Mammalian Cell Biotechnology in Protein Production

Mammalian Cell Biotechnology in Protein Production

Editors
Hansjörg Hauser · Roland Wagner

Walter de Gruyter · Berlin · New York 1997

Editors
Dr. Hansjörg Hauser
Gesellschaft für
Biotechnische Forschung mbH
Mascheroder Weg 1
D-38124 Braunschweig
Germany

Dr. Roland Wagner
Gesellschaft für
Biotechnische Forschung mbH
Mascheroder Weg 1
D-38124 Braunschweig
Germany

Cover illustration
Gravitational setting type bioreactors (see contribution p. 279)
With 132 figures and 52 tables

Library of Congress Cataloging-in-Publication Data

Mammalian cell biotechnology in protein production
editors, Hansjörg Hauser, Roland Wagner.
Includes bibliographical references and index.
ISBN (invalid) 03110134039 (alk. paper)
1. Proteins–Biotechnology. 2. Animal cell biotechnology.
3. Mammals–Cytology. I. Hauser, Hansjörg, 1949 – II. Wagner, Roland, 1956–.
TP248.65.P76M36 1997
660'.63–dc20 96-32908
CIP

Die Deutsche Bibliothek – Cataloging-in-Publication Data

Mammalian cell biotechnology in protein production
/ed. Hansjörg Hauser ; Roland Wagner. –
Berlin ; New York : de Gruyter, 1997
ISBN 3-11-013403-9
NE: Hauser, Hansjörg [Hrsg.]

∞ Printed on acid-free paper which falls within the guidelines of the ANSI to ensure permanence and durability.

Printed in Germany
Typesetting and Printing: Tutte Druckerei GmbH, Salzweg-Passau.
Binding: Lüderitz & Bauer GmbH, Berlin.
Cover Design: Hansbernd Lindemann, Berlin.

Preface

During the last decade, mammalian cell biotechnology has undergone rapid development in many aspects, ranging from cell biology to product recovery. Improvement of production techniques has been motivated by the license of such therapeutic candidates as tPA, factor VIII and erythropoietin which were especially pushed by the legendary progress in recombinant DNA technology. The idea of editing a book came into existence during several courses in "Animal Cell Technology" given at Gesellschaft für Biotechnologische Forschung (GBF)
here all scientific disciplines dealing with the successful production of recombinant glycoproteins were integrated. Due to the inter-disciplinary character, scientists with different educational backgrounds were able to discuss common problems which resulted in a fruitful exchange of ideas. However, a comprehensive book could not be assembled in a sensible way because mammalian cell biotechnology underwent turbulent developments during the eighties. In contrast, except for a few scientific disciplines (glycosylation, metabolic engineering, biosensor development, process modulation and genetic techniques), the development of fundamental methods for industrial application has been extensively accomplished. Anticipating that the technology acquired so far will be used for the next product generation, we were thoroughly convinced that it would be a good idea to collect and combine available information, placing it at the disposal of interested scientists and students in the form of a comprehensive handbook. Hence, our aim was to associate distinguished researchers from industry and academia in order to cover the respective subjects extensively.

The present book focuses on the different approaches and encompasses biological, biochemical and genetic aspects up to the recovery of the final end product by means of large scale production, with the emphasis on a fundamental explanation of the techniques and their practical success. The basic concept of this book is to integrate the most important disciplines in the field of manufacturing pharmaglycoproteins by cultured recombinant mammalian cell lines in one reference work and to serve as a guide to the original literature. As a result the book inevitably contains a broad spectrum of different subjects including virology, biochemistry, molecular genetics, cell biology, bioreactor technology, process simulation, validation, biophysics, hydrodynamics and legal aspects. It is not only meant for advanced students and doctoral researchers but also for experts working in the applied field who are interested in learning about the background of neighboring disciplines and the connections of this exciting scientific network. Based on the initial goal to make the book accessible to scientists with various educational backgrounds (e. g. bioengineers and molecular biologists), it is also emphatically recommended to scientists who are not directly involved in this field.

Due to the inter-disciplinary character of this book there was no alternative but to compile chapters provided by authors of the various fields. Apart from the different character of the articles, they have the advantage of having been written by outstanding experts in the respective research areas. Their willingness to impart their knowledge to their colleagues forms the character of Mammalian Cell Biotechnology in Protein Production, and their endeavor is gratefully acknowledged by the editors. In order to make it easier for the readership to understand the structure and use of this book, we decided to divide the chapters into different sections, ranging from the genetic manipulation of cells and the biological backgrounds to standard and new methods of protein production and product recovery. Finally, regulatory aspects and safety considerations are dealt with in so far as they concern industrial requirements. We are grateful to the publishers for their constant and diligent support. The editors entrust Mammalian Cell Biotechnology in Protein Production to the readership for frequent use and successful application and hope to have contributed a useful tool for improving discussions between scientists of different disciplines by providing a common platform open for critical considerations.

Braunschweig, in December 1996

Hansjörg Hauser
Roland Wagner

Contents

Contributors List

Hermann Allgaier
Dr. Karl Thomae
Birkendorfer Str. 65
D-88400 Biberach
Germany

Jeffrey J. Chalmers
Department of Chemical Engineering
The Ohio State University
140 West 19th Avenue
Columbus, OH 43210
USA

Devamita Chattopadhyay
Department of Chemical Engineering
The Ohio State University
Columbus, OH 43210
USA

Friedrich Dorner
Recombinant Cell Lines
Biomedical Research Center
IMMUNO AG
Uferstr. 15
A-2304 Orth
Austria

Jean-Marc Engasser
Institut National Polytechnique
de Lorraine
Laboratoire des Sciences du Génie
Chimique, CNRS
BP 172
F-54505 Vandœuvre-les-Nancy Cedex
France

Falko G. Falkner
Recombinant Cell Lines
Biomedical Research Center
IMMUNO AG
Uferstr. 15
A-2304 Orth
Austria

Ruth Freitag
Institut für Technische Chemie
Universität Hannover
Callinstr. 3
D-30167 Hannover
Germany

Miguel Garcia-Briones
Department of Chemical Engineering
The Ohio State University
Columbus, OH 43210
USA

Jean-Louis Goergen
Institut National Polytechnique
de Lorraine
Laboratoire des Sciences du Génie
Chimique, CNRS
BP 451
F-54001 Nancy Cedex
France

Hansjörg Hauser
Gesellschaft für Biotechnologische
Forschung
Dept. of Gene Regulation and
Differentiation
Mascheroder Weg 1
D-38124 Braunschweig
Germany

Michèle Himmelspach
Recombinant Cell Lines
Biomedical Research Center
IMMUNO AG
Uferstr. 15
A-2304 Orth
Austria

Alois Jungbauer
Institute of Applied Microbiology
University of Bodenkultur
Nußdorferlände 11
A-1190 Vienna
Austria

Brigitte Kaluza
Department of Cell Biology
Immunology/Oncology Division
Boehringer Mannheim GmbH
D-82377 Penzberg
Germany

Andre Lieber
Max-Delbrück-Centrum für
Molekulare Medizin
Cell Cycle Regulation and Gene
Substitution
Robert-Rössle-Str. 10
D-13122 Berlin
Germany

Annie Marc
Laboratoire des Sciences du Génie
Chimique, CNRS
BP 451
F-54001 Nancy Cedex
France

David Onions
Department of Veterinary Pathology
The University of Glasgow
Bearsden Road
GB-Glasgow G 61 1QH
Scotland

Volker Sandig
Max-Delbrück-Centrum für
Molekulare Medizin
Cell Cycle Regulation and Gene
Substitution
Robert-Rössle-Str. 10
D-13122 Berlin
Germany

Angela Savage
Chemistry Department
University College Galway
Galway
Ireland

Michaela Schäffner
Department of Cell Biology
Immunology/Oncology Division
Boehringer Mannheim GmbH
D-82377 Penzberg
Germany

Thomas Scheper
Institut für Technische Chemie
Universität Hannover
Callinstr. 3
D-30167 Hannover
Germany

Uwe Schlokat
Recombinant Cell Lines
Biomedical Research Center
IMMUNO AG
Uferstr. 15
A-2304 Orth
Austria

Friedrich Srienc
Department of Chemical Engineering
and Material Sciences
151 Amundson Hall
Minneapolis, Mn 55455-0132
USA

Michael Strauss
Max-Delbrück-Centrum für
Molekulare Medizin
Cell Cycle Regulation and Gene
Substitution
Robert-Rössle-Str. 10
D-13122 Berlin
Germany

Michiyuki Tokashiki
Teijin Limited
Biotechnology Research Laboratories
4-3-2 Asahigoaka
Hino-shi
Tokyo 191
Japan

Raghavan Venkat
Department of Chemical Engineering
The Ohio State University
Columbus, OH 43210
USA

Roland Wagner
Gesellschaft für Biotechnologische
Forschung mbH
Mascheroder Weg 1
D-38124 Braunschweig
Germany

Joachim Walter
Dr. Karl Thomae
Birkendorfer Str. 65
D-88400 Biberach
Germany

Ulrich H. Weidle
Dcpartment of Cell Biology
Immunology/Oncology Division
Boehringer Mannheim GmbH
D-82377 Penzberg
Germany

Manfred Wirth
Gesellschaft für Biotechnologische
Forschung
Abteilung Genexpression
Mascheroder Weg 1
D-38124 Braunschweig
Germany

Florian M. Wurm
Ecole Polytechnique Fédérale
Lausanne
Département Chimie
CH-1015 Lausanne
Switzerland

Seiichi Yokoyama
Teijin Limited
Biotechnology Research Laboratories
4-3-2 Asahigoaka
Hino-shi
Tokyo 191
Japan

1 Genetic Manipulation of Mammalian Cells

1.1 Heterologous Expression of Genes in Mammalian Cells

Hansjörg Hauser

1.1.1 Introduction

In this article I will review some of the basic features of expression in mammalian cells that are relevant to the specific manipulation of mammalian cells. Emphasis is put in the first instance on attaining a high enough level of expression to be able to isolate the product of the transferred gene.

The major advantage in using mammalian cells as an expression system for overproduction of recombinant proteins is the fact that these cells are able to carry out post-translational modifications as well as protein folding in an authentic manner. On the other hand, the bottleneck in the use of mammalian cells is the strength of expression. Whereas other expression systems often lead to extremely high yield expression in inexpensive growth media, mammalian cells require complex and relatively expensive media in order to keep them growing and secreting at a mediocre rate. In order to achieve an economically acceptable production process, genetic manipulation must result in very efficient expression of the relevant gene. This may be achieved either by constitutive or conditional expression of the gene whose product is of interest. In any case, the final expression should be as high as possible. Good expression values for secreted proteins are between 5 and 100 $\mu g/10^6$ cells/24 hours.

In order to achieve optimal expression, the engineering of mammalian cells includes the transfer of regulatory genes whose expression results in the overexpression of the gene of interest. The expression of these regulatory genes may require sophistic regulation, e. g. by external stimuli or intracellular signaling. The understanding of differential gene regulation is therefore a prerequisite for the successful development of high producer cell lines. A new approach concerns the use of heterologous expression principles in the mammalian hosts. These include the use of prokaryotic or yeast genes in a similar way as viruses, e. g. the vaccinia virus, with their own genes and regulators which use the mammalian cell as a host.

Our current knowledge is based on the studies of natural gene expression. Interestingly, highly specialized mammalian cells with the expression of certain genes, e. g. the β-globin production in erythroblasts or the immunoglobulin secretion, by B-cells, still represent the most efficient mammalian expression systems known so far. The principles of expression of these efficiently realized genes have to be understood and used for the construction of recombinant cell lines with advantageous properties.

Apart from the goal of achieving high expression in mammalian cells, it may be of interest to change some of the properties with respect to their specific abilities to modify proteins post-translationally. Post-translational modifications are carried out by a number of enzymatic activities. Often, cascades of enzymes act on the modification of one protein. Slight alterations which can be due to differences in species, differentiation state or mutagenic variation of enzymatic activities might lead to changes in the post-translational pattern of groups of proteins. First attempts to correct expression of genes responsible for such enzymatic activities have been successful. It is, however, obvious that these manipulations also require sophistically regulated gene expression.

1.1.2 Optimizing the Level of Gene Expression

The strength of protein expression in mammalian cells is a function of transcription, post-transcriptional events and translational efficiency (Wirth and Hauser, 1993). The basic parameters of expression of endogenous and transferred genes are summarized in Table 1.1.1.

Table 1.1.1 Parameters for expression strength of transfected genes

Transcriptional efficiency	– Influences from the chromosomal neighborhood of the transferred gene (promoters, enhancers, silencers, LCRs, SCSs, neighboring transcription in sense), structure of chromatin, CpG methylation islands; – Promoter/enhancer of the transferred gene – Copy number of the transferred gene
RNA processing and transport	– Capping, – splicing, – polyadenylation, – transfer to cytoplasm
mRNA turnover	– RNA sequences and peptide sequences of the translated RNA influencing mRNA stability; – shortening of the poly(A) tail – presence of the cap
Translation efficiency	– secondary structures, – cap, – modification of translation factors, – AUG context, – 5′UTR, – distant sequences, – binding of repressors of the mRNA, – general translation status of the cell, – poly (A) modification, – presence of antisense RNA, – translational masking
Protein stability	– Recognition by the ubiquitin-system (N-end amino acids, internal sequences)

The examination of expression levels of a large number of individual genes has demonstrated that post-transcriptional events and translation might play major roles in limiting the expression of a certain gene. The half-life of mRNAs in animal cells differs considerably. In contrast to many housekeeping mRNAs which do not show a significant turnover and are synthesized at a rate that allows doubling their amounts with each cell generation, most mRNA species are synthesized faster than is apparently required for maintenance of a steady state. Here, the degradation plays a major role in the availability of mRNAs for translation. The turnover of mRNAs for gene products which are only required for short time periods (cytokines, growth factors, development regulating genes) is rapid, and mRNA half-life is less than 30 minutes.

Control of post-transcriptional expression is mostly dictated by the individual gene. The current understanding of the processes involved in post-transcriptional events aims at avoiding elements in expression constructs which have a negative influence on the intended process. Unfortunately, our knowledge is not sufficiently developed to avoid all possible pitfalls concerning this expression level.

Translation is not only a function of the respective mRNA concentration in the cytoplasm. In the last decade it has become apparent that translation efficiency of mRNAs differs greatly and in some cases is tightly regulated. Again, this level of expression is governed by the sequence and structure of the individual gene. A number of mRNA features are known today to avoid inefficient translation and even to achieve regulation on this level.

1.1.2.1 Transcription Determining Elements

1.1.2.1.1 Transcription Initiation

The sequences immediately 5′ to the start of transcription are collectively referred to as the promoter (Fig. 1.1.1). The 30 or 40 nucleotides just 5′ to the start site of transcription initiation are regarded as the core promoter (Breathnach and Chambon, 1981). This sequence contains the mimimal elements necessary for transcription initiation. In most cases it concerns the TATA-box (25 to 35 nucleotides distance from the transcriptional start point) or initiatory elements which overlap the transcription initiation site (Roeder, 1991). It is composed of the core promoter and adjacent upstream activator sequences. The function of the promoter is to control the rate of initiation of transcription of the adjacent structural gene. The rate limiting step in transcription is the level of transcription initiation. Apart from premature termination and pausing which is observed during transcription of a few genes, the rate of transcriptional elongation is constant until the movement of RNA polymerase is stopped by termination signals. As a consequence, the density of polymerase molecules on a given gene reflects the frequency of initiation events. The promoter defines the 5′ end of the gene as well as a great part of its activity. The activity of the promoter might be further modified by enhancers and chromosomal elements which act from a distance on the promoter.

As far as it is known to date, all regulatory DNA elements contain sequence motifs, from a few to up to tens of nucleotides in length (Fig. 1.1.1). These sequence motifs can be regarded as the building blocks of genetic regulatory elements. They

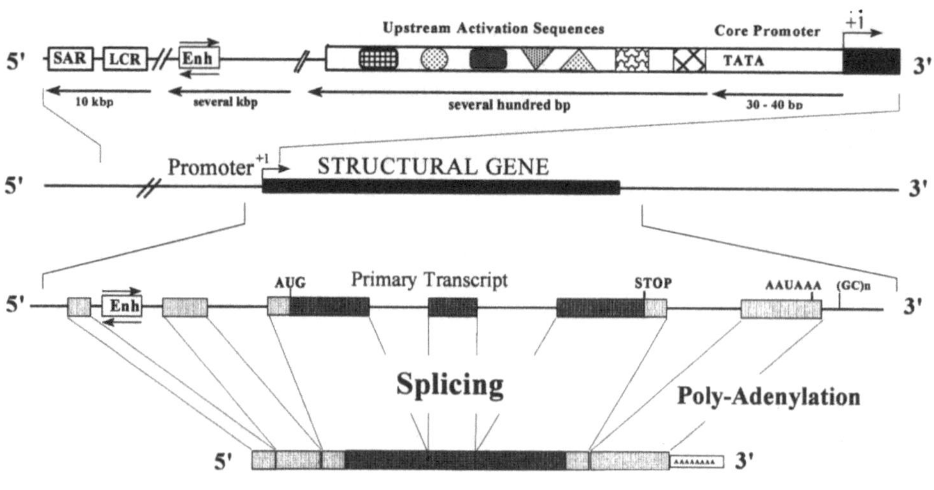

Fig. 1.1.1: Schematic representation of the eukaryotic transcription unit.
In the middle the transcription unit the structural gene and the promoter are indicated. The position marked +1 corresponds to the transcription start site. The promoter and further regulatory sequences are dissected into the components in the upper part of the figure. Enh: Enhancer sequences; SAR: scaffold attached regions; LCR: Locus Control Region. In the lower half of the figure the primary transcript and the mature are mRNA are shown. The initiator codon is indicated as AUG, the termination codon as stop. The signal for termination of transcription is indicated as AAUAAA. The poly(A) tail in the mRNA is indicated as AAAAAAAA.

represent the contact points of regulatory proteins. Natural promoters and enhancers are composed of a multitude of mostly different sequence motifs. This modularity means that a considerable degree of flexibility is available when deciding on the design of a promoter to be used for control of gene expression (Latchmann, 1991).

Elements regulating the activity of the core promoter are in most cases located immediately upstream and are collectively referred to as upstream activating sequences (UAS) Fig. 1.1.1). The UAS are usually found within several hundred nucleotides upstream from the promoter. Enhancers are relatively large sequence elements which are composed of DNA-sequence motifs, often identical to those found in the promoters. Enhancers are found proximal to promoter regions of genes, within introns and far upstream as well as downstream of the transcribed region. The distance between enhancers and the respective promoters can exceed the length of 10 kilobases. Normally, the enhancer stimulates the closest promoter. Today it is generally accepted that enhancers act from a distance by looping out interspersed DNA between the promoter and enhancer in order to get into physical contact with the transcription machinery of the promoter.

The proteins acting on the promoters can be divided into two classes: 1) Basal factors present in all cells include the RNA polymerase II with its co-factors. 2) Transcription regulatory factors interacting with the sequence motifs in the UAS and enhancers might be ubiquitous or tissue specific. They might also exist in an inactive but activatable fashion. The differential function of certain promoters in cells in different states of differentiation is explained by the quantitative and qualitative composition of DNA-binding factors and other proteins which indirectly interact with the DNA-binding factors.

Apart from positively acting elements, the existence of negatively acting DNA in UAS and enhancers has been demonstrated. These sequence elements which are interspersed in natural promoters are also recognized by the DNA-binding factors. The negative effect induced by these elements can be due to the binding of transcription repressing proteins as well as to the creation of unfavorable assemblies of transcription activating factors, thereby reducing the activity of the total promoter or enhancer.

Inducible promoters function by using two different mechanisms:

1. The cellular concentration of certain transcription factors defining the activity of a promoter is altered.
2. Modulatory events, like the binding of ligands to transcription factors might regulate their activity or binding to the respective DNA-sequences. Prominent examples include the binding of steroid hormones to their receptors (which are transcription factors by themselves) and the activation of their DNA-binding properties when heavy metal ions bind to certain transcription factors.

Various classes of DNA-binding proteins exist (Wingender, 1993). They differ in distinct features of their protein structure which account either for their interaction with DNA or for the interactions of the proteins in the formation of multimers. Dimerization such as is observed for *jun* and *fos* in the AP1 transcription factor complex is the most frequent feature of these factors. In the case where heterodimers can be formed there is a potential for a combinatorial increase in the possible regulatory elements formed. These provide a means for obtaining numerous combinations of individual proteins, thus resulting in the formation of heterodimers with potentially different functionalities (Lamb and McKnight, 1991). Proximity of the activation domains to the RNA polymerase II complex which is usually achieved by DNA-binding of the transcription factor activates the bound polymerase complex to start transcription. The DNA-binding factors usually contain a protein domain which allows sequence specific binding to DNA. In addition, they might contain activation or repression domains. (Hershbach and Johnson, 1993).

A limited number of DNA-binding proteins, a few hundred, have been suggested to interact directly or indirectly with the RNA polymerase II complex in order to stimulate its binding or activity.

DNA-binding factors that do not contain an activation domain exert their effects by proteins that allow a bridging between the DNA-binding factor and the polymerase II and thereby induce transcription initiation.

1.1.2.1.2 3′ End Formation and Polyadenylation

The poly(A) addition site usually precedes the end of the mRNA. This sequence which is the most conserved DNA recognition element in the eukaryotes is AATAAA. This sequence motif directs the cleavage of primary transcripts approximately 20 base pairs downstream of the poly(A) site. Most messenger RNAs do have a post-transcriptionally added poly(A) tract. The polyadenylation process, however, requires another, less conserved GU-rich, recognition sequence (T/G-box) which is located downstream of the cleavage site. This sequence, in collaboration with the poly(A) site leads to the endonucleolytic attack (Proudfoot, 1991). Termination of the transcription reaction takes place further downstream. The termination site is in most cases not known. The 3′ part of the transcript after AATAAA-mediated cleavage is degraded in the nucleus.

After cleavage of the primary transcripts by a nuclease about 20 bp downstream of the AATAA/A box poly(A) is added to the newly generated 3′ ends (polyadenylation). Polyadenylation is related to the stability and translatability of the mRNA. In the cytoplasm, the poly(A) end is successively shortened. Poly(A) ends with less than 27 adenosine residues are unstable in the cytoplasm (Munroe and Jacobsen, 1990; Jackson and Standart, 1990).

Some eukaryotic transcription units contain more than one poly(A) addition site. The differential choice of these sites is influenced by adjacent sequences and secondary structures. Several examples of tissue-specific usage of multiple poly(A) sites have been reported. In addition, some poly(A) sites are promiscuous, which leads to a partial read-through and the production of different with RNAs 3′ terminal ends. This may influence trancription from downstream located promoters. Transcription running through a 3′ located promoter can lead to occlusion of its activity (Emerman and Temin, 1986; Cullen et al., 1984).

Polyadenylation and signals for transcription termination are important elements of all expression vectors. cDNAs usually possess an AATAAA element located upstream the poly(A) tract. The T/G-box which is important for efficient polyadenylation is usually not contained in the cDNA sequence because the processing of the mRNA precursor occurs between the AATAAA- and the T/G-box (Proudfoot, 1991). Therefore, mRNAs derived from expression vectors are often incomplete or not polyadenylated at the authentic site. Another polyadenylation site and sequences influencing termination further downstream have to be supplied by the expression vector.

1.1.2.1.3 Chromatin

In contrast to prokaryotes the DNA of animal cells is associated with a number of nuclear proteins. It is wound on cores of histone proteins and condensed in the form of solenoids which again form higher orders of condensation (Fig. 1.1.2). Generally, decondensation of a DNA domain reflects its accessibility to transcription factors and activity of the genes located therein.

The nuclear DNA of eukaryotic cells is organized in chromatin domains with lengths between 5 and 200 kb. These domains contain loops with several genes or, more typically, singular genes. The nature of these domains seems to determine the extent of expression of the genes located therein (Gasser and Laemmli, 1987). A group of nuclear proteins (scaffold proteins) is involved in the structural organization of long chromosomal domains. The DNA-sites which interact with the nuclear scaffold proteins are called "scaffold associated regions" or "matrix associated regions" (SAR or MAR) (Fig. 1.1.3).

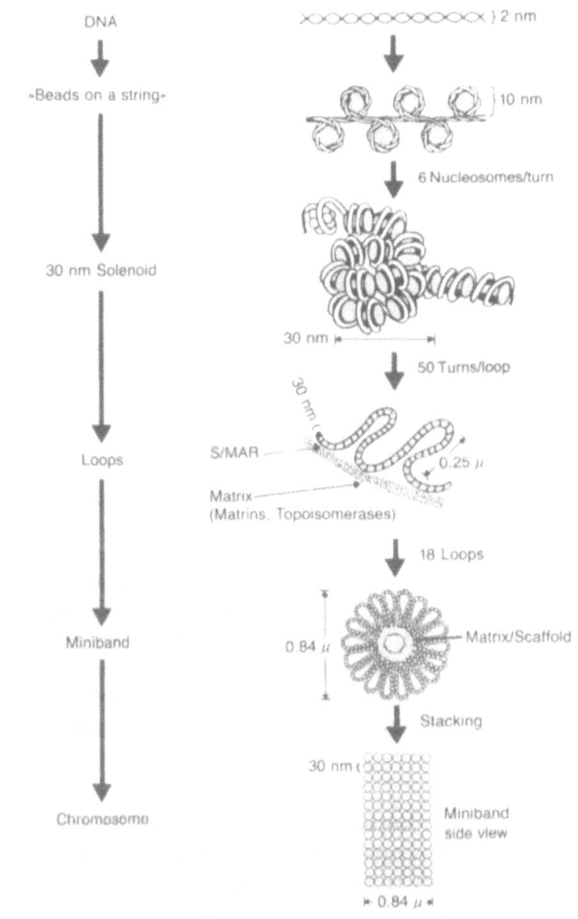

Base pairs per turn	Packaging ratio
10 b.p.	1
80 b.p. (160 b.p. per 2 turns)	6-7
1200 b.p. (per turn)	40
60000 b.p. (per loop)	680
1.1×10^{6} b.p. ± (per miniband)	1.2×10^{3}
18 Loops / miniband	1.2×10^{4}

Fig. 1.1.2: Organization of eukaryotic genomic DNA.
Double-stranded DNA is wound in two turns around a protein (histone) core. This leads to the "beads on a string" structure. The beads are called nucleosomes. They have a diameter of 100 Å. This is thought to be the typical structure of active chromatin allowing efficient transcription. The addition of a specialized histone (H1) to the links between the beads (nucleosome) leads to another level of winding of the "beads on a string", into the solenoid. This structure is typical for inactive chromatin domains (heterochromatin). The next level of organization includes functional chromatin domains which are represented as chromosomal loops fixed by SARs. During condensation to chromosomes important components of the scaffolds are retained.

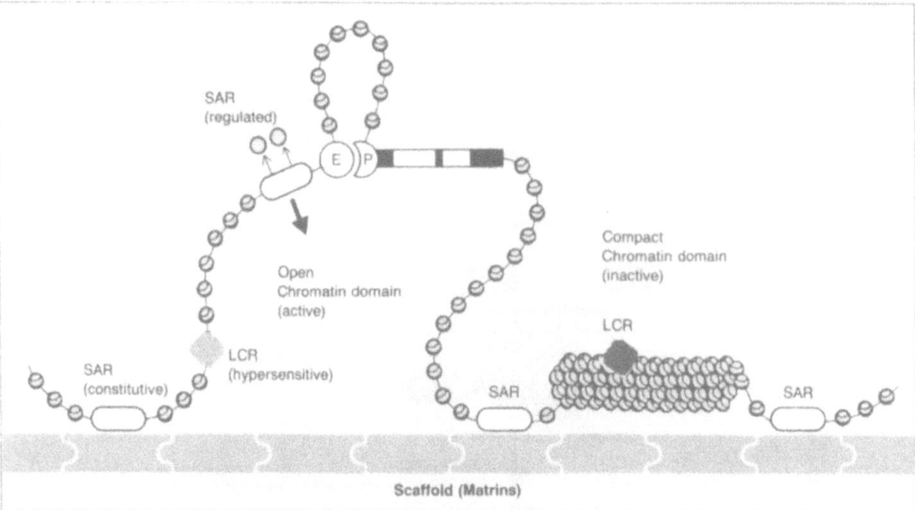

Fig. 1.1.3: Chromatin domains
Short DNA structures responsible for the attachment to the proteinaceous matrines of the nuclear scaffold exist within the chromatin which constitutes the general package form of DNA. Each chromosomal DNA is bordered by two such constitutive SAR sequences and isolated from the neighboring domain. Locus control regions (LCRs) or similar types of elements are responsible for the switches opening and closing (activation vs. inactivation) the domains. It is thought that the opening starts from the LCR and opening extends until it reaches the next SAR element. Decrease in the size of domains seems to be regulated by specific SAR elements frequently found associated with enhancer elements (E). All together these mechanisms serve to regulate the transcriptional strength from the promoter (P).

Scaffold-attached Regions

A-T rich sequences have been located at the borders of coding regions by electron microscopy. These sequences are closely associated with proteins of the inner "nuclear matrix" or "nuclear scaffold". In analogy to the cytoskeleton, the inner matrix (as opposed to the external lamina which forms the outer membrane and is the site of nuclear pores and transport of proteins) is thought to function by organizing the genome in the form of separate loops and to harbor the enzymes required for replication and transcription. Methods that involve mild detergents lead to the isolation of a proteinaceous network ("scaffold") which provides the high affinity for the putative "scaffold-attached regions" (SARs).

Topoisomerase II has been found to be a prototype scaffold protein which recognizes and also relaxes superhelical structures. Other components appear to be variants of histone H1 and a number of proteins that have been termed "matrins" in analogy to the lamins of the nuclear lamina. The first members of this class have been cloned and found to possess binding or activating domains that are otherwise typical of transcription factors.

SAR sequences do not demonstrate sequence homology and therefore the properties of SARs cannot be derived from mere sequence comparisons. Although all of them are A-T rich, this is not sufficient to provide these elements with the respective properties *in vitro* or *in vivo*. *In vitro*, these elements exhibit high binding affinities toward isolated scaffolds, and *in vivo*, they promote transcription. Both features appear to be linked and the biological effect appears to increase proportionally to the length. More refined analyses demonstrated that all SAR elements enable the separation of DNA strands under conditions that otherwise occur during transcription. Separation initiates at precisely-defined nucleation centers and proceeds until a highly G-C rich stretch is met. Nucleation sequences of this type were subsequently isolated in the form of 25 bp which were oligomerized in order to amplify strand separation. This approach resulted in relatively short (180 bp) SAR-type elements which were shown in many respects to resemble those at the domain borders (Bode et al., 1992).

Isolation and delimitation of SAR-DNAs revealed that SAR elements promote the activity of enhancers and promoters (Bode et al., 1995). They function in a cell type nonspecific manner and only when the DNA constructs have been integrated into the chromosomal DNA of the host. The expression of transcription units by flanking them with SAR elements strongly stimulates expression, depending on the nature of the promoter of the transcription unit. A further stimulation of this transcription can be achieved by adding butyrate to the cells. SAR-elements are now being used for improvement of expression plasmids (Klehr et al., 1991).

Locus-control-regions

In addition to a more or less uniform organization of DNA there are, within the genome, distinct nucleosome-free regions which are visualized most conveniently by their enhanced accessibility towards various nucleases. This so-called nuclease sensitivity can be subdivided into several levels. Super-hypersensitivity has been found at certain domain borders and is one of the features characterizing "locus-control regions" (LCRs). This upper level of sensitivity has been ascribed to displacement of nucleosomes by certain factors. Starting at a nucleation center, the initially localized opening of the domain progresses in one or both directions – a typical property of cooperative ("all-or-none") processes. Therefore, LCRs appear to be capable of opening individual domains and regulating their affinity.

LCRs are able to confer the tissue-specifity of expression on the coupled genes. Accordingly, the LCR of the β-globin domain mediates expression in erythroid tissues whereas the LCR of the lysozyme domain leads to the expression in macrophages.

1.1.2.1.4 Selection of General and Tissue Specific Gene Transcription Elements

From the description above, it is clear that efficient transcription of heterologous genes requires the use of strong promoter/enhancer combinations (Kriegler, 1990). Universal expression vectors therefore contain promoters which are efficiently used in a variety of cell lines. A series of natural promoters have been shown to issue a strong effect in a variety of cell types.

The strongest of these promoters/enhancers are of viral origin. Examples include the promoter/enhancer from the human Cytomegalovirus (CMV) as well as the murine Myeloproliferative virus (MPSV) (Boshart et al., 1985; Artelt et al., 1988). A combination of elements from two or even more distinct promoters in some cases have led to the construction of hybrid promoters with higher activity or an increased spectrum of cell and species specificity. Based on the availability of transcription factors in different cell types, the activity of all promoters strictly depends on the type of cell. Therefore, for certain cell types the use of specific promoters is superior to the use of generally strong promoters. The most prominent example is the immunoglobulin heavy chain enhancer/promoter which directs high level expression predominantly in B lymphocytic cell lines.

1.1.2.1.5 Synthetic Promoters

Artificial promoters which are based on known elements from natural promoters have been created. This required the detailed functional analysis of natural promoters. The resulting promoters may exhibit high expression, specific regulatability or new cell type specificity. One example concerns the construction of a strictly metal-inducible promoter by multimerization of a metal-responsive element from a metallothioneine promoter (McNeall et al., 1989).

In some cases inducible or constitutive expression systems require the presence of a high level of transacting factors , e. g. the glucocorticoid or estrogen receptor proteins. Cells which do not have these properties can be complemented by transfecting the respective expression constructs separately. As the genes for other transactivators are isolated they can be used to manipulate cell lines which lack these factors and render weak promoters highly active. Hippenmeyer und Highkin (1993) have engineered mammalian cells to produce high levels of heterologous proteins by expressing the Herpes Simplex Virus (HSV) VP 16 gene which strongly activates genes under the control of the HSV immediate early promoter. The knowledge of transcription factor protein structure will allow the engineering of novel potent transactivators which elicit the activity of specific promoters. This was successfully approached by fusing the DNA-binding moieties from different transcription factors to those encoding the hormone-binding/activation domains of steroid receptors rendering the activity of these transcription factors hormone-inducible (Eilers, et al., 1989).

The use of prokaryotic elements is an attractive way of maximizing gene expression and creating inducible expression systems in animals cells. A first step which showed that a bacterial control system could function in eukaryotic cells was performed by inserting DNA sequences which bind E. coli repressor proteins into functional animal cell promoters (Hu and Davidson, 1987). With another construct a cell was engineered to produce the respective repressor. Expression could be induced by inactivation of the repressor.

1.1.2.1.6 Regulatable Transcription

In certain circumstances, e. g. for the production of a toxic protein, time-restricted expression is required. The gene of interest is stably introduced into the host cells without perception of the negative effect from the gene product. After induction of specific gene expression the product is produced for a certain time. For the study of effects of certain gene products on cell physiology, conditional expression of the respective gene is often used.

A number of cellular and viral promoters/enhancers are known to respond to external stimuli by hormones, temperature, growth factors, viruses, etc. They contain short DNA sequences which are required for the response to external stimuli (response elements). Response elements have the same general characteristics as other promoter or enhancer elements. They are located in promoters as well as enhancers as single or multiple copies.

Among a number of approaches to achieve efficient regulation by external means, a few systems have been shown to be of value for a number of purposes. In these systems natural elements of the mammalian transcription machinery have been used in order to get the desired magnitude of regulation.

- The activation of the metallothioneine promoter by heavy metals has been optimized by multimerization of the DNA elements which are responsible for the heavy metal binding transcription factor. These promoters, although showing a certain basal level of constitutive transcription, are strongly inducible by heavy metal ions like cadmium and zinc. The ratio of induced and non-induced expression does not exceed the factor of 100. (Searle et al., 1985, Kaufman, 1990).
- The activation of steroid hormone receptors by their natural ligands has been applied to construct a regulatable expression system. Genes encoding fusion proteins which contain the hormone inducible domain of the steroid receptor and the DNA-binding domains of other factors which recognize certain promoter elements e.g. the GAL-binding sequence (motif from S. cerevisiae) were constructed. The gene encoding this transactivator fusion protein is stably established in the cell line of choice. The gene of interest is brought under the control of a synthetic promoter which is recognized by the fusion protein. Introduction of this expression construct into the transactivator expressing cell allows its strict regulation by adjustment of the concentration of the steroid hormone in the medium. (Eilers et al., 1989; Braselmann et al., 1993).
- The tetracycline operon from E. coli was multimerized and integrated in front of mammalian core promoters. Transactivation is achieved by a fusion protein consisting of the tetracycline operator as a DNA-binding moiety and a strong activator domain from the Herpes Simplex Virus transactivator protein VP16. Repression of this system is achieved by the addition of tetracycline or its derivatives which bind to the operator and thereby prohibit binding of the hybrid protein to the synthetic promoter. This system allows inducibility rates of up to five orders of magnitude (Gossen and Bujard, 1992).

A well-studied example of naturally regulated gene expression directed from a chromosomal regulatory entity is the chromosomal β-globin locus. (Fig. 1.1.4a). The five genes of this locus are expressed in a tissue specific and developmentally regulated fashion as the state of erythropoesis shifts during development (Evans et al., 1990). A regulatory region dominating the activity of the different genes within this locus lies further outside of the genes to be controlled. The regulatory DNA sequences that make up this "locus-control region" (LCR) contain binding sites for erythroid specific DNA-binding proteins as well as binding sites for numerous widely distributed transcription factors. The function of this locus is explained by the exertion of a polar effect on the expression of the different neighboring genes in a way that distal genes would become active during development by active repression of proximal genes (Fig. 1.1.4b). It was shown that the LCR confers position and copy number independent expression of genes with specific promoters placed under its control. The concept to use the LCRs for expression of heterologous genes in mam-

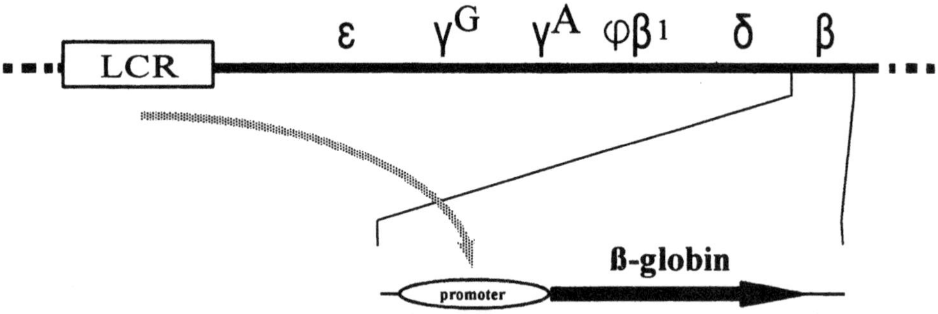

Fig. 1.1.4a: Schematic representation of the β-globin locus.
The locus control region (LCR) exerts its effect on the different globin genes during development, here shown as its effect on the β-globin promoter.

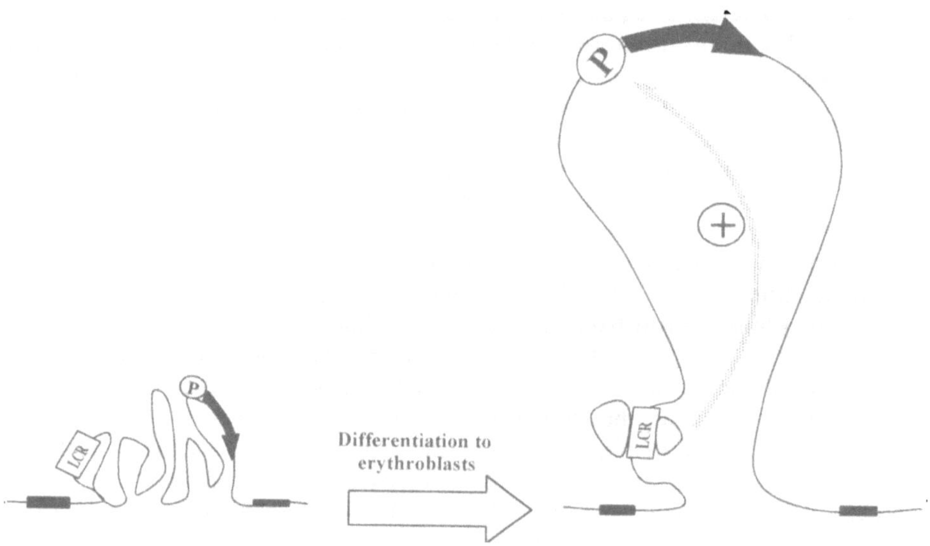

Fig. 1.1.4b: Schematic representation of the inactive and the active β-globin locus in erythroblasts of adults.
Scaffold attached regions (SAR) are given in the basement of the loops as black rectangles. The locus control region (LCR) is boxed, the β-globin gene illustrated by an arrow.

malian cells was elegantly realized (Needham et al., 1992). However, it suffers from the fact that only a limited number of LCRs have been described up to now. This reduces the extent of applications to a limited number of cell types.

In selecting an inducible vector-system for a particular gene, it is important to ensure that the inducing stimulus does not interfere with properties under study. It is also important to know the extent of induction which is required including the base level expression and the maximal achievable expression level.

1.1.2.1.7 Influence of Chromosomal State and Position of Transferred Genes
Transcriptional efficiency is not only the dominant but also the major level which determines the efficiency of expression of a certain gene. For this reason most researchers have put particular emphasis on the development of vectors which lead to high efficiency transcription. First, high-efficiency promoters and enhancers were selected and transcription inhibiting elements like silencers were removed. More recently, the chromosomal influence on integrated expression constructs was recognized to be of great importance for efficient expression. Consequently, strategies were developed allowing expression in chromosomal environments which favor transcription. One strategy is the creation of expression cassettes which stimulate the formation of active chromatin by making use of certain chromosomal elements with dominant effects on transcription (Klehr et al., 1992, Bode et al., 1995). A second strategy aims at targeting the expression construct to naturally existing chromosomal sites (a detailed description of thus technology is given in chapter 1.6.4). Recombinant retroviruses or parts thereof which integrate preferentially into such sites are used as probes. Known and characterized hyperactive sites might be targeted by homologous recombination (Karreman et al., 1996). The classical method still is to create a large number of cell clones harboring the transfected gene integrated randomly and to select those cell clones which show most efficient transcriptional activity of the gene of interest. The screening methods are described in detail in chapter 1.5.

The mode of integration of exogenously added DNA into the chromosomal DNA of the host cell is accomplished by two mechanisms, homologous recombination and nonhomologous (illegitimate) recombination. Illegitimate recombination does not depend on regions of homology between the DNA molecules and the mechanism is still obscure. Homologous recombination involves regions of homology between the host's DNA and the exogenous DNA (Lin et al., 1984). Directed integration into certain loci of a mammalian chromosome (gene targeting) is attempted via homologous recombination. However, the efficiency of this process is low and requires intensive screening (Sedivy and Joyner, 1992).

DNA methylation

In the DNA of higher animals, about 70 % of the cytosine residues in the 5′-CpG-3′ sequence are methylated whereas most other cytosine residues are not. During cell culture, methylated DNA sequences are usually inherited conservatively. Islands of CpG sequences are mostly found around the promoter areas of genes transcribed by RNA polymerase II. The methylation modification pattern is characterized by a full methylation of those genes that are inactive and undermethylation of those genes which are active (Eden and Cedar, 1994). The inverse correlation between DNA methylation and gene expression is striking and suggests that modification itself may play a direct role in the inhibition of transcription. Indeed, many individual endogenous genes can be activated from their repressed state by treatment with 5-azacytidine, an inducer of demethylation. This shows that methylation serves as a global suppressor of gene expression.

The influence of DNA modification on gene transcription is thought to be mediated through an effect of methyl moieties on DNA-protein interactions. DNA-methylation leads to the alteration of local protein recognition signals. Under certain conditions, DNA modification may directly inhibit the action of RNA polymerase itself. In addition to its direct effect on factor binding, DNA methylation may also operate by altering overall chromatin structure, thus indirectly influencing gene accessibility. In gene transfer, unmethylated sequences always integrate into the genome in a nuclease sensitive conformation. In contrast, *in vitro* methylated DNA adopts an inactive chromatin structure leaving the DNA resistant to nuclease digestion. DNA methylation influences chromatin structure in various ways, e.g., the methyl-CpG-binding protein, MeCp1, binds specifically to DNA-sequences that contain multiple symmetrically methylated CpGs. This protein is abundant in most cell lines and is probably one of the factors that act to inhibit transcription. Another way in which DNA methylation alters gene function is by modulation of DNA conformation which is independent of protein binding.

In the expression of heterologous genes, methylation plays a major role with respect to the stability of expression. Although it is not clear by which mechanism methylation of transferred genes is induced, the decrease of expression of transfected genes in the course of cell cultivation often correlates with *de novo* methylation of the respective promoters. The reason for these *de novo* methylations are not unterstood. It is, however, clear that methylation of transferred DNA depends on the integration site. It is tempting to speculate that the chromosomal DNA surrounding the transferred DNA exerts influences on its methylation pattern.

1.1.2.2 Post-Transcriptional Considerations of Gene Expression

Production of functional mRNA from the primary transcripts involves four major steps: Capping, splicing, polyadenylation and transport to the cytoplasm.

1.1.2.2.1 Capping

All RNA polymerase II transcribed RNAs formed in normal cells are capped. The cap consists of a terminal nucleotide, 7-methylguanylate (m^7G) in a 5′-linkage with the initial nucleotide of the mRNA chain. It is co-transcriptionally added to the primary transcript. The existence of the cap positively influences translation of cellular mRNAs as well as polyadenylation and splicing. With the exception of certain viral infections all cellular mRNAs transcribed by the RNA polymerase II are capped. The use of expression constructs with promoters which are not driven by RNA polymerase II requires specific attention, e.g., transcripts synthezised by the T7 polymerase are not capped and are therefore inefficiently translated.

1.1.2.2.2 Introns

Splicing of the primary transcripts starts on the nascent transcript and is completed after a maximum of 20 minutes after termination of transcription (Weiner, 1993). The intronic sequences are instable and are readily eliminated by degradation in the nucleus. Most higher eukaryotic genes contain introns (Fig. 1.1.1). When introduced into mammalian cells efficient expression of several genes has been shown to depend on the presence of an intron (Hamer and Leder, 1979; Buchman and Berg, 1988). Transcription, RNA processing and transport in the nucleus might be coupled in a spatially organized process (Mattaj, 1990). An intron is required for efficient expression from an immunoglobulin heavy chain promoter/enhancer, but the intron must not be a particular one. Furthermore, certain introns seem to be better in stimulating gene expression than the commonly used SV40 introns. In contrast, many genes do not require introns for mRNA formation when introduced into mammalian cells which is in line with the observation that some intron-less genes are efficiently expressed (Neuberger und Williams, 1988).

Many examples of differential processing of animal RNAs have been described. In some cases, the alternative splicing in one cell is optional, in other cases the choice of a certain splice is a cell-specific function (Maniatis, 1991). Differential splicing can lead to proteins which share a common N-terminus but differ in the C-terminal part. In the extreme case the resulting proteins are completely different.

Since the dependence of introns for efficient mRNA formation is usually not known the use of intron-containing vectors is still recommended. However, the use of introns can also create artifacts which are due to aberrant splicing. The SV40 small t-intron is one of the most commonly used introns in expression vectors. In vectors where the SV40 small t-intron is present in the 3′UTR, 5′splice donor sites derived from the cDNA genes together with the 3′splice acceptor site of the SV40 small t-intron are preferentially used resulting in deletions within the protein coding region (Huang and Gorman, 1990). 5′ splice donor sites in the cDNA could exist as a consequence of splice-junctions from former introns. Furthermore, ordinarily inactive splice consensus sequences (cryptic splice sites) are found in many genes. These may be utilized together with introns from vector sequences. Thus, the positioning of small introns in the 3′UTR of cDNA expression vectors should be avoided. If introns are used they should be located in the 5′UTR of the vector. The probability of aberrant splicing from this position is lower.

Fragments from different genes mediating transcription termination do not necessarily function very efficiently in heterologous surroundings. Promiscuous termination leads to mRNAs with different 3′ends. Run-through transcription can result in a series of difficulties including transcriptional interference on downstream promoters, instability of the oversized mRNA and hybrids with antisense RNA.

1.1.2.2.3 mRNA Stability

The half-life of an mRNA determines the length of time that a molecule can function to synthesize protein (Peltz et al., 1991; Cleveland and Yen, 1994). One of the sequence conditions found to be responsible for short half-lives of mRNAs encoding proto-oncogenes and cell growth factors is the repeated occurence of sequence AUUUA in the 3′UTR (Shaw and Kamen, 1986). The existence of these sequences in a certain mRNA species has been shown to be correlated to enhanced degradation of their poly(A) tract (Wilson and Treisman, 1988). However, AUUUA sequences alone are not always sufficient for mRNA destabilization. In some cases they act in concert with other RNA sequences (Laird-Offringa, 1992). mRNA secondary structures are also involved in regulation of the degradation rate. An interesting example concerns the transferrin receptor mRNA (reviewed by Müllner et al., 1989; Hentze, 1991).

1.1.2.3 Translational Control

Translation occurs on 80S ribosomes which are composed of a small (40S) and a large (60S) subunit. Translation begins with the stepwise assembly of a functional ribosome onto an mRNA. In brief, the cap structure at the 5′ end of the mRNA is bound by a large complex of proteins, including the cap binding-protein eIF-4E and the RNA helicase eIF-4a. According to Kozak, after binding the RNA, the small ribosomal subunit initiation-factor complex "scans" the 5′ untranslated region (5′ UTR) until it finds an AUG initiator codon in a favorable sequence context. (Kozak, 1986; Kozak, 1989b) However, this "scanning model" cannot accomodate a number of recent observations (for re-evaluation of translation initiation models see Sonenberg, 1993). Once an AUG is selected, the large ribosomal subunit joins the complex and peptide synthesis begins (Hershey, 1991). Similar to transcription, initiation of translation is the rate limiting step of protein synthesis.

Usually, the AUG codon next to the mRNA cap is used for initiation of translation. The initiation process is influenced by several aspects of mRNA structure.
1. The methylated cap structure (compare 1.1.2.2.1)
2. The primary sequence or context surrounding the AUG codon (see below)
3. The position of the AUG codon, i.e. whether it is the first AUG in the message
4. The leader length
5. Secondary structures upstream and downstream from the AUG codon.
6. 3′ UTR sequences that repress translation by modulating the length of the poly(A)tail (Richter, 1995).

Furthermore, evidence is increasing that other types of long distance cis-interactions between linear sequences and/or secondary structures with the initiator complex play a major role in the initiation efficiency of certain genes.

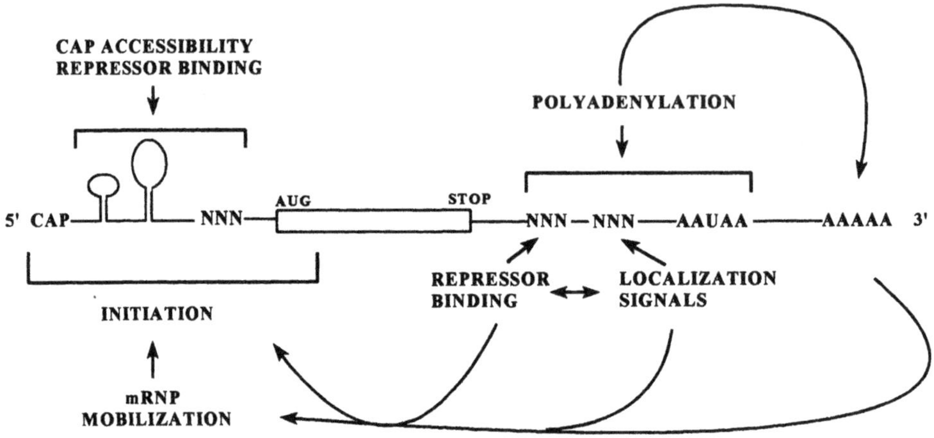

Fig. 1.1.5: Regulation of translation
The factors regulating translation initiation and their sites of action on the mRNA are depicted. The brackets indicate different independently acting processes which coordinately regulate the events (initiation and ribosome binding, polyadenylation) (according to Curtis et al, 1995).

A favorable context sequence around the AUG of numerous vertebrate mRNAs shows that two positions around the initiator AUG are most prominently conserved: position −3 which is often occupied by a purine residue, predominantly A and position +4 which mainly is a G residue. The primary sequence around the AUG determines the fidelity of initation. The sequence GCCG/AGCCACC*AUG*G was postulated to be an optimal sequence for translation (Kozak, 1986; Kozak, 1987a). Converting the context around the AUG of genes to be expressed in animal cells to the optimum 'Kozak sequence' should therefore increase their expression.

A number of genes produce two proteins from one mRNA. Here, the first AUG codon deviates from the consensus sequence in either or both positions −3 and +4. The result is "leaky scanning" in which some 40S ribosomes bypass the first AUG codon. Initiation occurs from the first and second AUG codon in these bifunctional mRNAs. Initiation at the second AUG may generate proteins from the same reading frame (protein isoforms) as well as from a different reading frame (unrelated proteins) (Kozak, 1991).

The sequence of the 5′UTR, also called the mRNA leader, affects translation efficiency. Attempts to pinpoint a motif in those sequences that mediate strong initiation have been unsuccessful. The apparent absence of a discrete effector motif and the fact that efficiently supporting translation initiation regions are deficient in G-residues underline the view that a moderately long, unstructured 5′ noncoding sequence may be necessary for efficient initiation of translation (Kozak, 1991b). However, the length by itself does not seem to be a dominant factor.

5′ UTR sequences which favor translation do exist, e.g., the Xenopus L.β-globin gene contains a 34 base translational enhancer (Falcone and Andrews, 1991). Connected to the body of reporter genes the translation efficiency is increased 10 to 300-fold in *in vitro* translation experiments (Falcone and Andrews, 1991) or up to 80-fold in *in vivo* investigations (Gallie et al., 1988).

The degree of secondary structure in the 5′ UTR is important for translation. Secondary structures in the 5′ UTR may hinder the binding step, scanning and AUG selection. Therefore, an unfolded 5′ leader should support these steps optimally, e.g., 5′ proximal secondary structures up to 15 bp from the cap impair translation (Godefroy-Colburn et al., 1985), and hairpin structures further downstream from the cap exceeding a stability of $\Delta G = -50$ kcal/mol severely affect translation.

1.1.2.3.1 Internal Translation Initiation

Picornaviruses harbor in their 5′UTR a cap-independent internal translation initiation element. The internal entry of ribosomes is mediated by a region termed internal ribosomal entry site (IRES). (Jackson, 1991; Simoes and Sarnow, 1991). Cap independent initiation has been reported for some cellular RNAs.

The potential for internal initiation on picornaviral 5′UTRs is exploited in eukaryotic expression vectors carrying polycistronic transcription units (Mountford and Smith, 1995). This development promises an improvement in the expression of homo- and heterodimeric proteins. In animal cells, most mRNAs are monocistronic and code for the synthesis of only one protein. The utilization of polycistronic transcription units is exceptional.

A tight linkage of expression is achieved by translation of the genes located on one single mRNA. Such vectors are referred to as bicistronic expression vectors

(Dirks et al., 1993; 1994). In several laboratories expression vectors carrying such bicistronic transcription units have been used to screen for high levels of foreign gene expression. In these constructs the gene of interest is contained in the first cistron and a selectable marker gene or a screening marker gene comprises the second cistron (compare chapter 1.5). Screening for high expression of the second cistron identified clones expressing high levels of the first cistron (Kaufman and Murtha, 1987; Boel et al., 1987; Wirth et al., 1990). In the absence of IRES elements insufficient reinitiation results in poor translation of the second cistron which is underexpressed if related to the expression of the first cistron (compare chapter 1.5). Expression levels for the second cistron are usually 10 to 200-fold lower if compared to the expression of the first cistron. Due to the low expression of the second cistron the strong selection pressure favors rearrangements in the transcription unit (Adam et al., 1991; Kaufman et al., 1991). Therefore, the use of neutral screening markers instead of selective markers is recommended (Wirth et al., 1990; 1991). Bicistronic expression vectors have been described for screening of high level expressing clones, in which the second underexpressed cistron was used as an indicator for the expression of the first cistron. Stoichiometric expression is desirable for proteins built of similar or heterologous subunits. Equimolar expression of both cistrons has now been achieved by using intercistronic regions mediation reinitiation or coupled initiation (Dirks et al., 1995).

1.1.2.3.2 Regulation of Mammalian mRNA Translation

Ferritin mediates iron uptake and detoxification. Its synthesis is translationally controlled by iron ion concentrations. It involves an intracellular protein, IRE-BP, capable of binding to a region of secondary structure (IRE) occurring once in the 5′ leader of ferritin mRNA. IRE-BP exists in two forms, depending on the iron concentration. Upon conditions of iron deprivation the high affinty form of IRE-BP is prevalent. Iron excess results in the low affinity IRE-BP. The affinity change is based on a change in the protein's redox state which is determined by the cell's iron status. The high affinity form binds to the 5′ IRE in ferritin mRNA and prevents its translation. (Klausner and Harford, 1989). This is the best documented example of translation regulation. Since the range of regulation capacity is lower that that achieved by regulated transcription systems, it has not yet been used in biotechnology. It should be mentioned that many more examples of translational regulation do exist but are not immediately obvious. In specific application they might cause problems.

1.1.3 Protein Sorting, Modification and Degradation

1.1.3.1 Protein Sorting

Mammalian cells are complex factories containing a number of compartments with specialized functions. Transport and sorting of newly synthesized proteins into these compartments is a well defined process. Protein transport and sorting are functions of the protein structure and are supported or modified by specific cellular factors.

In most cases the necessary information for delivery of proteins resides in the protein structure. The domains responsible for this information are called signal sequences and sorting sequences. Newly synthesized proteins are delivered by two routes to their final destination. One of the routes is the secretory pathway which is the most important one for biotechnological processes and includes the synthesis of the nascent peptide chain into the endoplasmatic reticulum (ER). From the ER the proteins are transported unidirectionally by vesicular transport to intermediate organelles and the Golgi apparatus. From the Golgi apparatus the proteins are delivered into secretory vesicles or vesicles with lysosomal destination. The second route encompasses the synthesis of proteins on free ribosomes into the cytoplasm and the delivery for organelles like the nucleus, the mitochondria or the peroxisomes. Since for biotechnological purposes membrane anchored and secreted proteins are the focus of interest, the second route is not discussed further.

The signal sequences responsible for the secretory route are in most cases N-terminal sequences which are cleaved off after translocation into the ER. They do not show an obvious homology on the level of amino acid sequence. Rather they show a common overall structure. The length of these signal peptides is between 13 and 30 amino acids. They have a net positive charge at the amino terminus, a core of at least 9 hydrophobic amino acids and a region for cleavage by the signal peptidase with small sized residues at positions -1 and -3 (with respect to the cleavage site). As shown by a number of experiments, signal sequences are functionally independent from the mature proteins as they work in different contexts.

Thus, it seems possible to direct intracellular proteins by engineering of their N-terminus into the secretory compartment. The secretory signal sequences seem to be dominant over other signals contained in the protein, which is probably based on the fact that it appears prior to the other types of signal or sorting sequences. Proteins with synthetic signal sequences are usually, but not always cleaved correctly. The main handicap, however, is the further transport through the secretory compartments. Only in very few cases was it possible to bring cellular or heterologous proteins to secretion. Specific mechanisms, mostly based on the protein folding pathway which resides within the secretory compartments might be responsible for eventual failure.

Integral membrane proteins contain in addition to the signal sequence for the ER a membrane anchored domain which is usually built of hydrophobic residues. This domain stops the transfer of the nascent protein and leads to the residence of the protein in the membrane. Integral membrane proteins might have one or more transmembrane domains with different or equally orientated N- or C-termini.

While secretion of normally cellular proteins by the addition of signal sequences is unpredictable, the engineering of proteins which are dedicated to the secretory pathway can be manipulated with regard to cell surface membrane anchorage versus secretion. Cell surface proteins which are usually anchored by a hydrophobic domain within the protein can be manipulated to be secreted by deletion of this domain. Conversely, in many cases it could be shown that the addition of anchor sequences to normally secreted proteins results in a cell surface membrane anchorage.

It was postulated that constitutive traffic to the plasma membrane requires no signals (Pfeffer and Rothman, 1987). Forward transport was assumed to be driven by bulk flow. However, this issue is still under debate and the final answer is not known today (Balch et al., 1994). Nevertheless, the maintenance of proteins in specific compartments of the secretory pathway is due to specific retention signals. A number of such signals have been identified, e.g. for retention in the ER (Munro and Pelham, 1986; Pelham, 1988).

1.1.3.2 Post-translational Modifications of Proteins

A number of covalent modifications of amino acids during or after translation have been uncovered. The function and the influence of these modifications on the properties of the relevant protein can be tremendous. In other cases, no obvious features of the modified proteins are recognizable if compared to the nonmodified ones. Basically, modifications may alter a protein by changing its tertiary structure, charge or hydrophobicity. Since there are no rules for the effects triggered by certain modifications on individual proteins it is difficult to discuss the relevance in a general manner.

The most important modification in secretory and membrane proteins in mammalian cells concerns glycosylation. Two types of glycosylation can be distinguished with respect to carbohydrate moiety, time point of addition, linkage to the protein and target amino acid. N-Glycosylation involves the *en-block* transfer of a precursor molecule to an asparagine by N-linkage of GlcNAc. The precursor is synthesized as a core unit of 14 saccharides. The asparagine residue is contained within the discrete recognition motif Asn-X-Ser/ Thr. During transport of the protein from the ER to the Golgi apparatus processing steps occur which include the addition and removal of certain residues in the carbohydrate chain. O-glycosylation of proteins is a post-translational modification which is carried out sequentially along the late ER and Golgi apparatus. O-glycans are only found linked to serines or threonines. There are no strict recognition sequences. For a detailed discussion of glycosylation compare Chapter 2.2.

The role of glycosylation for proteins cannot be defined straight forwardly. Carbohydrates have been shown to play a role in a multitude of biological processes such a protein folding, stability, targeting, clearance, as well as cell-matrix and cell-cell interactions (Varki, 1993). In most cases, glycosylation is the reason why a certain protein is produced in mammalian cells and not in other cells like yeasts or prokaryotes.

It is important to note that different mammalian cell lines producing the same glycoprotein give rise to different modifications. This seems to be dependent on the species from which the cell lines are derived and the level of differentiation. Even variants (clones) from the same cell line may synthesize carbohydrate side chains which are differently composed. Therefore, the choice of a producer cell line determines the detailed type of glycosylation of a certain protein. Often, glycoproteins from a single cell line are not homogeneously modified; a certain heterogeneity seen upon electrophoretic separation is produced from a single cell line, even if it is grown under identical conditions. Further alterations are obtained by changing culture conditions. Although this happens as well in nature, the production of pharmaceuticals requires reproducible processes and homogenous products.

1.1.3.3 Degradation of Proteins

With the exception of secreted proteins the overall concentration of a protein in a cell is a function of its rate of synthesis and degradation. Although there are several pathways of degradation found in the cell, the ubiquitin dependent degradation pathway seems to be the most important one (for review see Hershko and Ciechanover, 1992). The half-life of individual proteins in a cell varies considerably. Short-lived proteins have a half-life of less than 30 minutes. Short-lived proteins are typical signal transduction molecules like cell cycle controlling cyclins. Various proteins involved in constitutive maintenance of cellular architecture or constitutive functions have a half-life of more than a hundred hours. Again, a number of signals on proteins have been identified which determine the stability of a protein. Degradation of proteins is not localized to a certain cellular compartment. It happens in the cytosol as well as in the nucleus and the ER.

1.1.4 Choice and Improvement of Mammalian Cell Lines

1.1.4.1 Cell Lines for Biotechnological Use

Theoretically, any cell which is maintained in tissue culture can be manipulated to express foreign genes. Transient expression is a method which rapidly gives rise to high concentrations of the expressed products. Transient expression systems are therefore convenient to check the integrity of the expression construct and some properties of the derived product. Provided that only small amounts are required transient expression allows analysis of the nature of the gene product. Rapid results are achieved by transient expression in COS cells. However, vectors with specific properties are required to obtain optimal expression. Other host cell lines which lead to very efficient transient expression, but do not require specific features of the plasmid expression vector include human 293 and hamster BHK-21 cells. In both of these hosts efficient gene transfer and transcription but not episomal replication are responsible for high expression of the exogeneous DNA.

Depending on the cell line, the efficiency of gene transfer, expression and stability differs widely, e. g., expression from the same plasmid, when transferred to different, cell lines differs in order of magnitude, regardless of whether expression is measured in the transient phase or in stable cell clones. Although the nature of the expression plasmid with its control elements plays a major role for efficiency of expression, these great differences cannot be explained simply with the use of expression controlling sequences. It seems that certain cell lines allow the packaging for foreign DNA in a transcription favoring form of chromatin, whereas other cell lines induce the packaging of foreign DNA in less active chromatin. Alternatively, the activity of chromosomal sites for integration of foreign DNA may differ from cell line to cell line and by this determine expression strength. Although the phenomenon has been known for a long time, it is still not understood. The fact, however, remains that from hundreds of cell lines only a limited number show high expression of foreign genes.

Table 1.1.2. Properties of host cell lines for recombinant protein production

Property	**Examples**
Efficient expression of foreign genes	Transcription and post-transcriptional processing Protein synthesis Secretion
Post-translational modifications	e.g. Protein folding Glycosylation, Phosphorylation, ...
Stability of transgene expression	Genotypic (chromosomal stability) Phenotypic (CpG methylation, Chromosome structure alterations)
Absence of adventitious agents	Viruses Mycoplasms
Cultivation requirements	Serum, Growth factors Stabilizing proteins
Fermentation properties	Growth to high cell densities Resistance to shear forces Aeration processes Production of toxic compounds (e.g. lactic acid)
Production under fermentation conditions	High production at high cell density Reduced cell growth Protein requirements in the medium
Cell growth	Fast growth at low cell density Reduced growth at high cell density

Table 1.1.2 lists some other properties which cell lines for large scale production of recombinant proteins must have. Among those, stability of foreign gene expression and the absence of adventitious agents are key parameters for general acceptance or exclusion of certain lines for production of pharmaceutical proteins. Today, this exclusivity is no longer seen as strictly as it has been in the past, e.g., contaminating murine retroviruses are present in most murine myeloma cell lines. As a consequence, the purification procedure for the protein of interest must allow the reduction of such virus species for more than ten log-scales. In specific cases this might be so cost intensive that the process in total is not economically feasible. Taking all considerations together, only very few cell lines are acceptable for large scale production of pharmaceutical proteins (Tab. 1.1.3). Indeed, today most recombinant proteins are produced in CHO, SP2/0 and BHK-21 cells. This extremely small number results not only from rational considerations but mainly from the fact that control agencies, after having accepted a certain cell line as safe, give credit to further applications of processes based on the same cellular background.

Table 1.1.3.

Cell line	Species tissue	Ad-her-ence	Transfection method	Specific features	Prerequisites for high expression	Long term produc-tion	References
CHO dhfr	hamster	+	Calcium phosphate coprecipitation	dhfr	Gene amplification	+	Urlaub and Chasin, 1980; Kaufman and Sharp, 1982
BHK-21	hamster	+	Calcium phosphate coprecipitation	Efficient transient and stable expression	Gene amplification is not required	+	Wirth et al., 1988
Myeloma, e.g. SP2/O, 3558L P3-X63-AG8653	mouse	–	Lipofection, protoplast fusion, electroporation		Ig-promoter/enhancers are favorable	+	Oi et al., 1983
293	human	+	Calcium phosphate coprecipitation	Constitutive expression of Adenovirus E1a gene; extremely high transient expression; high expression in stable transferrants	Promoters, which are efficiently transactivated by E1a	+	Graham, et al., 1977; Walls et al., 1989

Cell clones resulting from stably transfected expression plasmids exhibit marked differences in expression levels of the introduced gene. These differences are predominantly caused by two reasons: different copy numbers and integration into more or less transcriptionally active sites. Good candidates for protein production are single clones which exhibit a stable and high-level expression of the foreign gene. This requires the screening for such candidates. Strategies and methods for screening overexpressing cell clones are described in chapter 1.5. Working with characterized single clones is preferred for many reasons. Clone mixtures are not homogenous with respect to their growth properties, expression levels and in some cases even in product quality. Particularly, nonproducing cells may overgrow the high-level producing cell clones if the product has adverse effects on cell growth.

1.1.4.2 Expression Vectors

Mammalian expression vectors normally contain prokaryotic elements which allow replication and selection in E. coli. They further harbor elements to mediate transcription initiation and termination (polyadenylation) in the cells dedicated for manipulation. Apart from these basic components mammalian expression vectors may contain a series of other elements which modulate transcription efficiency, post-transcriptional processes, replication, specific integration and gene amplification or helper function for delivery of the nucleic acids of the vector into the target cells.

For gene expression all steps up to transcription and eventually splicing are independent of the protein coding region and therefore can be adjusted by the vector construction, gene transfer method and selection. In contrast, control of expression for the subsequent levels is mostly dictated by the individual gene. Certain features of the transcribed mRNA like the 5′ UTR, the 3′ UTR or the context around the initiator AUG of a given gene may also determine its expression efficiency. Often, the mechanisms governing post-transcriptional regulation are not known or appear extremely complex. It should be kept in mind that at the RNA level not only linear sequences but also secondary structures play an important functional role. In many cases it is not possible to manipulate gene expression at the post-transcriptional level. Moreover, as coding sequences are involved in many post-transcriptional events manipulations in the nucleic acid sequence without changing the amino acid sequence of the gene product are required.

Many events controlling gene expression are partially or completely tissue specific. Therefore, expression parameters including transcriptional or post-transcriptional events exhibited in a certain host cell line cannot be expected to be identical in another cell line, even if the cell line originates from the same species and tissue.

Today, numerous expression vectors exist. More and more are commercially available. It is impossible to say which one is superior to others. It is clear, however, that the choice of the vector depends greatly on the cell line to be used and the way in which this cell is manipulated and how overexpressing cell clones are isolated.

1.1.4.3 Cell Growth and Apoptosis

In the past, most cell death in large scale culture was attributed to necrosis. Today, it is well known that mild environmental stress induces many cell types to undergo the genetically controlled process of "programmed cell death", also known as apoptosis.

Apoptosis is characterized by cell shrinkage, condensation of nuclear chromatin, nuclear fragmentation and rapid disappearance of cell debris after lysis. The biochemical hallmark of apoptosis is the fragmentation of DNA into oligomers of nucleosomal DNA, leading to the typical ladder. Apart from cytometrical analysis (microscopic and FACS), this ladder formation or the enzymatic addition of labeled nucleotides to free ends of the fragmented DNA serve as sensitive detection methods for apoptosis.

A number of agents, treatments and stress conditions lead to the induction of apoptosis. Specifically, the supply of nutrients, pH, shear forces, accumulation of waste products and the availability of oxygen play an important role in the decision whether a cell continues to live or undergoes apoptosis. In particular, medium depletion is a major cause of apoptosis for lymphoid cells. Certain serum factors, like IGF-1, are potent survival factors and preserve the cells from apoptotic death.

In large-scale culture for preparation of pharmaceuticals, apoptosis as well as necrosis are major obstacles. The existence of cell debris makes it more difficult to separate the product from the medium and the release of proteases can lead to degradation of the product (for review see Cotter and Al-Rubeai, 1995). Since the cessation of proliferation and the onset of apoptosis are often coupled it is difficult to reach an optimized production period at which maximal product formation takes place in nonproliferating cells. Therefore, compounds with anti-apoptotic function have been evaluated. Specifically, the use of aurintricarboxylic acid (ATA), zinc ions and antioxidants like vitamins C and E, have been evaluated (Helgason et al., 1993). The price of survival factors and cytokines is too high for commercial production.

1.1.4.4 Improvement of Cellular Properties

In order to fulfill the requested properties (see table 1.1.2), numerous cell lines have been screened. However, this approach was not very successful. Today, only a few cell lines show favorable properties for recombinant protein production (1.1.4.1). An alternative is to use these standard cell lines and isolate variants or mutants which exhibit the required properties. Indeed, this is possible and has been practiced with some success, e.g., variants concerning the glycosylation properties have been isolated from CHO and BHK-21 cells either after treatment with mutagenizing agents or as a result of spontaneous events. However, since all these events are random, screening is tedious and sometimes impossible. In most cases, loss of specific functions are achieved by this method. The gain of new functions is not usually possible. To overcome this problem, cell lines are genetically manipulated in order to achieve specific functions or metabolic properties. This approach is called metabolic design. A few examples are available which show the success of this approach, e.g., the introduction of a gene encoding glutamine synthetase is not only used as a selectable marker (see chapter 1.2) but renders the cells independent of exogenously added glutamine). Recently, the gene encoding α 2–6 sialyltransferase was transferred into BHK-21 and CHO cells which lack this enzyme. The transfectants secrete proteins which show α 2–6 linked Neuraminic acids linked to N-glycosyl-moieties (Grabenhorst et al., 1995).

References

Adam, M.A., Ramesh, N., Miller, A.D., and Osborne, W.R.A. (1991) Internal initiation of translation in retroviral vectors carrying picornavirus 5′ nontranslated regions. J. Virol, *65*. 4985–4990.

Artelt, P., Morelle, C., Ausmeier, M., Fitzek, M. and Hauser H. (1988) Vectors for efficient expression in mammalian fibroblastoid, myeloid and lymphoid cells via transfection or infection. Gene *68*. 213–219.

Balch, W.E., McCaffery, J.M., Plutner, H., Farquhar, M.G. (1994) Vesicular Stomatitis Virus Glycoprotein is sorted and concentrated during export from the Endoplasmic Reticulum. Cell *76*. 841–852.

Bode, J., Kohwi, Y., Dickinson, L., Joh, T., Klehr, D., Mielke, C. and Kohwi-Shigematsu, T. (1992) Biological significance of unwinding capability of nuclear Matrix associating DNAs.

Bode, J., Schlake, T., Ríos-Ramírez, M., Mielke, C., Stengert, M., Kay, V. and Klehr-Wirth, D. (1995) "Matrix-attached regions: Structural properties creating tanscriptionally active loci" in the International Review of Cytology series on "Structural and Functional Organization of the Nuclear Matrix", R. Berezney and K.W. Jeon, ed., Academic Press, 389–453.

Boel, E., Berkner, K.L., Nexoe, B.A. and Schwartz, T.W. (1987) Expression of a human pancreatic polypeptide precursors form a dicistronic mRNA in mammalian cells. FEBS Letters *219*. 181–188.

Boshart, M., Weber, F., Jahn, G., Dorsch-Häsler, K., Fleckenstein, B. and Schaffner, W. (1985) A very strong enhancer is located up-stream of an immediate early gene of human cytomegalovirus. Cell *41*. 521–530.

Braselmann, S., Graninger, P. and Busslinger, M. (1993) A selective transcriptional induction system for mammalian cells based on Gal4-estrogen receptor fusion proteins. Proc. Natl. Acad. Sci. USA. *90*. 1657–1661.

Breathnach, R. and Chambon, P. (1981) Organisation and expression of eukaryotic split genes coding for proteins. Annu. Rev. Biochem. *50*. 349–383.

Buchman, A.R. and Berg, P. (1988) Comparison of intron-dependent and intron-independent gene expression Mol. Cell. Biol. *8*. 4395–4405.

Cleveland, D.W. and Yen, T.J. (1994) Multiple determinants of eukaryotic mRNA stability. The New Biologist *1*. 121–126.

Cotter, T.G. and Al-Rubeai, M. (1995) Cell death (apoptosis) in cell culture systems. Tibtech *13*. 150–155.

Curtis, D., Lehmann, R., Zamore, P.D. (1995) Translational regulation in development. Cell *81*. 171–178.

Dirks, W., Wirth, M. and Hauser, H. (1993) Bicistronic transcription units for gene expression in mammalian cells. Gene *128*. 247–249.

Dirks, W., Schaper, F., Kirchhoff, S., Morelle, C. and Hauser, H. A multifunctional vector family for gene expression in mammalian cells. Gene *149*. 387–388.

Eden and Cedar, (1994) Role of DNA methylation in the regulation of transcription. Curr. Opin. Genet. Dev. *4*. 255–259.

Eilers, M., Picard, D., Yamamoto, K.R. and Bishop, J. (1989) Chimaeras of Myc oncoprotein and steriod receptors cause hormone-dependent transformation of cells. Nature *340*. 66–68.

Evans, T., Felsenfeld, G. and Reitman, M. (1990) Control of globin gene transcription. Annu. Rev. Cell. Biol. 95–124.

Falcone, D. and Andrews, D. W. (1991) Both the 5′untranslated region and the sequences surrounding the start site contribute to efficient initiation of translation in vitro. Mol. Cell. Biol. *11*. 2656–2664.

Gallie, D.R., Sleat, D.E., Watts, J.W., Turner, P.C. and Wilson, T.M.A. (1988) Mutational analysis of the tobacco mosaic virus 5′-leader for altered ability to enhance translation. Nucleic Acids Res. *16*. 883–893.

Gasser, S. and Laemmli, U.K. (1987) A glimpse at chromosomal order. Trends Genet. *3*. 16–22.

Godefroy-Colburn, T., Ravelonandro, M. and Pinck, L. (1985) Cap accessibility correlates with the initiation efficiency of alfalfa mosaic virus RNAs. Eur. J. Biochem. *147*. 549–552.

Gossen, M. and Bujard, H. (1992) Tight control of gene expression in mammalian cells by tetracycline-responsive promoters. Proc. Natl. Acad. Sci. USA. *89*. 5547–5551.

Grabenhorst, E., Hoffmann, A., Nimtz, M., Zettlmeißl, G. and Conradt, H.S. (1995) Construction of stable BHK-21 cells coexpressing human secretory glycoproteins and human Galβ1-4GlcNAc-R α2,6-sialyltransferase: α2,6-linked NeuAc is preferably attached to the Galβ1-4GlcNAcβ1-1Manα1-3-branch of biantennary oligosaccharides from secreted recombinant β-trace protein. Eur. J. Biochem. *232*, 718–725.

Graham, F.L., Smiley, J., Russel, W.C. and Naim, R. (1977) Characterictics of a human cell line transformed by DNA from human adenonines type 5. J. Gen. Virology *36*, 59–72

Hamer, D.H. and Leder, P. (1979) Splicing and the formation of stable RNA. Cell. *18*. 1299–1302.

Helgason, C.D. et al. (1993) Exp. Cell. Res. *206*. 302–310.

Hentze, M.W. (1991) Translational regulation of ferritin biosynthesis by iron: A review. Curr. Stud. Hematol. Blood Transfus. *58*. 115–126.

Hershbach, B.M. and Johnson, A.D. (1993) Transcriptional repression in eukaryotes, Annu. Rev. Biol. *9*. 479–509.

Hershey, J.W.B. (1991) Translational control in mammalian cells. Rev. Biochem. *60*. 717–755.

Hershko, A. and Ciechanover, A. (1992) The ubiquitin system for protein degradation. Annu. Rev. Biochem. *61*. 761–807.

Hippenmeyer, P. and Highkin, M. (1993) High level, stable production of recombinant proteins in mammalian cell culture using the Herpesvirus VP16 transactivator. Bio/Technology *11*. 1037–1041.

Huang, M.T.F. and Gorman, C.M. (1990) The simian virus 40 small-t intron, present in many common expression vectors, leads to aberrant splicing. Mol. Cell. Biol. *10*. 1805–1810.

Jackson, R.J. (1991) mRNA translation: Initiation without an end. Nature (London). *353*. 14–15.

Jackson, R.J. and Standart, N. (1990) Do the poly(A) tail and 3′ untranslated region control mRNA translation? Cell *62*. 15–24.

Karreman, S., Hauser, H. and Karreman, C. (1996) On the use of double FLP recognition targets (FRTs) in the LTR of retroviruses for the construction of high producer cell lines. Nucleic Acids Res. *24*, 1616–1624.

Kaufman, R.J., Davies M.V., Wasley, L.C. and Michnick, D. (1991) Improved vectors for stable expression of foreign genes in mammalian cells by use of the untranslated leader sequence from EMC virus. Nucleic Acids Res. *19*. 4485–4490.

Kaufman, R. (1990) Vectors used for expression in mammalian cells. Meth. Enzymol. *185*. 487–512.

Kaufman, R.J., Murtha, P. and Davies, M.V. (1987) Translational efficiency of polycistronic mRNAS and their utilization to express heterologous genes in mammalian cells. EMBO J. *6*. 187–193.

Klausner, R.D. and Harford, J.B. (1989) cis-trans models for post-transcriptional gene regulation. Science *246*. 870–872.

Klehr, D. Schlake, T., Maass, K. and Bode, J. (1992) Scaffold-attached regions (SAR elements) mediate transcriptional effects due to butyrate. Biochemistry *31*. 3222–3229.

Klehr, D., Maass, K. and Bode, J. (1991) Scaffold-attached regions from the human interferon beta domain can be used to enhance the stable expression of genes under the control of various promoters. Biochemistry *30*. 1264–1270.

Kozak, M. (1986) Point mutations define a sequence flanking the AUG initiator codon that modulates translation by eukaryotic ribosomes. Cell *44*. 283–292.

Kozak, M. (1987a) An analysis of 5'-noncoding sequences from 699 vertebrate messenger RNAs. Nucleic Acids. Res. *15*. 8125–8148.

Kozak, M. (1987b) Effects of intercistronic length on the efficiency of reinitiation by eucaryotic ribosomes. Mol. Cell. Biol. *7*. 3438–3445.

Kozak, M. (1989) Circumstances and mechanics of inhibition of translation by secondary structure in eucaryotic mRNAs. Mol. Cell. Biol. *9*. 5134–5142.

Kozak, M. (1991a) An analysis of vertebrate mRAN sequences: intimations of translational control. J. Cell. Biol. *115*. 887–903.

Kozak, M. (1991b) Effects of long 5' leader sequences on initiation by eukariotic ribosomes in vitro. Gene Expression *1*. 117–125.

Kriegler, M. (1990) Gene transfer and expression. A laboratory manual, Stockton Press.

Laird-Offringa, I. A. (1992) What determines the instability of c-myc proto-oncogene mRNA? BioEssays *14*. 119–124.

Lamb, P. and McKnight, S. L. (1991) Diversity and specificity in transcriptional regulation: the benefits of heterotypic dimerization. TIBS *16*. 417–422.

Latchman, D. S. (1991) Eucaroytic transcription factors, Academic Press Ltd., London.

Lin, F.-L., Sperle, K. and Sternberg, N. L. (1984) Model for homologous recombination during transfer of DNA into mouse L cells: role for DNA ends in the recombination process. Mol. Cell. Biol. *4*. 1020–1034.

Maniatis, T. (1991) Mechanisms of alternative pre-mRNA splicing. Science *251*. 33–34.

Mattaj, I. W. (1990) Splicing stories and poly (A) tales: An update on RNA processing and transport Curr. Opin. Cell. Biol. *2*. 528–538.

McNeall, J., Sanchez, A., Gray, P. P., Chesterman, C. N. and Sleigh, M. J. (1989) Hyperinducible gene expression from a metallothionein promoter containing additional metal-responsive elements. Gene *76*. 81–88.

Müllner, E. W., Neupert, B. and Kühn L. C. (1989) A specific mRNA binding factor regulates the iron-dependent stability of cytoplasmic transferrin receptor mRNA. Cell *58*. 373–382.

Mountford, P. S. and Smith, A. G. (1995) Internal ribosome entry sites and dicistronic RNAs in mammalian transgenes. Trends Genet. *11* (5). 179–184.

Munroe, D. and Jacobson, A. (1990) mRNA poly(A) tail, a 3' enhancer of translational initiation. Mol. Cell. Biol. *10*. 3441–3455.

Munro, S. and Pelham, H. R. B. (1986) An HSP 70-like protein in the ER: identity with the 78 kd glucose-regulated protein and immunoglobulin heavy chain binding protein. Cell *46*. 291–302.

Needham, M., Gooding, C. Hudson, K., Antonion, M., Groveld, F. and Hollis, M. LCR/MEL (1992) A versatile system for high-level expression of heterologous proteins in erythroid cells. Nucleic Acids Res. *20*. 997–1003.

Neuberger, M. S. and Williams, G. T. (1988) The intron requirement for immunoglobin gene expression is dependent upon the promoter. Nucleic Acids Res. *16*. 6713–6724.

Peltz, S. W., Brewer, G., Bernstein, P., Hart, P. A. and Ross, J. (1991) Regulation of mRNA turnover in eukaryotic cells. Crit. Rev. Euk. Gene Expression *1*. 99–126.

Pfeffer, S. R. and Rothman, J. E. (1987) Biosynthetic protein transport and sorting by the Endoplasmic Reticulum and Golgi. Annu. Rev. Biochem. *56*. 829–852.

Proudfoot, N. (1991) Poly(A) signals. Cell *64*. 671–674.

Richter, J. D. (1991) Translational control during early development. BioEssays *13*. 179–183.

Roeder, R. G. (1991) The complexities of eukaryotic transcription initiation: regulation of preinitiation complex assembly. TIBS *16*. 402–408.

Searle, P. F., Stuart, G. W. and Palmiter, R. D. (1985) Building a metal-responsive promoter with synthetic regulatory elements. Mol. Cell. Biol. *5*. 1480–1489.

Sedivy, J.M. and Joyner, A.L. (1992) Gene Targeting. Freeman and Co., New York
Shaw, G. and Kamen, R. (1986) A conserved AU sequence from the 3′ untranslated region of GM-CSF mRNA mediates selective mRNA degradation. Cell *46*. 659–667.
Simoes, E.A.F. and Sarnow, P. (1991) An RNA hairpin at the extreme 5′ end of the poliovirus RNA genome modulates viral translation in human cells. J. Virol. *65*. 913–921.
Sonenberg, N. (1993) Remarks on the mechanism of ribosome binding to eukaryotic mRNAs. Gene Expression *3*. 317–323.
Varki, A. (1993) Biological roles of oligosaccarides. All theories are correct. Glycobiology *3*. 97–130.
Walls, J.D., Berg, D.T., Yan, S.B. and Grimell, B.W. (1989) Amplification of multicistronic plasmids in the human 293 cell line and secretion of correctly processed recombinant human protein C. Gene *81*, 139–149.
Weiner, A.M. (1993) mRNA splicing and autocatalytic introns: distant cousins or the products of chemical determinism, Cell *72*. 161–164.
Wilson, T. and Treisman, R. (1988) Removal of poly(A) and consequent degradation of c-fos mRNA facilitated by AU-rich sequences. Nature (London). *336*. 396–399.
Wingender, E. (1993) Gene Regulation in Eukaryotes. VCH, Weinheim.
Wirth, M., Schumacher, L. and Hauser, H. (1991) Use of dicistronic transcription units for the correlated expression of two genes in mammalian cells in modern approaches to animal cell technology (Griffiths, B., Spier, R., and Meigner, R. Eds.) pp. 338–343, Butterworths.
Wirth, M. and Hauser, H. (1993) Genetic engineering of animal cells (ed. A. Pühler) pp. 1–82. VCH Publishes, Weinheim.
Wirth, M., Höxter, M., Morelle, C. and Hauser H. (1990) Bicistronic expression vectors facilitate screening for overexpressing mammalian cells in DECHEMA Biotechnology Conferences (Behrens, D., and Krämer, P. Eds.) Vol. 4, pp. 69–73. VCH Publishers, Weinheim.

1.2 Permanent Gene Expression in Mammalian Cells: Gene Transfer And Selection

Uwe Schlokat, Michèle Himmelspach, Falko G. Falkner and Friedrich Dorner*

During the last decade a wide variety of different gene transfer or 'transfection' methods have been developed to introduce foreign DNA into mammalian cells (Fig. 1.2.1). The chosen strategy will depend on cell type (e.g. primary versus permanent cells), growth characteristics (e.g. suspension versus adherent cells), and subsequent use of the engineered cell (e.g. scale-up of cell lines for large scale production of a desired protein versus cells to be manipulated for creating transgenic animals). Different strategies have proven to be more suitable than others for a given purpose.

In section 1.2.1 of this chapter the different gene transfer methods available to date will be discussed.

The ability of different cell lines to take up DNA varies widely. Generally, only a fraction of the cells manipulated to take up DNA at a given time actually is competent to do so. The amount of DNA taken up by individual cells in a given population also varies.

Once the DNA has been taken up by the cells the corresponding proteins will be *transiently* expressed until cellular nucleases have degraded the foreign DNA molecules. Depending on the cell line used, transient expression can last between a few hours and a few weeks.

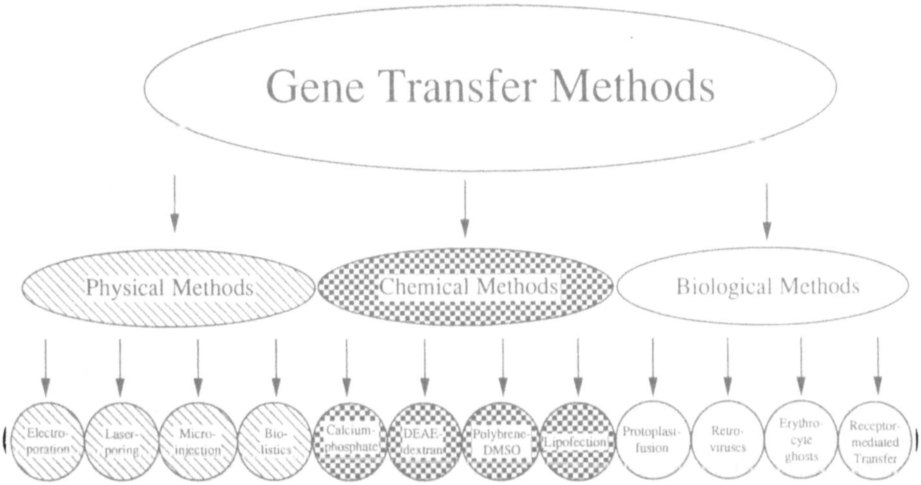

Fig. 1.2.1: A wide variety of gene transfer methods for different applications are available today. They are described in section 1.2.1

* Corresponding author

Using vectors with very strong promoters/enhancers (which ensure a high transcription rate) in conjunction with cells that easily take up DNA, expression levels as high as one microgram recombinant protein/10^6 cells × 24 hours can be achieved. This temporary gene expression might be improved even further by repeated transfection into the same cells on successive days (Ishikawa and Homcy, 1992), or by providing additional potent *trans*-acting factors such as Adenovirus E1a (Berg et al., 1992). Furthermore, cotransfection and subsequent expression of adenoviral VA genes might prevent the reduction of translational activity in the host cell during transient expression (Kaufman, 1985; Mathews and Shenk, 1991).

High level transient gene expression may also employ vectors capable of *replicating* within the transfected cells thereby increasing the template gene copy number. These vectors contain viral origins of replication and depend on the presence of the corresponding viral *trans*-acting replication factors in the transfected cells.

The most popular such system is based on Simian Virus 40 (SV40) origin of replication-containing vectors in conjunction with stably SV40 T-antigen expressing african green monkey derived CV-1 cells, termed COS (Kriegler, 1990a; Nakayama et al., 1992). If the target cells do not provide T-antigen permanently, recombinant protein (rprotein) and SV40 T-antigen-expressing plasmids can be cotransfected simultaneously (de Chasseval and de Villartay, 1992).

Thus, by *transient* expression, questions concerning functionality of the expression vector used and quality of the rprotein produced can quickly be resolved.

Episomal expression systems, analogous to the SV40 origin of replication/COS system, can also be applied to *longterm* (as opposed to *transient*) gene expression. However, rather than the SV40 origin of replication, genetic elements from viruses such as Epstein-Barr, Polyoma or Papova are used instead in the respective expression vectors (Groger et al., 1989; Heffernan and Dennis, 1991).

If an rprotein is to be expressed on a *preparative* scale most methods require its genetic information to be stably anchored within the cellular chromosomes in order to evade degradation and to ensure stable and equal distribution to the daughter cells. In order to quickly identify cells *stably and permanently* expressing rprotein, a gene harboring a *selectable marker* (e.g. a gene conferring resistance against some cytotoxic drug such as certain antibiotics) is usually co-introduced into the cell together with the gene coding for the rprotein.

Upon subsequent exposure to the cytotoxic agent only cells having taken up and expressing the selectable marker will survive. A certain fraction of these cells will also express the genes for the desired rprotein.

Presumably random recombination events in the cells can lead to the *amplification* of genes. If the copy number of the selectable marker gene is increased the host cell will be able to tolerate higher concentrations of the corresponding cytotoxic drug. Therefore, cells with very high copy numbers of the selection/amplification marker gene can be selected for by stepwise increases in the drug concentration in the medium.

Due to its physical linkage to the amplification marker gene at a given site in the chromosome, the expression vector for the rprotein is often co-amplified resulting in high yields of rprotein.

The different selection and amplification markers availaible to date will be reviewed in section 1.2.2 (Fig. 1.2.2).

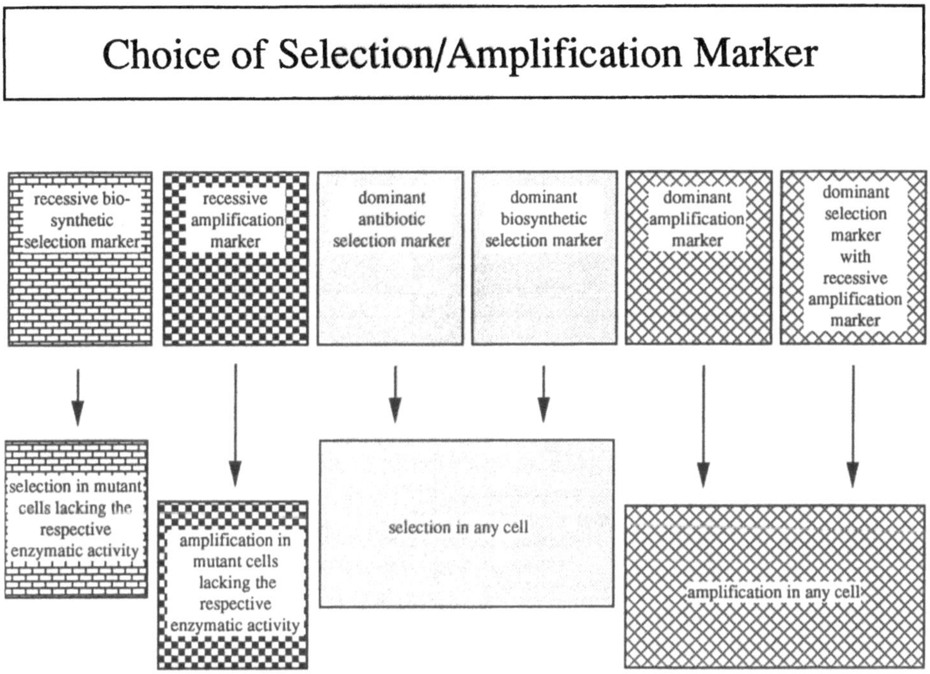

Fig. 1.2.2: The availability and suitability of different target cells and the demand for the amount of recombinant protein determines the choice of selection/amplification marker. The different selection and amplification markers available are introduced in section 1.2.2

Depending on the quantity of rprotein needed, the selection and amplification markers used, and the time available, different selection and amplification strategies have been developed (Fig. 1.2.3). These are described in section 1.2.3.

The various mammalian expression systems have led to high yield production of recombinant proteins. The prime restrictions associated with the available systems are the time consuming amplification procedure as well as the often insufficient post-translational modification systems of the host cells which can render large fractions of the produced rproteins inactive. Initial attempts to improve either are briefly discussed in section 1.2.4 (Fig. 1.2.4).

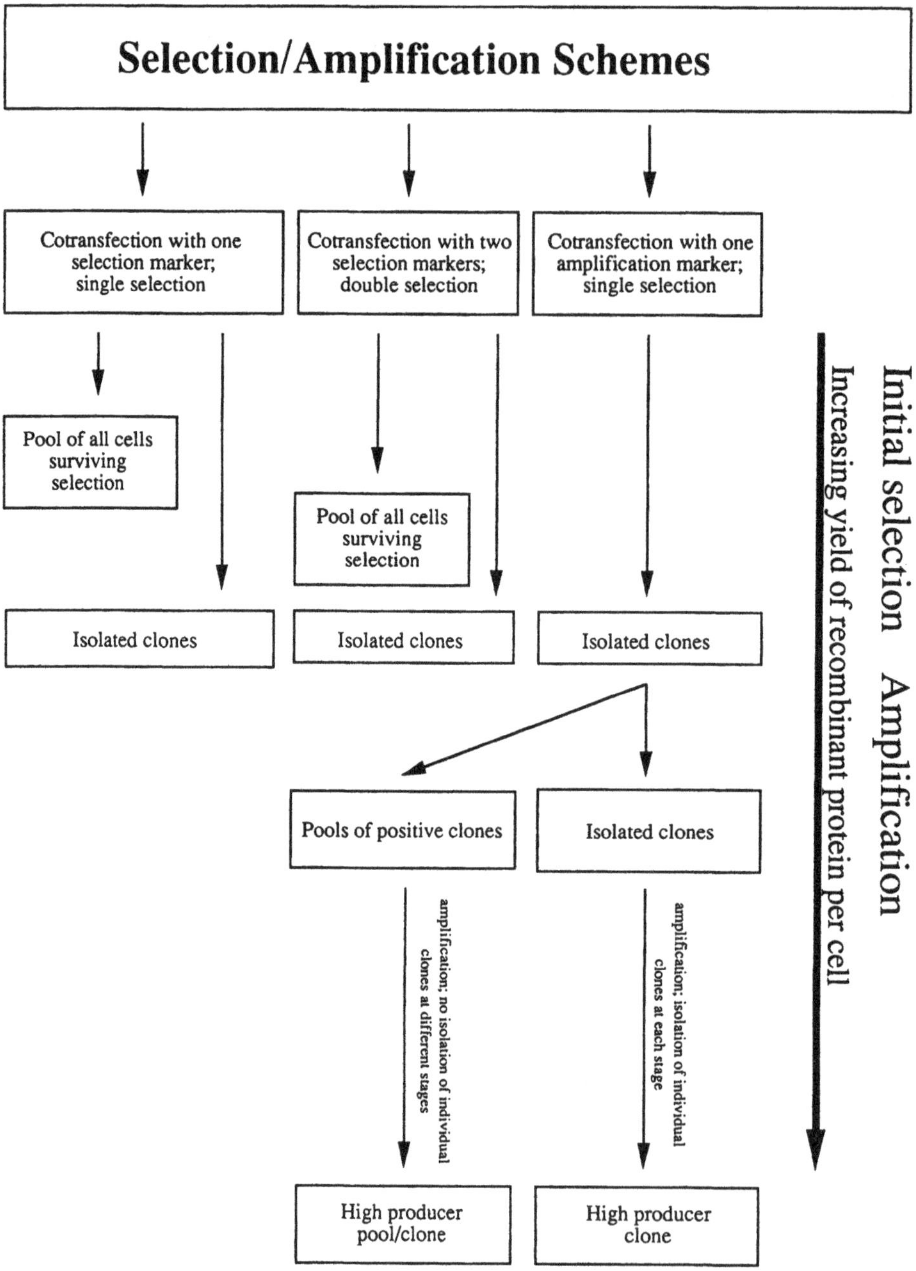

Fig. 1.2.3: Depending on the quantities of recombinant protein needed in what period of time, different selection/amplification strategies can be chosen. These are discussed in more detail in section 1.2.3

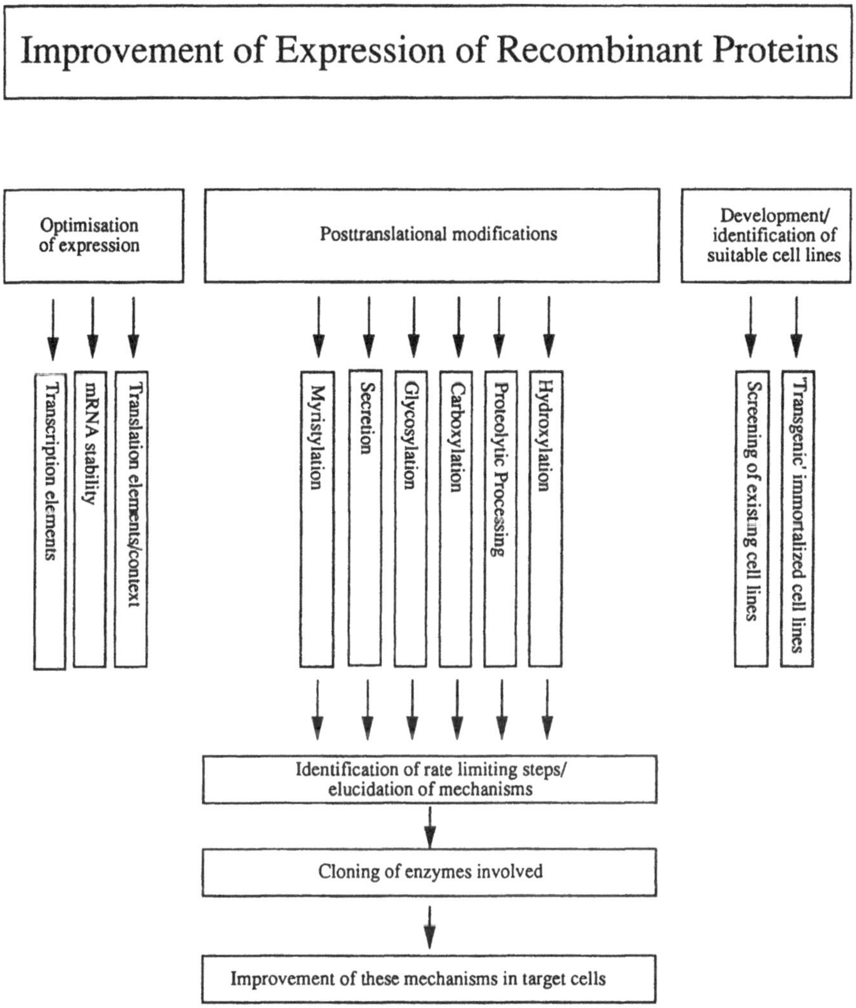

Fig. 1.2.4: Improvement of yield of fully functional recombinant proteins is a major concern of today's industrial cell technologists. Many different approaches are available, as outlined in section 1.2.4

1.2.1 Transfer of Foreign DNA into Mammalian Cell Lines

Nowadays a wide variety of techniques used to introduce foreign DNA into mammalian cells have become available (Fig. 1.2.1). Depending on the cells to be transfected the researcher generally has the choice between physical (electroporation, microinjection, laser poring and particle bombardment), biological (protoplast fusion, retrovirus mediated gene transfer, erythrocyte ghost fusion and receptor mediated gene delivery) and chemical (calcium phosphate coprecipitation, DEAE-dextran transfection, polybrene-DMSO and liposome mediated gene transfer) methods. While DEAE-dextran for transient and calcium phosphate coprecipitation for transient as well as stable transfections have become the most widely used methods, the other techniques also described below might provide significant advantages in specific cases.

1.2.1.1 Calcium Phosphate Coprecipitation Method

Originally described to increase the infectivity of adenoviral DNA, calcium phosphate coprecipitation has become the method of choice to introduce DNA into rapidly growing adherent tissue culture cells (Wigler et al., 1977; Maitland and McDougall, 1977). In contrast, anchorage independent cells generally are fairly poor recipients for calcium phosphate mediated DNA uptake (Okayama and Chen, 1991) even though the method might be considered in a few cases (Segal 1986a).

The method involves mixing calcium chloride, DNA and phosphate buffer at neutral pH, following which calcium phosphate DNA precipitates are formed. Methods whereby the precipitates are either formed prior to the application to the cells or within the petri dishes are available (Okayama and Chen, 1991). Once in the petri dish, the precipitates sink onto the cells and are subsequently taken up by phagocytosis. For many cell lines, transfection efficiency can be improved further by short exposures of the cells undergoing transfection to chemicals such as glycerol (Gorman et al., 1983; Frost and Williams, 1978), dimethyl sulfoxide (Stow and Wilkie, 1976), tubulin inhibitors (Faber and Eberle, 1976), lysosomal inhibitors such as chloroquine (Luthman and Magnusson, 1983), PEG (Shen et al., 1982), or sodium butyrate (Gorman and Howard, 1983). Even though parameters such as pH, duration of the precipitates on the cells, length of the additional chemical treatment, DNA concentration, DNA form, state of cell growth or temperature can influence the transfection efficiency, a fairly standardized procedure can be applied to a wide variety of different cell lines (Okayama and Chen, 1991). With commercial kits, such as the one from BRL, optimized systems are also available.

Calcium phosphate coprecipitation has proven to be the method of choice for the establishment of stable cell lines. Once in the cell, the transfected plasmids form large concatemers which can ultimately become stably integrated into the host cell's chromosomes by non-homologous recombination. Yet, only a small percentage (0.001–1%) of the transfected cells are stably transformed, while often 10% and more cells undergoing transfection take up DNA (Okayama and Chen, 1991). Recently, in situ formation (i.e. within the petri dish) of the calcium phosphate DNA precipitates has been found to dramatically improve stable transformation of various common fibroblastic and epithelial cell lines such as L, NIH 3T3, BHK, HeLa, CV1, NRK, C127 and CHO (Chen and Okayama, 1987; Chen and Okayama, 1988). With this modified method, up to 50% of the transfected cells were reported to become stable transformants (Chen and Okayama, 1987; Chen and Okayama, 1988). A very thorough investigation of the calcium phosphate DNA precipitate formation and its optimization has been published recently (Jordan et al., 1996).

1.2.1.2 DEAE-Dextran Mediated Gene Transfer

Similar to the calcium phosphate method, the polycationic diethyl-aminoethyl (DEAE)-dextran technique was originally used to enhance the infectivity of viral infections. Subsequently, the DEAE-dextran transfection method has been shown to be particularly suited for transient transfection assays (McCutchan and Pagano, 1968; Sompayrac and Dana, 1981). As with calcium phosphate coprecipitation, chemical shock treatment (e.g. with DMSO) significantly enhances transfection efficiency.

In contrast, the DEAE-dextran technique seems to be less suitable for stable transfection experiments (Milman and Herzberg, 1981). The DNA reportedly undergoes a higher mutation rate than DNA introduced into cells by other techniques (Calos et al., 1983, Razzaque et al., 1982; Ashman and Davidson, 1985). Since the DNA is taken up via endocytic processes DNA damage presumably results from its exposure to low pH or cellular nucleases. DEAE-dextran mediated gene transfer is used for some adherent cell lines such as COS cells but has been shown to be particularly valuable for transfection of suspension cells (for detailed protocols see Segal, 1986b).

1.2.1.3 Polybrene-DMSO Mediated Gene Transfer

Originally polybrene-DMSO mediated gene transfer had been described as a means to transfect chicken embryo fibroblasts with retroviral DNA (Kawai and Nishizawa, 1984). The synthetic polycation polybrene is thought to mediate the interaction of the DNA with the target cell membrane; subsequently, DMSO permeabilizes the membrane therefore facilitating internalization of the DNA into the cell.
The method was later adapted for the stable transformation of cell lines such as mouse L and NIH 3T3 cells, HeLa cells and a variety of cultured human fibroblasts at efficiencies on the order of 0,1 % (Morgan et al., 1986; Aubin et al., 1988; for a protocol, see Aubin et al., 1991). In particular, polybrene-DMSO mediated gene transfer has been reported to be extremely successful for the stable transformation of CHO cells (Chaney et al., 1986). About 10 % of the CHO cells exposed to transfection were observed to become stably transformed even with minute amounts of DNA (several nanograms).

1.2.1.4 Electroporation

When exposed to brief high voltage electric fields, cells are able to take up DNA from the surrounding medium (for a more general review on electroporation see Andreason and Evans, 1988, more detailed technical information can be obtained from Bio-Rad Laboratories). Simplified models postulate the transient creation of micropores in the cell membrane through which the DNA enters the cytosol thereby circumventing potential lysosome mediated DNA damage. Due to the latter effect, lower mutation frequencies than observed for the chemical gene transfer methods have been reported (Drinkwater and Klinedinst, 1986).

Similar to particle bombardment described further below, electroporation is particularly valuable when cells either resist efficient transfection by the more common chemical methods and/or grow in suspension (Potter et al., 1984; Toneguzzo and Keating, 1986; Sureau et al., 1986).

In addition to popular cell lines such as CV-1 (Chu et al., 1987), NIH 3T3 (Chu et al., 1987), COS (Callis et al., 1987; Cameron and Jennings, 1989), HeLa (Chu et al., 1987; Maxwell and Maxwell, 1988, Hiscott et al., 1988), and CHO (Lambert et al., 1990), many comparably exotic cells have been successfully electroporated (Chu et al., 1987; Maxwell and Maxwell, 1988; Andreason and Evans, 1989; Wilson et al., 1987a; Wilson et al, 1987b; Bishop and Frelinger, 1989; Klein et al., 1989; Sauer and Henderson, 1989; Sherman et al., 1989; Storkus et al., 1989; Brini et al., 1989; Sczakiel et al., 1989; Goldstein et al., 1989; Dawson et al., 1990; Basta et al., 1988; Leiden et al., 1989; Lanclos et al., 1989, Sorrentino et al., 1989).

While circular DNA is used for transient expression, linearized DNA is to be preferred for stable transfections (Spencer, 1991). Compared to the long concatemers frequently observed following calcium phosphate cotransfection, comparably few gene copies are inserted via electroporation (Toneguzzo et al., 1988). Since DNA fragments as large as 65 kb and more can be introduced into the cell by electroporation (Jastreboff et al., 1987) *in vitro* multimerization (ligation) of linearized plasmids prior to electroporation might lead to higher yields of rprotein in the resulting stable transformants (Spencer, 1991). Stable transformation efficiencies of up to 1 % have been observed in the cells subjected to electroporation (Chu et al., 1987).

Employing the Bio-Rad Gene pulser, the reader will find a good protocol how to perform stable transfections by electroporation in the article by Spencer (1991). Electroporation instruments from other companies, e.g. the Hoefer ProGenetor II or the Baekon 2000, might perform equally as well.

Unfortunately, the widespread use of electroporation as a means for standard gene transfer into mammalian cells has been counteracted by the time consuming necessity to individually determine the optimal electric field characteristics and other parameters for each given cell line (Andreason and Evans, 1989).

1.2.1.5 Protoplast Fusion

After stripping the outer cell wall bacteria can be fused to tissue culture cells by means of polyethylene glycol (Schaffner, 1980). Upon fusion the bacterial protoplasts empty their contents into the eukaryotic cells where the respective plasmidal expression vectors will subsequently be expressed. This method has turned out to be an efficient means to transiently screen plasmid cDNA expression banks for the presence of a desired rprotein (Caporale et al., 1990). Gene transfer efficiencies up to 1 % can be achieved. While protoplast fusion can result in the stable integration of large multimers into the cellular genome (de Saint Vincent et al., 1981), cotransformation of independent plasmids is difficult to achieve.

1.2.1.6 Lipid Mediated Gene Transfer (Lipofection)

The recent development of synthetic cationic lipids has greatly improved the conventional liposome mediated gene transfer technique, termed lipofection (Fraley et al., 1980; Wong et al., 1980). The new generation of liposome forming agents is represented by the Lipofectin reagent (e.g. available from BRL) composed of a mixture of DOTMA (N < 1-(2,3-dioleyloxy)propyl > N, N, N-trimethylammonium) and DOPE (dioleoyl phosphatidyl ethanolamine).

Lipofectin spontaneously and quantitatively complexes DNA (Felgner et al., 1987; Felgner and Ringold, 1989, for protocols see Felgner, 1991). Due to its positive electric charges this unilamellar DNA-lipid vesicle binds to and subsequently fuses with the anionic target cell membrane. This fusional DNA uptake bypasses temporary compartmentalization in lysosomes which is typical of conventional liposomes, thereby preventing likely damage of the DNA (Felgner and Ringold, 1989). Compared to most of the other transfection methods lipofectin mediated gene transfer is less sensitive to changes in the experimental conditions (Felgner, 1991). Adherent as well as anchorage independent cells can be lipofected with DNA as large as 130 kilobasepairs which represents a major advantage compared to the size constraints of conventional liposomes (Felgner, 1991).

A wide variety of primary cells and permanent cell lines (e.g. HeLa, HepG2, CHO, MRC5, COS, AtT20, NIH 3T3) have been lipofected to date. In many cases, transient, as well as stable, lipofection has significantly surpassed the traditional DEAE-dextran and calcium phosphate transfection efficiencies.

Further improvements resulted in the development of transfection reagents such as DOTAP (Boehringer Mannheim) or Transfectam (IBF). Apart from the absence of interference with physiological processes and absence of toxicity they offer the additional advantages of rapid internal degradation after DNA delivery and, in particular, they can be used in spite of the presence of serum in the media of lipofected cells.

Generally, a wide variety of Lipofectin-like reagents is nowadays available conferring improved 'transfectability' for individual cell types and lines.

1.2.1.7 Receptor Mediated Gene Delivery

Recently, a very elegant method termed receptor mediated gene delivery has been added to the repertoire of available gene transfer methods. DNA to be transferred into mammalian cells is complexed with a polycation-ligand conjugate such as polylysine-transferrin (Wagner et al., 1990; Zenke et al., 1990; Cotten et al., 1990; Wagner et al., 1991). While the polycation binds and condenses the DNA, the ligand serves to interact with its specific receptor on the surface of the target cell membrane. The DNA is subsequently taken up into the cell by endocytosis. Simultaneous employment of transcription/replication defective viruses in *trans* has further improved the efficiency of this method by facilitating the exit of the DNA from the endosomes into the cytoplasm (Cotten et al., 1992; Wagner et al., 1992). Nearly all of the cells undergoing gene transfer actually take up DNA, which can be as large as about 50 kb (Cotten et al., 1992). Receptor mediated gene transfer has thus been successfully used for transient and stable transfection experiments.

1.2.1.8 Microinjection

Foreign DNA can be directly introduced into the nucleus of the recipient cell by microinjection thereby avoiding potential damage in the cytoplasm or specific cellular compartments, e.g. endosomes (Capecchi, 1980; Anderson et al., 1980). Efficiency of DNA uptake is practically 100 % and stable integration has been reported to be as high as 20 % of the cells microinjected. The copy number of the genes introduced by microinjection can be controlled quite well. Unfortunately, equipment is very costly and extensive training of the experimentor is required. In spite of newly developed automated instruments (Pepperkok et al., 1988) only a rather limited number of cells can be injected. While therefore not common for the establishment of permanent cell lines microinjection has become the method of choice for the production of transgenic animals (De Pamphilis et al., 1988).

1.2.1.9 Particle Bombardment (Biolistics)

Particle bombardment originally was developed because of the need to efficiently introduce heterologous genes into plant cells (Sanford et al., 1987; for a general review on biolistics see Klein et al., 1992). DNA coated tiny microparticles approximately one micrometer in diameter are accelerated to velocities up to 4500 feet/second and directly delivered into the target cells. The commercially available biolistic particle delivery systems (e.g. the Bio-Rad PDS-1000/He system) rely on gold or tungsten microparticles whereas newly developed techniques employ ice particles. Once in the cell the latter simply dissolve leaving behind solely the DNA to be expressed (Brinegar et al., 1992).

To date, particle bombardment has been successfully used for transient expression as well as for the establishment of permanent recombinant cell lines from CHO (Fitzpatrick-McElligott, 1992), MCF-7 (a human mammary carcinoma line; Yang et al., 1990), NIH 3T3 (Zelenin et al., 1989) and T-lymphocyte cells (Fitzpatrick-McElligott, 1992). After bombardment, the viability of the target cells is surprisingly high with about 90 % survival rate. The widespread introduction of this technique for gene transfer into *standard* cell lines remains questionable, however. In addition to the high costs of the necessary equipment the bombardment conditions for different cell lines have to be individually and carefully optimized for a large number of factors affecting transformation efficiency (Klein et al., 1992).

Therefore, particle bombardment will prove valuable mostly for the introduction of DNA into specific cellular organelles (such as mitochondria) as well as into cells which are very hard to transfect otherwise (such as primary cells).

1.2.1.10 Retrovirus Mediated Gene Transfer

Due to their high infectivity and their broad host and tissue range retroviruses have proven suitable gene transfer vehicles for specific applications such as transgenic animal experiments (Jähner et al., 1985) and gene therapy (Naviaux and Verma, 1992; Hoeben et al., 1990).

Upon infection, the retroviral genome ultimately becomes stably integrated into the host cell chromosome. As with the other gene transfer methods, integration does not seem to be restricted to any specific target site; therefore, expression of rprotein from integrated viral genomes exhibits similar chromosomal position effects. Nonetheless, cells stably transformed by viral infection seem to express rprotein at consistently higher levels than their otherwise transfected counterparts.

The viral genome always integrates as a fully functional entity. All cells harboring integrated virus therefore express rprotein though at widely varying levels. In contrast, only a fraction of conventionally transfected cells express the rprotein due to *random* breakage of the expression vector prior to integration.

With few exceptions (Kozak and Kabat, 1990), the availability of amplification systems for conventionally transfected genes has precluded the common use of retroviruses for the establishment of standard production cell lines. However, recent developments of high titer and even amplifiable retroviral vectors might make retrovirus mediated gene transfer more attractive as a tool for the establishment of permanent high level production cell lines (Morgenstern and Land, 1990; Israel and Kaufman, 1990).

1.2.1.11 Laser Poring

When exposed to brief pulses of a laser beam tiny pores are transiently introduced into the cell membrane (Kurata et al., 1986; Tao et al., 1987). Heterologous DNA suspended in the surrounding medium can enter the cytoplasm through these holes. Stable transfection of tissue culture cells has been observed to be as high as 0.6 %. However, the complex and expensive equipment required will prevent the widespread introduction and use of this methodology for the establishment of standard permanent cell lines.

1.2.1.12. Erythrocyte Ghost Fusion

Recently it has become possible to remove the cytoplasmic contents of erythrocytes. The remaining 'ghosts' can be filled with heterologous DNA and the membranes resealed. Using polyethylene glycol, these vehicles can be fused to the recipient target cells thereby emptying their contents into the latter. This approach has successfully been applied to transient as well as stable gene transfer (Wiberg et al., 1983; Wiberg et al., 1986; Wiberg et al., 1987; Sowers, 1988; Sugawa et al., 1985). However, the complicated preparation of the ghosts as well as the difficulty of reproducibly packaging DNA into the ghosts have prevented its widespread use until now.

1.2.2 Selection and Amplification Markers

If an rprotein has to be expressed on a *preparative* scale most methods require its genetic information to be stably anchored within the cellular chromosomes in order to evade degradation and to ensure stable and equal distribution to the daughter cells. In order to quickly identify cells which have taken up DNA and which *stably and permanently* express rprotein a *selectable* marker gene is usually co-introduced into the cell together with the gene coding for the desired rprotein. Upon subsequent exposure to the selection pressure only cells expressing the selectable marker will survive. A certain *fraction* of these survivor clones will also express the rprotein coding genes which have been coincidently carried along like stowaways.

Selection can be either dominant or recessive (Fig. 1.2.2). *Dominant selection* can be applied to virtually any cell type. Typical dominant selection markers are bacterial proteins which mediate resistance against certain antibiotics.

Recessive selection requires the target cell to be deficient in the respective enzymatic activity. Recessive selection markers usually play crucial roles in cellular biochemical pathways such as nucleotide metabolism. Cells can synthesize nucleotides by two pathways: either by *de novo* synthesis or by conversion of nucleoside precursors to the corresponding nucleotides via the *salvage* route.

If *de novo* synthesis is inhibited, e. g. due to the lack of a crucial enzyme, the cells can still grow by utilizing nucleoside precursors in the salvage pathway. Upon omission of these precursors the cells would die unless an intact gene for the defective enzyme in the *de novo* synthesis pathway has been experimentally provided.

For selection, the cells are split into selective medium 24 to 48 hours after gene transfer. Upon exposure to the selection pressure cells having taken up the selection marker gene, inserted it into a chromosome (usually by non-homologous recombination) and ultimately expressing the selectable marker will survive. Most selection protocols require the cells to grow rapidly for optimal selection. About two weeks post-transfection, small colonies resistant to selection will be visible in the petri dish. A certain fraction of these cell clones also express the genes for the desired rprotein. Coexpression mainly results from the formation of large concatemers of randomly linked rprotein and selectable marker genes prior to co-integration into the chromosome.

If only selectable marker genes become integrated, no rprotein is produced. Breakage of the expression vectors *within* the rprotein transcription unit at concatemer formation leads to integration of non-functional rprotein genes and, thus, no rprotein is expressed either.

Chromosomal positioning effects, such as integration into transcriptionally highly active sites of chromosomes modulate expression levels of rprotein dramatically. Therefore, at *initial* selection, expression levels of rprotein do not necessarily correlate with rprotein gene copy number (in contrast, at *amplification*, close correlation between rprotein gene copy number and expression levels are usually observed; see below).

Compared to the antibiotic based selection markers, the 'biochemical' selection markers often carry the additional advantage of being useful as *amplification* markers: cells which have increased the selection marker gene copy number due to random recombination events can be identified by their resistance to increased concentrations of otherwise toxic substrate analogues. Often these cells have coamplified the rprotein encoding gene due to its physical linkage to the amplification marker gene. Stepwise successive increases of the cytotoxic drug concentration therefore lead to cells with high yields of rprotein (for a detailed discussion of the biochemical mode of action of the cellular selection/amplification markers see Kaufman, 1990a; for excellent collections of selection/amplification protocols see Kriegler, 1990b; Kaufman, 1990b).

The selection/amplification marker gene can either be located on the same plasmid as the rprotein gene as a separate transcription unit (Lenz and Weidle, 1990), or, alternately, be part of a *dicistronic* messenger RNA (Kaufman et al., 1988; Kaufman et al., 1987, Balland et al., 1988). In the latter case, the second cistron may be translated via an internal ribosome binding site (Kaufman et al., 1991; Davies and Kaufman, 1992) (see also chapter 1.1.2.3.1 this volume).

If the selection marker gene is located on a *different* plasmid, both plasmids have to be transfected simultaneously (Scahill et al., 1983). In this case, a 10 to 100: 1 ratio of expression plasmid harboring the rprotein gene versus the selection marker gene containing plasmid should be used (Kaufman et al., 1985). Thus, the fraction of rprotein-producing cell clones out of all cell clones resulting from a given stable transfection can be improved.

Below, the more common selection and amplification markers useful for the establishment of rprotein expressing cell lines will be described. Less widely used markers rwith selection/amplification potential, such as adenylate deaminase (Debatisse et al., 1986), uridine monophosphate synthetase (Kanalas and Suttle, 1984), inosine monophosphate 5′-dehydrogenase (Huberman et al., 1981), mutant hypoxanthine-guanine phosphoribosyltransferase (Brennand et al., 1982), mutant thymidine kinase (Roberts and Axel, 1982), thymidylate synthetase (Rossana et al., 1982), ribonucleotide reductase (Sabourin et al., 1981), arginosuccinate synthetase (Su et al., 1981), HMG coenzyme A reductase (Luskey et al., 1983), N-acetylglucosaminyl transferase (Criscuolo and Krag, 1982), threonyl-tRNA synthetase (Gantt et al., 1981) and Na^+/K^+-ATPase (Pauw et al., 1986) will not be discussed.

1.2.2.1 Dominant Antibiotic Selection Markers

Antibiotic selection markers confer a dominant resistance against antibiotics. Only cells having taken up such a marker gene will be able to survive and grow in medium containing the corresponding antibiotic. The antibiotic marker genes most commonly used for the establishment of permanent cell lines encode the proteins *neomycin phosphotransferase* (Colbere-Garapin et al., 1981), *hygromycin B phosphotransferase* (Gritz and Davies, 1983) and *puromycin acetyltransferase* (Vara et al., 1986).

Neomycin and hygromycin B phosphotransferases confer resistence against certain aminoglycoside antibiotics. The neomycin phosphotransferase gene was originally identified on the bacterial transposon Tn5 (Beck et al., 1981). Experimentally, neomycin phosphotransferase is used in combination with the antibiotic G418 (Geneticin). G 418 blocks translation by interacting with 80S ribosomes.

After transfection, an antibiotic concentration of 400 to 800 µg G 418/ml medium is suitable for selection of resistant clones in most standard cell lines (Colbere-Garapin et al., 1981; Blasquez et al., 1989; Wirth et al., 1988; for a detailed protocol on neomycin and hygromycin selection see Santerre et al., 1991).

The antibiotic hygromycin B causes mistranslation due to interference with translocation. Phosphotransferases which confer resistance against hygromycin B have been found in Streptomyces as well as E. coli (Gritz and Davies, 1983). The hygromycin phosphotransferase gene from the latter has successfully been used as a selection marker gene for the establishment of permanent cell lines (Parkinson et al., 1990; Giordano and McAllister, 1990). Suitable concentrations for selection usually are in the range of 100 to 200 µg hygromycin B/ml medium.

Puromycin causes premature release of the nascent polypeptide chain during translation. Resistance against puromycin is mediated by a Streptomyces derived acetyltransferase. A puromycin concentration of 10 µg/ml medium is toxic to most standard cell lines and routinely used for selection (Vara et al., 1986; Wirth et al., 1988).

Less common, but equally useful as a dominant selection marker in higher cells might be the bacterial *bleomycin resistance conferring gene product* (Genilloud et al., 1984), which mediates resistance against the closely related antibiotics bleomycin (Genilloud et al., 1984) and phleomycin (Mulsant et al., 1988; Leiting and Noegel, 1991; Prentice and Kingston, 1992). These antibiotics are believed to cause scission of DNA strands.

1.2.2.2 Dominant Biosynthetic Selection Markers

Since tryptophan is an essential amino acid, mammalian cells die upon omission of exogenous tryptophan. By transfection of the bacterial *tryptophan synthetase* gene tryptophan prototrophy can be installed in mammalian cells in indole containing medium.

Histidinol dehydrogenase, another bacterial enzyme, can also be used as a selection marker in mammalian cells in the absence of histidine and the presence of histidinol.

Both of the described enzymes exert a dominant mode of action in mammalian cells (Hartman and Mulligan, 1988).

1.2.2.3 Recessive Biosynthetic Selection Markers

In the nucleotide salvage pathway, *thymidine kinase* converts deoxyuridine and thymidine into deoxyuridine monophosphate and thymidine monophosphate, respectively. When *de novo* synthesis of purines and thymidine is blocked by aminopterin, thymidine kinase defective cells die unless supplied with exogenous thymidine kinase genes (Wigler et al., 1977; Littlefield, 1966; Colbere-Garapin et al., 1979). In addition, the purine salvage pathway has to be fed with hypoxanthine, and thymidine has to be supplied in the medium as a substrate for thymidine kinase (such as in HAT medium).

Thymidine kinase deficient cells can readily be identified by their ability to grow in bromodeoxyuridine-containing medium which otherwise is toxic to the cell.

When *de novo* synthesis is blocked and thymidine is in short supply, thymidine kinase has been reported to be amplifiable (Roberts and Axel, 1982) as well as useful even as a dominant selection marker (Mercola et al., 1980).

Hypoxanthine-guanine phosphoribosyltransferase can function as a recessive selection marker by restoring the purine salvage pathways in accordingly deficient cells

(Jolly et al., 1983). Analogous to thymidine kinase selection, *de novo* synthesis has to be blocked by using HAT medium.

Adenine phosphoribosyltransferase (Murray et al., 1984; Lowy et al., 1980; Stambrook et al., 1984) converts adenine to adenosine monophosphate in the purine salvage pathway. In cells lacking this enzymatic activity, adenine phosphoribosyltransferase functions as a recessive selection marker when *de novo* purine synthesis is blocked by azaserine and sufficient adenine is simultaneously provided in the medium.

1.2.2.4 Recessive Amplification Markers

The prototype and most common recessive amplification marker is the *wild type dihydrofolate reductase*. Dihydrofolate reductase converts folate to tetrahydrofolate, a crucial precursor molecule of glycine anabolism as well as thymidine and purine *de novo* biosynthesis.

Employing a sophisticated mutation/selection procedure, Urlaub and Chasin isolated a CHO cell line lacking a functional endogenous dihydrofolate reductase gene (Urlaub and Chasin, 1980). This cell line can only be propagated when the medium is supplemented with necessary salvage route components (e.g. adenosine, deoxyadenosine and thymidine).

Upon omission of the supplements, these dihydrofolate reductase deficient CHO cells die unless experimentally provided with an exogenous dihydrofolate reductase cDNA in an appropriate expression vector.

For amplification, the medium is further supplemented with varying concentrations of the folate analogue methotrexate (amethopterin) which binds and, therefore, inhibits dihydrofolate reductase. At a given methotrexate concentration only cells with cellular dihydrofolate reductase molecules numerous enough to outnumber the methotrexate concentration in the cell will survive. Increased dihydrofolate reductase concentrations result from previous recombination events which in turn lead to an increased dihydrofolate reductase gene copy number (up to several thousand copies per cell; Crouse et al., 1983). Sequential increases in methotrexate concentration thus eventually lead to highly methotrexate resistant cells with highly amplified dihydrofolate reductase gene copy numbers. The amplification scheme is aided by the tremendously inhibitory effect of methotrexate even at extremely low, i.e. nanomolar, concentrations. Therefore, multiple rounds of amplification can be performed (up to a millimolar methotrexate concentration) thereby dramatically increasing the dihydrofolate reductase gene copy number.

Ornithine decarboxylase is the first and indispensable enzyme in polyamine biosynthesis and, therefore, essential to cell growth. A CHO mutant cell line deficient in ornithine decarboxylase avtivity (CHO C55.7; Steglich and Scheffler, 1982; Pilz et al., 1990) can be propagated in the presence of putrescine in the cell culture medium. Integration of heterologous ornithine decarboxylase genes (Moshier et al., 1990; Hölttä et al., 1989; Grens et al., 1989) restores putrescine independent cell growth. Selection for amplified cells is achieved by means of increasing amounts of difluoromethylornithine, which inhibits ornithine decarboxylase (Chiang and McConlogue, 1988).

The *CAD protein* catalyzes the three initial steps in *de novo* uridine monophosphate biosynthesis. This multifunctional protein contains the enzymatic activites carbamoyl-phosphate synthase, aspartate transcarbamoylase and dihydroorotase. CHO D20 cells which are deficient in the three enzymatic activities described can be propagated in uridine-containing medium (Patterson and Carnright, 1977). Prototrophy is restored by transfection of a copy of the genomic CAD gene (de Saint Vincent et al., 1981).

Since the aspartate transcarbamoylase moiety is inhibited by phosphonoacetyl aspartic acid, amplified cells can be selected for by sequentially increased concentrations of this compound (Wahl et al., 1984).

The E. coli aspartate transcarbamoylase gene has also been used as an amplification marker in CHO cells lacking aspartate transcarbamoylase activity (Ruiz and Wahl, 1986).

1.2.2.5 Dominant Amplification Markers

The identification of dominant amplifiable markers has removed the limiting need for appropriate mutant mammalian cells. Several such amplification markers have been identified to date: xanthine-guanine phoshoribosyltransferase, adenosine deaminase, multiple drug resistance, asparagine synthetase, glutamine synthetase, and dihydrofolate reductase mutants.

Mutant dihydrofolate reductase genes have been isolated from mouse cell lines which exhibit increased resistance to methotrexate (i.e. a lower affinity to methotrexate; Simonsen and Levinson, 1983; McIvor and Simonsen, 1990). These mutant dihydrofolate reductase molecules were subsequently found to carry an arginine rather than a leucine at amino acid position 22 in one case, and a tryptophan rather than a phenylalanine at position 31 in another. They confer a dominant drug resistance to the target cells, whereby the action of endogenous (methotrexate sensitive) dihydrofolate reductase becomes negligible.

However, due to the insensitivity of the mutant dihydrofolate reductase molecules to methotrexate, the corresponding gene can not be amplified as high as the wild type dihydrofolate reductase gene.

Another widely used dominant amplification marker is *glutamine synthetase*. Glutamine synthetase converts glutamate and ammonia to glutamine. In addition to its role as an amino acid, glutamine is necessary for nitrogen detoxification as well as purine biosynthesis. Omission of glutamine and the presence of low concentrations of methionine sulphoximine in the cell culture medium, which inhibits endogenous glutamine synthetase, leads to cell death. Only cells which acquire additional copies of the glutamine synthetase gene will survive.

Subsequent amplification is achieved by stepwise increases in methionine sulphoximine concentration (Cockett et al., 1990; Bebbington et al., 1992).

The E. coli *xanthine-guanine phosphoribosyltransferase* converts xanthine to xanthine monophosphate. In mammalian cells this reaction in the salvage pathway is very inefficient and, therefore, negligible. In order to use the E. coli enzyme as a dominant marker, the conversion of inosine monophosphate to xanthine monophosphate in the *de novo* pathway has to be blocked. This is achieved by inhibiting inosine monophosphate dehydrogenase with mycophenolic acid (Mulligan and Berg, 1981).

Cells with amplified exogenous xanthine-guanine phosphoribosyltransferase genes are identified by their ability to grow in medium containing reduced amounts of xanthine (Chapman et al., 1983; Israel et al., 1989).

A 170 kilodalton cellular membrane glycoprotein mediates *resistance against multiple cytotoxic drugs (mdr)* such as colchicine, adriamycin and actinomycin D. The mdr protein (also called P-glycoprotein) is an energy-dependent multidrug efflux pump transporting the cytotoxic agents out of the cell (Gottesman and Pastan, 1988). Distinct P-glycoprotein mutations might confer preferential resistance or sensitivity to one or another drug (Choi et al., 1988).

Mdr cDNA copies can be used as dominant amplification marker genes (Gros et al., 1986; Kane et al., 1988; Kane et al., 1989). Cells with amplified mdr gene copy numbers are identified by their increased resistance to colchicine (Kane et al., 1988; Kane et al., 1989) or adriamycin (Gros et al., 1986). However, the degree of amplification might be rather limited since cells with comparably few copies of the mdr gene already exhibit very high drug resistance (Gros et al., 1986).

At cytotoxic concentrations of adenosine or cytotoxic analogues thereof (e.g. 9-β-D-xylofuranosyl adenine, xyl-A) *adenosine deaminase* converts these agents to ultimately nontoxic compounds. Xyl-A causes premature chain termination during DNA as well as RNA synthesis. Therefore, it efficiently exerts its toxic effects even in confluent and stagnant cells.

Increasing amounts of 2′-deoxycoformycin, which binds to and inhibits adenosine deaminase, are used to identify cells with amplified adenosine deaminase genes.

The use of xyl-A as a selective agent in cells with amplified adenosine deaminase expression levels, however, is complicated by the frequent loss of adenosine kinase activity which also leads to xyl-A resistant cell growth (Kellems et al., 1989). Therefore, an alternative, more sophisticated selection/amplification protocol has been developed which selects for the simultaneous presence of adenosine kinase and deaminase (Yeung et al., 1983).

Due to its low endogenous expression levels, adenosine deaminase can be used as a dominant marker (Israel and Kaufman, 1990; Kaufman et al., 1986; Yeung et al., 1985; Kaufman et al., 1989).

Mammalian *asparagine synthetase* synthesizes asparagine from glutamine and aspartic acid. Asparagine synthetase can be inhibited by the glutamine analogue albizzin or the aspartate analogue β-aspartyl hydroxamate. Inhibition of endogenous asparagine synthetase at low albizzin concentrations and omission of asparagine from the culture medium allow dominant selection for heterologous asparagine synthetase expression. Cells with amplified asparagine synthetase gene copy numbers are identified by their increased resistance to albizzin (Andrulis et al., 1987).

A bacterial asparagine synthetase utilizes ammonia rather than glutamine as the amide donor. In the presence of albizzin which solely inhibits the cellular asparagine synthetase, and in the absence of asparagine the bacterial enzyme provides an even more convenient marker (Cartier et al., 1987; Cartier and Stanners, 1990). In this case, β-aspartyl hydroxamate is used for amplification.

1.2.3 Selection and Amplification Strategies

Different expression strategies can be chosen depending on the quantities of rprotein needed in a particular period of time (Fig. 1.2.3).

If only qualitative or analytical questions are to be addressed it usually suffices to look at *pools of selected clones*. Even though only a certain fraction of these clones will express the rprotein, the latter is usually easily detectable by means of immunological tests (Western Blotting, ELISA) or activity assays (if available). If a higher sensitivity is required, *in vivo* radioactive labeling of the proteins and subsequent immunoprecipitation of the rprotein from cytoplasmic extract or supernatant can be performed (depending on the type of protein). Usually, in this case, the rproteins can be assayed for about three weeks post-transfection (*see chapter by Wirth, this volume for identification of cell clones with high level gene expression*).

If the rprotein is only poorly expressed or if cell lines with high yields of rprotein are to be established, *single cell clones* have to be isolated and screened for expression. While cell pools usually yield several hundred nanograms rprotein/10^6 cells × 24 hours, cell clones thereof can produce up to 10 micrograms (Grinnell et al., 1990).

Cell clones can be isolated by means of cloning cylinders or trypsin-EDTA saturated cotton swabs. For industrial laboratories automated devices capable of isolating large numbers of clones by limited dilution and even performing subsequent rprotein identification are available (Steindl at al., 1990).

An elegant method to create and rapidly identify initial high producing clones has been described recently (Wirth et al., 1988; Page and Sydenham, 1991). This approach involves the cotransfection of two (rather than one) different selectable marker genes on separate plasmids, together with the gene coding for the rprotein.

Cells at a given transfection event take up varying amounts of DNA. Upon *double* selection cells having taken up *several* plasmids (rather than one or only a few) are predominantly selected for. The pool of these cells, as well as isolated clones thereof, often express significantly higher levels of rprotein when compared to cell lines established according to the single selection procedure. Useful selection marker combinations with this approach include neomycin phosphotransferase in combination with puromycin acetyltransferase, dihydrofolate reductase or hygromycin phosphotransferase.

If high level production of rprotein for industrial purposes is required, the respective heterologous genes in the host cell line need to be amplified. Particularly at early stages of amplification, the transfected DNA can be relatively unstable in the host genome leading to gradual loss of rprotein expression. Subcloning of high yield clones at given stages of the amplification might prove helpful in order to stabilize the heterologous genetic information in the host cell genome. In addition, maintenance of selection pressure over a prolonged period of time also is important (*see chapter by Wurm, this volume for stability of amplified cell clones*).

An interesting amplification scheme that obviates time consuming screening and subcloning at each amplification step has been developed by Kaufman et al. (Kaufman et al., 1986). Pools of initially positive clones are amplified over several successive rounds before screening for individual high producers within the high producing pools is performed.

Generally, amplification by means of recessive amplification markers such as dihydrofolate reductase or ornithine decarboxylase seems to obtain higher gene copy numbers compared to dominant amplification markers. On the other hand, it is more convenient to employ dominant markers which can be directly used in virtually any cell type. Therefore, indirect strategies have been developed to use recessive markers even in cells which already express the corresponding enzymatic activity endogenously.

In one strategy, a strong promoter is linked to the recessive amplification marker gene to be transfected. In the case of wild type dihydrofolate reductase, initial moderately increased methotrexate resistance of the transfected cells is thus due to high level dihydrofolate reductase expression rather than to amplification (Murray et al., 1983). This strategy can also be successfully applied to other recessive amplification markers (Chiang and McConlogue, 1988).

Cotransfer with a dominant antibiotic selection marker can also be used to employ the advantages of a recessive amplification marker in cells already exhibiting this enzymatic activity endogenously. This has been successfully demonstrated for wild type dihydrofolate reductase in combination with hygromycin (Walls et al., 1989) or neomycin phosphotransferases (Wirth et al., 1988). Upon transfection, the initial selcction is based either on resistance to the antibiotic alone (Walls et al., 1989) or can be based on antibiotic resistance in combination with moderate methotrexate concentrations already at this stage (Wirth et al., 1988). Subsequently, amplification is solely based on resistance against increased methotrexate concentrations.

An analogous strategy can be applied to amplification via *metallothionein.* Upon cotransfer of the metallothionein gene with a dominant selection marker gene such as the neomycin phosphotransferase gene, amplified cells can be identified by means of increased resistance against heavy metal ions such as cadmium or zinc (Beach and Palmiter, 1981; Friedman et al., 1989).

In either case, the transfected amplification marker genes are often preferentially amplified over their endogenous counterparts (Kim and Wold, 1985).

In specific applications, selection of cells which express the desired rprotein can be performed without the need of a cotransferred selection marker. If fluorescent nontoxic compounds are available which interact with the expressed rprotein (such as substrate analogues, ligands or specific antibodies), high level producing cells can sometimes be selected by a *fluorescent activated cell sorter* (Kaufman et al., 1978; McClelland et al., 1984; Kavathas and Herzenberg, 1983b). Likewise, cells with amplified rprotein genes can sequentially be selected (Kavathas and Herzenberg, 1983a).

An important tool in developmental research as well as for the identification of cells which have incorporated exogenous DNA by *homologous* recombination is provided by *positive/negative selection* (Mansour et al., 1988).

Positive selection allows the identification of cells which have taken up exogenous DNA, such as thymidine kinase deficient cells. These cells can grow in HAT medium only after the acquisition of a foreign tk gene. By *negative selection*, cells which have taken up the selection marker gene are selectively killed, such as thymidine kinase positive cells. These will die upon incorporation of the cytotoxic thymidine analogue bromodeoxyuridine.

The positive/negative selection markers hypoxanthine-guanine phosphoribosyltransferase, adenine phosphoribosyltransferase and thymidine kinase are *recessive* and require mutant cells deficient in the respective marker activity (von der Lugt et al., 1991). Recently a neomycin phosphotransferase-thymidine kinase fusion protein has been constructed which allows *dominant* positive/negative selection (Schwartz et al., 1991).

1.2.4 Future Prospects

The availability of highly efficient gene transfer methods in combination with suitable amplification systems has enabled the researcher to express recombinant proteins at high yield in practically every cell line chosen.

Because yield is no longer the prime restriction, attention will be focused on the development of expression systems which reduce or even obviate the time consuming amplification procedure. Targeting of foreign genes into transcriptionally highly active regions of the host chromosome represents one possibility (Mannix, 1991; Wurm et al., 1996); another approach includes the optimized combination of the available regulatory elements for transcription (Friedman et al., 1989).

Furthermore, coexpression of several recombinant proteins at predetermined rates within a given cell will become feasible. Finally, attempts to identify cells with amplified heterologous genes on a basis other than their increased resistance to cytotoxic agents will be increasingly performed (Kavathas and Herzenberg, 1983a; Rogers et al., 1989).

Nowadays, the most severe restriction to high yield expression stems from the inefficiency of many cells to perform complex post-translational modifications of recombinant proteins. In order to be fully functional, many recombinant proteins need to be extensively modified post-translationally. For example, blood factors such as the anticoagulant Protein C must be carboxylated, hydroxylated, glycosylated, proteolytically processed and properly secreted (Grinnell et al., 1990). In spite of high yield expression, insufficient cellular modification systems often render the recombinant protein inactive (Kaufman et al., 1986). Identification and/or development of more suitable cell lines (Pavirani et al., 1989), elucidation of the modification mechanisms involved (Dorner et al., 1987; Wallin and Martin, 1988), and cloning of the corresponding modification and processing enzymes will ultimately further improve the yield of fully active recombinant proteins (Fig. 1.2.4), either by coexpression *in vivo* or by protein-chemical downstream processing steps *in vitro* (Bristol et al., 1994; Wasley et al., 1993; Schloka et al., 1996; Fischer et al., 1995; Rehemtulla et al., 1993; Grabenhorst et al., 1995; Drews et al., 1995).

Acknowledgements

We wish to thank M. Duschek, G. Mohr, A. Preininger, A. Rauscher and V. Stichler for their enthusiasm in developing and establishing cell lines which produce a wide variety of recombinant proteins at high yield. In addition, critical reading of the manuscript by N. Barrett is gratefully acknowledged.

References

Anderson, F. W., Killos, L., Sanders-Haigh, L., Kretschmer, P. J. and Diacumakos, E. G. (1980) Replication and expression of thymidine kinase and human globin genes microinjected into mouse fibroblasts. Proc. Natl. Acad. Sci. USA *77*, 5399–5403.

Andreason, G. L. and Evans, G. A. (1988) Introduction and expression of DNA molecules in eucaryotic cells by electroporation. BioTechniques *6*, 650–660.

Andreason, G. L. and Evans, G. A. (1989) Optimization of electroporation for transfection of mammalian cell lines. Analyt. Biochem. *180*, 269–275.

Andrulis, I., Chen, J. and Ray, P. (1987) Isolation of human cDNAs for asparagine synthetase and expression in jensen rat sarcoma cells. Mol. Cell. Biol. *7*, 2435–2443.

Ashman, C. R. and Davidson, R. L. (1985) High spontaneous-mutation frequency of BPV shuttle vector. Somat. Cell. Mol. Genet. *11*, 499–504.

Aubin, R. J., Weinfeld, M. and Paterson, M. C. (1988) Factors influencing the efficiency and reproducibility of polybrene-assisted gene transfer. Somat. Cell. Mol. Genet. *14*, 155–167.

Aubin, R., Weinfeld, M. and Paterson, M. C. (1991) Polybrene/DMSO-assisted gene transfer. In: Gene Transfer and Expression Protocols (Murray, E. J., ed.) pp. 35–44, Humana Press, Clifton, New Jersey.

Balland, A., Faure, T., Carvallo, D., Cordier, P., Ulrich, P. Fournet, B., de la Salle, H., Lecocq, J.-P. (1988) Characterisation of two differently processed forms of human recombinant factor IX synthesised in CHO cells transformed with a polycistronic vector. Eur. J. Biochem. *172*, 565–572.

Basta, P. V., Sherman, P. A. and Ting, J. P. Y. (1988) Detailed delineation of an interferon-γ-responsive element important in human HLA-DRA gene expression in a glioblastome multiform line. Proc. Natl. Acad. Sci. USA *85*, 8618–8622.

Beach, L. and Palmiter, R. (1981) Amplification of the metallothionein-1 gene in cadmium-resistant mouse cells. Proc. Natl. Acad. Sci. USA *78*, 2110–2114.

Bebbington, C. R., Renner, G., Thomson, S., King, D., Abrams, D. and Yarranton, G. T. (1992) High-level expression of a recombinant antibody from myeloma cells using a glutamine synthetase gene as an amplifiable selectable marker. Bio/Technology *10*, 169–175.

Beck, E., Ludwig, E. A., Auerswald, R., Reiss, B. and Schaller, H. (1981) Nucleotide sequence and exact localization of the neomycin phosphotransferase gene from transposon Tn5. Gene *19*, 327–336.

Berg, D. T., McClure, D. B. and Grinnell, B. W. (1992) E1a-responsive mammalian host/vector system for the stable high-level expression of secreted proteins. Nucl. Acids Res. *20*, 5485–5486.

Bishop, G. A. and Frelinger, J. A. (1989) Haplotype-specific differences in signaling by transfected class II molecules to a Ly-1$^+$ B-cell clone. Proc. Natl. Acad. Sci. USA *86*, 5933–5937.

Blasquez, V. C., Xu, M., Moses, S. C. and Garrard, W. T. (1989) Immunoglobulin κ gene expression after stable integration. J. Biol. Chem. *264*, 21183–21189.

Brennand, J., Chinault, A. C., Konecki, D. S., Melton, D. W. and Caskey, C. T. (1982) Cloned cDNA sequences of the hypoxanthine/guanine phosphoribosyltransferase gene from a mouse neuroblastoma cell line found to have amplified genomic sequences. Proc. Natl. Acad. Sci. USA *79*, 1950–1954.

Brinegar, C., Gilmore, J., Johnson, M. and Walker, N. (1992) Methods and apparatus for ice-mediated introduction of substances into biological material in a non-lethal manner. Int. Patent Appl. No. PCT/US91/05022. Int. Publ. No. WO 92/01802.

Brini, A. T., Harel-Bellan, A., Korner, M. and Farrar, W. L. (1989) Detection of enhancer binding proteins recognizing the human immunodeficiency virus long terminal repeat by in situ gel retardation. Biochem. Biophys. Res. Comm. *160*, 268–275.

Bristol, J.A., Freedman, S.J., Furie, B.C. and Furie, B. (1994) Profactor IX: The propeptide inhibits binding to membrane surfaces and activation by factor XIa. Biochemistry *33*, 14136–14143.

Callis, J., Fromm, M. and Walbot, V. (1987) Expression of mRNA electroporated into plant and animal cells. Nucl. Acids Res. *15*, 5823–5831.

Calos, M.P., Lebkowski, J.S. and Botchan, M.R. (1983) High mutation frequency in DNA transfected into mammalian cells. Proc. Natl. Acad. Sci. USA *80*, 3015–3019.

Cameron, F.H. and Jennings, P.A. (1989) Specific gene suppression by engineered ribozymes in monkey cells. Proc. Natl. Acad. Sci. USA *86*, 9139–9143.

Capecchi, M.R. (1980) High efficiency transformation by direct microinjection of DNA into cultured mammalian cells. Cell *22*, 476–488.

Caporale, L.H., Chartrain, N., Tocci, M. and DeHaven, P. (1990) Protoplast fusion in microtiter plates for expression cloning in mammalian cells: demonstration of feasibility using membrane bound alkaline phosphatase as a reporter enzyme. Gene *87*, 285–289.

Cartier, M. and Stanners, C.P. (1990) Stable, high-level expression of a carcinoembryonic antigen-encoding cDNA after transfection and amplification with the dominant and selectable asparagine synthetase marker. Gene *95*, 223–230.

Cartier, M., Chang, M. and Stanners, C. (1987) Use of the Escherichia coli gene for asparagine synthetase as a selective marker in a shuttle vector capable of dominant transfection and amplification in animal cells. Mol. Cell. Biol. *7*, 1623–1628.

Chaney, W.G., Howard, D.R., Pollard, J.W., Sallustio, S. and Stanley, P. (1986) High frequency transfection of CHO cells using polybrene. Somat. Cell. Mol. Genet. *12*, 237–244.

Chapman, A.B., Costello, M.A., Lee, F. and Ringold, G.M. (1983) Amplification and hormone-regulated expression of a mouse mammary tumor virus-eco gpt fusion plasmid in mouse 3T6 cells. Mol. Cell. Biol. *3*, 1421–1429.

Chen, C. A. and Okayama, H. (1988) Calcium phosphate-mediated gene transfer: a highly efficient transfection system for stably transforming cells with plasmid DNA. BioTechniques *6*, 632–638.

Chen, C. and Okayama, H. (1987) High efficiency transformation of mammalian cells by plasmid DNA. Mol. Cell. Biol. *7*, 2745–2752.

Chiang, T. and McConlogue, L. (1988) Amplification and expression of heterologous ornithine decarboxylase in Chinese hamster cells. Mol. Cell. Biol. *8*, 764–769.

Choi, K.H., Chen, C.J., Kriegler, M. and Roninson, I.B. (1988) An altered pattern of cross resistance in multi-drug-resistant cells results from spontaneous mutations in the mdr-l (p-glycoprotein) gene. Cell *53*, 519–529.

Chu, G., Hayakawa, H. and Berg, P. (1987) Electroporation for the efficient transfection of mammalian cell with DNA. Nucl. Acids Res. *15*, 1311–1326.

Cockett, M.I., Bebbington, C.R. and Yarranton, G.T. (1990) High level expression of tissue inhibitor of metalloproteinases in Chinese hamster ovary cells using glutamine synthetase gene amplification. Bio/Technology *8*, 662–667.

Colbere-Garapin, F., Chousterman, S., Horodniceanu, F., Kourilsky, P. and Garapin, A.C. (1979) Cloning of the active thymidine kinase gene of herpes simplex virus type 1 in Escherichia coli K–12. Proc. Natl. Acad. Sci. USA *76*, 3755–3759.

Colbere-Garapin, F., Horodniceanu, F., Kourilsky, P. and Garapin, A. (1981) A new dominant hybrid selective marker for higher eucaryotic cells. J. Mol. Biol. *150*, 1–14.

Cotten, M., Längle-Rouault, F., Kirlappos, H., Wagner, E., Mechtler, K., Zenke M., Beug, H. and Birnstiel, M.L. (1990) Transferrin polycation-mediated introduction of DNA into human leukemic cells – stimulation by agents that affect the survival of transfected DNA or modulate transferrin receptor levels. Proc. Natl. Acad. Sci. USA *87*, 4033–4037.

Cotten, M., Wagner, E., Zatloukal, K. Phillips, S., Curiel, D.T. and Birnstiel, M.L. (1992) High-efficiency receptor-mediated delivery of small and large (48 kilobase) gene constructs using the endosome-disruption activity of defective or chemically inactivated adenovirus particles. Proc. Natl. Acad. Sci. USA *89*, 6094–6098.

Criscuolo, B.A. and Krag, S.S. (1982) Selection of tunicamycin-resistant Chinese hamster overy cells with increased N-acetylglucosaminyltransferase activity. J. Cell Biol. *94*, 586–591.

Crouse, G.F., McEwan, R.N. and Pearson, M.L. (1983) Expression and amplification of engineered mouse dihydrofolate reductase minigenes. Mol. Cell. Biol. *3*, 257–266.

Davies, M.V. and Kaufman, R.J. (1992) The sequence context of the initiation codon in the encephalomyocarditis virus leader modulates efficiency of internal translation initiation. J. Virol. *66*, 1924–1932.

Dawson, C.W., Rickinson, A.B. and Young, L.S. (1990) Epstein-Barr virus latent membrane protein inhibits human epithelial cell differentiation. Nature *344*, 777–780.

de Chasseval, R. and de Villartay, J.-P. (1992) High level transient gene expression in human lymphoid cells by SV40 large T-antigen boost. Nucl. Acids Res. *20*, 245–250.

De Pamphilis, M.L., Herman, S.A., Martinez-Salas, E., Chalifour, L.E., Wirak, D.O., Cupo, D.Y. and Miranda, M. (1988) Microinjecting DNA into mouse ova to study DNA replication and gene expression and to produce transgenic animals. BioTechniques *6*, 662–680.

de Saint Vincent, B.R., Delbruck, S., Eckhart, W., Meinkoth, J., Vitto, L. and Wahl, G. (1981) The cloning and reintroduction into animal cells of a functional CAD gene, a dominant amplifiable genetic marker. Cell *27*, 267–277.

Debatisse, M., Hyrien, O., Petit-Koskas, E., de Saint-Vincent, B.R. and Buttin, G. (1986) Segregation and rearrangement of coamplified genes in different lineages of mutant cells that overproduce adenylate deaminase. Mol. Cell. Biol. *6*, 1776–1781.

Dorner, A.J., Bole, D.G. and Kaufman, R.J. (1987) The relationship of N-linked glycosylation and heavy chain-binding protein association with the secretion of glycoproteins. J. Cell. Biol. *105*, 2665–2674.

Drews, R., Paleyandra, R.K., Lee, T.K., Chang, R.R., Rehemtulla, A., Kaufman, R.J., Drohan, W.N. and Lubon, H. (1995) Proteolytic maturation of protein C upon engineering the mouse mammary gland to express furin. Proc. Natl. Acad. Sci USA *92*, 10462–10466.

Drinkwater, N.R. and Klinedinst, D.K. (1986) Chemically induced mutagenesis in a shuttle vector with a low-background mutant frequency. Proc. Natl. Acad. Sci. USA *83*, 3402–3406.

Faber, F.E. and Eberle, R. (1976) Effects of cytochalasin and alkaloid drugs on the biological expression of herpes simplex virus type 2 DNA. Exp. Cell Res. *103*, 15–22.

Felgner, P.L. (1991) Cationic liposome-mediated transfection with lipofectin reagent. In: Gene Transfer and Expression Protocols (Murray, E.J., ed.) pp. 81–89, Humana Press, Clifton, New Jersey.

Felgner, P.L. and Ringold, G.M. (1989) Cationic liposome-mediated transfection. Nature *337*, 387–388.

Felgner, P.L., Gadek, T.R., Holm, M., Roman, R., Chan, H.W., Wenze, M, Northtop, J.P., Ringold, G.M. and Danielsen, M. (1987) Lipofectin: a highly efficient, lipid-mediated DNA-transfection method. Proc. Natl. Acad. Sci. USA *84*, 7413–7417.

Fischer, B.E., Schlokat, U., Mitterer, A., Reiter, M. Mundt, W., Turecek, P.L., Schwarz, H.P. and Dorner, F. (1995) Structural analysis of recombinant von Willebrand factor produced at industrial scale fermentation of transformed CHO cells co-expressing recombinant furin. FEBS Lett. *375*, 259–262.

Fitzpatrick-McElligott, S. (1992) Gene transfer to tumor infiltrating lymphocytes and other mammalian somatic cells by the particle bombardment method. Bio/Technology *10*, 1036–1040.

Fraley, R., Subramani, S., Berg, P. and Papahadjopoulous, D. (1980) Introduction of liposome-encapsulated SV40 DNA into cells. J. Biol. Chem. *255*, 10431–10435.

Friedman, J.S., Cofer, C.L., Anderson, C.L., Kushner, J.A., Gray, P.P., Chapman, G.E., Stuart, M.C., Lazarus, L., Shine, J. and Kushner, P.J. (1989) High expression in mammalian cells without amplification. Bio/Technology *7*, 359–362.

Frost, E. and Williams, J. (1978) Mapping temperature-sensitive and host-range mutations of adenovirus type 5 by marker rescue. Virology *91*, 39–50.

Gantt, J.S., Bennett, C.A., and Arfin, S.M. (1981) Increased levels of threonyl-tRNA synthetase in a borrelidin-resistant Chinese hamster ovary cell line. Proc. Natl. Acad. Sci. USA *78*, 5367–5370.

Genilloud, O., Garrido, M.C. and Moreno, F. (1984) Transposon Tn5 carries a bleomycin resistance determinant. Gene *32*, 225–233.

Giordano, T.J. and McAllister, W.T. (1990) Optimization of the hygromycin B resistance-conferring gene as a dominant selectable marker in mammalian cells. Gene *88*, 285–288.

Goldstein, S., Fordis, C.M. and Howard, B.H. (1989) Enhanced transfection efficiency and improved cell survival after electroporation of G2/M-synchronized cells and treatment with sodium butyrate. Nucl. Acids Res. *17*, 3959–3971.

Gorman, C.M. and Howard, B.H. (1983) Expression of recombinant plasmids in mammalian cells is enhanced by sodium butyrate. Nucl. Acids Res. *11*, 7631–7648.

Gorman, C., Padmanabhan, R. and Howard, B.H. (1983) High efficiency DNA-mediated transformation of primate cells. Science *221*, 551–553.

Gottesman, M.M. and Pastan, I. (1988) The multidrug transporter, a double-edged sword. J. Biol. Chem. *263*, 12163–12166.

Grabenhorst, E., Hoffmann, A., Nimtz, M., Zettlmeissl, G. and Conradt, H.S. (1995) Construction of stable BHK-21 cells coexpressing human secretory glycoproteins and human Gal(β1-4)GlcNAc-R α2,6-sialyltransferase. Eur. J. Biochem. *232*, 818–725.

Grens, A., Steglich, C., Pilz, R. and Scheffler, I.E. (1989) Nucleotide sequence of the Chinese hamster ornithine decarboxylase gene. Nucl. Acids Res. *17*, 10497.

Grinnell, B.W., Walls, J.D., Gerlitz, B., Berg, D.T., McClure, D.B., Ehrlich, H., Bang, N.U. and Yan, S.B. (1990) Native and modified recombinant protein C: function, secretion, and posttranslational modifications. Adv. Appl. Biotechnol. Series *11*, 30–63.

Gritz, L. and Davies, J. (1983) Plasmid-encoded hygromycin B resistance: the sequence of hygromycin B phosphotransferase gene and its expression in Escherichia coli and Saccharomyces cerevisiae. Gene *25*, 179–188.

Groger, R. K., Morrow, D. M. and Tykocinski, M. L. (1989) Directional antisense and sense cDNA cloning using Epstein-Barr virus episomal expression vectors. Gene *81*, 285–294.

Gros, P., Neriah, Y.B., Croop, J.M. and Housman, D.E. (1986) Isolation and expression of an complementary DNA that confers multidrug resistance. Nature *323*, 728–731.

Hartman, S. and Mulligan, R. (1988) Two dominant-acting selectable markers for gene-transfer studies in mammalian cells. Proc. Natl. Acad. Sci. USA *85*, 8047–8051.

Heffernan, M. and Dennis, J.W. (1991) Polyoma and hamster papovavirus large T antigen-mediated replication of expression shuttle vectors in chinese hamster ovary cells. Nucl. Acids Res. *19*, 85–92.

Hiscott, J., Wong, A., Alper, D. and Xanthoudakis, S. (1988) Trans activation of type l interferon promoters by simian virus 40T antigen. Mol. Cell. Biol. *8*, 3397–3405.

Hoeben, R.C., van der Jagt, C.M., Schoute, F., van Tilburg, N.H., Verbeet, M.P., Briet, E., van Ormondt, H. and van der Eb, A.J. (1990) Expression of functional factor VIII in primary human skin fibroblasts after retrovirus-mediated gene transfer. J. Biol. Chem. *965*, 7318–7323.

Höltta, E., Hirvonen, A., Wahlfors, J., Alhonen, L., Jänne, J. and Kallio, A. (1989) Human ornithine decarboxylase(ODC)-encoding gene: cloning and expression in ODC-deficient CHO cells. Gene *83*, 125–135.

Huberman, E., McKeown, C.K. and Friedman, J. (1981) Mutagen-induced resistance to mycophenolic acid in hamster cells can be associated with increased inosine 5′-phosphate dehydrogenase activity. Proc. Natl. Acad. Sci. USA *78*, 3151–3154.

Ishikawa, Y. and Homcy, C. J. (1992) High efficiency gene transfer into mammalian cells by a double transfection protocol. Nucl. Acids Res. *20*, 4367.

Israel, D. I. and Kaufman, R. J. (1990) Retroviral-mediated transfer and amplification of a functional human factor VIII gene. Blood *75*, 1074–1080.

Israel, N., Chenciner, N., Houlmann, C. and Streeck, R. E. (1989) An expression vector for high-level protein synthesis in Vero cells. Gene *81*, 369–372.

Jähner, D., Haase, K., Mulligan, R. and Jaenisch, R. (1985) Insertion of the bacterial gpt gene into the germ line of mice by retroviral infection. Proc. Natl. Acad. Sci. USA *82*, 6927–6931.

Jastreboff, M.M., Ito, E., Bertino, J.R. and Narayanan, R. (1987) Use of electroporation for high molecular weight DNA-mediated gene transfer. Exp. Cell Res. *171*, 513–517.

Jolly, D., Okayama, H., Berg, P., Esty, A., Filpula, D., Bohlen, P., Johnson, G., Shively, J. Hunkapillar, T. and Friedmann, T. (1983) Isolation and characterization of a full-length expressible cDNA for human hypoxanthine posphoribosyltransferase. Proc. Natl. Acad. Sci. USA *80*, 477–481.

Jordan, M., Schallhorn, A. and Wurm, F.M. (1996) Transfecting mammalian cells: optimization of critical parameters affecting calcium phosphate precipitate formation. Nucl. Acids Res. *24*, 596–601.

Kanalas, J.J. and Suttle, D.P. (1984) Amplification of the UMP synthase gene and enzyme overproduction in pyrazofurin-resistant rat hepatome cells. Molecular cloning of a cDNA for UMP synthase. J. Biol. Chem. *259*, 1848–1853.

Kane, S.E., Reinhard, D.H., Fordis, C.M., Pastan, I. and Gottesman, M.M. (1989) A new vector using the human multidrug resistance gene as a selectable marker enables overexpression of foreign genes in eucaryotic cells. Gene *84*, 439–446.

Kane, S.E., Troen, B.R., Gal, S., Veda, K., Pastan, I. and Gottesman, M.M. (1988) Use of cloned multidrug resistance gene for coamplification and overproduction of major excreted protein, a transformation-regulated secreted acid protease. Mol. Cell. Biol. *8*, 3316–3321.

Kaufman, R.J. (1985) Identification of the components necessary for adenovirus translational control and their utilization in cDNA expression vectors. Proc. Natl. Acad. Sci. USA *82*, 689–693.

Kaufman, R.J. (1990a) Use of recombinant DNA technology for engineering mammalian cells to produce proteins. In: Large-scale mammalian cell culture technology (Lubiniecki, A. S., ed.) pp. 15–69, Marcel Dekker Inc., New York and Basel.

Kaufman, R.J. (1990b) Selection and coamplification of heterologous genes in mammalian cells. Meth. Enzym. *185*, 537–566.

Kaufman, R.J., Bertino, J.R. and Schimke, R.T. (1978) Quantitation of dihydrofolate reductase in individual parental and methotrexate-resistant murine cells. J. Biol. Chem. *253*, 5852–5860.

Kaufman, R.J. Davies, M.V., Wasley, L.C. and Michnik, D. (1991) Improved vectors for stable expression of foreign genes in mammalian cells by use of the untranslated leader sequence from EMC virus. Nucl. Acids Res. *19*, 4485–4490.

Kaufman, R.J., Murtha, P. and Davies, M.V. (1987) Translational efficiency of polycistronic mRNAs and their utilization to express heterologous genes in mammalian cells. EMBO J. *6*, 187–193.

Kaufman, R.J., Murtha, P., Ingolia, D.E., Yeung, C.Y. and Kellems, R.E. (1986) Selection and amplification of heterologous genes encoding adenosine deaminase in mammalian cells. Proc. Natl. Acad. Sci. USA *83*, 3136–3140.

Kaufman, R.J., Wasley, L.C. and Dorner, A.J. (1988) Synthesis, processing, and secretion of recombinant human factor VIII expressed in mammalian cells. J. Biol. Chem. *263*, 6352–6362.

Kaufman, R.J., Wasley, L.C., Davies, M.V., Wise, R.J., Israel, D.I. and Dorner, A.J. (1989) Effect of von Willebrand factor coexpression on the synthesis and secretion of factor VIII in Chinese hamster ovary cells. Mol. Cell. Biol. *9*, 1233–1242.

Kaufman, R.J., Wasley, L.C., Furie, B.C., Furie, B. and Shoemaker C.B. (1986) Expression, purification, and characterization of recombinant γ-carboxylated factor IX synthesized in Chinese hamster ovary cells. J. Biol. Chem. *261*, 9622–9628.

Kaufman, R.J., Wasley, L.C., Spiliotes, A.J., Gossels, S.D., Latt, S.A., Larsen, G.R., and Kay, M.R. (1985) Coamplification and coexpression of human tissue-type plasminogen activator and murine dihydrofolate reductase sequences in Chinese hamster ovary cells. Mol. Cell. Biol. *5*, 1750–1759.

Kavathas, P. and Herzenberg, L.A. (1983a) Amplification of a gene coding for human T-cell differentiation antigen. Nature *306*, 385–387.

Kavathas, P. and Herzenberg, L.A. (1983b) Stable transformation of mouse L cells for human membrane T-cell differentiation antigens, HLA and β2-microglobulin: Selection by fluorescence-activated cell sorting. Proc. Natl. Acad. Sci. USA *80*, 524–528.

Kawai, S. and Nishizawa, M. (1984) New procedure for DNA transfection with polycation and dimethyl sulfoxide. Mol. Cell. Biol. *4*, 1172–1174.

Kellems, R.E., Johnstone, E.M., Ward, K.E. and Little, S.P. (1989) Adenosine deaminase: a dominant amplifiable genetic marker for use in mammalian cells. In: Genetics and molecular biology of industrial microorganisms (Hershberger, C.L., Queener, S.W. and Hegeman, G., eds.) pp. 215–255, American Society for Microbiology, Washington DC.

Kim, S.K. and Wold, B.J. (1985) Stable reduction of thymidine kinase activity in cells expressing high levels of anti-sense RNA. Cell *42*, 129–138.

Klein, D.P., Jongstra-Bilen, J., Ogryzlo, K., Chong, R. and Jongstra, J. (1989) Lymphocyte-specific Ca^{2+}binding protein LSP1 is associated with the cytoplasmic face of the plasma membrane. Mol. Cell. Biol. *9*, 3043–3048.

Klein, T.M., Arentzen, R., Lewis, P.A. and Fitzpatrick-McElligott, S. (1992) Transformation of microbes, plants and animals by particle bombardment. Bio/Technology *10*, 286–291.

Kozak, S. L. and Kabat, D. (1990) Ping-pong amplification of a retroviral vector achieves high-level gene expression: human growth hormone production. J. Virol. *64*, 3500–3508.

Kriegler, M. (1990a) Gene transfer and expression: a laboratory manual, pp. 25–29, Stockton Press, New York.

Kriegler, M. (1990b) Gene transfer and expression: a laboratory manual, pp. 103–113, Stockton Press, New York.

Kurata, S.I., Tsukakoshi, M., Kasuya, T. and Ikawa, Y. (1986) The laser method for efficient introduction of foreign DNA into cultured cells. Exp. Cell Res. *162*, 372–378.

Lambert, H., Pankov, R., Gauthier, J. and Hancock, R. (1990) Electroporation-mediated uptake of proteins into mammalian cells. Biochem. Cell Biol. *68*, 729–734.

Lanclos, K.D., Michael, S.K., Gu, Y.C. Howard, E.F., Stoming, T.A. and Huisman, T.H. J. (1989) Transient chloramphenicol acetyltransferase expression of the Gγ globin gene 5′-flanking regions containing substitutions of C to T at position−158, G to A at position−161, and T to A at position−175 in K562 cells. Biochim. Biophys. Acta *1008*, 109–112.

Leiden, J.M., Karpinski, B.A., Gottschalk, L. and Kornbluth, J. (1989) Susceptibility to natural killer cell-mediated cytolysis is dependent of the level of target cell class I HLA expression. J. Immunol. *142*, 2140–2147.

Leiting, B. and Noegel, A.A. (1991) The ble gene of Streptoalloteichus hindustanus as a new selectable marker for Dictyostelium discoideum confers resistance to phleomycin. Biochem. Biophys. Res. Commun. *180*, 1403–1407.

Lenz, H. and Weidle, U. (1990) Expression of heterobispecific antibodies by genes transfected into producer hybridoma cells. Gene *87*, 213–218.

Littlefield, J.W. (1966) The use of drug-resistant markers to study the hybridization of mouse fibroblasts. Exp. Cell Res. *41*, 190–196.

Lowy, I., Pellicer, A., Jackson, J., Sim, G.K., Silverstein, S. and Axel, R. (1980) Isolation of transforming DNA: cloning the hamster aprt gene. Cell *22*, 817–823.

Luskey, K.L., Faust, J.R., Chin, D.J., Brown, M.S. and Goldstein, J.L. (1983) Amplification of the gene for 3-hydroxy-3-methylglutaryl coenzyme A reductase, but not for the 53-kDa protein, in UT–1 cells. J. Biol. Chem. *258*, 8462–8469.

Luthman, H. and Magnusson, G. (1983) High efficiency polyoma DNA transfection of chloroquine treated cells. Nucl. Acids Res. *11*, 1295–1308.

Maitland, N.J. and McDougall, J.K. (1977) Biochemical transformation of mouse cells by fragments of herpes simplex virus DNA. Cell *11*, 233–241.

Mannix, C.J. (1991) A productive mix of cultures. Tibtech *9*, 406–407.

Mansour, S.L., Thomas, K.R. and Capecchi, M.R. (1988) Disruption of the proto-oncogene int-2 in mouse-embryo-derived stem cells: a general strategy for targeting mutations to non-selected genes. Nature *336*, 348–352.

Mathews, M.B. and Shenk, T. (1991) Adenovirus-associated RNA and translation control. J. Virol. *65*, 5657–5662.

Maxwell, I.H. and Maxwell, F. (1988) Electroporation of mammalian cells with a firefly luciferase expression plasmid: kinetics of transient expression differ markedly among cell types. DNA *7*, 557–562.

McClelland, A., Kuhn, L.C. and Ruddle, F.H. (1984) The human transferrin receptor gene: genomic organization, and the complete primary structure of the receptor deduced from a cDNA sequence. Cell *39*, 267–274.

McCutchan, J.H. and Pagano, J.S. (1968) Enhancement of the infectivity of simian virus 40 deoxyribonucleic acid with diethylaminoethyl-dextran. J. Natl. Cancer Inst. *41*, 351–357.

McIvor, R.S. and Simonsen, C.C. (1990) Isolation and characterization of a variant dihydrofolate reductase cDNA from methotrexate-resistant murine L5178Y cells. Nucl. Acids Res. *18*, 7025–7032.

Mercola, K.E., Stang, H.D., Browne, J., Salser, W. and Cline, M.J. (1980) Insertion of a new gene of viral origin into bone marrow cells of mice. Science *208*, 1033–1035.

Milman, G. and Herzberg, M. (1981) Efficient DNA transfection and rapid assay for thymidine kinase activity and viral antigenic determinants. Somat. Cell Genet. *7*, 161–170.

Morgan, T.L., Maher, V.M. and McCormick, J.J. (1986) Optimal parameters for the polybrene-induced DNA transfection of diploid human fibroblasts. In Vitro Cell. Dev. Biol. *22*, 317–319.

Morgenstern, J.P. and Land, H. (1990) Advanced mammalian gene transfer: high titre retroviral vectors with multiple drug selection markers and a complementary helper-free pakkaging cell line. Nucl. Acids Res. *18*, 3587–3596.

Moshier, J.A., Gilbert, J.D., Skunca, M., Dosescu, J., Almodovar, K.M. and Luk, G.D. (1990) Isolation and expression of a human ornithine decarboxylase gene. J. Biol. Chem. *265*, 4884–4892.

Mulligan, R.C. and Berg, P. (1981) Selection for animal cells that express the Escherichia coli gene coding for xanthine-guanine phosphoribosyl transferase. Proc. Natl. Acad. Sci. USA *78*, 2072–2076.

Mulsant, P., Gatignol, A., Dalens, M. and Tiraby, G. (1988) Phleomycin resistance as a dominant selectable marker in CHO cells. Somat. Cell. Mol. Genet. *14*, 243–252.

Murray, A., Drobetsky, E. and Arrand, J. (1984) Cloning of the complete human adenine phosphoribosyltransferase gene. Gene *31*, 233–240.

Murray, M.J., Kaufman, R.J., Latt, S.A. and Weinberg, R.A. (1983) Construction and use of a dominant, selectable marker: A Harvey Sarcoma Virus dihydrofolate reductase chimera. Mol. Cell. Biol. *3*, 32–43.

Nakayama, N., Yokota, T. and Arai, K. (1992) Use of mammalian cell expression cloning systems to identify genes for cytokines, receptors, and regulatory proteins. Curr. Op. Biotech. *3*, 497–505.

Naviaux, R.K. and Verma, I.M. (1992) Retroviral vectors for persistent expression in vivo. Curr. Op. Biotech. *3*, 540–547.

Okayama, H. and Chen, C. (1991) Calcium phosphate mediated gene transfer into established cell lines. In: Gene Transfer and Expression Protocols (Murray, E. J., ed.) pp. 15–21, Humana Press, Clifton, New Jersey.

Page, M.J. and Sydenham, M.A. (1991) High level expression of the humanized monoclonal antibody campath-1 in Chinese hamster ovary cells. Bio/Technology *9*, 64–68.

Parkinson, J.F., Grinnell, B.W., Moore, R.E., Hoskins, J., Vlahos, C.J. and Bang, N.U. (1990) Stable expression of a secretable deletion mutant of recombinant human thrombomodulin in mammalian cells. J. Biol. Chem. *265*, 12602–12610.

Patterson, D. and Carnright, D.V. (1977) Biochemical genetic analysis of pyrimidine biosynthesis in mammalian cells: I. isolation of a mutant defective in the early steps of de novo pyrimidine synthesis. Somat. Cell Genet. *3*, 483–495.

Pauw, P.G., Johnson, M.D., Moore, P., Morgan, M., Fineman, R.M., Kalka, T. and Ash, J.F. (1986) Stable gene amplification and overexpression of sodium- and potassium-activated ATPase in HeLa cells. Mol. Cell. Biol. *6*, 1164–1171.

Pavirani, A., Skern, T., Le Meur, M., Lutz, Y., Lathe, R., Crystal, R.G., Fuchs, J.-P., Gerlinger, P. and Courtney, M. (1989) Recombinant proteins of therapeutic interest expressed by lymphoid cell lines derived from transgenic mice. Bio/Technology *7*, 1049–1054.

Pepperkok, R., Schneider, C., Philipson, L. and Ansorge, W. (1988) Single cell assay with an automated capillary microinjection system. Exp. Cell Res. *178*, 369–376.

Pilz, R.B., Steglich, C. and Scheffler, I.E. (1990) Molecular and genetic characterization of an ornithine decarboxylase-deficient Chinese hamster cell line. J. Biol. Chem. *265*, 8880–8886.

Potter, H., Weir, L. and Leder, P. (1984) Enhancer-dependent expression of human kappa immunoglobulin genes introduced into mouse pre-B lymphocytes by electroporation. Proc. Natl. Acad. Sci. USA *81*, 7161–7165.

Prentice, H.L. and Kingston, R.E. (1992) Mammalian promoter element function in the fission yeast Schizosaccharomyces pombe. Nucl. Acids Res. *20*, 3383–3390.

Razzaque, A., Mizusawa, H. and Seidman, M.M. (1982) Rearrangement and mutagenesis of a shuttle-vector plasmid after passage in mammalian cells. Proc. Natl. Acad. Sci. USA *80*, 3010–3014.

Rehemtulla, A., Roth, D.A., Wasley, L.C., Kuliopulos, A., Walsh, C.T., Furie, B., Furie, B.C. and Kaufman, R.J. (1993) In vitro and in vivo functional characterization of bovine vitamin K-dependent γ-carboxylase expressed in Chinese hamster ovary cells. Proc. Natl. Acad. Sci. USA *90*, 4611–4615.

Roberts, J.M. and Axel, R. (1982) Gene amplification and gene correction in somatic cells. Cell *29*, 109–119.

Rogers, P., Fisher, R. and Guyden, J. (1989) Neutral selection of transfected mammalian cells using tissue plasminogen activator gene expression. Gene *81*, 151–158.

Rossana, C., Rao, L.G. and Johnson, L.F. (1982) Thymidylate synthetase overproduction in 5-fluorodeoxyuridine-resistant mouse fibroblasts. Mol. Cell. Biol. *2*, 1118–1125.

Ruiz, J.C. and Wahl, G.M. (1986) Escherichia coli aspartate transcarbamylase: a novel marker for studies of gene amplification and expression in mammalian cells. Mol. Cell. Biol. *6*, 3050–3058.

Sabourin, C.L.K., Bates, P.F., Glatzer, L., Chang, C.C., Trosko, J.E. and Boezi, J.A. (1981) Selection of amphidicolin-resistant CHO cells with altered levels of ribonucleotide reductase. Somat. Cell Genet. *7*, 255–268.

Sanford, J.C., Klein, T.M., Wolf, E.D. and Allen, N. (1987) Delivery of substances into cells and tissues using a particle bombardment process. J. Part. Sci. Technol. *5*, 27–37.

Santerre, R.F., Walls, J.D. and Grinnell, B.W. (1991) Use of vectors to confer resistance to antibiotics G418 and hygromycin in stably transfected cell lines. In: Gene Transfer and Expression Protocols (Murray, E. J., ed.) pp. 245–256, Humana Press, Clifton, New Jersey.

Sauer, B. and Henderson, N. (1989) Cre-stimulated recombination at loxP containing DNA sequences placed into the mammalian genome. Nucl. Acids Res. *17*, 147–161.

Scahill, S.J., Devos, R., van der Heyden, J. and Fiers, W. (1983) Expression and characterization of the product of a human interferon cDNA gene in Chinese hamster ovary cells. Proc. Natl. Acad. Sci. USA *80*, 4654–4658.

Schaffner, W. (1980) Direct transfer of cloned genes from bacteria to mammalian cells. Proc. Natl. Acad. Sci. USA *77*, 2163–2167.

Schlokat, U., Fischer, B., Herlitschka, S., Antoine, G., Preininger, A., Mohr, G., Himmelspach, M., Kistner, O., Falkner, F.G. and Dorner, F. (1996) Production of highly homogeneous and structurally intact recombinant von Willebrand Factor multimers by Furin mediated propeptide removal *in vitro*. Biotechn. Appl. Biochem., in press.

Schwartz, F., Maeda, N., Smithies, O., Hickey, R., Edelmann, W., Skoultchi, A. and Kucherlapati, R. (1991) A dominant positive and negative selectable gene for use in mammalian cells. Proc. Natl. Acad. Sci. USA *88*, 10416–10420.

Sczakiel, G., Döffinger, R. and Pawlita, M. (1989) Testing for electrotransfection parameters by use of the fluorescent dye lucifer yellow CH. Anal. Biochem. *181*, 309–314.

Segal, S. (1986a) Calcium phosphate transfection of nonadherent and adherent cells with purified plasmids. In: Basic Methods in Molecular Biology (Davis, L.G., Dibner, M.D. and Battey, J.F., eds.) pp. 286–289, Elsevier Science Publishing Co., Inc., New York, New York.

Segal, S. (1986b) DEAE dextran-mediated transfection of nonadherent and adherent mammalian cells. In: Basic Methods in Molecular Biology (Davis, L.G., Dibner, M.D. and Battey, J.F., eds.) pp. 290–292, Elsevier Science Publishing Co., Inc., New York, New York.

Shen, Y., Hirschhorn, R.R., Mercer, W.E., Surmacz, E., Tsutsui, Y., Soprano, K.J. and Baserga, R. (1982) Gene transfer: DNA microinjection compared with DNA transfection with a very high efficiency. Mol. Cell. Biol. *2*, 1145–1154.

Sherman, P.A., Basta, P.V., Moore, T.L., Brown, A.M. and Ting, J.P.Y. (1989) Class II box consensus sequences in the HLA-DR alpha gene: transcriptional function and interaction with nuclear proteins. Mol. Cell. Biol. *9*, 50–56.

Simonsen, C.C. and Levinson, A.D. (1983) Isolation and expression of an altered mouse dihydrofolate reductase cDNA. Proc. Natl. Acad. Sci. USA *80*, 2495–2499.

Sompayrac, L.M. and Dana, K.J. (1981) Efficient infection of monkey cells with DNA of simian virus 40. Proc. Natl. Acad. Sci. USA *78*, 7575–7578.

Sorrentino, B., Ney, P., Bodine, D. and Nienhius, A.W. (1989) A 46 base pair enhancer sequence within the locus activating region is required for induced expression of the gamma-globin gene during erythroid differentiation. Nucl. Acids Res. *18*, 2721–2731.

Sowers, A.E. (1988) Fusion events and nonfusion contents mixing events induced in erythrocyte ghosts by an electric pulse. Biophys. J. *54*, 619–626.

Spencer, S.C. (1991) Electroporation technique of DNA transfection. In: Gene Transfer and Expression Protocols (Murray, E.J., ed.) pp. 45–52, Humana Press, Clifton, New Jersey.

Stambrook, P., Dush, M., Trill, J. and Tishfield, J.A. (1984) Cloning of a functional human adenosine deaminase phosphoribosyltransferase (APRT) gene: identification of a restriction-fragment-length polymorphism and preliminary analysis of DNAs from APRT-deficient families and cell mutants. Somat. Cell. Mol. Genet. *4*, 359–367.

Steglich, C. and Scheffler, I.E. (1982) An ornithine decarboxylase-deficient mutant of Chinese hamster ovary cells. J. Biol. Chem. *257*, 4603–4609.

Steindl, F., Jungbauer, A., Assadian, A., Blüml, G., Gruber, G., Buchacher, A., Purtscher, M., Hohenwarter, O., Tauer, C., Polacsek, K., Maiwald, G., Predl, R., Uhl, K., Katinger, H., Rubenzer, P., Jambor, F., Livingston, A. Donhauser, T., Puchegger, K. and Atzler, J. (1990) GCSS – General cell screening system. In: Synopsis 1989 (Institut für Angewandte Mikrobiologie, ed.) pp. 79–84, Institut für Angewandte Mikrobiologie, Universität für Bodenkultur, Vienna, Austria. Commercial information on GCSS by SLT Labinstruments GmbH, Untersbergstr. 1, 5082 Gröding/Salzburg, Austria.

Storkus, W.J., Alexander, J., Payne, J.A., Dawson, J.R. and Cresswell, P. (1989) Reversal of natural killing susceptibility in target cells expressing transfected class I HLA genes. Proc. Natl. Acad. Sci. USA *86*, 2361–2364.

Stow, N.D. and Wilkie, N.M. (1976) An improved technique for obtaining enhanced infectivity with herpes simplex virus type 1 DNA. J. Gen. Virol. *33*, 447–458.

Su, T.-S., Bock, H.-G.O., O'Brien, W.E. and Beaudet, A.L. (1981) Cloning of a cDNA for argininosuccinate synthetase mRNA and study of enzyme overproduction in a human cell line. J. Biol. Chem. *256*, 11826–11831.

Sugawa, H., Uchida, T., Yoneda, Y., Ishiura, M. and Okada, Y. (1985) Large macromolecules can be introduced into cultured mammalian cells using erythrocyte membrane vesicles. Exp. Cell Res. *159*, 410–418.

Sureau, C., Romet-Lemonne, J.L., Mullins, J.I. and Essex, M. (1986) Production of hepatitis B virus by a differentiated human hepatoma cell line after transfection with cloned circular HBV DNA. Cell *47*, 37–47.

Tao, W., Wilkinson, J., Stanbridge, E.J. and Berns, M.W. (1987) Direct gene transfer into human cultured cells facilitated by laser micropuncture of the cell membrane. Proc. Natl. Acad. Sci. USA *84*, 4180–4185.

Toneguzzo, F. and Keating, A. (1986) Stable expression of selectable genes introduced into human hematopoietic stem cells by electric field-mediated DNA transfer. Proc. Natl. Acad. Sci. USA *83*, 3496–3499.

Toneguzzo, F., Keating, A., Glynn, S. and McDonald, K. (1988) Electric field-mediated gene transfer: characterization of DNA transfer and patterns of integration in lymphoid cells. Nucl. Acids Res. *16*, 5515–5532.

Urlaub, G. and Chasin, L.A. (1980) Isolation of Chinese hamster cell mutants deficient in dihydrofolate reductase activity. Proc. Natl. Acad. Sci. USA *77*, 4216–4220.

van der Lugt, N., Maandag, E.R., te Riele, H., Laird, P.W. and Berns, A. (1991) A pgk::hprt fusion as a selectable marker for targeting of genes in mouse embryonic stem cells: disruption of the T-cell receptor δ-chain-encoding gene. Gene *105*, 263–267.

Vara, J.A., Portela, A., Ortin, J. and Jimenez, A. (1986) Expression in mammalian cells of a gene from Streptomyces alboniger conferring puromycin resistance. Nucl. Acids Res. *14*, 4617–4624.

Wagner, E., Cotten, M., Mechtler, K., Kirlappos, H. and Birnstiel, M.L. (1991) DNA-binding transferring conjugates as functional gene-delivery agents-synthesis by linkage of polylysine or ethidium homodimer to the transferrin carbohydrate moiety. Bioconjugate Chem. *2*, 226–231.

Wagner, E., Zatloukal, K., Cotten, M., Kirlappos, H., Mechtler, K., Curiel, D.T. and Birnstiel, M.L. (1992) Coupling of adenovirus to transferrin-polylysine/DNA complexes greatly enhances receptor-mediated gene delivery and expression of transfected genes. Proc. Natl. Acad. Sci. USA *89*, 6099–6103.

Wagner, E., Zenke, M., Cotten, M., Beug, H. and Birnstiel M.L. (1990) Transferrin-polycation conjugates as carriers for DNA uptake into cells. Proc. Natl. Acad. Sci. USA *87*, 3410–3414.

Wahl, G., de Saint Vincent, B.R. and DeRose, M.L. (1984) Effect of chromosomal position on amplification of transfected genes in animal cells. Nature *307*, 516–520.

Wallin, R. and Martin, L.F. (1988) Early processing of prothrombin and factor X by the vitamin K-dependent carboxylase. J. Biol. Chem. *263*, 9994–10001.
Walls, J.D., Berg, D.T., Yan, S.B. and Grinnell, B.W. (1989) Amplification of multicistronic plasmids in the human 293 cell line and secretion of correctly processed recombinant human protein C. Gene *81*, 139–149.
Wasley, L.C., Rehemtulla, A., Bristol, J.A. and Kaufman, R.J. (1993) PACE/Furin can process the vitamin K-dependent pro-factor IX precursor within the secretory pathway. J. Biol. Chem. *268*, 8458–8465.
Wiberg, F.C., Sunnerhagen, P. and Bjursell, G. (1986) New, small circular DNA in transfected mammalian cells. Mol. Cell Biol. *6*, 653–662.
Wiberg, F.C., Sunnerhagen, P. and Bjursell, G. (1987) Efficient transient and stable expression in mammalian cells of transfected genes using erythrocyte ghost fusion. Exp. Cell Res. *173*, 218–231.
Wiberg, F.C., Sunnerhagen, P., Kaltoft, K., Zeuthen, J. and Bjursell, G. (1983) Replication and expression in mammalian cells of transfected DNA; description of an improved erythrocyte ghost fusion technique. Nucl. Acids Res. *11*, 7287–7302.
Wigler, M., Silverstein, S., Lee, L.S., Pellicer, A., Cheng, V.C. and Axel, R. (1977) Transfer of purified herpes virus thymidine kinase gene to cultured mouse cells. Cell *11*, 223–232.
Wilson, A.K., Claypool, W.D. and de Lanerolle, P. (1987a) The use of electroporation to insert antibodies to myosin light chain kinase into rat alveolar macrophages. Biophys. J. *51*, 214a.
Wilson, A.K., Claypool, W.D. and de Lanerolle, P. (1987b) Incorporation of affinity-purified antibodies to myosin light chain kinase by electroporation inhibits macrophage motility. Fed. Proc. *46*, 1097.
Wirth, M., Bode, J., Zettlmeissl, G. and Hauser, H. (1988) Isolation of overproducing recombinant mammalian cell lines by a fast and simple selection procedure. Gene *73*, 419–426.
Wong, T.K., Nicolau, C. and Hofschneider, P.H. (1980) Appearence of beta-lactamase activity in animal cells upon liposome-mediated gene transfer. Gene *10*, 87–94.
Wurm, F.M., Johnson, A., Ryll, T., Koehne, C., Scherthan, H., Glaab, F., Lie, Y.S., Petropoulos, C.J. and Arathoon, W.R. (1996) Gene transfer and amplification in CHO cells – efficient methods for maximizing specific productivity and assessment of genetic consequences. Proc. N.Y. Acad. Sci., in press.
Yang, N.S., Burkholder, J., Roberts, B., Martinell, B. and McCabe, D. (1990) In vivo and in vitro gene transfer to mammalian somatic cells by particle bombardment. Proc. Natl. Acad. Sci. USA *87*, 9568–9572.
Yeung, C.-Y., Ingolia, D.E., Roth, D.B., Shoemaker, C., Al-Ubaidi, M.R., Yen, J.Y., Ching, C., Bobonis, C., Kaufman, R.J. and Kellems, R.E. (1985) Identification of functional murine adenosine deaminase cDNA clones by complementation in Escherichia coli. J. Biol. Chem. *260*, 10299–10307.
Yeung, C.-Y., Riser, M.E., Kellems, R.E. and Siciliano, M.J. (1983) Increased expression of one of two adenosine deaminase alleles in a human choriocarcinoma cell following selection with adenine nucleosides. J. Biol. Chem. *258*, 8330–8337.
Zelenin, A.V., Titomirov, A.V. and Kolesnikov, V.A. (1989) Genetic transformation of mouse cultured cells with the help of high-velocity mechanical DNA injection. FEBS Lett. *244*, 65–67.
Zenke, M., Steinlein, P., Wagner, E., Cotten, M., Beug, H. and Birnstiel M.L. (1990) Receptor-mediated endocytosis of transferrin polycation conjugates – an efficient way to introduce DNA into hematopoietic cells. Proc. Natl. Acad. Sci. USA *87*, 3655–3659.

1.3 Vectors for Gene Transfer and Expression in Animal Cells

Volker Sandig, Andre Lieber and Michael Strauss

1.2.1 Introduction

The introduction of foreign genes into animal cells has become an invaluable tool for the investigation of both eukaryotic gene function and regulation within the last 15 years. Moreover, high level expression of particular genes in animal cells is of growing interest to biotechnology as an alternative way of producing proteins which are impossible to express in a biologically active form in bacteria. In contrast to the latter, animal cells are more difficult to transfect and it is even harder to establish cell clones expressing high levels of the desired protein. There are numerous physical and chemical means to introduce a cloned gene into a recipient cell which are more or less efficient. Eventually, the transfected gene will become integrated into the genome of the recipient cell and will be expressed to a certain extent depending on the chromosomal location. Efficient gene transfer vehicles as well as expression vectors are required to overcome the limitations of classical transfection of cloned genes or plasmid vectors.

Viruses are known to introduce their genome efficiently into recipient cells. That is why most of the well-characterized animal viruses have been tested for their ability to transfer and to express foreign genes in animal cells. Many of them have proven to be useful for various purposes whereas others seem to be more restricted in their usefulness. In general, viral vectors are the most efficient means to transfer genes into particular cell types. Some of them are also efficient in integrating the transferred genes, others persist extrachromosomally. The suitability of the existing vector types for different purposes depends on several parameters which are difficult to compare from the literature. Therefore, we give here a brief description of the different types of vectors and, at the end, compare their suitability for different applications.

1.3.2 Papova Viral Vectors

The papova viruses SV40, polyoma virus, BK and JC virus are the smallest representatives of the DNA tumor viruses and are similar in their genome organization but different in their host range (Tooze, 1981). Their genome size is about 5 kb and accomodates, in overlapping reading frames, 5 or 6 genes. Half of the genome, the so-called early region, codes for the T antigens. The other half, the late region, codes for the three viral coat proteins. The regulatory region is located within a short stretch of about 300–400 nucleotides between these two regions. This contains not only the strong early and late promoters and enhancers but also, between them, the origin of replication (Tooze, 1981). SV40 replicates in monkey cells and to a lesser extent in human fibroblasts, polyoma virus replicates in mouse and less in other rodent cells, BK virus and JC virus replicate only in human cells (Tooze, 1981; Fried and Prives, 1986).

SV40 was the first vector used for the transfer of foreign genes into mammalian cells (Goff and Berg, 1976; Hamer et al., 1977) and remained the only one for several years. In initial studies, early and late region replacement vectors were developed. The foreign gene was inserted by replacing either the early or the late viral region using the respective viral promoter to drive the expression. In their pioneering studies on the expression, of influenza viral hemagglutinin, Gething and Sambrook could demonstrate that both types of replacement vectors are efficient in foreign gene expression reaching levels of 2 µg of protein per 10^6 cells in 24 hours (Gething and Sambrook, 1981; Gething and Sambrook, 1982). However, the disadvantage of these vectors is the requirement of a helper virus to compensate for the missing function in virus replication. Thus, this system is a lytic one and, therefore, only suitable for short term expression. Nevertheless, it was used by many groups for several years for research purposes.

As mentioned above, there is a short and well-defined regulatory sequence in the papova viral genome which also harbors the origin of replication. This sequence from SV40 was cloned into plasmid vectors and transfected into cells expressing the large T antigen of SV40, resulting in extremely high levels of replication (Myers and Tjian, 1980), which finally kills the cells after about two weeks. Thus, the origin of replication is the only sequence which is needed in cis and large T antigen is required in trans to allow for replication of plasmid vectors. CV-1 monkey cells, stably transformed with an origin-defective SV40 genome and therefore expressing large T antigen, were established as host cells for the replication of origin-plasmids (Gluzman, 1981). These plasmids are actually replicated to copy numbers of up to 400,000 per cell (Mellon et al., 1981). The famous COS cells are still frequently used to overexpress genes under the control of either one of the viral promoters present in the origin region or other strong promoters. Expression levels are comparable to those obtained with the viral replacement vectors with the advantages that this system is easier to handle and allows expression for at least ten days before the cells are killed by the exhaustive synthesis of DNA.

Similar origin-vector/helper cell systems were established from polyoma virus (Muller et al., 1983) but are less frequently used because of lower levels of replication and expression. This basic principle was also adapted for other viruses (see below). The replicating vector system based on the COS cells was later used to establish a cloning protocol for genes in mammalian cells (Aruffo and Seed, 1987). A complete cDNA library can be cloned in an origin-vector and transfected into COS cells, where small fractions of recombinants are established in individual cells. After identification by expression of cells harboring the cDNA of interest, these cells are selectively isolated and the plasmid DNA is recovered. Following transformation of E. coli, plasmid pools can be transfected again. By repeating the transfection/recovery procedure three to four times it is possible to isolate the cDNA clone of interest (Aruffo and Seed, 1987; Hamann et al., 1993). In summary, the origin-vector system in combination with the COS cells is a convenient and versatile expression system for research purposes aimed at short-term overexpression or cloning of particular genes. Since origin-containing plasmid vectors are replicated efficiently, every successfully transfected cell will express the gene of interest. There is no recent review of this system. For the reader interested in the history of viral vectors we recommend (Gluzman, 1982; Sambrook, 1987; Strauss et al., 1986).

1.3.3 Vectors Derived from Papilloma Viruses

Papilloma viruses were originally classified as members of the papovavirus family (Melnick et al., 1974) because they also have a closed circular double-stranded DNA genome which is complexed with histones, condensed into nucleosomes, and encapsulated in an icosahedral virion. However, molecular genetics studies have shown that the papillomaviruses constitute a distinct group of viruses. Their genomes are 50% larger (~ 7900 bp) than SV40 and polyoma virus, and there are virtually no similarities in the sequences. Most papillomaviruses have a single host and grow only in differentiated cutaneous or mucosal epithelium at specific anatomical sites (Broker and Botchan, 1986).

There are papillomaviruses of diverse kinds in various vertebrates, including more than forty subtypes in humans (Pfister et al., 1986). They are all known to persist in the form of their episomal DNA in their natural host tissues, the differentiated keratinocytes of epithelia and they are the causative agents of warts (Broker and Botchan, 1986). The only papillomavirus genome which has been used extensively as a vector up to now is the bovine papillomavirus subtype 1 (BPV). The transforming and replication functions of this virus have been studied in great detail. Whereas the early genes E5 and E6 are responsible for malignant transformation, the E1 gene is associated with the replication function (Broker and Botchan, 1986). However, some genes seem to have overlapping functions, e.g. E6/E7 can induce immortalization and plays a role in maintaining high copy numbers of extrachromosomal plasmids of the cloned viral genome (Lusky and Botchan, 1985). It has been shown that 69% of the viral genome including the early genes and the origin of replication are sufficient to transform mouse fibroblasts and also to maintain the recombinant plasmid in an episomal state (Lowy et al., 1980). Whereas viral DNA persists in the natural host cell at copy numbers between 50 and 200, lower copy numbers were observed for recombinant plasmids based on pBR322. After removing a so-called poison sequence from pBR322 (Lusky and Botchan, 1981), stably replicating BPV vectors could be generated (DiMaio et al., 1982). Attempts to separate the replication functions into cis and trans sequences turned out to be difficult (Strauss et al., 1986). Thus, it has not been possible yet to establish an origin-vector system like the COS system of SV40. The 69% early region fragment of BPV cloned into small plasmid vectors is still the vector of choice for many applications.

It is noteworthy that it turned out to be difficult to include selectable marker genes in the BPV vector to allow for selection of the vector-harboring cells. Experiments using the herpes virus tk gene (Lusky et al., 1983), the E. coli gpt gene (Law et al., 1982), or the mouse dihydrofolate reductase gene (Breatnach, 1984) showed that the vectors had undergone rearrangements and integration into cellular DNA in selected colonies. However, vectors including the neo gene were maintained as multicopy plasmids upon selection with standard concentrations of Geneticin and little rearrangement was observed (Matthias et al., 1983; Meneguzzi et al., 1984). Niwa et al. (1991) have generated a vector based on the 69% BPV fragment and a mutant neo gene driven by a weak promoter to express the human interleukin-2 gene. Using high concentrations of Geneticin, transfectants containing high vector copy number (greater than 300) were selected. In these transfected mouse or CHO cell clones the vector sequences were found to be integrated into the host chromosomes. Stable maintenance of the integrated copies as well as high-level expression of IL-2 was observed (Niwa et al., 1991). Similar integration and subsequent amplification of the vector sequences was observed earlier using the metallothionein gene as a selectable marker and cadmium as the selective agent (Karin et al., 1983). Apparently the strength of the selective pressure is critical for either episomal replication or stable integration of the recombinant molecules.

Various genes which are of interest for therapeutic purposes have been expressed using BPV vectors and stable cell lines could be generated which produce the protein of interest at relatively high levels. For instance, the human growth hormone was expressed from the metallothionein promoter in a BPV vector at levels up to 5 μg/ml/day (Pavlakis and Hamer, 1983). Even the surface antigen of hepatitis B virus, which was difficult to express at high levels using other expression systems, was continously secreted at levels of 10 μg/ml/day for a period of 85 days (Hsiung et al., 1984). Thus, this system is a fairly reliable one for long-term expression at moderate or high levels and it is, therefore, useful for large-scale production of pharmaceutical proteins and antigens for vaccination (Madej et al., 1992).

1.3.4 Vectors Derived from Epstein-Barr Virus

Epstein-Barr virus (EBV) is a member of the Herpes virus family. These large DNA viruses share the ability to infect target cells either by a latent or a lytic mechanism. In the latent state, the DNA genome is maintained in an episomal form and only a subset of viral genes is expressed. The host range of EBV is restricted to human B lymphocytes. EBV was discovered as the virus causing Burkitts lymphoma, the most common tumor affecting children in some parts of East Africa. B lymphocytes that are latently infected with EBV are efficiently immortalized.

Multiple copies of the EBV genome persist as closed circles within the host cells. The viral genome is replicated once per cell cycle employing the cellular replication machinery (Adams, 1987). This process requires only two viral elements: the latent origin of replication acting in cis and the trans-acting nuclear protein EBNA-1. This simple mechanism provided the basis for the development of shuttle plasmids containing viral and bacterial origins allowing replication in human cells as well as in E. coli (Yates et al., 1985). EBNA-1 can be supplied either by a chromosomal copy of replication defective virus, e.g. as in Raji cells (Yates et al., 1984), or by the gene localized on the plasmid itself. Because EBNA-1 has no immortalizing or transforming properties, plasmid maintenance and cell immortalization are well separated.

The shuttle system can be applied to a wide range of human and monkey cell lines, including hepatocytes. Neither the expression of EBNA-1 nor the presence of multiple copies of episomes seem to interfere with a set of hepatocyte specific functions (Lutfalla et al., 1989)
In most cell lines, low levels of EBNA-1 are sufficient to maintain replication. A cryptic promoter within the bacterial β lactamase gene can be used to drive EBNA-1 expression (Yates et al., 1985; Haver et al., 1989; Peterson and Legerski, 1991). Higher levels of EBNA-1 expression may enhance the transfection efficiency in some cell lines, but these cells often enter a crisis after prolonged cultivation (Vidal et al., 1991).

The major advantages of this system are the high rates of stable gene transfer, relatively high levels of expression and the easy recovery of episomal DNA. Since no integration process is required, stable transfer can be achieved in up to 50% of the cells if transfection is carried out by electroporation in the G2/M phase (Teshigawara and Kasutra, 1992). This makes the system suitable for expression cloning of even low abundant cDNAs (Peterson and Legerski, 1991; Jalanko et al., 1988). EBV-based shuttle vectors were compared with those derived from SV40 for the purpose of cDNA cloning and were found to be superior with regard to transfection efficiencies and expression levels (James et al., 1989). EBV-derived vectors in combination with a strong promoter/ enhancer, represent excellent tools for high level foreign gene expression in human cells. This may be important for the production of authentic human glycoproteins. Compared to their integrated equivalents, the average expression levels obtained with an EBV vector are 10–100 times higher (Jalanko et al., 1988). Stable expression of influenza hemagglutinin of up to more than 10 $\mu g/10^6$ cells was obtained in HeLa cells using an EBV vector and the CMV promoter. Fewer clones have to be analyzed to find a high producer because position effects important for integrated genes do not play a role in the episomal system. However, the persistence of extrachromosomal DNA is strictly dependent on the continuous presence of selection pressure.

Although high stability of clones over a period of more then 100 days has been described (Jalanko et al., 1988), integration of the vector resulting in a decrease of expression cannot be ruled out.

Several features make EBV-derived vectors useful for studying induced mutations in human cells. The frequency of spontaneous gene mutations in the episomes is comparable with those in human chromosomes. If genes are used as a target which can be analyzed or selected for in E. coli, mutation frequencies can easily be determined after plasmid rescue (Haase et al., 1989).

Another potential application of EBV vectors which can be envisaged is the analysis of promoter activities independent from chromosomal localization. The potential usefulness of the system for gene therapy, e.g. in lymphocytes, has not been evaluated yet.

1.3.5 Adenoviral Vectors

Like the papovaviruses, the adenoviruses have served as excellent tools for the investigation of the molecular mechanisms of gene expression and malignant transformation (Doerfler, 1983–1984). There are several subtypes of adenoviruses in humans and animals. Their effects on the host cells range from highly oncogenic (human Ad12) to non-oncogenic (human Ad2 and Ad5). Some of the early genes have been studied in detail. The E1a products are the first viral proteins synthesized after infection. They function as regulators of viral and cellular gene expression, as well as of replication. The E1a region is sufficient for immortalization of primary cells and it cooperates with E1b in malignant transformation. The generation of a cell line expressing E1a and E1b, called 293, (Graham et al., 1977) allowed for the establishment of E1-replacement vectors. In addition, the E3 region turned out to be non-essential for viral growth in cultured cells and could, therefore, be replaced by foreign sequences.

Another important element of the adenoviral genome is the major late promoter (MLP) which drives the expression of most of the late genes. Despite differential splicing of the late transcripts, they are all linked at their 5' ends to a so-called tripartite leader sequence, consisting of three parts fused together by splicing, which are responsible for the very efficient translation. The MLP, together with the tripartite leader sequence, can be used to drive foreign gene expression (Logan and Shenk, 1984; Berkner and Sharp, 1985).

On the basis of in-depth knowledge about the adenoviral life cycle and its molecular biology, the design of adenoviral vectors seems to be rather simple. In general, foreign genes without a promoter can be placed into the E1 or E3 regions downstream of the respective promoter. However, they can also be inserted into both regions, together with a heterologous promoter, in either orientation. The MLP, the early SV40 and the CMV promoter were successfully used. The genomes of adenoviruses are large ($\sim$ 36 kb) and helper-independent vectors can accomodate up to 8.3kb of foreign sequences if large parts of E1 and E3 are replaced or deleted (Bett et al., 1994).

The insertion of foreign sequences can be done either directly into the cloned viral genome or indirectly by homologous recombination *in vivo* following transfection. The direct approach is very difficult because there are only a few unique restriction sites within the viral genome. For the indirect approach, subgenomic fragments spanning the respective region for insertion are cloned in plasmid vectors and the foreign DNA is cloned into this sequence. Mainly, the left end of the viral genome (map units 0 to 17 representing 17% of the genome) is cloned into a plasmid and the foreign sequence is then placed between map units 1 and 9, replacing most of E1. The resulting recombinant is transfected together with viral genomic DNA, lacking the left-terminal end up to map unit 4, into 293 helper cells which provide the missing E1 functions in *trans*. The viral genome is provided either as part of a large bacterial plasmid or as a restriction fragment of purified viral DNA (Bett et al., 1994; Schaak et al., 1995). The overlapping sequences from m.u. 9 to 17 are sufficient to allow for recombination. Recombinant viruses can be harvested from the transfected cells. The frequency of recombinants can vary between 5 and 90%. Plaque purification is required for generation of pure recombinant virus stocks (Berkner, 1992).

Heterologous proteins were expressed at moderate or high levels using adenoviral vectors including T antigens of SV40 and polyoma virus, thymidine kinase of Herpes simplex, DHFR, CAT, factor IX, and ribonucleotide reductase (Solnick, 1981; Gluzman et al., 1982; Thummel et al., 1983; Yamada et al., 1985; Manseur et al., 1985; Manseur et al., 1986; Berkner and Sharp, 1984; Massie et al., 1986; Berkner et al., 1987 Schaffhausen et al., 1987; Davidson and Hassel, 1987). The highest expression levels obtained yield approximately 100–200 µg protein from a confluent 10 cm plate of 293 cells (Berkner, 1992).

The ability of adenovirus to infect different target cells efficiently has been used to generate stably transduced cell lines with several selectable marker genes including neo, tk, and DHFR. However, the most interesting application in this regard is the efficient immortalization of different cell lines by transduction of large T antigen of SV40, preferentially using an origin-defective early region of SV40 (van Doren and Gluzman, 1984).

Recombinant adenoviruses are also attractive for vaccination purposes. Various serotypes have been adapted as potential vaccine vectors. The surface antigen gene of hepatitis B virus was successfully transferred into different cell lines. An E3-deficient vector carrying the gene for the HBV envelope protein was inoculated into hamsters intranasally (Morin et al., 1987). The vector replicated in the lung and production of antibodies to the antigen was detected. The potential for vaccine development has also been explored for the HIV envelope protein (Hung et al., 1988), measles virus hemagglutinin (Alkatib and Briedis, 1988), and Herpes Simplex Virus gB glycoprotein (Johnson et al., 1988). All three antigens were expressed from the same viral vector (Solnick, 1981). This result is very promising for the development of an oral vaccine.

A new area of adenovirus research was initiated when these vectors were applied to the development of gene therapy strategies. The major advantages of adenoviruses compared to retroviral vectors are the ability to transfer genes in a relatively stable manner into resting cells, extremely high virus titers of purified virus stocks (10^{10}–10^{11} pfu/ml) and high infectivity *in vitro* and *in vivo*. In addition, they have some tropism for epithelial cells. Thus, lung epithelia, liver parenchyma, and intestine parenchyma are attractive targets (Stratford-Perricaudet et al., 1990). Different marker genes and therapeutic genes have been delivered into animal model systems. The first clinical protocol for the treatment of cystic fibrosis is based on the use of an adenoviral vector carrying the CFTR gene which has previously been applied successfully in rat (Crystal et al., 1994). The major limitation for the general application of adenoviral vectors for gene therapy results from the short-term mode of expression. Despite the lack of E1 function in the target cell, large amounts of viral proteins are produced along with the therapeutic protein, cytotoxic T cells are attracted and the infected cells are eliminated. In order to reduce immunogenicity, genes from early regions E2 or E4 have been conditionally inactivated or deleted from the virus (Englhardt et al., 1994; Armentano et al., 1995). However, the immunizing effect of the virus may also be used to its advantage in protocols involving short-term expression, e.g. for the transfer of suicide genes into cancer cells (Shu-Hsia et al., 1994).

1.3.6 Retroviral Vectors

Retroviruses are natural vectors of foreign genes. They are well known for their ability to pick up, in the course of their replication, cellular sequences called proto-oncogenes. These viruses have an RNA genome which is reverse-transcribed in the host cell and integrates into the genomic DNA of the infected cell (Varmus and Brown, 1989). The integrated provirus always has the same order of genes: LTR-gag-pol-env-LTR. In addition to the efficient infection of various cell types, the retroviruses offer the possibility of integrating one copy of a gene in a defined manner into the genome. Most retroviral vectors are derived from murine or avian retroviruses. The murine Moloney-MuLV was the source for almost all existing retroviral vector systems (Egilitis and Anderson, 1988; McLachlin et al., 1990). Since the essential components for viral packaging were discovered, vector systems have been developed consisting of the replication-deficient minimal vector construct and of packaging cell line contributing the retroviral structural proteins in *trans* (Mann et al., 1983; Miller, 1990).

Since retroviral vectors appeared to be ideal transfer vehicles for gene therapy much effort was made toward the improvement of safety and stability. Both vectors and packaging cell lines were improved by introducing various modifications and mutations that would reduce or abolish the ability of the vector DNA to recombine with the helper sequences within the packaging lines to generate infectious recombinant viruses (Miller and Buttimore, 1986; Bosselman et al., 1987; Markowitz et al., 1988a; Markowitz et al., 1988b; Dougherty et al., 1989; Morgenstern and Land, 1990). These modifications did not have much effect on the efficiency of gene transfer or on the level of expression.

Most attempts to improve the level of expression of inserted genes concerned the orientation of the foreign gene within the vector and the choice of suitable promoters. Whereas the original vectors with the foreign gene under control of the viral LTR promoter gave reasonable levels of expression (10–500 ng protein/10^6 cells/ml/24 hours), the insertion of heterologous promoters rarely resulted in higher levels. However, the use of heterologous promoters offered the possibility of achieving tissue specific expression. To this end, hybrid genes were inserted into the vector in antisense orientation to prevent overriding by transcription from the LTR (Guild et al., 1988).

Most retroviral constructs contain two heterologous genes, a marker gene confering a selectable advantage upon infected cells, and the gene of interest. The two genes can either be expressed from the same retroviral promoter (Gilboa et al., 1985) or from different promoters (Gilboa et al., 1985; Levine et al., 1991). In the first case two different mRNA molecules are normally generated due to the natural splicing mechanism of the retrovirus. Dicistronic vectors were developed which produce higher titer virus, permit the insertion of larger genes, and show a higher stability of expression of the transferred genes (Levine et al., 1991). At least one of the selectable markers, the neomycin-resistance gene, was shown to have a negative effect upon the expression of the second gene (Artelt et al., 1991). Similar effects cannot be ruled out for other selectable marker genes. Therefore, in the most recent vectors the use of a selectable marker gene was avoided (Dranoff et al., 1993). These vectors rely on the high efficiency of infection and on the detection of the desired gene product.

Retroviral vectors are a good choice for efficient transfer and integration of a single copy of a particular gene into a target cell, resulting in moderate but stable expression levels for several months (McLachlin et al., 1990). They are also superior for *in vitro* gene therapy approaches using dividing cells such as hematopoietic stem cells (Mulligan, 1993). However, they are not suitable for *in vivo* gene therapy into resting cells (Miller et al., 1990). In biotechnological applications they cannot compete with other vector types regarding high-level expression but the stability of expression might be advantageous for some purposes. For a review on retroviral vectors see Miller (1992).

1.3.7 Vaccinia Virus Vectors

Vaccinia virus is the best-studied member of the poxvirus family because of its world-wide use as a vaccinating agent against smallpox. Poxviruses are the largest of all animal viruses. A double-stranded genome of 100,000 to 300,000 bp enables them to be largely independent from host metabolic pathways. They are the only vertebrate viruses known to carry their own transcription system, including DNA-dependent RNA polymerase. Replication and transcription of the genome takes place in the cytoplasm.

The host range of vaccinia virus is restricted to avian and mammalian cells. A set of about 100 early genes, expressed immediately after virus entry into the cytoplasm, and about 100 late genes following after the onset of DNA-replication allow for a high speed of protein synthesis and virus multiplication (for review see Moss, 1990). These features make vaccinia virus a powerful tool for short-term gene expression.

Insertion of foreign DNA requires site specific recombination due to the large size of the genome. Recombinant viruses are generated by transfection of plasmids carrying the gene of interest flanked by non-essential viral sequences into cells infected with vaccinia virus. Usually, the gene is hooked up to early or late viral promoters (P7.5 or P11) and subsequently introduced into the vaccinia thymidine kinase gene (Mackett et al., 1982). Inactivation of the tk gene by recombination with the plasmid allows for selection of recombinants in tk-deficient cell lines and reduces the virulence of these viruses (Buller et al., 1985). Improved methods for isolation of recombinants have been described (Chakrabarti et al., 1985; Isaacs et al., 1990). The vaccinia expression system is only suitable for cDNA expression because splicing has not been observed, for either early or late vaccinia transcripts. DNA fragments of up to 25 kb can be accomodated.

Foreign gene expression can be enhanced considerably by using an appropriately designed vaccinia virus expressing the RNA polymerase of bacteriophage T7. The gene of interest is cloned downstream from a T7 promoter, introduced into the cells by transfection and transcribed by T7 RNA polymerase in the cytoplasm of cells infected with the recombinant vaccinia virus (Fuerst et al., 1986). The efficiency of gene expression in this hybrid system is due to the high catalytic activity and to the very stringent promoter specificity of the T7 RNA polymerase. In this form, the hybrid system is well-suited for cloning of genes from cDNA libraries based on the identification of the biological function of the gene product within single clones. It has been successfully applied to the cloning of plasma membrane transporters (Blakely et al., 1990; Usdin et al., 1991; Karschin, 1993). In contrast to expression cloning in COS or EBV-immortalized cells, the transfection rates reach nearly 100% because transport to the nucleus is not required. Protein levels are high enough to allow for detection of expression in pools of 5000–10000 recombinants. Vaccinia-assisted expression cloning is applicable to various cell types due to the broad host range of the virus and to all plasmid libraries, established in vectors containing either T7, T3 or SP6 promoters. Vaccinia viruses expressing the respective phage polymerases have been generated and show comparable properties (Rodriguez et al., 1990; Usdin et al., 1993).

Large-scale production of recombinant proteins in the vaccinia/T7 RNA polymerase system was limited by the transfection procedures until a second virus was used to deliver the target gene (Fuerst et al., 1987). In the advanced system, the expression cassette is also amplified by virus replication. Although amounts of specific RNA were extremely high, comprising 30% of the total steady-state level of RNA in the cytoplasm, the protein levels were comparably moderate. This was due to inefficient capping of T7 transcripts by the virus-encoded enzyme. To overcome this problem, the 5'UTR from encephalomyocarditis virus was cloned between the T7 promoter and the target gene, facilitating cap-independent translation initiation (Elroi-Stern et al., 1989). Vaccinia virus recombinants generated in this way represent the system of choice for batch production of authentic mammalian proteins. These vectors are 35–70 times more efficient than the original vaccinia system (Dougherty et al., 1989) and 200-fold compared with the BPV system (Lefkowitz et al., 1990). The amount of recombinant protein can equal 10% of the total cellular protein.

Probably the most exiting application of vaccinia virus vectors is their use as a recombinant live vaccine. The concept is based on several unique features of the virus: It has a successful history as an immunizing agent. Its efficacy as a vaccine strain to eliminate the disease of smallpox with minimal side effects is well-documented. Vaccinia virus is very stable under normal environmental conditions. It has been proposed that one inoculation is sufficient to introduce immunity. Since one virus particle can carry several genes, protection against subsequent challenge by the different infectious agents is possible (for reviews see Tartagalia et al., 1990; Moss, 1991). The safety in using these viruses was increased by generation of highly attenuated viruses. Most of these viruses lose the capacity for DNA replication in mammalian cells which results in a considerable decline of late gene expression. In the case of strain MVA, proven to be avirulent even in immunosupressed animals, the block occurs after replication. Therefore, this strain is well-suited for antigen expression driven from late promoters (Sutter and Moos, 1992). The development of vectors for veterinary and medical vaccines, based on other poxviruses, is in progress (Taylor et al., 1992). An excellent review by Moss (Moss, 1992) is highly recommended.

1.3.8 Alphavirus Vectors

Alphaviruses are among the most simple enveloped animal viruses. They have a broad host range. In their natural environment they shuttle between arthropods and vertebrates, including humans, causing mild to severe diseases. Two members of the genus, Sindbis and Semliki forest virus, have been studied in detail. As gene expression vectors they appeared in the literature only a few years ago. However, due to some unique features and the ease of vector construction they are becoming increasingly popular.

The alphavirus genome, a plus-strand RNA molecule (a typical mRNA), is translated immediately after cell entry to generate the viral replicase. Nothing more than this enzyme and replication signals at both ends of the RNA are required for RNA amplification, which takes place in the cytoplasm of the cell. The intermediate minus-strand RNA serves as a template for the synthesis of full-length genomic RNA and a subgenomic transcript coding for the structural proteins (Schlesinger and Schlesinger, 1990).

In the simplest form of a vector only the replication system of the virus is used. The sequence of interest is cloned into a plasmid, which serves as a template for *in vitro* transcription by phage polymerases (Rice et al., 1987). The transcript coding for the gene of interest, which is controlled by the subgenomic RNA promotet, and the replicase gene, is then transfected by standard methods. After the RNA has entered the cytoplasm, active replication starts, resulting in 10^5 RNA molecules per cell. Efficient translation signals lead to high-level production of heterologous protein, reaching up to 3% of the cell protein (Xiong et al., 1989). The system is well-suited as a vaccine vector due to the elevated protein levels and sustained expression *in vivo* compared to non-treplicative RNA or DNA gene transfer (Johanning et al., 1995). The self-limiting mode of replication as well as the lack of recombination or integration events make the system safe for human application.

Instead of naked RNA, recombinant viral particles may be used to infect cells more efficiently. For this purpose, a helper RNA is cotransfected into BHK cells which codes for the structural proteins and the replication signals but lacks the packaging signal. Whereas high titer virus stocks can be produced, as little as 10^2 particles injected into the mouse spleen result in a strong CTL response (Zhou et al., 1994).

To further improve biosafety of the vector a mutation has been introduced into one of the spike proteins resulting in production of an uncleaved form of the protein. The resulting particles are not infectious unless they are first treated with an exogenous protease (Berglund et al., 1993).

Both Sindbis and Semliki forest virus-based vectors have recently been used to overexpress proteins in a wide variety of cell lines (Lundstrom et al., 1994; Szekly et al., 1993). Applications are similar to that of vaccinia virus. The new vectors seem to be equally efficient but easier to generate and to handle.

1.3.9 Baculovirus Vectors

One decade after their development Baculovirus vectors have proven to be the most powerful, versatile and popular eukaryotic expression system available. These viruses were first mentioned as the agent causing polyhedrosis in silkworms. A long record of research initiated to find a cure for the disease led to broad knowledge about the biology and genetics of this virus family infecting various arthropod species. Baculoviruses are large enveloped viruses harboring a double-stranded DNA genome of 80–200 kbp which is packaged into rod-shaped nucleocapsids. During infection of insect cells, two types of virus particles are formed. Whereas extracellular particles (ECV) are released by budding during the first day of infection causing virus spread over the insect larva, the other type (Occluded Virus = OV) is assembled later and remains inside the cell nucleus in large proteinous capsules, called polyhedra. The polyhedrin which is the main component of these capsules has a molecular weight of 29 kD. It accumulates during the very late phase of infection (3.–6. day p.i.) to high levels, amounting to about 30% of cellular protein. Polyhedra protect the virions from decay in the environment but the most abundant proteins involved in the occlusion process, polyhedrin and p10, are not essential for the virus life cycle under cell culture conditions (for review see Blissard and Rohrmann, 1990). Therefore, the loci of these genes are excellently suited for insertion of foreign genes and allow for high-level expression.

As the genome of the virus is rather large, recombinant viruses have to be generated by homologous recombination between viral DNA and a transfer vector plasmid which are cotransfected into insect cells. For the most extensively studied baculovirus strain, Autographa californica nuclear polyhedrosis virus (AcNPV), a wide variety of transfer plasmids have been developed. Initially heterologous genes were inserted at the polyhedrin locus next to the transcriptional start site of the promoter (Smith et al., 1983). However, the polyhedrin promoter can be replaced by other late or very late baculovirus promoters which are transcribed by the virally encoded RNA polymerase. Generally, the highest level of expression (up to 50% of the cellular protein) is obtained with the very late promoters, p10 and polyhedrin. Since these promoters extend over the transcriptional start site, most recent transfer vectors include viral sequences also covering the translational start site (Possee and Howard, 1987) or even more downstream sequences (Luckow and Summers, 1988). Unfortunately, the alteration of the polyhedrin ATG start codon to ATT in these vectors does not completely abolish translation from this site, resulting in a fraction of polyhedrin fusions (Beames et al., 1991).

Several proteins can be expressed at the same time by replacement of more than one non-essential gene or by insertion of expression cassettes in a dispensable region of the genome (Urakarwa et al., 1989; Wang et al., 1991; Belyaev and Roy, 1993). If the polyhedrin gene is left intact in these vectors, occluded viruses are produced which are suitable for propagation in insect larvae or for improved pesticide development (Steward et al., 1991). For a long time the most time consuming part in the process of vector construction was the isolation of recombinant viruses after transfection, occuring with a frequency of 0.5–2%. Plaque assays were used to distinguish recombinants with a deletion in the polyhedrin gene from wild type viruses with the latter forming cloudy polyhedrin containing plaques compared to the clear plaques of recombinant viruses. Putative recombinants had to be replaqued three times and the presence of the foreign gene had to be demonstrated since spontaneous mutations could occur.

Virus isolation was simplified later by using PCR and FACS methods (Peng et al., 1993) or by cloning the lacZ gene coding for β galactosidase into the virus genome. Such viruses can be detected on Xgal plates by forming blue plaques whereas recombinant plaques do not stain due to the disruption of the lacZ gene. An alternative solution was to raise the ratio between recombinants and nonrecombinants. A first substantial improvement was achieved by cloning a unique restriction site in the vicinity of the polyhedrin gene (Kitts et al., 1990). Linearized baculovirus DNA is inefficient in infection, but when used in cotransfections together with a plasmid vector, recombinant viruses are obtained with a frequency of about 30%.

A real breakthrough was achieved by generation of linearized virus DNA deleted in an essential gene at the polyhedrin locus (Kitts and Possee, 1993). The gene has to be replaced by recombination with a plasmid vector carrying the missing gene in order to form a viable virus. This method of recombinant selection became the method of choice, since it leads to recombination frequencies close to 100%. It is now routinely used in many laboratories and is available from commercial sources.

In addition to the ease in use and the rapid scaling up, the high expression levels and the native conformation of the expressed proteins are the major advantages of the baculovirus system. Generally, expression is 20–250 times higher than in mammalian cells (Gong, 1993). Most of the co- and post-translational modifications including N- and O-linked glycosylation, carboxymethylation, phosphorylation, fatty acid acylation, amidation and signal peptide cleavage can be carried out by the insect cells (Luckow, 1991). Although glycosylations are not identical with those occuring in mammalian cells, biological functions and activities of recombinant proteins are normally not affected. To allow for complete protein modification a weaker late promoter should control the foreign gene rather than one of the very late promoters (Sridhar et al., 1992).

Baculoviruses seem to be safe for use due to their host restriction, their inability to survive in the environment without polyhedra, and the absence of transforming genes. However, non-occluded baculoviruses were recently shown to enter cultured mamalian liver cells very efficiently. If the foreign gene inserted into the virus is driven by a mammalian promoter, high-level gene expression takes place in hepatocytes (Hofmann et al., 1995). The "infection" process is not a productive one since neither viral gene expression nor DNA replication has been observed. This phenomenon may be useful in asigning a new gene therapy vector specific for liver cells. It shows, on the other hand, that care has to be taken in the construction of viruses expecially if toxic gene products are to be expressed.

1.3.10 Comparison of Vector Systems

Since the aim of this review is the discussion of the usefulness of different vector systems for the application in biotechnology it is essential to compare the most important properties of the systems available. We have attempted to do so and summarize the main characteristics in table 1.3.1. It seems useful to classify the vector systems according to their productivity in relation to their use as either short-term or long-term expression systems. The status of the vector within the host cell (ex trachromosomal vs. integrated) might be of importance for some purposes.

Trying to give advice to the beginner in the field of foreign gene expression or to biotechnologists looking for the most suitable system for their purposes, we can clearly state that the baculovirus vector/insect cell system comprises the easiest tool to obtain several milligrams of the protein of interest from a few liters of cell culture. In cases where the mode of glycosylation really matters, the vaccinia/T7-hybrid system is a very good alternative. However, if long-term expression at moderate or high levels is the priority, the EBV- or BPV episomal vectors should be prefered. The latter are comparable in their yields with the co-amplification systems based on selectable genes like metallothionein or DHFR (see chapter 1.4). Stable long-term expression at moderate levels can also be achieved from plasmid vectors with very strong promoters or with bacteriophage promoters in conjunction with their respective RNA polymerase (Lieber et al., 1993). The general advantage of all viral systems compared with integrated plasmid vectors is that tedious colony screenings for high producers are not required.

1.3.11 Conclusions

We have tried to summarize the state of the art in the field of viral vectors, in particular for application in biotechnology. With this goal in mind, it was impossible to go into the details of most systems. However, we hope that this introduction to the field will help the interested reader to select a system most suitable for his or her purpose. For a detailed discussion of the properties of the individual systems we must refer to more specialized review articles. Detailed reviews of several viral and other expression systems have been published. Several systems, e.g. the adeno-associated virus, were not discussed because their suitability for biotechnology has not yet been shown. The most important achievements in the field of foreign gene expression in animal cells were connected with extensive studies on the molecular biology of viruses. Thus, it seems not unlikely that other viral systems will be explored for the application to foreign gene expression. It also seems possible that improved cell-free expression systems will soon compete with the viral systems.

References

Adams, A. (1987) Replication of latent Epstein-Barr virus in Raji cells. J Virol 61, 1743–1746.

Alkatib G. and Briedis, D.J. (1988) High-level eucaryotic in vivo expression of biologically active measles virus hemagglutinin by using an adenovirus type 5 helper-free vector system. J. Virol. 62, 2718–2727.

Armentano, D., Sookdeo, C.C., Hehir, K.M., Gregory, R.J., St George, J.A., Prinz, G.A., Wadsworth, S.C. and Smith, A.E. (1995) Characterisation of an adenovirus gene transfer vector containing an E4 deletion. Human Gene Therapy 6, 1343–1353.

Artelt, P., Grannemann, R., Stocking, C., Friel, J., Bartsch, J. and Hauser, H. (1991) The prokaryotic neomy cin-resistance-encoding gene acts as a transcriptional silencer in eukaryotic cells. Gene 99, 249–254.

Aruffo, A. and Seed, B. (1987) Molecular cloning of a CD28 DNA by a high-efficiency Cos cell expression system. Proc. Natl. Acad. Sci. USA 84, 8573–8577.

Beames, B., Braunagel, S., Summers, M.D. and Lanford, R.E. (1991) Polyhedrin initiator codon altered to AUU yelds unexpected fusion protein from a baculovirus vector. Biotechniques 11, 378–383.

Belyaev, A.S. and Roy, P. (1993) Development of baculovirus triple and quadruple expression vectors: Coexpression of three or four bluetongue virus proteins and the synthesis of blueton gue virus-like particles in insect cells. Nucl. Acid Res. 21, 1219–1223.

Berglund, P., Sjöberg, M., Atkins, G.J., Sheahan, B.J., Garoff, H. and Liljeström, P. (1993) Semliki Forest virus expression system: production of conditionally infectious recombinant particles. Bio Technology 11, 916–920.

Berkner, K.L. (1992) Expression of Heterologous Sequences in Adenoviral Vectors, Current Topics in Microbiology and Immunology 158, 39–66.

Berkner, K.L. and Sharp, P.A. (1984) Expression of dihydrofolate reductase and of the ad jacent E1b region in an Ad5-dihydrofolate reductase recombinant virus. Nucleic Acids Res. 12, 1925–1941.

Berkner, K.L. and Sharp, P.A. (1985) Effect of the tripartite leader on synthesis of a non-viral protein in an adenovirus 5 recombinant. Nucleic. Acids Res. 13, 841–857.

Berkner, K.L., Schaffhausen, B.S., Roberts, T.M. and Sharp, P.A. (1987) Abundant expression of polyomavirus middle T antigen and dihydrofolate reductase in an adenovirus recombinant. J. Virol. 61, 1213–1220.

Bett, A.J., Haddara, W., Prevec, L. and Graham, F.L. (1994) An efficient and flexible system for construction of adenovirus vectors with insertions or deletions in early regions 1 and 3. Proc. Natl. Acad. Sci. USA 91, 8802–8806.

Blakely, D.R., Clarc, A.J., Rudnick, G. and Amara, S.G. (1990) Vaccinia-T7 RNA polymerase expression system: evaluation for the expression cloning of plasma membrane transporters. Anal. Biochem. 194, 302–308.

Blissard, G.W. and Rohrmann, G.F. (1990) Baculovirus diversity and molecular biology. Annual Review of Entomology 53, 127–155.

Bosselman, R.A., Hsu, R.-Y., Bruszewski, J., Hu, S., Martin, F. and Nicolson, M. (1987) Replication-defective chimeric helper proviruses and factors affecting generation of competent virus: Expression of moloney murine leukaemia virus structural genes via the metallothionein promoter. Mol. Cell. Biol 7, 1797–1806.

Breathnach, R. (1984) Selective amplification in methotrexate-resistant mouse cells of an artificial dihydropholate reductase transcription unit making use of cryptic splicing and polyadeny lation. EMBO J. 3, 901–908.

Broker, T.R. and Botchan M. (1986) Papillomaviruses: Retrospectives and Prospectives. Cancer Cells Cold Spring Harbor Lab. 4, 17–36.

Buller, R.M.L., Smith, G.L., Cremer, K., Notkins, A.L. and Moss, B. (1985) Decreased virulence of recombinant vaccinia virus expression vectors is associated with a thymidin-kinase negative phenotype. Nature 173, 813–815.
Chakrabarti, S., Brechtling, S.K. and Moss, B. (1985) Vaccinia virus expression vector: Coexpression of ß-galactosidase provides visual screening of recombinant virus plaques. Mol. Cell. Biol. 5, 3403–3409.
Crystal, R.G., McElvaney, N.G., Rosenfeld, M.A., Chu, C.S., Mastrangeli, A., Hay, J.G., Brody, S.L., Jaffe, H.A., Eissa, N.T. and Danel, C. (1994) Administration of an adenovirus containing the human CFTR cDNA to the respiratory tract of individuals with cystic fibrosis. Nature Genet. 8 1, 42–51.
Davidson, D. and Hassell, J.A. (1987) Overproduction of polyoma virus middle T antigen in mammalian cells through the use of an adenovirus vector. J. Virol. 61, 1226–1239.
DiMaio, D., Triesman, R., and Maniatis, T. (1982) A bovine papilloma-virus vector which propagates as an episome in both mouse and bacterial cells. Proc. Natl. Acad. Sci. USA 79, 4030–4034.
Doerfler, W., ed. (1983–1984) The molecular biology of adeno viruses, vols 1–3. Springer, Berlin Heidelberg New York (Current topics in microbiology and immunology, vols 109–111).
Dougherty, J.P., Wisniewski, R., Yang, S., Rhode, B.W. and Temin, H.M. (1989) New retrovirus helper cells with almost no nucleotide sequence homology to retroviral vectors. J. Virol. 63, 3209–3212.
Dranoff, G., Jaffee, E., Lazenby, A., Golumbek, P., Levitsky, H., Brose, K., Jackson, V., Hamada, H., Pardoll, D. and Mulligan, R.C. (1993) Vaccination with irradiated tumor cells engineered to secrete murine granulocyte-macrophage colony-stimulating factor stimulates potent, specific, and long-lasting anti-tumor immunity. Proc. Natl. Acad. Sci. 90, 3539–3543.
Eglitis, M.A. and Anderson, W.F. (1988) Retroviral vectors for introduction of genes into mammalian cells. BioTechniques 6, 608–614.
Elroi-Stein, O., Fuerst, T.R. and Moss, B. (1989) Cap-independent translation of mRNA conferred by encephalomyocarditis virus 5'sequence improves the performance of the vacciniavi rus/bacteriophage-T7 hybrid system. Proc. Natl. Acad. Sci. USA 86, 6126–6130.
Engelhardt, J.F., Ye, X., Doranz, B. and Wilson, J.M. (1994) Ablation of E2a in recombinant adenovirus improves transgene persistence and decreases immune response in mouse liver. Proc. Natl. Acad Sci.USA 91, 6196–6200.
Fried, M. and Prives, C. (1986) The Biology of Simian Virus 40 and Polyomavirus. Cancer Cells, Cold Spring Harbor Lab. 4, 1–16.
Fuerst, T.R., Earl, P.L. and Moss, B. (1987) Use of a hybrid vaccinia virus -T7 RNA polymerase system for expression of target genes. Mol. Cell. Biol. 7, 2538–2544.
Fuerst, T.R., Niles, E.G., Studier, F.W. and Moss, B. (1986) Eucaryotic transient expression system based on recombinant vaccinia virus that synthesizes bacteriophage T7 RNA polymerase. Proc. Natl. Acad. Sci. USA 83, 8122–8126.
Gething, M.-J. and Sambrook, J. (1981) Cell surface expression of influenza hemagglutinin from a cloned DNA copy of the RNA gene. Nature 293, 620–625.
Gething, M.-J. and Sambrook, J. (1982) The expression of the influenza virus hemagglutinin gene from SV40-HA recombinants. In: Eukaryotic Viral Vectors (Gluzman, Y. ed.). Cold Spring Harbor Laboratory, New York pp. 29–33.
Gilboa, E., Eglitis, M.A., Kantoff, P.W. and Anderson, W.F. (1986) Transfer and expression of cloned genes using retroviral vectors. BioTechniques 4, 504–512.
Gluzman,Y. (1981) SV40-transformed simian cells support the replication of early Sv40 mutants. Cell 23, 175–182.
Gluzman, Y. (1982) Eucaryotic viral vectos. Cold Spring Habor Laboratory, New York.

Gluzman Y., Reichl, H. and Solnick, D. (1982) Helper-free adeno virus type-5 vectors. In: Eukaryotic viral vectors (Gluzman, Y., ed.), Cold Spring Harbor Laboratory, New York, pp 187-192.

Goff, S.P. and Berg P. (1976) Construction of hybrid viruses containing Sv40 and lambda phage DNA segments and their propagation in cultured monkey cells. Cell 9, 695-705.

Gong, T. (1993) Overview over insect cell culture and BEVS. JRH Biosciences.

Graham, F.L., Smiley, J., Russel, W.C. and Nairn, R. (1977) Characteristics of a human cell line transformed by DNA from human adenovirus type 5. J. Gen. Virol. 36, 59-72.

Guild, B.C., Finer, M.H., Housman, D.E. and Mulligan, R.C. (1988) Development of retrovirus vectors useful for expressing genes in cul tured murine embryonal cells and hematopoietic cells in vivo. J. Virol. 62, 3795-3801.

Hamann, J., Fiebig, H. and Strauss, M. (1993) Expression cloning of the early activation antigen CD69, a type II integral membrane protein with a c-type lectin domain. J. Immunol 150, 4920-4927.

Hamer, D.H., Davoli, D., Thomas Jr., C.A. and Fareed, G.C. (1977) SV40 carrying an E.coli suppressor gene. J. Mol. Biol. 112, 155-182.

Haase, S.B., Heinzel, S.S., Krysan, P.J. and Calos, M.P. (1989) Improved EBV shuttle vectors. Mutation Res.

Haver, C.A., Getty, R.R. and Tykocinski, M.L. (1989) Nucleic Acid Res. 13, 1989-2002.

Hofmann, C., Sandig, V., Jennings, G., Rudolf, M., Schlag, P. and Strauss, M. (1995) Efficient gene transfer into human hepatocytes by baculovirus vectors. Proc. Natl. Acad. Sci. USA 92, 10099-10103.

Hsiung, N., Fitts, R., Wilson, S., Milne, A. and Hamer, D. (1984) Efficient hepatitis B surface antigen using a bovine papillomavirus-metallothionein vector. J. Mol. Appl. Genet. 2, 497-506.

Hung, P.P., Morin, J.E., Lubeck, M.D., Barton, J.E., Molnar-Kimber, K.L., Mason, B.B. Dheer, S.K., Jarocki-Witek, V., Kostek, B., Zandle, G., Conley, A.J. and Davis, A.R. (1988) Recombinant adenovirus as a vehicle for the HBV surface antigen or HIV enve lope protein genes. In: Human retroviruses, cancer and AIDS: approaches to prevention and therapy. Liss, New York, 349-361.

Isaacs, S.N., Kotwal, G.J. and Moss, B. (1990) Reverse guanine phosphoribosyltransferase selection of recombinant vaccinia viruses. Virology 178, 626-630.

Jalanko, A., Kallio, A. and Ulmanen, I. (1988) Comparision of mammalian cell expression vectors with and without an EBV-replicon. Arch-Virol. 103, 157-166.

Jalanko, A., Kallio, A., Ruohonen-Letho, M., Soderlund, H. and Ulmanen, I. (1988) An EBV-based mammalian cell expression vector for efficient expression of cloned coding sequences. Biochem. Biophys. Acta 949, 206-212.

James, M.R., Stary, A., Daya-Grosjean, L., Drougard, C. and Sarasin, A. (1989) Comparative study of Epstein-Barr virus and SV40-based shuttle-expression vectors in human repair deficient cells. Mutation Res. 220, 169-185.

Johanning, F.W., Conry, R.M., LoBuglio, A.F., Wright, M., Sumerel, L.A., Pike, M.J. and Curiel, D.T. (1995) A sindbis virus mRNA polynucleotide vector achieves prolonged and high level heterologous gene expression in vivo. Nucl. Acid Res. 23, 1495-1501.

Johnson, D.C., Ghosh-Choudhury, G., Smiley, J.R., Falls, L. and Graham, F.L. (1988) Abundant expression of herpes simplex virus glycoprotein gB using an adenovirus vector. Virology 164, 1-14.

Karin, M., Cathala, G. and Nguyen-Hun, M.C. (1983) Expression and regulation of a human metallothionein gene carried on an autonomously replicating shuttle vector. Proc. Natl. Acad. Sci. USA 80, 4040-4044.

Karschin, A. (1993) Heterologuus expression of the membrane proteins that control cellular excitability. EXS 63, 31-479.

Kitts, P.A. and Possee, R.D. (1993) A Method for production of recombinant baculovirus expression vectors at high frequency. Biotechniques 14, 810–817.

Kitts, P.A., Ayres, M.D. and Possee, R.D. (1990) Linearization of baculovirus DNA enhances the recovery of recombinant virus expression vectors. Nucl.Acid Res. 18, 5667–5672.

Law, M.-F., Howard, B., Sarver, N. and Howley, P.M. (1982) Expression of selective traits in mouse cells transformed with a BPV DNA-derived hybrid molecule containing E. coli gpt. In: Eucaryotic viral vectors (Gluzman, ed.), Cold Spring Harbor Laboratory, New York 79–85.

Lefkowitz, E.J., Pattnaik, A.K. and Ball, L.A. (1990) Complementation of vesicular stomatitis virus glycoprotein G mutant with wild type protein expressed from either a bovine papilloma virus or a vaccinia virus vector system. Virology 187, 373–383.

Levine, F., Yee, J.K. and Friedmann, T. (1991) Efficient gene expression in mammalian cells from a dicistronic transcriptional unit in an improved retroviral vector. Gene 108, 167–174.

Lieber, A., Sandig, V., Bähring, S., Sommer, W. and Strauss, M. (1993) Stable high level gene expression in mammalian cell by phage T7 RNA polymerase. Methods in Enzymol. 217, 66–74.

Logan J. and Shenk, T. (1984) Adenovirus tripartite leader sequence anhances translation of mRNAs late after infection. Proc. Natl. Acad. Sci. USA 81, 3655–3659.

Lowy, D.R., Dvoretzky, I., Shober, R., Law, M.-F., Engel, L. and Howley, P.M. (1980) In vitro tumorigenic transformation by a defined sub-genomic fragment of bovine papilloma virus DNA. Nature 287, 72–74.

Luckow, V.A. (1991) Cloning and expression of heterologous genes in insect cells with baculovirus vectors. In: Recombinant DNA technology and Ap plications (Procop, A., Bajpai, R.K. and Ho, C., eds.), McGraw-Hill.Inc. 97–152.

Luckow, V.A. and Summers, M.D. (1988) Trends in the development of baculovirus expression vectors. Bio/Technology 6, 47–55.

Lundstrom, K., Mills, A., Buell, G., Allet, E., Adami, N. and Liljeström, P. (1994) High-level expression of the human neurokinin-1 receptor in mammalian cell lines using the Semliki Forest virus expression system. Eur. J. Biochem. 224, 917–921.

Lusky, M. and Botchan, M. (1981) Inhibitory effect of specific pBR322 DNA sequences upon SV40 replication in simian cells. Nature 293, 79–81.

Lusky, M. and Botchan M.R. (1985) Genetic analysis of bovine papillomavirus type 1 trans-acting replication factors. J. Virol. 53, 955.

Lusky, M., Berg, L., Weiner, H., and Botchan, M. (1983) Bovine papilloma-virus contains an activator of gene expression at the distal end of the early transcription unit. Mol. Cell. Biol. 3, 1108–1122.

Lutfalla, G., Armbruster, L., Dequin, S. and Bertolotti, R. (1989) Construction of well-differentiated human hepatoma cells and of appropriate Epstein-Barr virus based shuttle vectors. Gene 76, 27–39.

Mackett, M., Smith, G.L. and Moss, B. (1982) Vaccinia virus: A selectable eucaryotic cloning vector. Proc. Natl. Acad. Sci. USA 79, 7415–7419.

Madej, M., Conway, M.J., Morgan, A.J., Sweet, J., Wallace, L., Qualtiere, L.F., Arrand, J.R. and Mackett, M. (1992) Purification and characterization of Epstein-Barr virus gp340/220 produced by a bivine papillomavirus virus expression vector system. Vaccine 10, 777–782.

Mann, R., Mulligan, R.C. and Baltimore, D. (1983) Construction of a retrovirus packaging mutant and its use to produce helper-free defective retrovirus. Cell 33, 153–159.

Mansour, S.L., Grodzicker, T. and Tjian, R. (1985) An adenovirus vector system used to express polyoma virus tumor antigens. Proc. Natl. Acad. Sci. USA 82, 1359–1363.

Mansour, S.L., Grodzicker, T. and Tjian, R. (1986) Downstream sequences affect transcription initiation from the adenovirus major late promoter. Mol. Cell. Biol. 6, 2684–2694.

Markowitz, D., Goff, S. and Bank, A. (1988a) A safe packaging line for gene transfer: Separating viral genes on two different plasmids. J. Virol. 62, 1120–1124.
Markowitz, D., Goff, S. and Bank, A. (1988b) Construction and use of a safe and efficient amphotropic packaging cell line. Virology 167, 400–406.
Massie, B., Gluzman, Y. and Hassell, J.A. (1986) Construction of a helper-free recombinant adenovirus that expresses polyomavirus large T antigen. Mol. Cell. Biol. 6, 2872–2883.
Matthias, P.D., Bernard, H.U., Scott, A., Brady, G., Hashimoto-Gotoh, T. and Schütz, G. (1983) A bovine papillomavirus vector with a dominant resistance marker replicates extrachromosomally in mouse and E. coli cells. EMBO J. 2, 1487–1492.
McLachlin, J.R., Cornetta, K., Eglitis, M.A. and Anderson, W.F. (1990) Retroviral-mediated gene transfer. Prog. Nucleic Acid. Res. Mol. Biol. 38, 91–135.
Mellon, P., Parker, V., Gluzman, Y. and Maniatis, T. (1981) Identification of DNA sequences required for transcription of the human alpha1-globin gene in a new SV40 host-vector system. Cell 27, 279–288.
Melnick, J.L., Allison, A.C., Butel, J.S., Eckhart, W., Eddy, B.E., Kit, S., Levine, A.J., Meles, J.A.R., Pagano, J.S., Sachs, L. and Vonka, V. (1974) Papovaviridae. Intervirology 3, 106.
Meneguzzi, G., Binetry, B., Grisoni, M. and Cozin, F. (1984) Plasmidial maintenance in rodent fibroblasts of a BPVI-pBR322 shuttle vector without immediate oncogenic transformation of the recipient cells. EMBO J. 3, 365–371.
Miller, A.D. (1990) Retrovirus packaging cells. Human gene therapy 1, 5–14.
Miller, A.D. (1992) Retroviral Vectors. In: Current Topics in Microbiology and Immunology, Viral Expression Vectors (Muzyczka, N., ed.), pp 1–24, Springer Verlag, Berlin Heidelberg.
Miller, A.D. and Buttimore, C. (1986) Redesign of retrovirus packaging cell lines to avoid recombination leading to helper virus production. Mol. Cell. Biol. 6, 2895–2902.
Miller, D.G., Adam, M.A. and Miller, A.D. (1990) Gene transfer by retrovirus vectors occurs only in cells that are actively replicating at the time of infection. Mol. Cell. Biol. 10, 4239–4242.
Morgenstern, J.P. and Land, H. (1990) Advanced mammalian gene transfer: high titre retroviral vectors with multiple drug selection markers and a complementary helper-free packaging cell line. Nucleic Acids Res. 18, 3587–3596.
Morin, J.E., Lubeck, M.D., Barton, J.E., Conley, A.J., Davis, A.R. and Hung, P.P. (1987) Recombinant adenovirus induces antibody response to hepatitis B virus surface antigen in hamsters. Proc. Natl. Acad. Sci. USA 84, 4626–4730.
Moss, B. (1990) Poxviridae and their replication. In: Virology, Second Edition (Fields, B. N., ed.), 2079–2111.
Moss, B. (1991) Vaccinia virus: a tool for research and vaccine development. Science 252, 1662–1667.
Moss, B. (1992) Vaccinia virus vectors In: Current Topics in Microbiology and Immunology, Viral Expression Vectors (Muzyczka, N., ed.), pp 1–24 Springer Verlag, Berlin Heidelberg.
Muller, W.J., Mueller, C.R., Mes, A.-M. and Hassel, J.A. (1983) Polyomavirus origin for DNA replication comprises multiple gene tic elements. J. Virol. 47, 586–599.
Mulligan, R.C. (1993) The basic science of gene therapy. Science 260, 926–932.
Myers, R.M. and Tjian, R. (1980) Construction and analysis of SV40 origins defective in tumor antigen binding and DNA replication. Proc. Natl. Acad. Sci. USA 77, 6491–6493.
Niwa, H., Yamamura, K. and Miyazaki, (1991) Efficient selection for high-expression transfectants with a novel eukaryotic vector. Gene 108, 193–199.
Pavlakis, G.N. and Hamer, D.H. (1983) Regulation of a metall othionein-growth hormone hybrid gene in bovine papilloma virus. Proc. Natl. Acad. Sci. USA 80, 397–401.
Peng, S., Sommerfelt, M.A., Berta, G., Berry, A.K., Kirk, K.L., Hunter, E. and Sorscher, E.J. (1993) Rapid purification of recombinant baculovirus using fluorescence-activated cell sorting. Biotechniques 14, 274–277.

Peterson C. and Legerski, R. (1991) High frequency transformation of human repair deficient cell lines of human repair deficient cell lines by an Epstein-Barr virus based cDNA expression vector. Gene 107, 279–284.
Pfister, H., Krubke, J., Dietrich, W., Iftner, T. and Fuchs, P.G. (1986) Classification of the papillomaviruses-mapping the genome. Ciba Found. Symp. 120, 3.
Possee, R.D. and Howard, S.C. (1987) Analysis of the polyhedrin gene promoter of Autographa californica nuclear polyhedrosis virus. Nucl. Acid Res. 15, 10233–10248.
Rice, C.M., Levis, R., Strauss, J.H. and Huang, H.V. (1987) Production of infectious RNA transcripts from Sindbis virus cDNA clones: Mapping of lethal mutations, rescue of temperature sensitive marker, and in vitro mutagenesis to generate defined mutants. J. Virol. 61, 3809–3819.
Rodriguez, D., Zhou, Y., Rodriguez, R.K., Durbin, R.K., Jimenez, V., McAllister, W.T. and Esterban, M. (1990) Regulated expression of nuclear genes by T3 RNA polymerase and lac repressor using recombinant vaccinia virus vectors. J. Virol. 64, 4815–4857.
Sambrook, J. (1987) Vectors for high level exression of proteins in mamalian cells. 10. Anniversary series of Focus.
Schaak, J., Langer, S. and Guo, X. (1995) Efficient selection of recombinant adenovirus by vectors that express ß-galactosidase. J. Virol. 69, 3920–3923.
Schaffhausen, B.S., Bockus, B.J., Berkner, K.L., Kaplan, D. and Roberts, T.M. (1987) Characterization of middle T antigen expressed by using an adenovirus expression ssytem. J. Virol. 61, 1221–1225.
Schlesinger, S. and Schlesinger, M.J. (1990) Replication of Toogaviridae and Flavaviridae. In: Fields Virology (Fields, B.N. and Knipe, D.M., eds.), Raven Press, New York, pp. 697–711.
Shu-Hsia, C., Shine, H.D., Goodman, J.C., Grossman, R.G. and Woo, S.L.C. (1994) Gene therapy for brain tumors: Regression of experimental gliomas by adenovirus-mediated gene transfer in vivo. Proc. Natl. Acad. Sci. USA 91, 3054–3057.
Smith, G.E., Summers, M.D. and Fraser, M.J. (1983) Production of human ß interferon in insect cells infected with a baculovirus expression vector. Mol. Cell. Biol. 3, 2156–2165.
Solnick, D. (1981) Construction of an adenovirus-SV40 recom binant producing SV40 T antigen from an adenovirus late promoter. Cell 24, 135–143.
Sridhar, P., Panda, A.K., Pal, R., Talwar, G.P. and Hasnain, S.E. (1992) Temporal nature of the promoter and not relative strength determines the expression of an extensively processed protein in a baculovirus system. FEBS Letters 315, 282–286.
Steward, L.M.D., Hirst, M., Lopez-Ferber, M., Merrywether, A.T., Cayley, P.J. and Possee, R.D. (1991) Construction of an improved baculovirus insecticide containing an insect specific toxin gene. Nature 352, 85–88.
Stratford-Perricaudet, L.D., Levrero, M., Chasse, J.-F., Perricaudet, M. and Briand, P. (1990) Evaluation of the transfer and expression in mice of an enzyme-encoding gene using a human adenovirus vector. Hum. Gene Ther. 1, 241–256.
Strauss, M., Kiessling, U. and Platzer, M. (1986) Gene transfer in animal cells. I. Techniques, vectors and selective markers. Biol. Zbl. 105, 209–243.
Sutter, G. and Moss, B. (1992) Nonreplicating vaccinia vector efficiently expresses recombinant proteins. Proc. Natl. Acad. Sci. USA 89, 10847–10851.
Szekely, L., Jin, P., Jang, W.Q., Rosen, A., Wiman, K.G., Klein, G. and Ringgertz, N. (1993) Position dependent nuclear accumulation of retinoblastoma(Rb) protein during in vitro myogenesis. J. Cell Physiol. 155, 313–322.
Tartagalia, J., Pincus, S. and Paoletti, E. (1990) Poxvirus based vectors as vaccine candidates. Crit Rev. Immunol. 10, 13–30.
Taylor, J., Weinberg, R., Tartagalia, J., Richardson, C., Alkhatib, G., Briedis, D., Appel, M., Norton, E. and Paoletti, E. (1992) Nonreplicating viral vectors as potential vaccines:

recombinant canarypoxvirus expressing measles virus fusion and hemagglutinin glycoproteins. Virology 187, 321–328.

Teshigawara, K. and Kasutra, Y. (1992) A simple and efficient mammalian gene expression system using an EBV-based vector transfected by electroporation in G2/M phase. Nucleic Acid Res. 20, 2607.

Thummel, C., Tjian, R., Hu, S. H. and Grodzicker, T. (1983) Translational control of SV40T antigen expressed from the adeno virus late promoter. Cell 33, 544–464.

(1) Tooze, J., (1981) Molecular Biology of Tumor Viruses. Part 2: DNA Tumor Viruses. 2nd Edn., Cold Spring Harbor Laboratory, New- York.

Urakarwa, T., Ferguson, M., Minor, P. D., Cooper, J., Sulivan, M., Almond, Y. W. and Bishop, D. L. H. (1989) Synthesis of immunogenic, but noninfectious poliovirus particles in insect cells by a baculovirus expression vector. J. Gen. Virol. 70, 1453–1463.

Usdin, T. B., Braunstein, M. J., Moss, B. and Isaacs, S. N. (1993) SP6 RNA polymerase containing vaccinia virus for rapid expression of cloned genes in tissue culture. Biotechniques 14, 222–224.

Usdin, T. B., Mezey, E., Chen, C., Braunstein, M. J. and Hoffman (1991) Cloning of the cocaine-sensitive bovine dopamine transporter. Proc. Natl. Acad. Sci. USA 88, 11168–11171.

van Doren, K. and Gluzman, Y. (1984) Efficient transformation of human fibroblasts by adenovirus-simian virus 40 recombinants. Mol. Cell. Biol. 4, 1653–1656.

Varmus, H. and Brown, P. (1989) Retroviruses. In: Mobile DNA (Berg, D. E. and Howe, M. M., eds.), American Society for Microbiology, Washington, D.C. 53–108.

Vidal, M., Wrighton, C., Eccles, S., Burke, J. and Grosveld, F. (1990) Differences in human cell lines to support stable replication of Epstein-Barrvirus-based vectors. Biochem. Biophys. Acta 1048, 171–177.

Wang, X., Ooi, B. G. and Miller, K. L. (1991) Baculovirus expression vectors for multible gene expression and for occluded virus production. Gene 100, 131–137.

Xiong, C., Levis, R., Shen, P., Schlesinger, W. S., Rice, C. M. and Huang, H. V. (1989) Sindbis virus: an efficient, broad host range vector for gene expression in animal cells. Science 243, 1188–1191.

Yamada, M., Lewis, J. A. and Grodzicker, T. (1985) Overproduction of the protein product of a nonselected foreign gene carried by an adenovirus vector. Proc. Natl. Acad. Sci. USA 82, 3567–3571.

Yates, Y. L., Warren, N. and Sugden, B. (1985) Stable replication of plasmids derived from Epstein-Barr virus in a variety of mammalian cells. Nature 313, 812–815.

Yates, Y. L., Warren, N., Reisman, D. and Sugden, B. (1984) A cis acting element from Epstein-Barr viral genome that permits stable replication of recombinant plasmids in latently infected cells Proc. Natl. Sci. USA 81, 3806–3810.

Zhou, X., Berglund, P., Zhao, H., Liljeström, P. and Jondal, M. (1994) Generation of cytotoxic and humoral immune response by nonreplicative recombinant Semliki Forest virus. Proc. Natl. Acad. Sci. USA , 3009–3013.

1.4 Aspects of Gene Transfer and Gene Amplification in Recombinant Mammalian Cells

Florian M. Wurm

1.4.1 Introduction

Mammalian genomes are exceptionally plastic. A reflection of this is the finding that some DNA sequences within mammalian genomes, whether they are endogenous or transfected, can be induced and/or selected for amplification to high copy numbers. This phenomenon has been very useful for the expression and production of proteins of scientific and pharmaceutical value, since increased copy numbers of genes often correlate with both high levels of messenger RNA and efficient protein synthesis. Experimentally induced gene amplification has been used successfully to express a large number of recombinant proteins, particularly in immortalized Chinese Hamster Ovary (CHO) cells (Ringold et al., 1981; Kaufman and Sharp, 1982; Kaufman et al., 1986; Weidle et al., 1988; Wurm et al., 1986). The amplified DNA sequences are observable by light microscopy as homogeneously staining regions or extended chromosomal regions (HSR or ECR) (Biedler and Spengler, 1976; Nunberg et al., 1978; Stark et al., 1989; Pallavicini et al., 1990; Gudkov and Kopnin, 1987; Milbrandt et al., 1983). However, ECRs do not represent the only form of amplified sequences in mammalian cells. In some cell lines cultivated in the presence of folate antagonists or other drugs, elements which are extrachromosomal and therefore genetically unstable, have been observed by light microscopy. These elements are called double minute chromosomes (DMs) and have paired, acentric circular structures (Brown et al., 1981; Carroll et al., 1988; Vitek, 1987). In addition, autonomously replicating circular precursors, smaller in size than DMs, have been detected which also are thought to be intermediates for chromosomally amplified DNA (Carroll et al., 1988; Ruiz and Wahl, 1988). Unlike DMs, these autonomous episomes cannot be seen by light microscopy. Usually, DMs and ECRs do not coexist within the same cell (Cherif et al., 1989). Evidence indicates that both the small episomes and the larger DMs have the capacity to integrate into a chromosome of the host, thereby forming the more stable ECRs. (Stark et al., 1989; Gudkov and Kopnin, 1987; Carroll et al., 1988; Ruiz and Wahl, 1988; Brown et al., 1981; Ruiz and Wahl, 1990; Levan and Levan, 1982; Trent et al., 1986). It is not apparent at present why some cells exhibit mainly the formation of DMs, whereas other cells, CHO cells being the prime example, exhibit the formation of ECRs quite readily upon certain selection procedures.

After a brief introduction of the primary cell line for the discussion of gene transfer and amplification in mammalian cells I will describe the path of the transfected DNA and the sequence of events in cells when they are exposed to conditions which result in selective amplification of the transferred DNA. A major part of this review will be a discussion of the elegant work by Wahl and co-workers (Stark et al., 1989; Ruiz and Wahl, 1988; Ruiz and Wahl 1990; Stark and Wahl, 1984) who established evidence for the possible relationship of episomal structures containing plasmid sequences and chromosomally amplified DNA. Finally, I will present some data and discuss work from my laboratory and those of others, in particular work by Ma, Hamlin and co-workers (Milbrandt et al., 1983; Ma et al., 1993; Hamlin et al., 1991) that address aspects of chromosomal events during amplification and the genetic implications (and resulting productivities) when cell populations containing amplified sequences are cultivated over extended periods.

1.4.2 History of CHO Cells as Hosts for the Production of Recombinant Proteins

It is fortunate that one of the most popular mammalian hosts for recombinant protein production (CHO) is also the choice for studies regarding the phenomena discussed here. The reason for this coincidence may be that other cell lines which exhibit the formation *and persistance* (under selective condition) of DMs are impractical hosts for recombinant protein expression, since rapid loss of gene copies occurs when the culture conditions are non-selective (Brown et al., 1981; Carroll et al., 1988). In contrast, stable, highly amplified sequences in the form of ECRs have been observed in CHO cells and other cells which have been cultivated without selective agents for many months (Weidle et al., 1988; Pallavicini et al., 1990; Stark and Wahl, 1984). Though not studied as extensively as CHO cells with respect to the chromosomal status of amplified sequences, Baby Hamster Kidney (BHK) cells (Schmid et al., 1991), mouse myeloma cells (Hendricks et al., 1988; Hendricks et al., 1989) and recently, human lymphoblastoid cells (Namalwa) (Okamoto et al., 1990) also have been used as hosts for production of recombinant proteins. The selection protocols and expression vectors used with these cell hosts are similar to those typically used with CHO cells. BHK, myeloma and Namalwa cells and their genomes react as CHO cells do to the transfer of exogenous DNA and the application of selective procedures, as far as the limited data available allows one to conclude.

The popularity of CHO cells as hosts for the introduction of exogenous DNA and their use for the synthesis of encoded proteins is based mainly on two sets of considerations. The first encompasses a number of technical, practical issues which are outlined in more detail below and in chapter 1.4.3. The other set of considerations is based on safety discussions which were held for many years and started several decades ago when the first biological products (i.e. vaccines and later, natural interferons) were developed on the basis of primary and transformed mammalian cells. An extremely important goal of both the manufacturers of recombinant protein pharmaceuticals and the regulatory agencies controlling the marketing approval of novel therapeutics is to minimize eventual risks associated with the use of immortalized recombinant mammalian cell hosts.

CHO cells were established for the study of somatic cell genetics in 1957 by T. Puck (for review see Puck, 1985). These cells have been the object of many studies since then and are very well characterized with respect to a variety of aspects, including karyotype, chromosome structures, gene mapping, culture conditions and media requirements. Upon the emergence of modern DNA-technology, one particular derivative of the original CHO cells (see below) was found to be easy to work with and was used in a number of pioneering gene transfer experiments (Ringold et al., 1981; Kaufman and Sharp, 1982). It was possible to sub-cultivate these cells after the identification of genetically altered candidate clones producing the protein(s) of interest under increasingly selective stringencies. This resulted in the generation of even higher productive, "amplified" subclones. Another important practical consideration of industrial scale production within the pharmaceutical industry was the finding that CHO can be adapted quite easily to grow in suspension. In the mid-eighties scientists and engineers at Genentech developed a 10000 L production process for rtPA, using "deep tank" technology with suspension adapted CHO cells and their special needs in mind (Arathoon and Birch, 1986).

CHO cells are considered to be a "safe" host for the production of proteins of therapeutic value which, in most cases, will be administered parenterally to human patients. Human pathogenic viruses like Polio, Herpes, Hepatitis B, HIV, Measles, Adenoviruses, Rubella and Influenza do not replicate in these cells. Thus, the risk of a viral adventitious agent being involuntarily carried along with the product of interest can be considered extremely low. Wiebe et al., tested a total of 44 human pathogenic viruses for replication in CHO cells and found only 7 (Reo 1, 2, 3, Mumps and Parainfluenza 1, 2, 3) which were able to infect these cells (Wiebe et al., 1989). These safety considerations were thoroughly discussed among scientists both from regulatory agencies and from the biopharmaceutical sector. One can only speculate whether or not an immortal human cell substrate would have received the approval for production and marketing of a protein therapeutic. Finally then in 1987, recombinant human tissue plasminogen activator (tPA) "Activase®", was the first therapeutic protein from a "transformed" mammalian cell line to receive approval for marketing.

Thus, "Activase®" tPA opened the door for many other protein therapeutics to be produced in CHO cells. Due to complex processing and modification requirements, many proteins of value in human therapy have to be produced in a mammalian cell. At present (early 1993), only three other recombinant proteins produced in an immortal mammalian cell host have received approval for marketing: (i) human erythropoeitin, a growth promoting factor for red blood cells (1989), (ii) recombinant Factor VIII for patients with reduced clotting capacity (1992) and (iii) a recombinant Hepatitis B vaccine, approved for use in France (1992). However, a much larger number of proteins derived from recombinant CHO cells are in the process of clinical evaluation and the number of approved biologics from recombinant CHO cells will very likely increase rapidly in the near future. Examples are: various versions of HIV gp120 (a candidate AIDS vaccine), human DNase (a drug for the treatment the pulmonary problems in cystic fibrosis), humanized anti-Her-2 antibody (for the treatment of certain breast cancers), soluble tumor necrosis factor-receptor immunoadhesin (for treatment of septic shock), and human nerve growth factor (for treatment of peripheral neuropathy).

1.4.3 The DHFR/Methotrexate/CHO Expression System: A Multilayer Selection System for High Level Expression of Recombinant Genes

In 1980 Urlaub and Chasin developed a CHO cell line (Urlaub and Chasin, 1980) that was created by mutagenesis of the original cell line K1 established by Puck (Kao and Puck, 1968). As intended by the investigators, this mutant line exhibited a complete inactivation and/or removal of it's two alleles encoding Dihydrofolate Reductase (DHFR). Growth of these cells is dependent therefore on supplementation of the culture medium with hypoxanthine and thymidine. The amino acid glycine had to be added as well, since, like the original K-1 cell line, the DHFR minus line was glycine dependent as well. This cell line quickly became the host of choice in transfections which employed the use of expression vectors carrying a functional DHFR gene and a secondary gene of interest (either provided on the same or on second, co-transfected plasmid). In selective medium (without glycine, hypoxanthine and thymidine; "GHT-minus") DHFR positive clones can be easily identified after about two weeks (see Fig. 1.4.1). Usually, many of the recombinant DHFR positive clones also have been found to express the gene of interest (Ringold et al., 1981; Kaufman and Sharp, 1982; Wurm et al., 1986; Wurm et al., 1992).

DHFR vectors also have been utilized in cell lines which still contain functional endogenous DHFR genes. However, with such cells, modified selection procedures are required. For example, a second marker gene (Schmid et al., 1991; Chiang and McConlogue, 1988; Kaufman et al., 1986) or a mutant marker DHFR gene can be used. Also, combinations of several markers have been employed successfully (Wirth et al., 1988).

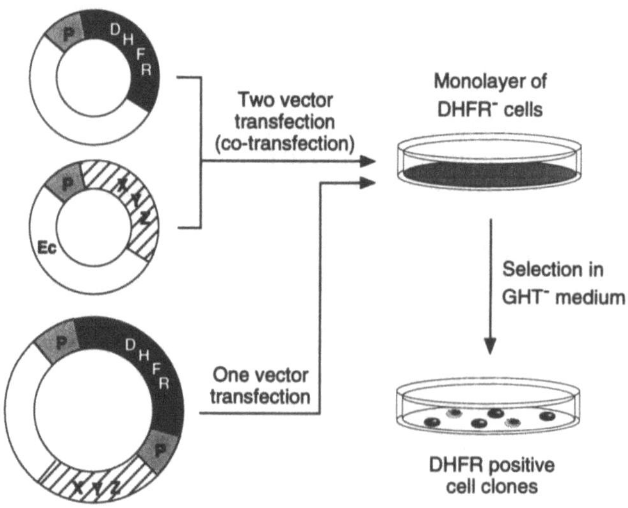

Fig. 1.4.1: Transfection and co-transfection of DHFR containing expression vectors into CHO cells deficient in endogenous DHFR activity. Selection of recombinant clones in GHT minus medium (GHT minus: medium lacks glycine, hypoxanthine and thymidine).

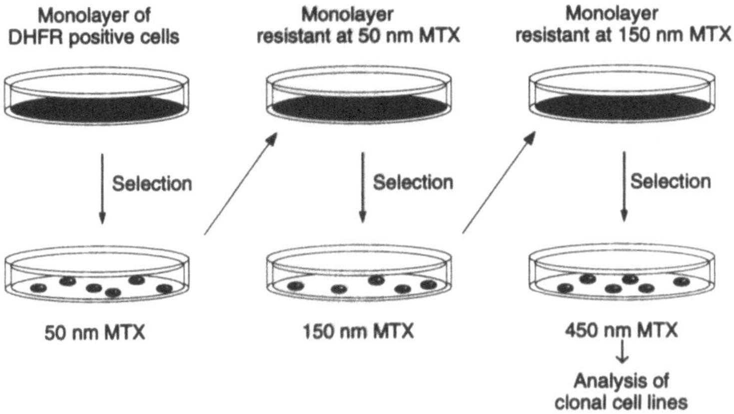

Fig. 1.4.2: Incremental increase in methotrexate (MTX) concentration in culture and selection of subpopulations of cell clones resistant to high levels of methotrexate (nm: nanomolar).

DHFR activity can be blocked completely by the folate derivative methotrexate (MTX) (Camargo and Cervenka, 1980). It can be added to the cell culture medium and is taken up readily by the cultivated cells. Thus the exposure of recombinant cell clones to MTX provides opportunities for additional, multiple layers of selection. Depending on the protocol employed, incrementally increased concentrations (2–3-fold for each step) from 10 nanomolar to several hundred micromolar MTX can be utilized. At each of the concentration levels applied, subpopulations of cells can be identified, which have increased DHFR activity as compared to their parent generation of cells (Fig. 1.4.2). In principle, high DHFR activity in cells can be the final result of either one of two genetic phenomena: a) high level of expression from a single or a few plasmid molecules integrated into the host genome or b) amplification of plasmids that express at relatively low levels. In fact, the latter phenomenon, i.e. amplification, seems to be quite frequent in transformed cell lines. In CHO cells and in other cell lines, as mentioned above, amplification is associated with the formation of marker chromosomes characterized by long homogeneously staining regions, HSRs or ECRs, which contain the transfected DNA (Weidle et a., 1988; Nunberg et al., 1978; Pallavicini et al., 1990; Milbrandt et al., 1983; Kaufman et al., 1985). Using a stepwise incremental selection procedure with MTX, a number of cell lines have been developed which have hundreds or even thousands of copies of the transfected plasmid constructs (Wurm et al., 1986; Nunberg et al., 1978; Pallavicini et al., 1990; Milbrandt et al., 1983; Kaufman et al., 1985; Goeddel et al., 1983; Smith et al., 1987; Capon et al., 1989). Some of these cell lines express the desired protein at surprisingly high levels with peak productivities reported in the several hundreds of milligrams per liter in standard cell culture production processes (Wood et al., 1990), Fouser et al., 1992; Brand et al., 1993).

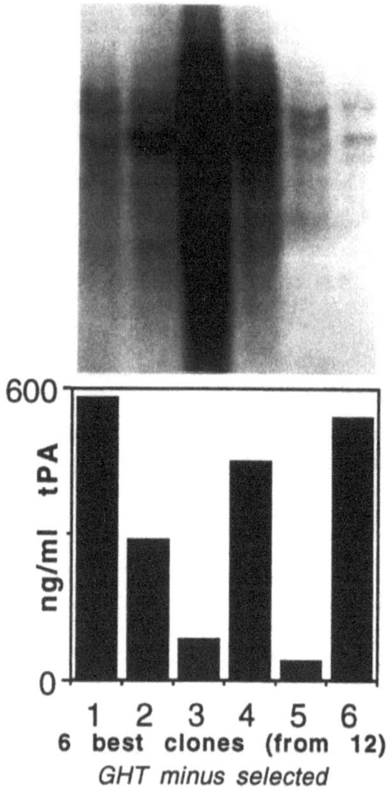

Fig. 1.4.3: Southern analysis and product expression of 6 best producer (clonal) cell lines established upon selection in GHT minus medium upon cotransfection with a DHFR vector and a tPA expression vector.
a) upper part of figure: Southern type DNA hybridization or restricted genomic DNA of each of 6 best expressors. Hybridization was performed with a rtPA specific probe only. b) lower part of figure: Expression levels for recombinant tissue plasminogen activator (rtPA) expression in nanogram per milliliter of the same 6 best clones selected and expanded to individual cell lines.

1.4.4 Linearization of Plasmid DNA is a Prerequisite for Integration

Graham and van der Eb (Graham and van der Eb, 1973) showed that the transfer of DNA into mammalian cells in culture can be achieved by exposing cells to naked DNA complexed into micro-precipitates of calcium phosphate. This simple method (termed calcium phosphate transfection) has become very popular. Typically, osmotic shock is part of the transfection procedure. Osmotic shock is thought to promote the transport of individual DNA/calcium phosphate complexes across the cell membrane perhaps by promoting endocytosis. Up to 90 % of cells subjected to optimized transfection conditions take up complexed DNA. However, in addition to the transit across the cells outer membrane, the transport of DNA from the cytoplasm to the nucleus seems to be a significant barrier to gene transfer as well. Only 1–5 % of transfected cells have detectable calcium phosphate complexes within the nucleus (Loyter et al., 1982). Moreover, the number of emerging colonies upon transfection by the calcium phosphate technique is even lower: usually, only between 0.01 % and 0.5 % of the cells give rise to recombinant cell colonies. It seems that the transfer of a sufficient amout of DNA into the nucleus and the modification of this DNA by nuclear enzymes to enable the integration of at least a few functional plasmid molecules into a receiving chromosome is by no means an efficient process. It has been speculated that the association of the DNA with calcium phosphate might protect against nuclease attack during the transport to the nucleus. However, one has to assume that at some point in time these complexes become dissociated and that nuclear endo- and exonucleases have access to the „naked"? DNA. In fact it has been shown that circular plasmid DNA becomes linearized within hours inside the nucleus – an essential step for integration into the linear DNA backbone of a chromosome (Finn et al., 1989). The mechanism by which the DNA is transported to the nucleus and finally to the site of chromosomal integration is not known yet and little is also known about the integration of non-homologous DNA molecules into the chromosomal DNA.

With respect to the site of integration there is no reason to believe that there are chromosomes within the CHO genome into which transfected DNA molecules integrate preferentially. Again, only limited and sporadic data is available at present on the sites of primary integration into chromosomes of CHO cells or other cells. The question arises whether there is a preference for transfected DNA to integrate into transcriptionally active regions of chromosomes, which represent only a small fraction of mammalian genomes. So far, no conclusive data has been presented giving an answer to this question. However, the observed variability with respect to the expression levels of recombinant proteins when analyzing independently established clonal cell lines upon transfection seems to be indicative of random integration loci – assuming that (chromosomal) position effects would be mainly responsible for the wide range of expression levels found.

It is interesting to note, in the context of integration sites, that clonal sublines from cells selected to grow in high levels of methotrexate showed ECRs on various chromosomes with no preference for a particular chromosome or chromosome site (i.e. median or terminal) (Pallavicini et al., 1990). In light of recent data on gene amplification mechanisms (to be discussed below) it is not possible to draw conclusions from these findings regarding primary integration site(s).

In addition to calcium phosphate transfection, more recently, electroporation and liposome mediated transfection techniques have been developed for transfer of DNA into mammalian cells. These techniques appear to result in higher transfection efficiencies (Chu et al., 1987; Barsoum, 1990) as compared to calcium phosphate mediated DNA transfer. Such observations can be generalized only with some caution, since all these techniques and their success rates are highly variable and depend on many diverse parameters which can be optimized for each individual protocol

(Chen and Okayama, 1987). It might be that various techniques, resulting in different transfection efficiencies, i.e. the number of colonies arising in selective medium following transfection, would also differ in the amount of DNA integrated into the genome per recombinant cell. This by itself could have profound effects on the number of initially integrated molecules and, subsequently, on the achievable expression levels per transformant.

Linearization of closed circular plasmid DNA seems to be a prerequisite for integration. When dependent on host enzyme activity single and double strand cleavage of the plasmid DNA is expected to occur at single or multiple (random) loci. Depending on the overall size of the plasmid molecule and the size of the DHFR gene cassette relative to the vector, a significant number of molecules would be linearized within the DHFR sequence which clearly would jeopardize the functionality of this plasmid. Obviously, molecules whose DHFR expression cassette would be interrupted or deleted by the linearization process would not be able to transfer a functional marker gene to the host cells and such cells would be eliminated through the selection procedure. Therefore, opening the circular plasmid molecule by restriction enzyme digestion with appropriate enzymes prior to transfection has been recommended for improvement of transfection efficiencies with protocols which are used for integration of plasmid DNA into chromosomal DNA. Specific restriction enzyme digestions have been performed with plasmids containing fragments and sequences which are homologous to parts of the receiving genome and whose function it was to improve the rate of homologous recombination events for the integration of the transferred DNA into certain target regions of the genome (Jasin and Berg, 1988; Smithies et al., 1985; Zheng and Wilson, 1990).

Recently, we have used vectors with sequence elements which were derived from CHO DNA and which might participate in integration events involving homologous sequences. The sequence elements employed are derived from endogenous retroviral sequences residing at high copy numbers within the genome of CHO cells (Wurm et al., 1992; Anderson et al., 1990); Anderson et al., 1991). We found that the presence of these fragments promoted higher transfection efficiencies and improved expression levels when a vector containing a gene of interest (recombinant tissue plasminogen activator or CD4-IgG) was co-transfected (Wurm et al., 1992). Restriction of a modified DHFR vector containing a fragment of a defective C-type sequence with an enzyme which opens the plasmid in the center of the retroviral sequence, i.e. within the region of homology, resulted in a large number of clones that express the recombinant protein at very high levels (Fig. 1.4.4). Southern analysis of these high expressor clones revealed that the copy number of the integrated plasmid sequences was much lower than the copy numbers in clones derived from control transfections which did not employ retroviral sequences. We concluded from these results that the presence of retroviral sequences in transfection cocktails mediated introduction of the recombinant DNA into regions of the genome of high transcription rate, possibly marked by individual members of the residing retroviral sequence families (Wurm et al., 1994).

The majority of recombinant cell clones emerging from a calcium phosphate-transfection seem to have received more than just a single copy of the transfected plasmid DNA molecules (Kaufman and Sharp, 1982; Milbrandt et al., 1983). We have verified multi-copy integrations in clonal cell lines selected in GHT minus

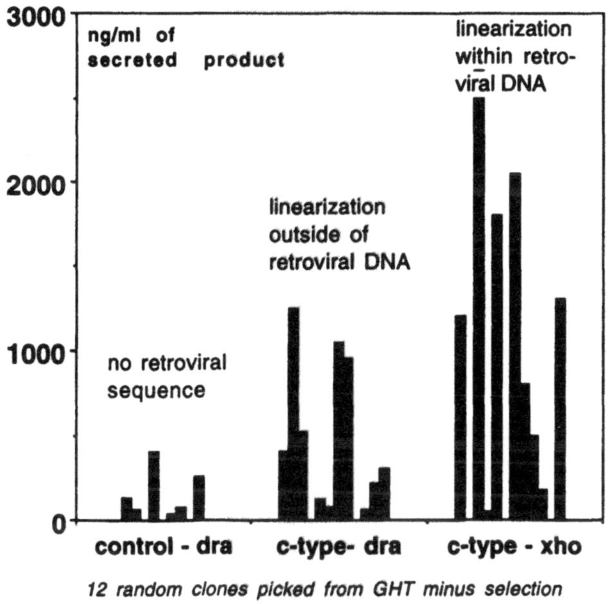

Fig. 1.4.4: Expression analysis of 3 independent co-transfections, with 12 clonal cells lines established randomly for each of the transfections. Transfections were done with linearized plasmids, with and without the presence of a CHO-C-type retroviral sequence.
control-dra:
co-transfection of standard expression vectors for DHFR and tPA, both linearized by restriction within the ampicillin resistance gene.
c-type-dra:
co-transfection of the same tPA expression vector together with a DHFR plasmid containing a 6.6 kb CHO-derived C-type sequence. This plasmid was linearized in the region of the ampicillin resistance gene.
c-type-xho:
co-transfection using the same pair of plasmids as in b), but the DHFR vector was linearized within the center of the 6.6kb CHO-C-type sequence. The bars represent the individual tPA expression levels of 12 randomly chosen clones per transfection.

medium and have found in fact a wide variation in copy number in those primary clones, as is shown in Fig. 1.4.3 (upper part). Based on the intensity of the hybridization signal exhibited and analyzed by computer assisted densitometry, we estimated the copy numbers of 4 best clones (from 12 established and expanded), expressing rtPA, to vary about one hundredfold. In fact, this phenomenon constitutes the basis for the high success rate of co-transfections of DHFR plasmids and genes of interest cloned into a second expression vector. Again, it is not well understood which mechanisms drive the co-integration of co-transfected plasmids (provided in circular or linear form) into the same chomosomal locus. However this seems to occur in the majority of cases, at least in those clones which co-express the plasmid-derived DHFR gene and the gene of interest. One can speculate that

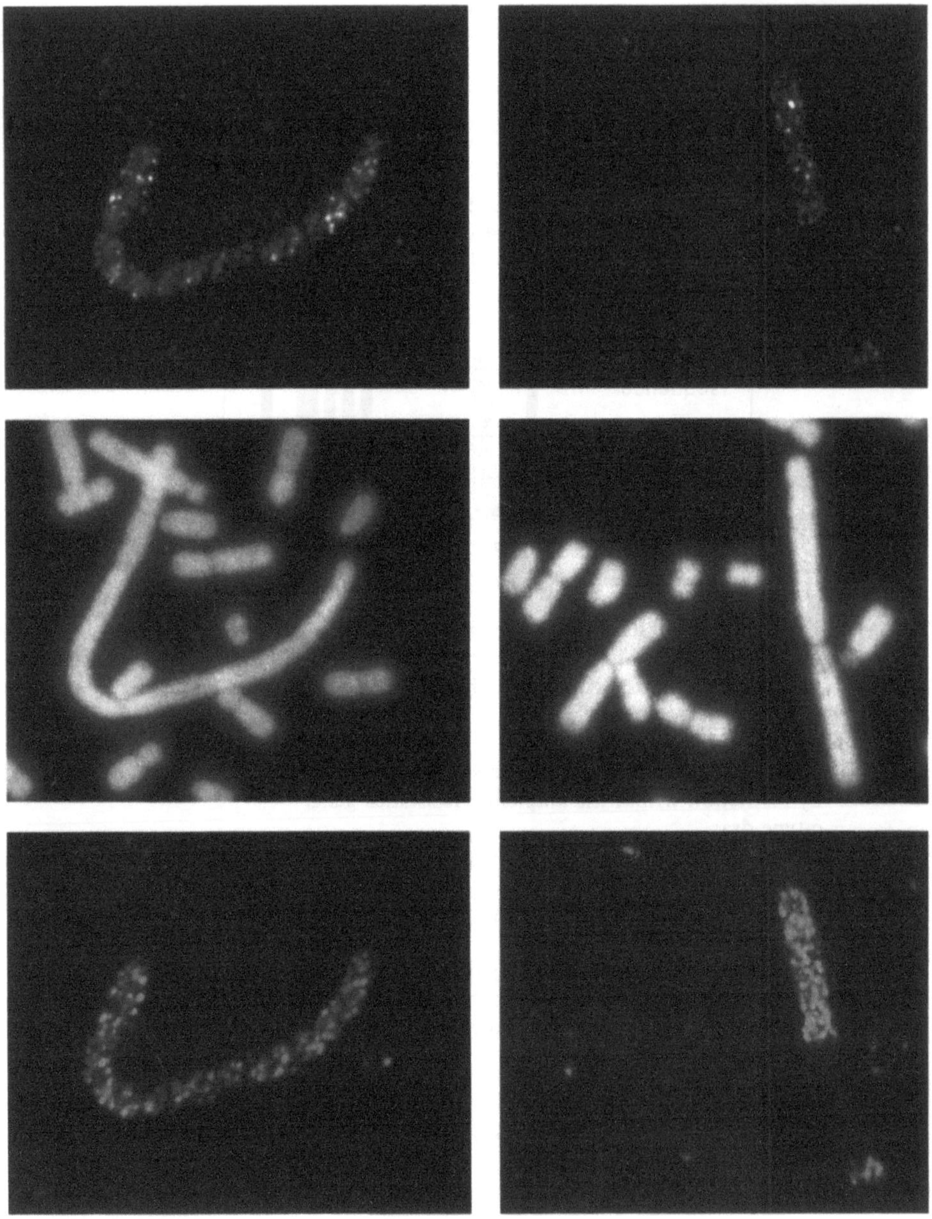

Fig. 1.4.5: Double probed fluorescence in situ hybridization of two metaphases obtained from highly amplified cell line expressing the mouse c-myc gene, initially established by co-transfection of two plasmids. Center: metaphase spreads show exceptionally extended chromosomes. Top: Hybidization signal obtained with probes specific for the DHFR plasmid. Bottom: Hybridization signal obtained with probe specific for the c-myc plasmid, showing fluorescence over same regions of chromosomes.

the hundreds or thousands of molecules of DNA being carried into the nucleus might both become targets for attack by nucleases as well be exposed to nuclear (repair?) ligase activity. It has been established that mammalian cells have abundant end-joining capabilities (Wilson et al., 1982) and cotransfected DNA becomes physically and genetically linked (Perucho et al., 1980). This phenomenon seems to be independent of regions of homology between the ligated partners. Ligation of linearized plasmid molecules would generate multimeric chains of derivative molecules which contain the DNA sequences of the original plasmid mixture, in a random order, and in proportion to the initial molar ratios. Indeed, co-transfection protocols usually recommend the use of a plasmid cocktail consisting of a molar ratio of 1:5 to 1:20 of the DHFR plasmid and the plasmid carrying the gene of interest. The idea is, of course, to optimize expression of the gene of interest while providing sufficient expression of the selectable marker gene.

Support for the concept of co-integration of co-transfected DNAs also comes from our own studies using fluorescence *in situ* hybridization (FISH). Using double probed chromosomes of cells showing ECRs, with two differently labeled DNAs representing the two co-transfected plasmids, allowed for fluorescence signals with two clearly separate wave lengths. As is shown in Fig. 1.4.5 a and b, the differential fluorescence signal indicates the presence of the individual plasmid sequences over the same stretch of chromosomal DNA of each of those highly elongated chromosomes, visualized in the center of the composite.

1.4.5 Chromosome Segments with Plasmid Sequences are Excised, Circularize and Form Replicating Episomes which Finally Reintegrate into the Host Genome

Integration and subsequent expression of transfected DNA in mammalian cells can be considered to be the outcome of many complex and poorly understood steps. When protocols for "transient transfections" are used, up to 90 % of subjected cells take up and express the transfected DNA possibly residing inside the cells as episomes. In contrast, stably transfected cell colonies develop – in selective media – from less than one percent of the cells which had been exposed to the DNA initially. This indicates that a cell which "succeeds" in sustaining expression of the transfected DNA (and thereby divides and forms a colony) represents a rare exception in comparison to most cells which succumb to the selection procedure. FISH studies on metaphase chromosomes of cell clones selected immediately after transfection without the exposure to MTX indicate the existence of a single integration site per cell (Wurm and Pallavicini, 1990). Therefore, according to the data obtained so far, it appears that integration of plasmid DNA into chromosomal DNA is a rare and most likely a singular event.

The mechanisms responsible for the amplification of sequences and for their behavior within mammalian genomes have been the subject of intensive research for many years. It would be impossible to do justice to all the studies published over two decades. Rather I refer the interested reader to key papers and reviews which are cited in the literature list. Using broad strokes, I hope to summarize the latest understanding of these phenomena. Most studies have been done with transformed (i.e. permanent), recombinant and non-recombinant cell lines using a number of antiproliferative agents for the selection and subcultivation of mutants. Among the many antiproliferative agents which have been used, MTX is a unique one – for several reasons. As I mentioned before, MTX is an efficient inhibitor of the enzyme DHFR. The latter is a member of the family of enzymes which are responsible for the synthesis of certain amino acids and nucleotides. DHFR provides precursors for the synthesis of adenosine and thymidine. Blocking DHFR activity with MTX either completely or partially lowers the intracellular pool of these components, because no alternative pathway for synthesis of these components is available. It has been shown that a low intracellular level of thymidine can result in the misincorporation of uridine monophosphate instead of thymidine monophosphate during DNA-replication. This can trigger the (single strand) excising activity of repair enzymes and might be one of the contributing factors for increased recombinogenic activity. Such an induced plasticity might result in chromosomal rearrangements and, consequently, in amplification due to gross chromosomal changes (Goulian et al., 1980). In fact, the occurrence of single stranded gaps within chromosomes has been demonstrated in cells which have been exposed to MTX. Cytogenetically, these structures have been documented as fragile sites (Barbi et al., 1984; Mondello et al., 1994; Raffetto et al., 1979). Thus, when added to the culture medium of cells, MTX not only kills cells with insufficient DHFR-activity, but may also account for an increased rate in recombinogenic events, which could contribute to amplification of afflicted chromosomal regions.

Let me return to the work of many groups who have tried over the last 10 years or so to unravel the numerous and complex steps involved in chromosomal amplification. In particular, Wahl and collaborators have contributed significantly to the understanding of the early events during integration and amplification of recombinant DNA whereas Ma and Hamlin and coworkers (Milbrandt et al., 1983; Ma et al., 1993; Hamlin et al., 1991) have addressed in their studies events which seem to occur somewhat later in the sequence of phenomena. Recently, Wahl and coworkers have analyzed in detail transformants of transfected CHO cells which displayed high frequency amplification upon exposure to high levels of MTX. They use the term "hyperamplifyable" to describe the phenotype of such cells (Ruiz and Wahl, 1988). In these cells, representing about 3 % of the initial recombinant transfectants, integration of the transfected DNA into a (random?) chromosomal site resulted in its inverted duplication along with (inverted) duplication of host sequences of one of the two arms of the chromosomal DNA at the insertion site (Fig. 1.4.6).

The overall size of the duplication was more than 70 kb and thus proved to be much larger than the donated plasmid DNA. Convincing evidence was presented that the primary transformants grown *in the absence of MTX* underwent rapid copy number changes. This is most likely due to the generation of rather large episomal elements (larger than 750 kb but too small to detected by light microscopy) containing multiple copies of the structure outlined in Fig. 1.4.6, (inverted duplication of chromosomal DNA with the plasmid DNA in its center). It has been proposed that one of the reasons for the almost instantaneous emergence of extrachromosomal

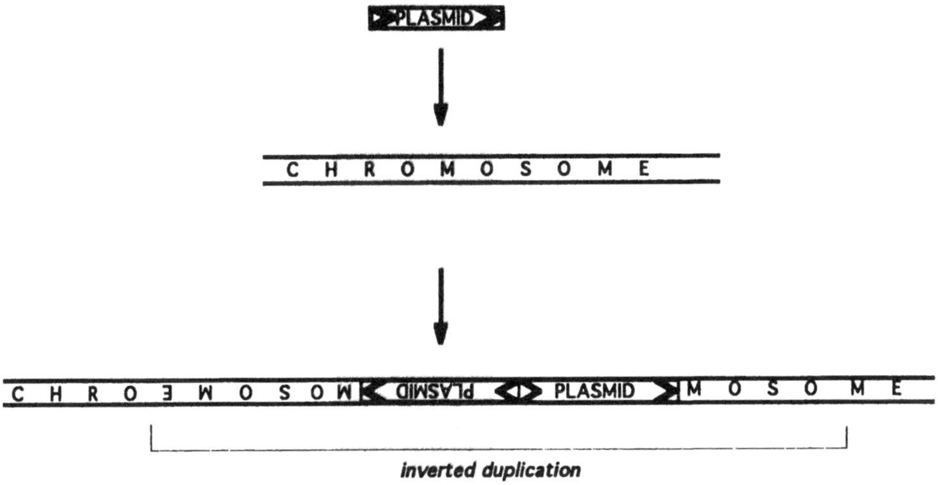

Fig. 1.4.6: Simplified model for the formation of an inverted duplication after gene insertion (according to Ruiz and Wahl, 1988).

elements in some, if not most, transfected cells, is chromosomal instability at or around the site of initial integration of the donated plasmid DNA.

So far, one has to assume that the site of integration is random (Folger et al., 1982). There are no data available at present to assume a preferential site(s) within CHO chromosomes. Also, there is no evidence to indicate a preference for integration of foreign DNA into transcriptionally active regions over non-coding DNA within the genome. Classic genetic studies however, have shown that not all regions of the chromosomes are equally recombinogenic. One way to explain such position effects is to invoke the existence of recombinogenic hot spots equivalent to the *Escherichia coli* chi sites. The rapid emergence of extrachromosomal elements from such sites would provide cells with a hyperamplifiable phenotype and with a high degree of flexibility in response to the environment in which they are kept. If higher copy numbers of such extrachromosomal elements were the reason for a growth advantage over cells with single or low copy numbers, transient stabilization of this early amplification could occur, as long as the selective condition (lack or low level of nucleoside precursors) exists. The exposure of cells, exhibiting such a wide range of expression of the selective gene due to variations in copy number, to low and moderate levels of methotrexate, would result in a further increase of the overall copy number of the transferred DNA within the selected subpopulations.

A separate FISH cytogenetic study done by the group of Wahl (Windle et al., 1991) utilizing cells containing a stable, single chromosomally integrated DHFR gene (different from the ones discussed above) provided a number of interesting insights. Exposure of these cells to MTX for only eight to nine cell population doublings resulted in the generation of subclones exhibiting, in the majority of cells, a wide variation in copy number and in structure of multiple, randomly distributed *extrachromosomal* elements. Evidently, MTX exposure of cells containing a single DHFR copy in a (stable) chromosomal site results in frequent excision of DHFR containing DNA from the chromosome. Another result of this was the generation of extrachromosomal elements which fluctuate in copy number and structure to a large extent. This fluctuation in copy number parallels the observations with respect to the early emergence of submicroscopic elements in hyperamplifiable, freshly transfected cells discussed above (Ruiz and Wahl, 1988).

Prolonging MTX exposure of the same stable cell line containing a chromosomally integrated DHFR gene to a 30 to 35 day period resulted in a different result. When a number of resistant clones were analyzed using FISH, the majority of cells (98 %) of each of the lines showed *intrachromosomal* hybridization in clustered amplicons (tightly arranged stretches of DNA containing plasmid sequences). Copy number (as measured by fluorescene intensity) and chromosomal sites of these amplicons varied from clone to clone quite dramatically (up to 30-fold), but, in contrast, were relatively consistent within each individual clone (only a 1.2-fold variation). These results demonstrated copy number and amplicon location heterogeneity among subclones selected in a single 75 nanomolar MTX selection step. They were generated by *extrachromosomal* intermediates that preceded the formation of the stable chromosomal amplicons detected at the 30–35 cell population doublings.

A very interesting and highly enlightening article on the early events during amplification of the endogenous DHFR locus in CHO-K1 cells was published recently by Ma et al., (1993). The main findings of these studies, employing FISH extensively on various populations of cells generated by exposure to MTX can be summarized as follows. The initial event for amplification in CHO-K1 cells was found to be a duplication of the endogenous DHFR locus by two versions of sister chromatid fusions. One type of fusion occurs after symmetric chromosome breaks (the two chromatids of a chromosome break at the same locus). The fusion of the two frayed ends (of equal length) of the two chromatids would "heal" the breakage prior to the next mitosis. Alternatively, asymmetric breaks of the two chromatids within the chromosome carrying the endogenous DHFR gene would fuse. In both cases, chromosomes with two centromeres and a duplication of the DHFR locus emerge. Such dicentric chromosomes are, due to the forces exerted during mitotic separation of the two centromeres, subject to further breakage events, facilitating and mediating in this manner additional rounds of chromatid fusion and, possibly, further duplication events. This mechanism, the "bridge/breakage/fusion model" was suggested previously to explain the high frequency of dicentric chromosomes in cells with amplified DHFR copies (Kaufman and Sharp, 1982). This model, based on the original observation by the late Nobel laureate Barbara McClintock during the forties on genetic instability in corn (McClintock, 1941), would explain why the early amplicons are most often located on the same chromosome arm as the parental single-copy locus, but very often far away from that locus. It also provides an explanation for the clustering (sometimes recognizable as multiple bands in *in situ* hybridization) of the amplified sequences.

It is unlikely that the process of sister chromatid fusion would be restricted to the amplification of endogenous DHFR genes only. Similar events will occur therefore with transfected and chromosomally integrated DNA sequences as well, including DNA which has been initially amplified via the excision, episome expansion and reintegration pathway, as suggested by Wahl, Stark and coworkers (Stark et al., 1989; Carroll et al., 1988; Ruiz and Wahl, 1988). Such mechanisms would then contribute to the distribution of the amplified DNA to other chromosomes, as well as to additional amplification (via duplication) of amplified clusters.

1.4.6 CHO Cells with Highly Amplified, Chromosomally Localized DNA Sequences are the Result of Long-Term Exposure to Incrementally Increased Concentrations of Methotrexate

Seventy-five nanomolar is a relatively low concentration of methotrexate in culture media, when compared to the concentrations quoted in most publications describing the establishment of highly productive CHO cells. Some cell lines have been established, which were resistant to more than one hundred micromolar methotrexate (Wurm et al., 1986). However the establishment of such highly resistant cell lines is a time consuming enterprise. Considering the plasticity of the eukaryotic genome, such resistance to MTX is most likely the result of additional chromosomal rearrangements affecting copy number and structure of the amplified sequences. Cell populations have to be subcultivated continuously for many weeks, with incremental elevation of the MTX concentration in the culture medium. Usually several periods of poor growth and extensive cell death will occur during this process. Only a fraction of the cell population survives, indicating the selection of distinct subpopulations of cells which express DHFR at elevated levels, mostly due to further selective DNA amplification (Wurm et al., Kaufman et al., 1985; Brand et al., 1993).

In comparison to lines selected at nanomolar MTX concentrations, cell lines resistant to micromolar concentrations show even more elongated chromosomal structures which hybridize to probes representing the transfected DNA (Kaufman and Sharp, 1982; Pallavicini et al., 1990). Examples of these structures are shown in Fig. 1.4.7. They are, most likely, equivalent to the regions originally described by Biedler and Spengler (1976) as homogeneously staining regions (HSRs). Some of the chromosomes containing a large number of more or less tightly arranged repetitive sequence elements homologous to the transfected DNA differ dramatically from "normal" CHO chromosomes. This raises questions about the long- and short-term stability of such aberrant chromosomes during continuous subcultivation and in consequence, the production of recombinant proteins based on genes residing within these structures.

The general trend toward more excessive chromosomal amplification in cell lines selected at higher levels of MTX is established in the literature. However, individual primary clones isolated at the minimal level of stringency (in GHT minus medium) vary in their ability to respond to methotrexate induced amplification. Some clones would not give rise to subclones surviving in higher levels of methotrexate even after extended exposure to low levels of methotrexate. Others would do so quite readily. These might be the ones defined by Wahl and coworkers as having a hyper-amplifiable phenotype. Also, it is interesting to note in this context, that two groups have isolated sequence elements from mouse and CHO DNA, respectively, which,

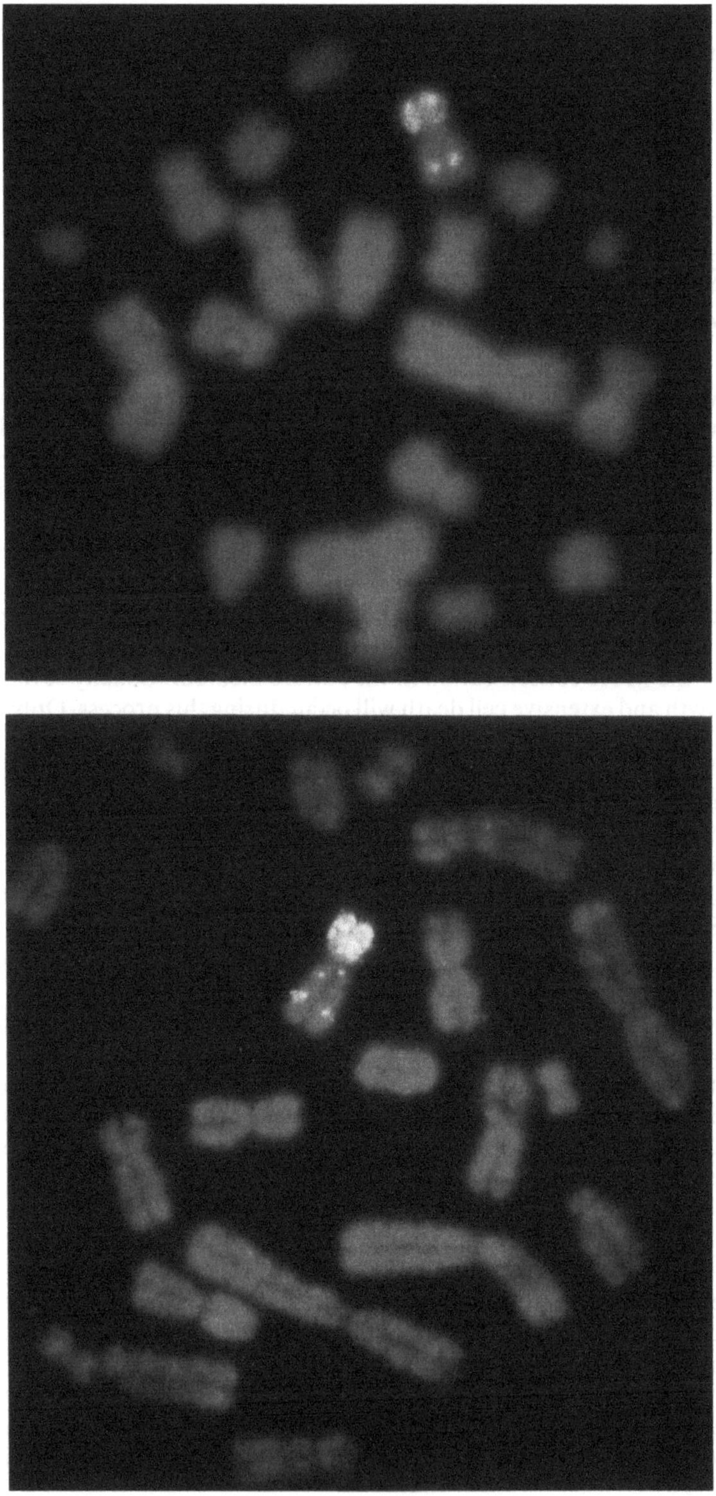

Fig. 1.4.7a

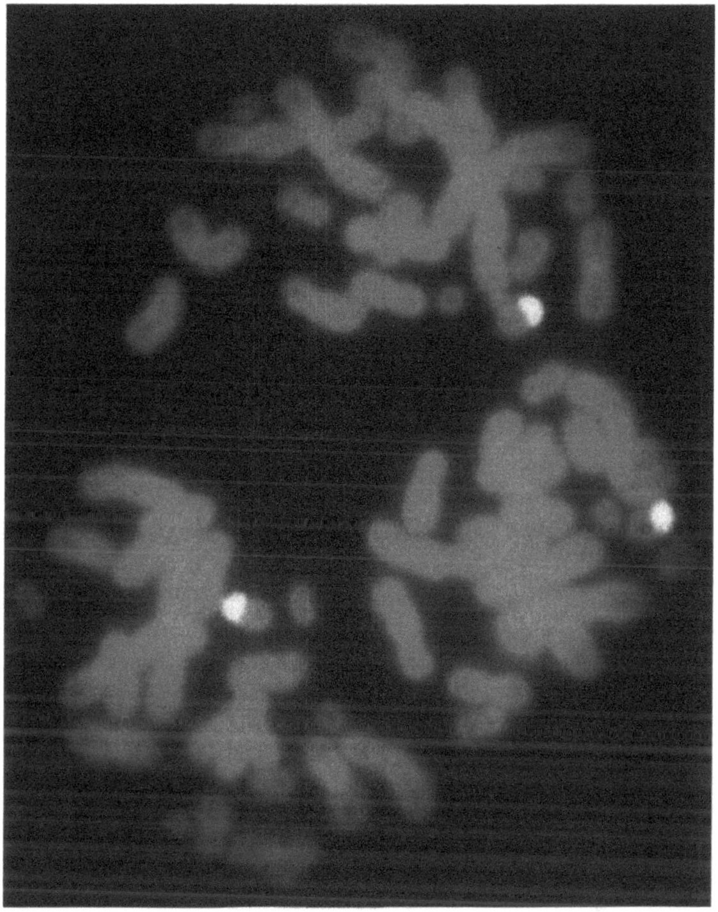

Fig. 1.4.7b

Fig. 1.4.7: FISH of metaphase preparations of various recombinant CHO cell lines hybridized with DNA probes homologous to recombinant sequences a) cell line producing recombinant tPA. b) cell line producing a truncated version of the human CD4 receptor. c) cell line expressing the mouse cy-myc protein. d) cell line expressing the mouse c-myc protein cultured in the presence of MTX.

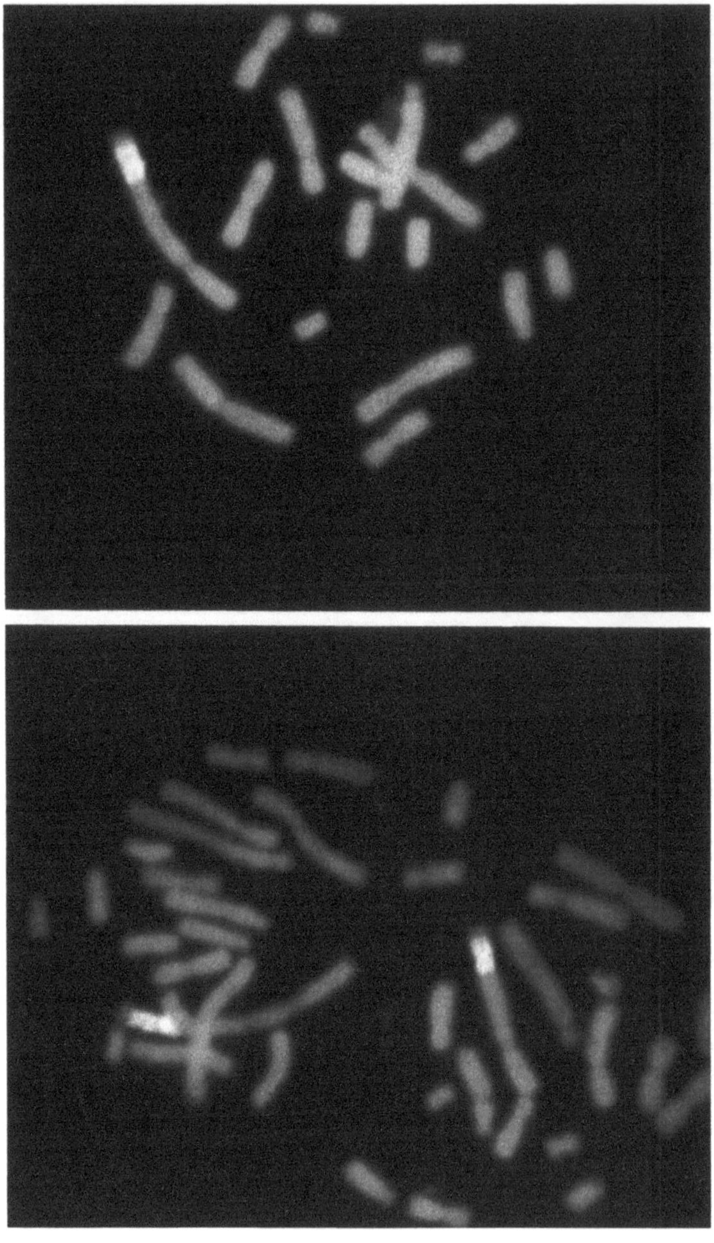

Fig. 1.4.7c

when integrated into expression vectors, facilitate the chromosomal amplification of the transfected DNA (Wegner et al., 1989; Zastrow et al., 1989; McArthur and Stannners, 1991). In both cases, the DNA is characterized by unique AT-rich repetitive elements. One can envision then that, occasionally, the surrounding genomic DNA of the integrated plasmid sequences has structural and sequence characteristics which support the amplification machinery. Evidence for effects mediated by the

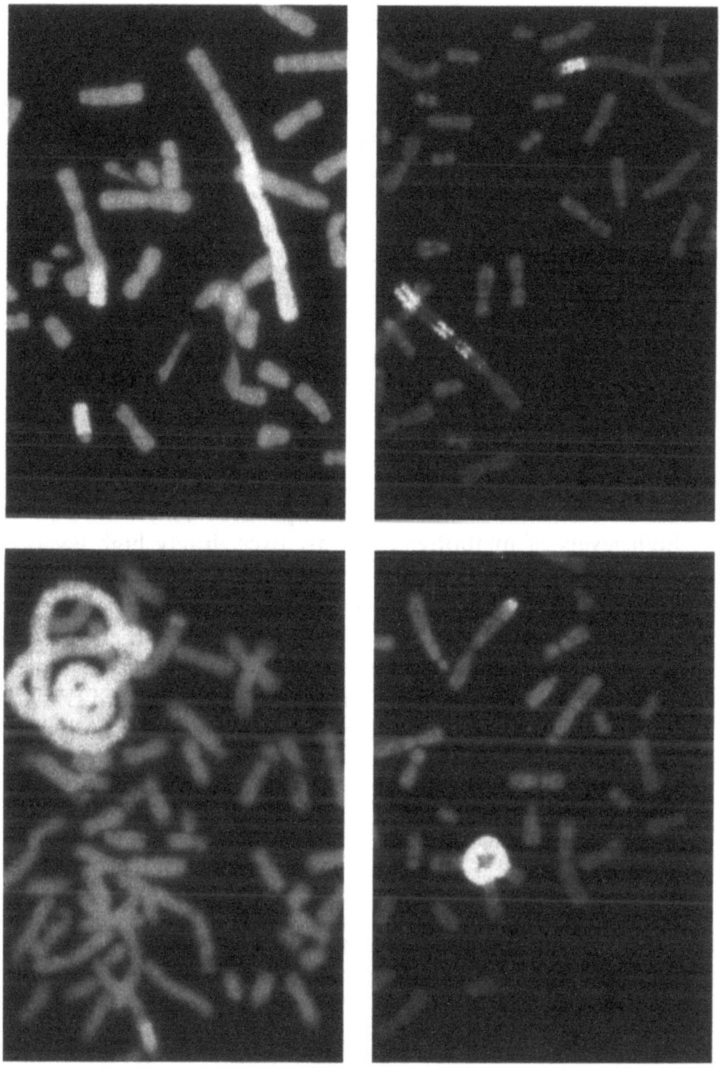

Fig. 1.4.7d

chromosomal domains on amplification of integrated plasmid sequences comes from a paper by Carroll et al., (1987). These authors showed evidence for the involvement of a mammalian replication origin within an episome containing transfected DNA. Also, findings by Windle et al., (1991) support the involvement of host sequences in the facilitation and/or generation of excisional episome formation following plasmid DNA integration. Episomes containing such signal sequences derived from the receiving genome which facilitate replication and amplification could then, upon MTX selection, increase in copy number rapidly and eventually reintegrate in multimeric arrangements into chromosomal DNA. We have found, as mentioned before, quite a striking variation of copy numbers of integrated plasmid sequences in primary

(GHT minus selected) clones (Fig. 1.4.3). The top part of the figure shows the hybridization pattern obtained with DNA from 6 rtPA producing CHO cell lines (clones) when probed with the rtPA expression cassette. The intensity of the signal is judged to be correlated with the copy number of the integrated DNA molecules. Such variations may have profound effects on the degree of responsiveness to MTX treatment and thus achievable amplification.

Another reason for the observed ranges in expression of different clones from the same transfection, *excluding* copy number effects, might be the influence of the locus into which the primary integration event occurred on transcription rates of surrounding DNA. Some differences in expression might be the result of varying influences exerted by different chromosomal environments into which the transfected DNA integrated. If, for example, the transfected DNA becomes integrated into a chromosomal region where clusters of highly transcribed endogenous genes reside, the transcription rate of the inserted DHFR gene would likely be higher. In contrast, lower transcription rates would be expected when the same gene becomes inserted into a region which represents an extended cluster of middle or highly repetitive, noncoding sequence elements. In the first case, progeny cells would be resistant to moderate or high levels of methotrexate. Only exceedingly high levels of methotrexate would be able to slow down the growth of these cells which express high levels of DHFR from a few favorably inserted copies. The selection of spontaneous mutants with further increased expression levels would be rather inefficient at low levels of MTX. This influence would be maintained even when episome formation occurs since the size of those episomes exceeds that of the plasmid DNA by several hundredfold. In addition, the episomes have been shown to include large regions of the original integration locus (Carrol et al., 1987). The type of DNA at the integration locus might even influence the rate of episome formation and the extent of copy number amplification in such episomal DNA molecules prior to reintegration.

It becomes clear from the above discussion that a multitude of diverse structures are formed during the initial integration and primary amplification events, as well as during periods of stepwise selection in MTX. If one's goal is to obtain a reliable source of cells for the stable expression of protein(s) of interest, one should wonder about the genetic stability of these structures, both at the level of the individual cell as well as at the level of the cell populations generated during selection. Usually, to minimize the risk of heterogeneity, "young" cell populations generated from the expansion of a colony on a 100 mm plate or of a single cell in a multiwell plate would be evaluated and then expanded to generate a frozen cell bank. We have done extensive studies addressing the question of genetic stability of clones and non-cloned cell populations using FISH.

1.4.7 The Presence or Absence of Methotrexate in the Culture Media Affects the Overall Copy Number of Amplified Sequences Within a Cell Population

The genetic stability of recombinant sequences in transfected CHO cell lines is an important issue both for the researcher requiring a stable model system for gene expression studies as well as for the manufacturer who wants to produce a pharmaceutically important protein in these cells. Recombinant cell populations are usually established as clonal progeny of isolated colonies or single cells. Nevertheless, the inherent plasticity of eukaryotic genomes, as well as subtle selective pressures applied during continuous cultivation, may affect the stability of the transfected and integrated DNA sequences. Hence, verification of integration and amplification of recombinant sequences is important. This may be performed using a variety of DNA-based techniques, including Southern DNA blot analysis, cytogenetic analysis of ECRs and chromosomal *in situ* hybridization. Southern DNA blot analysis provides little information about subpopulation heterogeneity and rearrangement of amplified sequences and gives no information on chromosomal location and number of integration sites. ECRs are detectable as rather large, uniformly staining regions and are recognized as such only when they constitute more than 10 % of the total length of the chromosome.

Recently, several groups have reported the detection of unique gene sequences in chromosome mapping experiments using FISH with chemically-modified non-isotopic probes. These techniques offer high sensitivity with markedly improved speed and spatial resolution (Lawrence et al., 1988; Lichter et al., 1990; Lichter et al., 1988; Pinkel et al., 1986). My colleagues and I have used this powerful technique with a number of recombinant CHO cells which, in part, were established for the production of candidate protein pharmaceuticals. Our goal was to analyze the degree of amplification and distribution of recombinant sequences in CHO cells which had been established for reliable growth in moderate to high levels of MTX. We performed the analysis of these cytogenetic parameters on cells which were subcultivated both in the presence and absence of MTX.

It is controversial whether the continued presence of a selective agent (i.e. MTX) in long-term cultures of recombinant CHO cells is required for production stability. Whereas some studies suggest MTX is required for stable production of recombinant proteins (Weidle et al., 1988), others indicate that continued culture of clonal cell lines in MTX may not be necessary (Brown et al., 1981). In fact, in contrast to the generally accepted belief, we have shown, by means of FISH, both in clonal and in non-clonal cell lines, that continuous exposure to MTX at the same concentration under which the cell populations were initially established, promotes genetic *instability* at the chromosomal level. I will describe in section 1.4.9 in more detail the observed phenomena. Essentially, it was found that the presence of MTX in the culture medium results in the generation of mostly *unstable*, novel chromosomal rearrangements (Pallavicini et al., 1990).

The first of our FISH analysis studies was done with cell lines expressing the mouse c-myc gene in an inducible manner. The advantage of this system is that the experimental observations can be attributed solely to the status (expression level, chromosomal location, degree of amplification) of the DHFR selector gene and the inhibitor MTX. When these cells are cultivated at 37 °C, the inducible *Drosophila melanogaster* heat shock protein 70 promoter is in an off-status. In one of these lines, containing more than 2000 copies of the transfected DNA, no expression of c-myc RNA and protein was observed at 37 °C. The continuous exposure to MTX resulted in a slightly higher copy number of plasmid DNA, as compared to a population of cells which had been cultivated in the absence of any selective pressure for at least 4 weeks. The difference was about twofold, as indicated by DNA Southern hybridizations. The expression of the c-myc RNA (upon induction by shifting the cells to 43 °C for 1 hour) was elevated accordingly (Wurm et al., 1986; Pallavicini et al., 1990). A similar observation with a constitutive promoter driving tPA expression was reported by Weidle et al. (Weidle et al., 1988) who showed both a reduction in copy number and a gradual decline of productivity in cells grown without MTX for 40 to 60 days. Interestingly, during the remaining observation time, from 50 to about 100 days in the absence of MTX, no further decline in productivity was observed. I will discuss below in chapters 1.4.8 and 1.4.9 data obtained in our laboratory by FISH which might explain these observations and shed some light on the underlying cytogenetic mechanisms.

1.4.8 Unique Patterns of Chromosomally Integrated Amplified Sequences in Clonal Recombinant Cell Lines

The loci of integration and the degree of amplification of recombinant sequences in clonal cells are thought to vary between individual clones and thus may be unique descriptors for identification of recombinant cell lines. FISH analysis with chemically-modified probes complementary to the transfected DNA can then be used to label the integrated, amplified DNA sequences.

Following our initial observation of a drop in copy number in cells containing the mouse c-myc gene when cultivated in the absence of MTX, we extended our studies with this line and also analyzed the chromosomal integration pattern of transfected sequences in two additional CHO cell lines. The three clonal cell lines studied under this program expressed one of the three following proteins: 1) human rtPA (Goeddel et al., 1983), 2) a truncated version of the human CD4 receptor (Smith et al., 1987) and 3) the murine c-myc protein (Wurm et al., 1986).

The three cell lines had been established by transfecting CHO-DUKX cells with DNA encoding the genes of interest and containing a DHFR expression construct. Either individual clones or pools of clones were cultivated for extended periods in media containing stepwise increasing concentrations of MTX. In each case, a single individual clone expressing the desired level of recombinant protein was expanded for long-term growth and production. FISH was performed using biotinylated probes complementary to the recombinant sequences. Prior to hybridization, the cells were grown for at least four subcultivations (14 days) in the absence of MTX.

Three different integration patterns (site, degree) of amplified sequences were found. The FISH signals in the rtPA expressing cell line were localized in one midsize submetacentric chromosome (probably either chromosome 3, 4, or the X-chromosome). The signals covered, as a bright band of fluorescence, most of the length of the short arm and showed additional, much weaker fluorescence spots over the center of the chromatids in the long arm (Fig. 1.4.7). For the cell line expressing the truncated CD4 molecule, a strong signal was found over the telomeric end of a small acrocentric chromosome (possibly a derivative of either chromosome 5, 6, or 7) which covered about one third of its' total length (Fig. 1.4.7b). Finally, cells expressing the mouse c-myc protein exhibited a bright band of hybridization in the center of the long arm of a chromosome we assume is number 2 (Fig. 1.4.7c). These unique integration patterns were found in 95–99% of all metaphases from each cell line.

Subsequent to the first analysis, we cultivated these cell lines in the absence of methotrexate for extended periods and occasionally performing FISH analysis. We found the chromosomal structures described above to be stable within the observation period (a minimum of 60 days and in one case 160 days). These observations indicated a high degree of genetic stability of chromosomally amplified sequences in the absence of methotrexate.

1.4.9 MTX Induced Heterogeneity of Amplified Sequences

A number of reports describe multiple chromosomal aberrations and a wide variety of diverse structures from cell to cell when mouse and human cells were subcultivated in the presence of MTX (Raggetto et al., 1979; Mondello et al.; 1984; Vitek, 1987). In an attempt to explain some of the effects observed both in cell culture and *in vivo* (upon chemotherapy of cancer patients), Goulian and coworkers (Goulian et al., 1980) reported, as mentioned in chapter 1.4.5, that MTX exposure results in an intracellular depletion of the thymidine pool. This, in turn, leads to frequent misincorporation of uridine monophosphate into replicating DNA. The presence of the ribonucleotide within the DNA backbone might be the cause for increased repair synthesis along the cells chromosomes. Consequently it might be responsible for elevated recombinogenic activity as well. Our own observations with respect to the dynamics of chromosomally amplified sequences in CHO cells led us to believe that the phenomenon described by Goulian et al., might be a contributing factor.

In fact, we have shown that continuous cultivation of cells in media containing high concentrations of MTX is associated with rearrangement and variable amplification of transfected sequences within almost half of the cells of a population expressing c-myc (Wurm et al., 1986; Pallavicini et al., 1990). Several types of integration patterns were found in these cells: 1) chromosomes with highly extended regions of hybridizing sequences, 2) cells containing transfected DNA integrated into multiple sites on an individual chromosome or 3) into multiple chromosomes, 4) chromosomes joined at amplified regions, 5) circular chromosomes, 6) cells with small derivative chromosomes or fragments and, 7) cells containing a single "normal" integration characteristic of the particular cell line (Fig. 1.4.7a). Examples of variant hybridization patterns in metaphases obtained from cells producing recombinant c-myc cultured in the presence of MTX are shown in Figure 1.4.7d. Variant hybridization patterns were visible in approximately 40 % of c-myc CHO metaphases. Similar results were observed with another recombinant CHO cell line producing a chimeric CD4-immunoglobulin protein (Kaufman et al., 1985). When grown in the presence of MTX, CD4-IgG producing cells showed substantial heterogeneity in both chromosomal location and copy number of amplified sequences coding for CD4-immunoglobulin (CD4-IgG).

To explain the discrepancy between the stable, unique structures described in chapter 1.4.8 and the variable, heterogeneic hybridization signals outlined here, one has to assume that the cell population would go through a transition phase when MTX is removed from the culture medium. We were able to find cytogenetic support for the existence of such a transition phase. Fig. 1.4.8a summarizes the numerical analysis of variant and "normal" hybridization pattern observed within the c-myc producing cell line when cultivated in the absence of MTX. The "normal" hybridization pattern is defined as a single, unique signal which had been identified as characteristic of this cell line. We use the term "master integration" for these hybridization signals. An "abnormal" or variant integration would be found in cells which show hybridization signals different and/or additional to the one defined as the master integration. When these CHO cells were grown for 40 days without MTX, the frequency of multiple and unusual hybridization patterns gradually declined, whereas the percentage of cells carrying the master integration increased from about 50 % to about 96 % after 40 days. Correspondingly, the frequency of cells with additional and unusual (more extended) hybridization signals declined. We verified that this type of population drift occurred with other CHO cells lines, both cloned and non-cloned, carrying highly amplified plasmid sequences (Fig. 1.4.8b). Again, the percentage of cells carrying a single, unique master integration increased from about 40 % to almost 80 % in 12 to 15 days of culture in two clonal cell lines (A, B) isolated from a culture containing 500 nanomolar MTX. Another "polyclonal" population (C) established from cells selected at 250 nanomolar MTX had been subcultivated for 50 days without this drug and showed an increase from 52 % to 69 % of cells containing a single hybridization signal. Since this population of cells was polyclonal it was not possible to identify a master integration. In fact, we suspected that there might be cells in this population which carried more than just a single hybridization signal per cell and, instead, two signals on the same chromosome. Such a "double" integration, similar to the triplication of the c-myc band seen in one of the metaphase spreads in Fig. 1.4.7d, might be the result of a major segment duplication within that chromosome and, a chromosome with such a structure might meet the stability criteria of a master integration. This view is supported by the finding that the frequency of cells with two integrations per cell increased upon the removal of methotrexate from the culture medium.

Of course, the unstable, unusual chromosome structures described above contribute to the overall copy number of transfected DNA within a cell population. Hence the loss of these structures over time will decrease the number of recombinant gene sequences harbored within the cell population. Weidle et al., (1988) analyzed the relative copy number of clonal cell lines cultivated in the absence of MTX by Southern DNA blot hybridization analysis and observed a gradual decline in intensity of the hybridization signal. As I pointed out above, while these authors also recorded a corresponding decline of productivity over the initial 40 to 50 days of

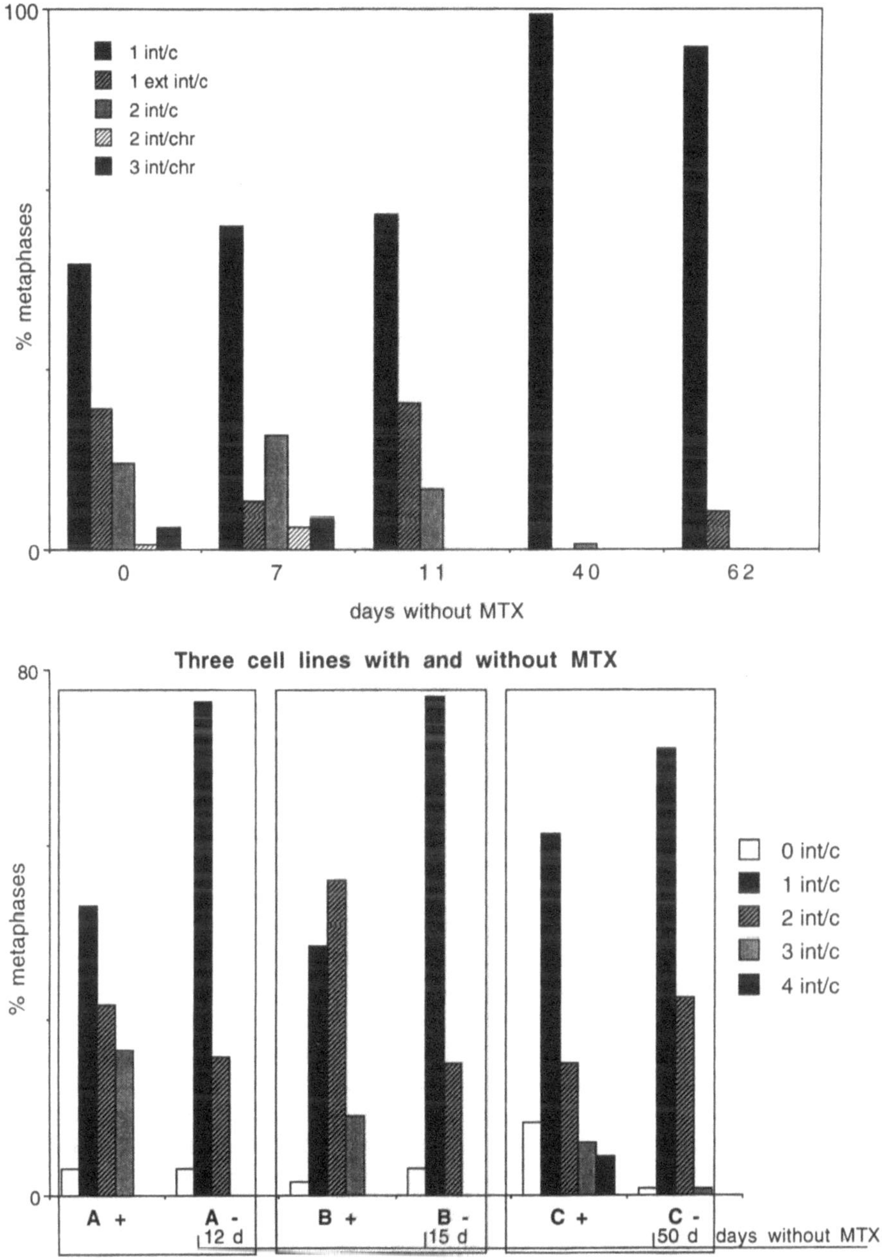

Fig. 1.4.8: Distribution of integraton patterns of amplified recombinant sequences in CHO cell lines (int/c + Integration per cell, ext.int/c + extended integration per cell, int/chr. + integration per chromosome)
a) CHO cell line adapted to 320 μM MTX and expressing the mouse c-myc gene
b) Two clonal cell lines adapted to 500 nM MTX (A and B), and a heterogeneous cell line adapted to 250 nM MTX (C) expressing CD4-IgG (−/+ = without and with MTX).

analysis, the productivity remained stable for another 50 days. It seems that a similar phenomenon to the one observed with FISH in our case might have occurred in these cell lines of Weidle and co-workers as well. A gradual elimination from the cell population of cells carrying highly amplified, but genetically unstable chromosomes might have occurred. Along with that a steady increase in the number of cells which contain a stable master integration but may have an overall lower degree of amplification would result.

1.4.10 Does Chromosomal Heterogeneity of Integrated Plasmid Sequences Within Cell Populations Grown in the Absence of Selection Result in Gradual Loss of Productivity (due to Overgrowth by „Low Producer" Cells)?

A stable, reliable source of cells is an essential prerequisite for large scale production of recombinant proteins for therapeutic use. Several precautions are typically implemented to minimize the opportunity for genetic drift and variability with respect to productivity. These include the establishment of large repositories of the recombinant cells shortly after a producer cell line is identified (master and working cell banks). Also, time limits for subcultivation of thawed cells to production levels usually are recommended. Cell growth and recombinant protein productivity rates are usually measured at several intervals during culture. However, these assessments have failed to establish a convincing relationship between recombinant protein production, proliferation rate, and genetic stability of integrated recombinant sequences. Although high copy numbers and extended regions of amplified sequences are often positively correlated with high expression levels in CHO cells, it is not clear whether extensive competition of cellular resources may have adverse consequences on cell proliferation. Such a competitive relationship between growth and productivity has been discussed in the past and more recently by Gu et al., (1992).

To test the hypothesis of a possible correlation between high levels of expression and reduced cell growth rates, we created a cell line with both heterogeneous integration loci and heterogeneous amplification levels of the transfected plasmid sequences. Obviously, the assumption is that cells with low and high copy numbers of integrated plasmid sequences would also have low and high levels of productivity. In view of what has been discussed so far, I am aware of this being an oversimplification. However, when considering large numbers of clones, the general trend might still hold true. Long term perfusion fermentation of a heterogeneous population of cells was expected to be the „best" condition for overgrowth of more rapidly proliferating, low- or nonproducer-subpopulations. Such a phenomenon has been observed in monoclonal-antibody producing hybridomas and, by inference, it might occur with recombinant CHO cells as well.

Approximately 300 recombinant CHO colonies selected in GHT minus medium after transfection with an expression vector for a chimeric CD4-IgG molecule (Capon et al., 1989) were pooled and subcultivated in increasing concentrations of MTX over a period of about 3 months. The resulting cell lines were maintained in media containing 250 nanomolar MTX while adapting them to suspension growth over ten subcultivations. FISH with probes specific for the transfected CD4-IgG plasmid showed that the population was heterogeneous in both number and loci of integration sites and in the degree of amplification of transfected sequences. Each metaphase contained at least one, and often two or more hybridization signals with no specific integration pattern. The amount of recombinant sequences per cell (estimated by probe-linked fluorescence) varied substantially and we hoped recombinant protein production per cell would vary accordingly. We used this heterogeneous cell population to investigate the relationship between protein production and cell growth.

Cells were cultivated in a 2 L vessel equipped for long term perfusion fermentation. This bioreactor configuration allowed the supply of up to 10 volumes of fresh medium per day while the same volume of spent medium could be removed from the vessel. The cells were grown in this system in the absence of MTX for more than 90 days. Various cultivation variables (i. e. glucose consumption, oxygen uptake rate, lactate production, etc.) were monitored and adjustments were made on a daily basis to maximize growth rates. Assuming an inverse correlation between cell growth and productivity as stated in the beginning of this chapter, faster proliferation of cells would produce lower levels of recombinant protein (and possibly fewer copies of recombinant sequences). Thus, a gradual drift to lower specific recombinant protein productivity should occur within this particular cell population cultivated under these conditions.

During the culture period, daily samples were taken for the assessment of cell density and levels of recombinant protein concentration using a CD4-IgG enzyme linked immunoassay. The specific productivities (pg/cell day) of recombinant protein, shown in Fig. 1.4.9, varied considerably during the measurement period. During

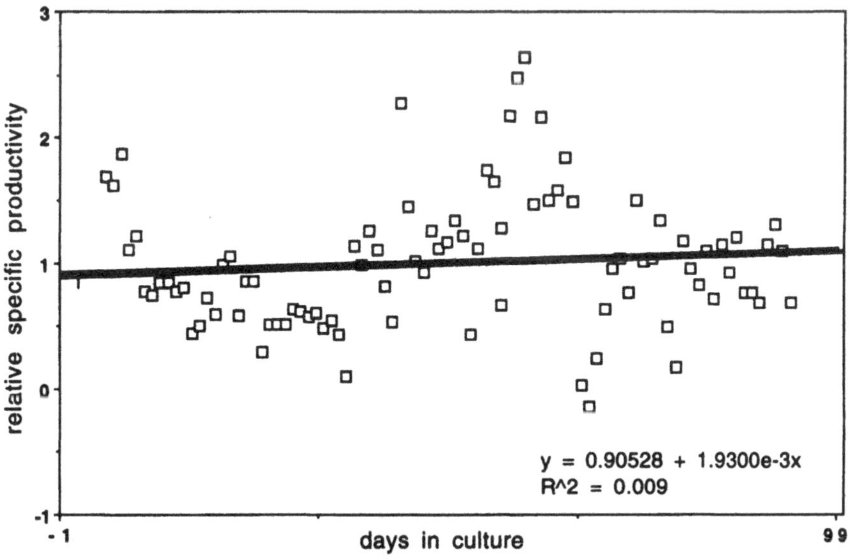

Fig. 1.4.9: Perfusion cultivation of CD4-IgG expressing CHO cells over almost 100 days. Evaluation of relative specific productivity.

this long term perfusion experiment we observed two periods of very high densities, i.e. 55×10^6 viable cells/ml and 115×10^6 viable cells/ml, when multiple volumes of fresh medium were perfused through the reactor. Also, occasional operational errors or equipment failures resulted in dramatic cell losses, reducing the number of viable cells in the reactor to less than 5 %. These changes in environmental conditions resulted not only in variability of growth rates and cell densities but also influenced specific recombinant protein productivities. They were lowest at times of declining viabilities and cell death. However, regression analysis of the overall protein production rates did not indicate any decline. Rather a slight increase of specific recombinant protein productivity was observed throughout the culture period.

The distribution, amplification and localization of recombinant sequences in this heterogeneous population of cells was measured at the beginning (day 15) and toward the end (day 80) of the culture period using FISH. There was no apparent difference between the distribution patterns of the recombinant sequences in cells measured at these time intervals (not shown). These data indicate both cytogenetic and protein production stability of this heterogeneous population of cells cultivated continuously for almost 100 days (> 140 cell population doublings). Surprisingly, they do not support the assumed concept of an inverse correlation between growth rate and productivity.

1.4.11 Summary and Conclusions

I have attempted in this article to summarize and highlight most of the presently known aspects of gene transfer into and amplification within one of the most popular mammalian cell hosts, DHFR deficient Chinese hamster ovary cells (CHO). Expression vectors containing the desired gene of interest are used in combination with a functional DHFR expression cassette either on the same or on a separate vector. Selection and identification of recombinant cells is performed using the DHFR phenotype. It is assumed that integration into a random chromosomal site within the nucleus is a primary, but rare, and as such, singular event. Recombinant cell clones might vary with respect to the stability within the primary locus of integration. Also, profound effects on transcription rates might be exerted by the DNA at the integration site. Integration of plasmid DNA seems to be accompanied by an inverted duplication both of plasmid DNA and large stretches of neighboring chromosomal DNA. Hyperamplifiable phenotypes have been observed which frequently excise the region of the inverted duplication. Such excised episomes have been found to be circular, self replicative, submicroscopic elements which undergo rapid copy number fluctuations. Under stringent, selective conditions, for example, when the DHFR inhibitor methotrexate is added to the culture medium, subpopulations of cells can be selected which, during the initial phase of selection, have a higher copy number of these episomal structures. In CHO cells, upon prolonged exposure to MTX, reintegration into the chromosomes occurs frequenty. Double minute chromosomes (DMs), larger derivatives of the episomes, observed in other cell lines more frequently than in CHO cells, usually do not persist in CHO cells.

Integrated episomes in CHO cells give rise to homogeneously staining regions (HSRs) or extended chromosomal regions (ECRs). These ECRs contain several hundreds if not thousands of copies of the initially transfected DNA. They undergo fluctuations in copy number, chromosomal locus and length when cells are continuously exposed to MTX. In clonal cell lines, which have been cultivated for several weeks in the absence of MTX, very specific types of recombinant integration structures are detected in greater than 90 % of cells, which can be used as identifying markers for each clonal cell line (master integrations). These structures are genetically quite stable and have been observed to be maintained in cell populations for up to 170 days. In the presence of MTX, master integrations might still be present in about 40 to 50 % of the cells, but other, more variable, more extended ECRs were observed in the remainder of the cells. These types of ECRs appear to be genetically unstable since they disappear from the population of cells when these are cultivated in the absence of MTX. They contribute to the overall copy number of recombinant genes within a population of cells and thus may also contribute to overall productivity. Thus, removal of methotrexate may result in an initial drop in productivity. In one experiment, performed with a polyclonal population of cells containing a high degree of variability with respect to chromosomal amplification of integrated plasmid sequences, we could not detect an inverse correlation between cell growth and copy number. In other words, during long term perfusion culture, cells with a lower copy number of inserted genes did not overgrow the remainder of the cells during an observation period of 100 days.

In spite of many unanswered questions and, possibly, some concern about the degree of ignorance we have to accept for the time being, CHO cells have been extremely successful as a reliable and robust substrate for large, even industrial scale quantities of high value proteins. Also, no other cell line is available that has been studied to the same extent as CHO cells have with respect to a multitude of parameters and characteristics. Though many details have to be unraveled in the future about gene transfer and amplification in mammalian cells, CHO cells surely will lead the way to a better understanding of these complex mechanisms. The knowledge acquired has provided us, and will continue to do so, with better means to use CHO cells safely and reliably both for the production of valuable pharmaceuticals as well for the study of gene transfer and expression of foreign genes in mammalian hosts.

Acknowledgments

I am very thankful to my colleagues Drs. Avi Ashkenazi, Marshall Dinowitz, Chris Petropoulos, Mary Burke Sliwkowski (all at Genentech) and Dr. Leonard Hayflick (University of California, San Francisco) for critically reading the manuscript and improving it through helpful comments and suggestions. I also want to express my gratitude towards Dr. Robert Arathoon for encouragement and continuing support and many enlightening discussions.

References

Anderson, K.P., Lie, Y., Low, M.-A., Williams, S.R., Fennie, E.H., Nguyen, T.P., Wurm, F.W. (1990) Presence and transcription of intracisternal A-particle-related sequences in CHO cells. J. Virology 64, 2021–2032.

Anderson, K.P., Low, M.-A., Lie, Y.S., Keller, G.-A., Dinowitz, M. (1991) Endogenous origin of defective retroviruslike partides from a recombinant Chinese hamster ovary cell line. Virology 181, 305–311.

Arathoon, W.R. and Birch, J.R. (1986) Large-scale cell culture in biotechnology. Science 232, 1390–1395.

Barbi, G., Steinbach, P. and Vogel, W. (1984) Nonrandom distribution of methotrexate-induced aberrations on human chromosomes. Detection of further folic acid sensitive fragile sites. Hum. Genet. 68, 290–294.

Barsoum, J. (1990) Introduction of stable high-copy-number DNA into Chinese hamster ovary cells by electroporation. DNA and Cell Biology 9, 293–300.

Biedler, J.L. and Spengler, B.A. (1976) A novel chromosome abnormality in human neuroblastoma and antifolate-resistance chinese hamster cell lines in culture. J. Natl. Cancer Inst. 57, 3, 683–689.

Brand, H., Froud, S.J., Metcalfe, H.K., Onadipe, A., Shaw, A. and Westlake, A.J. (1993) Selection strategies for highly productive recombinant mammalian cell lines. In: Animal Cell Technology, Products of Today, Prospects for Tomorrow (Berthold, W., MacDonald, C. et al., eds) (in press).

Brown, P.C., Beverley, S.M. and Schimke, R.T. (1981) Relationship of amplified dihydrofolate reductase genes to double minute chromosomes in unstably resistant mouse fibroblast cell lines. Mol. Cell. Biol. 1, 1077–1083.

Camargo, M. and Cervenka J. (1980) Pattern of chromosomal replication in synchronized lymphocytes. 1. Evaluation and application of methotrexate block. Human Genetics. 54, 47–53.

Capon, D., Chamow, S.M., Mordenti, J., Marsters, S.A., Gregory, T., Mitsuya, H., Byrn, R.A., Lucas, C., Wurm, F.M., Groopman, J.F., Broder, S. and Smith, D.H. (1989) Designing CD4 immnoadhesins for AIDS therapy. Nature 337, 525–531.

Carrol, S.M., Gaudray, P., DeRose, M.L., Emery, J.F., Meinkoth, J.L., Nakkim, E., Subler, M., Von Hoff, D.D., Wahl, G.M. (1987) Characterization of an episome produced in hamster cells that amplify a transfected CAD gene at high frequency: Functional evidence for a mammalian replication origin. Mol. Cell. Biology 7, 1750–1750.

Carroll, S.M., DeRose, M.L., Gaudray, P., Moore, C.M., Needham-Vandevanter, D.R., Von Hoff, D.D. and Wahl, G.M. (1988) Double minute chromosomes can be produced from precursors derived from a chromosomal deletion. Mol. Cell. Biol. 8, 1525–1533.

Chen, C. and Okayama, H. (1987) High efficiency transformation of mammalian cells by plasmid DNA. Mol. Cell. Biol. 7, 2745–2752.

Cherif, D., Lavialle, C., Modjitahedi, N., Le Coniat, M., Berger, R. and Brison, O. (1989) Selection of cells with different chromosomal localizations of the amplified c-myc gene during in vivo and in vitro growth of the breast carcinoma cell line SW 613-S. Chromosoma 97, 327–333.

Chiang, T.-S. and McConlogue (1988) Amplification and expression of heterologous ornithin decarboxylase in Chinese hamster ovary cells. Mol. Cell. Biol. 8, 764–769.

Chu, G., Hayakawa, H. and Berg, P. (1987) Electroporation for the efficient transfection of mammalian cells with DNA. Nucl. Acids Res. 15, 1311–1326.

Finn, G.K., Kurz, B.W., Cheng, R.Z. and Shmookler, R.J. (1989) Homologous plasmid recombination is elevated in immortally transformed cells. Mol. Cell. Biol. 9, 4009–4017.

Folger, K.R., Wong, E.A., Wahl, G., Capecchi, M.R. (1982) Patterns of integration of DNA microinjected into cultured mammalian cells: evidence for homologous recombination between injected plasmid molecules. Mol. Cell. Biol. 2, 1372–1387.

Fouser, A.L., Swanberg, S.L, Lin, B.-Y., Benedict, M., Kelleher, K., Dumming, D. A. and Riedel, G.E. (1992) High level expression of a chimeric antiganglioside GD2 antibody: Genomic kappa sequences improve expression in cos and CHO cells. Biotechnology 10, 1121–1127.

Goeddel, D.V., Kohr, W.J., Pennica, D. and Vehar, G.A. (1983) Human tissue plasminogen activator, pharmaceutical compositions containing it, processes for making it, and DNA and transformed intermediates therof. United Kingdom Patent 2, 119,804.

Goulian, M., Bleile, B. and Tseng, B.Y. (1980) Methotrexate-induced misincorporation of uracil into DNA. Proc. Natl. Acad. Sci. USA 77, 1956–1960.

Graham, F.L. and van der Eb, A.J. (1973) A new technique for the assay of infectivity of human adenovirus 5 DNA. Virology 52, 456–467.

Gu, M.B., Kern, J.A., Todd, P. and Kompala, D.S. (1992) Effect of amplification of dhfr and lac Z genes on growth and β-galactosidase expression in suspension cultures of recombinant CHO cell.

Gudkov, A.V. and Kopnin, B.P. (1987) Gene amplification and multidrug resistance. Sov. Sci. Rev. D. Physicochem. Biol. 7, 95–136.

Hamlin, J.L., Leu, T.-H., Vaughn, J.P., Ma, C. and Dijkwel, P.A. (1991) Amplification of DNA sequences in mammalian cells. Prog. Nucl. Acid Research and Mol. Biol. 41, 203–239.

Hendricks, M.B., Banker, M.J. and McLaughlin, M. (1988) A high efficiency vector for expression of foreign genes in myeloma cells. Gene 64, 43–51.

Hendricks, M.B., Luchette, C.A. and Banker, M.J. (1989) Enhanced Expression of an immunoglobulin based vector in myeloma cells mediated by coamplification with a mutant dihydrofolate reductase gene. Biotechnology 7, 1271–1274.

Jasin, M. and Berg, P. (1988) Homologous integration in mammalian cells without target gene selection. Genes and Development 2, 1353–1363.

Kao, F.-T. and Puck, T.T. (1968) Genetics of somatic mammalian cells VII. Induction and isolation of nutritional mutants in Chinese hamster cells. Proc. Natl. Acad. Sci. USA 60, 1275–1281.

Kaufman, R.J. and Sharp, P. (1982) Amplification and Expression of Sequences Cotransfected with a Modular Dihydrofolate Reductase Complementary DNA Gene. J. Mol. Biol. 159, 601–621.

Kaufman, R.J., Murtha, P., Ingolia, D.E., Yeung, C.-Y. and Kellems, R.E. (1986) Selection and amplification of heterologous genes encoding adenosin deaminase in mammalian cells. Proc. Natl. Acad. Sci. USA 83, 3136–3140.

Kaufman, R.J., Wasley, L.C., Furie, B.C. and Shoemaker, C.B. (1986) Expression, purification and characterization of recombinant Y-carboxilated factor IX synthesized in Chinese hamster ovary cells. J. Biol. Chem. 261, 9622–9628.

Kaufman, R.J., Wasley, L.C., Spiliotes, A.J., Gosseis, S.D., Latt, S.A., Larsen, G.R., Kay, R.M. (1985) Coamplification and coexpression of human tissue-type plasminogen activator and murine dihydrofolate reductase in Chinese hamster ovary cells. Mol. Cell. Biol. 5, 1750–1759.

Lawrence, J.B., Vilnave, C.A. and Singer, R.H. (1988) Sensitive, high-resolution chromatin and chromosome mapping in situ: Presence and orientation of two closely integrated copies of LBV in a lymphoma line. Cell 52, 51–61.

Levan, G. and Levan, A. (1982) Transitions of double minutes into homogeneously staining regions and C-bandless chromosomes in the SFWA tumor. In: Gene Amplification (Schimke, R.T., ed.), Cold Spring Harbor Laboratory, Cold Spring Harbor, N.Y., pp 91–97.

Lichter, P., Chieh-Ju, C.T., Call, K., Hermanson, G., Evans, G.A., Housman, D. and Ward, D. (1990) High-resolution mapping of human chromosome 11 by in sity hybridization with cosmid dones. Science 247, 64–69.

Lichter, P., Cremer, T., Chieh-Ju, C.T., Watkins, P.C., Manuelidis, L. and Ward, D.C. (1988) Rapid detection of human chromosome 21 aberrations by in sity hybridization. Proc. Natl. Acad. Sci USA 85, 9664–9668.

Loyter, A., Scangos, G.A. and Ruddle, F.H. (1982) Mechanisms of DNA uptake by mammalian cells: Fate od exogenously added DNA monitored by the use of fluorescent dyes. Proc. Natl. Acad. Sci. USA 79, 422–426.

Ma, C., Martin, S., Trask, B. and Hamlin, J.L. (1993) Sister chromatid fusion initiates amplification of the dihydrofolate reductase gene in Chinese hamster cells. Genes and Development 7, 605–620.

McArthur, J.G. and Stanners, C.P. (1991) A genetic element that increases the frequency of gene amplification. J. Biol. Chem. 266, 6000–6005.

McClintock, B. (1941) The stability of broken ends of chromosomes in Zea mays. Genetics 26, 234–282.

Milbrandt, J.D., Azizkhan, J.C. and Hamlin, J.L. (1983) Amplification of a cloned Chinese hamster dihydrofolate reductase gene after transfer into a dihydrofolate reductase-deficient cell line. Mol. Cell. Biol. 3, 1274–1282.

Mondello, C., Giorgi, R. and Nuzzo, F. (1984) Chromosomal effects of methotrexate on cultured human lymphocytes. Mutation Research 139, 67–70.

Mondello, C., Giorgi, R. and Nuzzo, F. (1984) Chromosomal effects of methotrexate on cultured human lymphocytes. Mutation Reserach 139, 67–70.

Nunberg, J.H., Kaufman, R.J., Schimke, R.T., Urlaub, G., Chasin, L.A. (1978) Amplified dihydrofolate reductase genes are localized to a homogeneously staimng region of a single chromosome in a methotrexate-resistant Chinese hamster ovary cell line. Proc. Natl. Acad. Sci. USA 75, 5553–5556.

Okamoto, M., Nakayama, C., Nakai, M. and Yanagi, H. (1990) Amplification and high level expression of a cDNA for human granulocyte-macrophage colony-stimulation factor in human lymphoblastoid Namalwa cells. Biotechnology 8, 550–553.

Pallavicini, M.G., DeTeresa, P.S:., Rosette, C., Gray, J.W. and Wurm, F.M. (1990) Effects of Methotrexate (MTX) on Transfected DNA Stability in Mammalian Cells. Mol. Cell. Biol. 10, 401–404.

Perucho, M., Hanahan, D. and Wigler, M. (1980) Genetic and physical linkage of exogenous sequences in transformed cells. Cell 22, 309–317.

Pinkel, D., Straume, T. and Gray, J.W. (1986) Cytogenetic analysis using quantitative, high sensitivity fluorescence hybridization. Proc. Natl. Acad. Sci. USA 83, 2934–2938.

Puck, T.T. (1985) Development of the Chinese hamster ovary (CHO) cell for use in somatic cell genetics. In: Molecular Cell Genetics (Gottesman, M. M., ed.), pp 37–64.

Raffetto, G., Parodi, S., Faggin, P. and Maconi, A. (1979) Relationship between cytotoxicity and induction of sister-chromatid exchanges in mouse foetal cells exposed to several doses of carcinogenic and non-carcinogenic chemicals. Mutation Research 63, 335–343.

Raggetto, G., Parodi, S., Faggin, P. and Maconi, A. (1979) Relationship between cytotoxicity and induction of sister-chromatid exchange in mouse foetal cells exposed to several doses of carcinogenic and non-carcinogenic chemical. Mutation Research 63, 335–343.

Ringold, G., Dieckmann, B. and Lee, F. (1981) CO-Expression and amplification of Dihydrofolate Reductase cDNA and the Escherichia coli XGPRT gene in Chinese hamster ovary cells. J. of Molecular and Applied Genetics 1,165–175.

Ruiz, J.C. and Wahl, G.M. (1988) Formation of an inverted duplication can be an initial step in gene amplification. Mol. Cell. Biol. 8, 4302–4313.

Ruiz, J.C. and Wahl, G.M. (1990) Chromosomal destabilization during gene amplification. Mol. Cell. Biol. 10, 3056–3066.

Schmid, G., Zilg, H. and Johannsen, R. (1991) Development of a low-serum growth medium for production of a human glycoprotein by recombinant BHK cells. In: Production of Biologicals from Animal Cells in Culture (Spier, Griffiths, Meignier, eds.) pp 144–147.

Smith, D.H., Byrn, R.A., Marsters, S.A, Gregory, T., Groopman, J.E. and Capon, D.J. (1987) Blocking of HIV-1 infectivity by a soluble, secreted form of the CD4 antigen. Science 328, 1704–1707.

Smithies, O., Gregg, R.G., Boggs, S.S., Koralewski, M.A. and Kucherlapati, R.S. (1985) Insertion of DNA sequences into the human chromosomal betaglobin locus by homologous recombination. Nature 317, 230–234.

Stark, G.R. and Wahl, G.M. (1984) Gene amplification. Annual Review of Biochemistry 53, 447–491.

Stark, G.R., Debatisse, M., Giulotto, E. and Wahl, G.M. (1989) Recent Progress in Understanding Mechanisms of Mammalian DNA Amplification. Cell 57, 901–908.

Trent, J.M., Meltzer, P., Rosenblum, M., Harsh, G., Kinzler, K., Marchal, R., Flinberg, A. and Vogelstein, B. (1986) Evidence for rearrangement, amplification and expression of c-myc in a human glioblastoma. Proc. Natl. Acad. Sci. USA 83, 470–473.

Urlaub, G. and Chasin, L.A. (1980) Isolation of Chinese hamster cell mutants deficient in dihydrofolate reductase activity. Proc. Natl Acad. Sci. USA 77, 4216–4220.

Vitek, J.A. (1987) Similarity in dynamics of single and double minute chromosomes incidence and number of chromosomal aberrations during longterm treatment of a human cell line with methotrexate. Neoplasma 34, 665–670.

Wegner, M., Zastrow, G., Klavinius, A., Schwender, S., Müller, F., Luksa, H., Hoppe, J., Wienberg, J. and Grummt, F. (1989) Cis acting sequences from mouse rDNA promote plasmid DNA amplification and persistence in mouse cells: implication of HMG-I in their function. Nucl. Acids Research 17, 9909–9932.

Weidle, U.H., Buckel, P. and Wienberg, J. (1988) Amplified expression constructs for human tissue-type plasminogen activator in Chinese hamster ovary cells: instability in the absence of selective pressure. Gene 66, 193–203.

Wiebe, M.E., Becker, F., Lazar, R., May, L., Casto, B., Semense, M., Fautz, C., Garnick, R., Miller, C., Masover, G., Bergman, D. and Lubiniecki, A. S. (1989) A multifaceted approach to assure that recombinant tPA is free of adventitious virus. In: Advances in Animal Cell Biology and Technology for Bioprocesses (Spier, Griffitlis, Stephenne, Crooy, eds.), pp 68–71.

Wilson, J.H., Berget, P.B. and Pipas, J.M. (1982) Somatic cells efficiently join unrelated DNA segments end to end. Mol. Cell. Biol. 2, 1258–1269.

Windle, B., Draper, B.W., Yin, Y., O'Gorman, S. and Wahl, G.M. (1991) A central role for chromosome breakage in gene amplification, deletion formation, and amplicon integration. Genes and Development 5, 160–174.

Wood, C.R., Dorner, A.J., Morris, G.E., Alderman, E.M., Wilson, D., O'Hara, Jr., R.M. and Kaufman, R.J. (1990) High level synthesis of immunoglobulins in Chinese hamster ovary cells. J. Immunology 145, 3011–3016.

Wurm, F.M. and Pallavicini, M.G. (1990) unpublished observations.

Wurm, F.M., Gwinn, K.A. and Kingston, R.E. (1986) Inducible overexpression of the mouse c-myc protein in mammalian cells. Proc. Natl. Acad. USA 83, 5414–5418.

Wurm, F.M., Johnson, A., Etcheverry, M.T., Lie, Y.S., Petropoulos, C.J. (1994) Retrotargeting: Use of defective retroviral DNA fragments to improve recombinant protein production in mammalian cells in: Spier, R.E., Griffiths, J.B., Berthold, W.: Animal cell technology – products of today, propspects for tomorrow, p. 24–29.

Wurm, F.M., Johnson, A., Lie, Y., Etcheverry, T.M. and Anderson, K.P. (1992) Host cell derived retroviral sequences enhance transfection and expression efficiency in CHO cells.

In: Animal Cell Technology: Developments, Processes and Products. (Spier, Griffiths, MacDonald, eds.), pp 35–41.
Zastrow, G., Koehler, U., Müller, F., Klavinius, A., Wegner, M., Wienberg, J., Weidle, U. and Grummt, F. (1989) Distinct mouse DNA sequences enable establishment and persistence of plasmid DNA polymers in mouse cells. Nucleic Acids Research 17, 1867–1879.
Wirth, M., Bode, J., Zettlmeissl, G. and Hauser, H. (1988) Isolation of overproducing recombinant mammalian cells by a fast and simple selection procedure. Gene 73, 419–426.
Zheng, H. and Wilson, J. H. (1990) Gene targeting in normal and amplified cell lines. Nature 344, 170–173.

1.5 Isolation of Recombinant Cell Clones Exhibiting High-Level Expression of the Introduced Gene

Manfred Wirth

1.5.1 Introduction

Stable transfer of expression plasmids into animal cells has become a widely distributed method for the generation of recombinant animal cells producing a foreign protein. Although productivity levels range below the values which are achieved by *E. coli* or yeast expression systems, the biotechnological application has gained more and more importance. This may be due to the fact that animal cells are able to perform several post-translational modifications, which are necessary for some proteins to function properly, and efficiently secrete proteins into the medium. Due to economic reasons, the generation of cell lines that stably produce high levels of the foreign protein over a long period is desirable. Expression levels mainly depend on parameters which influence the transcription efficiency, translation efficiency or the mRNA stability of the expression unit on the transfected plasmid (see chapter 1.1). The transcriptional efficiency is determined by the interaction between the promoter/enhancer unit with cellular transcription factors, the copy number and the site of integration of the plasmid into the host chromosome. Therefore, the knowledge which promoter/enhancer unit is most active in the chosen cell line, which 5'UTR is favorable for optimal translation and which 3'UTR serves for optimal RNA processing and stability is of crucial importance for the construction of the optimal expression plasmid. In addition to the individual plasmid components, the chromosomal state of the plasmid after integration is of considerable importance. Out of the two factors which influence expression on this level, the plasmid copy number and the integration locus, the copy number can be influenced by a mechanism which is termed 'amplification'. The phenomenon of amplification and routes to high copy numbers via gene amplification are described in chapter 1.4. The second factor, the chromosomal site of integration, cannot be influenced well at present but promising approaches are now in development (gene targeting). The widely used ,non-directed' gene transfer allows plasmids to integrate randomly and with different copy numbers into the genome and generates a population of cell clones which differ in their expression levels as well as in their stability of expression. As shown in Fig. 1.5.1, for example, mouse L cells, which have been transfected with a gene for a secreted blood plasma protein, vary considerably with respect to their expression levels. The shape of this distribution is typical and seems to occur in general: Some clones do not express the transfected gene or express it below the detection level of the method. A broad middle class shows a minor variation in expression. A small number of clones belong to the upper class and exhibit expression clearly above the average level. In Fig. 1.5.1 expression of the best expressing clone is only 3-fold above average. However, in general the isolation of highlights in expression clearly depends on the number of clones which have been screened. From

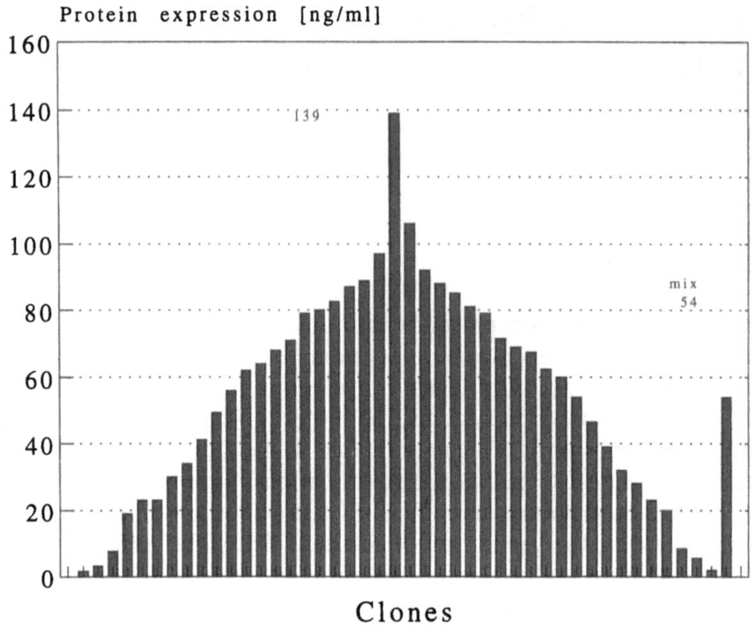

Fig. 1.5.1: Variation in the expression level of transfected cell clones.
Mouse L cells were stably transfected with an expression plasmid carrying the human metallothioneine promoter and a cDNA for a human therapeutic protein. Clones arising from the selection procedure were analyzed for protein expression by a specific ELISA.

the authors experience the expression levels of super-producers range from 5 to 30-fold above the average expression measured with the pool of clones. Comparisons of different cell lines and gene transfer methods show that the variation is not restricted to certain cell lines, and is commonly found in many cell types and with several different methods of gene transfer (Wirth et al., 1993). To isolate clones with economically satisfying production levels ($> 5\ \mu g/10^6$ cells in 24 hours) a tedious and time-consuming subcloning and testing procedure is required. Depending on the protein detection method weeks to months have to be spent until a favorable cell clone is isolated. It is not surprising that efforts have been made in the last few years to simplify this screening procedure. This contribution will focus on newly developed methods for screening for high-level producing clones.

1.5.2 Direct Screening Methods

1.5.2.1 Methods Which Directly Estimate the Expression of the Product

Before genetic manipulation of cells to be screened for protein production, a simple detection procedure has to be established which allows quantification of expression in an acceptable time. Many detection methods for specific proteins are available.

These include Western blotting, RIAs and assays which detect the biological activity of a given protein. ELISA assays optimally fit the requirements of ease and speed. Preceding ELISA, the transfected cells are seeded into microtiter plates to give rise to single cell clones using the limiting dilution procedure. If the protein is secreted, which is the case for many proteins of commercial interest, supernatants can be directly taken for analysis of expression levels. ELISA assays usually take 3–4 hours and several 96-well plates can be handled at the same time allowing the screening of several hundred clones.

1.5.2.2 The Filter Assay for Protein-Secreting Cells

Considerably more cell clones per time unit can be analyzed by a procedure designated as filter immunoassay. The procedure dates to a paper published by

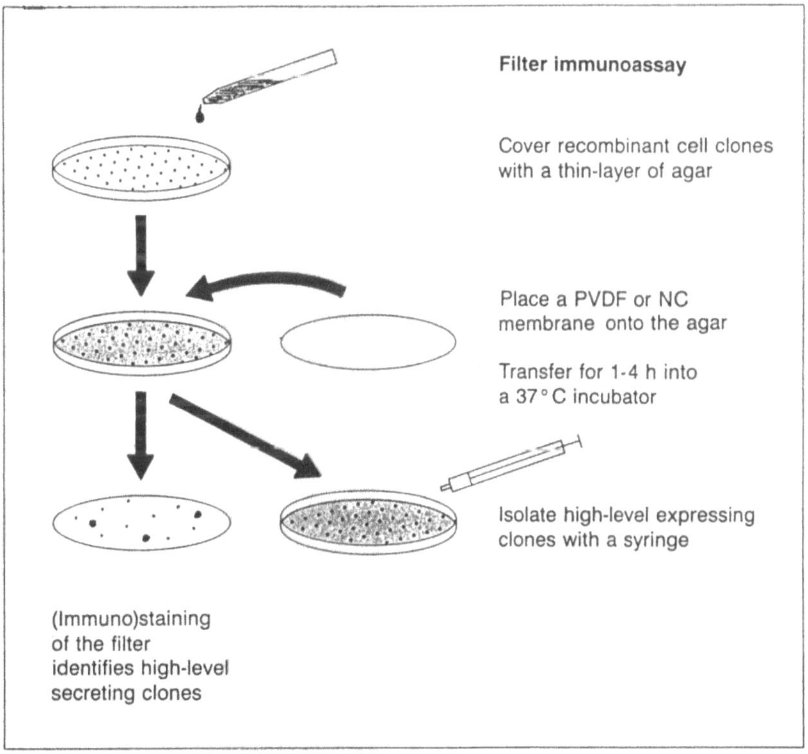

Fig. 1.5.2: The filter immunoassay.
Proteins, which are secreted from cell clones on the plate, bind to an overlaid membrane and can be detected by indirect staining methods. Spot intensities on the membrane correlate with the specific productivity of the corresponding cell clone. Individual cell clones are picked with a syringe (adapted from Dirks et al. 1994. Applications of expression vectors containing bicistronic transcription units in mammalian cells. In: Cell culture in Pharmaceutical Research, Vol. 11, p. 260, Springer, Berlin).

McCracken and Brown, who used this procedure for the isolation of Fao (rat hepatoma) cell mutants unable to secrete bovine serum albumin (McCracken and Brown, 1984). The original procedure has been refined and tailored to the isolation of high-level secreting cell clones (Wirth et al., 1990; Walls and Grinnell, 1990; Walls et al., 1989). The procedure is schematically depicted in Figure 1.5.2. Cell clones on tissue culture plates arising after the primary selection procedure are covered with thin layers of agar. Secreted proteins diffuse through the agar and can bind to a nitrocellulose or nylon membrane lying on the agar layer. After 1–4 hours the membranes are removed and can be stained by indirect immunoreaction for the presence of the expected protein. If the protein is a stable enzyme, it can be detected directly through its enzymatic activity provided that a chromogenic reaction product is available. Small intense spots on the filters identify high-level secretors on the master plates. These clones simply can be picked with a syringe. To date, many recombinant animal cells have been screened using this filter technique. These include CHO and BHK-21 cells secreting human antithrombin III (Wirth et al., 1990), BHK-21 cells secreting human IL-2 and its mutants or BHK-21 cells secreting a shortened form of the human placental alkaline phosphatase, human protein C and tissue plasminogen activator (Walls and Grinnell, 1990; Walls et al., 1989). With a recommended density of 100–200 clones per 10 cm plate, between 1000 and 2000 clones can be screened easily within one day. The procedure can be used for identification of high-level secreting cell clones which have been generated without any gene amplification. In addition, it is applied to facilitate the time-consuming amplification screening. For instance, in a particular gene amplification strategy the pools of CHO or 293 cell clones are subjected to several rounds of amplification screening (in pools) and are finally screened for the highest-expressing subclones (Walls et al., 1989). Another amplification strategy favors subcloning after each round of amplification. This procedure can be simplified by making use of the filter immunoassay after each amplification step.

1.5.2.3 FACS-Analysis for Cells Expressing the Protein Intracellularly or on the Cell Surface

Several proteins of fundamental or pharmaceutical interest are not secreted into the extracellular space, but remain in intracellular compartments or are displayed at the plasma membrane. Examples encompass the immunoglobulin binding protein (BiP) and the protein disulfide isomerase which are ER resident proteins or enzymes of the protein glycosylation pathway, and are found in the ER and Golgi compartments. Macromolecules which are stacking to or spanning the plasma membrane often function as receptors for extracellular signals like hormones and cellular markers. They might form transport channels or display a discrete enzymatic activity. Isolation of the pure expression product requires a more complex purification procedure because it is necessary to remove cellular contaminants like cellular proteins or membrane components. A multi-step purification is always accompanied by a yield reduction of the desired product. Futhermore, several membrane proteins have the property of aggregating in the absence of membranes and require detergents to remain soluble. Self aggregation is mediated by the transmembranal part of the respective protein. Provided that biological function is not impaired, access to a large amount of pure product is possible if protein mutants are expressed in the cells which lack membrane spanning domains or retention signals and which are therefore efficiently secreted into the medium. A now classical example is the 'soluble form of CD4' which has gained interest in HIV research in the last years in a stabilized form as a fusion protein with the F_c-part of an immunoglobulin chain (Capon et al., 1989; Byrn et al., 1990; Gregersen et al., 1990; Zettlmeissl et al., 1990). Another example is the expression of a secreted deletion mutant of human thrombomodulin, a cell membrane protein which plays a central regulatory role in the protein C anti-coagulant pathway (Parkinson et al., 1990). Neutral endopeptidase (also known as enkephalinase or CD10) is a type II membrane protein and has been expressed as a fully active soluble protein after deletion of its membrane spanning domain and fusion to a foreign signal sequence (Lemay et al., 1989).

Often, not the purified product but the recombinant cell line itself with the protein expressed internally or on its surface is of biological value. For instance, in retroviral packaging cell lines the high-level expression of the retroviral gagpol and env genes is desirable to provide optimal packaging for retroviral vector RNA. Cell lines expressing high levels of a viral or a signal receptor are helpful for the study of virus replication or the action of extracellular signaling agents. Investigations are facilitated in such a recombinant cell line if it can be compared to the original cell. Example cell lines encompass recombinant mouse L cells carrying the IL-2 receptor (Beckers et al., 1988), human 293 kidney cells expressing the CD4 protein (Reil et al., 1994) (these cells are infectable with HIV-1) and cells expressing the dopamine or the acetylcholin receptor (Todd et al., 1989).

In general, for identification and isolation of cells with high-level surface expression flow cytometry can be used. A prerequisite is the availability of a nontoxic, fluorescence agent with affinity to the protein of interest. The reagent may be a fluorescence labeled ligand, substrate analogue or antibody. For several years, FACS (fluorescence activated cell sorting) screening (Kavathas, 1990) and methods like panning (Wysacki and Sato, 1978; Seed and Aruffo, 1987) or isolation with magnetic beads have been used for gene cloning. This is accomplished by transfection of cDNA expression libraries or genomic DNA libraries into cells and screening for the presence of the expected receptor genes. Many receptor genes and lymphocyte surface marker genes have been isolated in the past with the help of expression screening using one of the above methods. Unlike panning or the isolation with magnetic beads, FACS analysis allows a qualitative and a quantitative determination of cell surface expression of the respective protein. FACS analysis also allows the separation of cells exhibiting different expression levels, rendering the method useful for screening for high-level producer cells. For identification of high-level expressing cell clones by FACS analysis, suspended cells are stained with an antibody directed against the respective surface molecule. The antibody itself or a second antibody recognizing the first antibody is coupled to a fluorescent dye line FITC, phycoerythrin or rhodamin. Cell suspensions then are loaded and small droplets are created which contain less than one cell per droplet. The droplets are directed through a laser beam leading to excitation of the fluorescent marker. The FACS machine operates in a way that excitation results in a negative charging of the droplet. The droplets are deviated by an electric field according to the amount of their negative surface charge which is proportional to the number of specific protein molecules expressed on the surface of the cell. To enrich for a population of cells exhibiting high-level expression of a surface molecule, usually 2 to 3 consecutive rounds of FACS sorting have to be performed. Cells exhibiting the brightest fluorescence are separated from the pool. Usually 1 to 5 % of the cells from the whole cell pool are separated, but single clone sorting can be applied, too. One out of several typical expression profiles of FACS analysis is shown in Figure 1.5.3. Hamster CHO cells were stably transfected with an expression plasmid carrying the K^bgene, which codes for a protein of the mouse histocompatibility locus. Cell pools were analyzed for the presence of K^b on the surface of the transfectants (compare A and B) and high-level expressing cell clones were isolated by sorting out single cell clones. FACS analysis of one such subclone is depicted in C.

Several publications describe the use of FACS sorting for isolation of high-level expressing cell clones. Kavathas and Herzenberg (1983) isolated mouse L-cell transformants expressing the human T-cell differentiation antigen Leu-2. Sixfold repetition of sorting and regrowth of the brightest 0.1–0.3 % of the fluorescence-tagged cells resulted in a cell population staining 40 times brighter than the starting populations. Beckers et al., (1988) isolated by FACS-sorting subpopulations of L-cells expressing the native and mutant form of the alpha-subunit of the human Il-2 receptor (Tac). Using a fluorescinated monoclonal anti-Tac antibody, several-fold enrichment for high-level expressing cells was achieved by repeated sorting of the top 5 % of highly fluorescinating cells. Strair and coworkers (1988) described a series of retroviral vectors in which cell surface antigens like the transferrin receptor or Leu-1 were incorporated as a selectable marker. Application of the FACS-technique and direct staining methods allowed the rapid detection of cotransfected packaging cell lines producing high titers of recombinant retroviruses. After three sortings for the brightest 10 % of the population a seven- to eightfold increase in fluorescence accompanied by an equivalent increase in virus production was measured in the selected population if compared to those of the starting population. Each sort selected for cells with a high level of cell surface antigen expression and virus production.

Wirth et al. (1990) screened for CHO cells transfected with a bicistronic expression vector carrying the human antithrombin III gene and the K^b marker together with an expression plasmid harboring the mouse DHFR cDNA. By alternating amplification screening with methotrexate and surface marker screening by FACS a population of cells could be derived which exhibited high levels of K^b expression. Due to a coupling of the genes, antihrombin III production was also high in these cells (see chapter 1.5.3).

Flow cytometry is possible with molecules residing inside the cell, too. Johnston et al. (1983) used FACS for selection of amplified DHFR sequences in CHO cells. Staining was performed with a membrane-permeable, fluorescent methotrexate, which binds tightly to the dihydrofolate reductase (DHFR) thereby inhibiting the protein's activity. Multiple rounds of sorting accumulated cells with 50-fold increase in DHFR amplification. Kaufman and Murtha (1987) used a similar method to isolate a subpopulation of DHFR expressing cells after transient transfection. In conclusion, flow cytometry is a potent method for the qualitative and quantitative analysis of cell surface expression. However, for isolation of high-level expressing

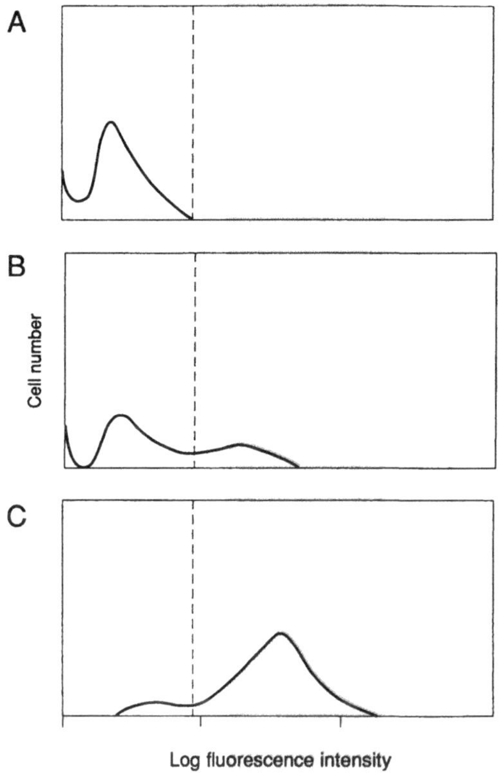

Fig. 1.5.3: FACS sorting of transfected cells for isolation of clones with high-level cell surface expression.
A: CHO control cells, B: CHO cells transfected with an expression plasmid for K^b antigen C: CHO subclone after FACS sorting. The fluorescence profile may vary depending on the cell line, transfection and gene. Overlapping and non-overlapping fluorescence profiles can occur and the positive population can exhibit a more heterogenous staining pattern resulting in the occurence of a broader curve. The bar represents the threshold which is set in a way that cells with fluorescence above the bulk of the control cells can be counted.

cells clones the time requirement for the repeated sorting and the cost for the equipment have to be seriously considered when choosing this method. As possible alternatives, methods using magnetic beads and panning were mentioned.

1.5.3 Indirect Screening Methods

The methods in the preceding paragraph identify favorable recombinant cell clones by measuring directly the amount of the specific protein. This direct screening is preferred if simple and rapid detection methods are available for the protein. If such methods are missing, the indirect screening for high-level expression represents the method of choice and is preferred to the 'brute force' application. For the indirect screening methods the expression of a marker gene is used to identify cell clones with high-level expression of the desired product. This marker gene is cotransferred together with the gene of interest and codes for a protein that is easily detectable or to be screened for. The quality of genetic linkage of the marker gene and the gene of interest is crucial for the efficiency of this procedure. Two methods that differ in the method of linkage are presented in the following.

1.5.3.1 Double Selection

Stable integration of a transfected gene into mammalian cells is not a very efficient process. To detect this rare event a marker is co-introduced which is often a selectable gene and this marker is selected for (compare chapter 1.2). Due to formation of a 'transgenome' during the integration process, genes become linked to form multicopy concatamers. Cells expressing the marker gene then also have incorporated the gene of interest. As markers, genes coding for antibiotic resistance are often chosen. Most cells arising from the selection procedure show the expression of the gene of interest – with a certain distribution of expression levels (Fig. 1.5.1). Several attempts have been made to isolate high-producing cell clones simply by increasing the selective pressure (e.g. raising the concentration of the antibiotic drug used for selection). However, these experiments failed. Two factors may be responsible for this effect: insufficient linkage of the genes and the occurence of cell clones exhibiting a nonspecific resistance to the drug. Nonspecific resistances are not correlated to increased gene expression of the introduced selectable marker gene, but are evoked by other mechanisms. It has been observed, for example, that upon high concentrations either the import of drugs can become restricted or their intracellular concentration is decreased by specifically enhanced export. A well-investigated example for the last case is the multidrug resistance gene (mdr) which has gained importance in cancer therapy and as a selectable marker. The mdr gene codes for a protein called P protein which is responsible for the energy dependent efflux of drugs like adriamycin, colchicine and actinomycin D and which confers to the cell the phenotype of multidrug resistance. To avoid the occurance of such or similar nonspecific resistance a screening strategy termed combined selection or double selection was developed (Wirth et al., 1988; Zettmeißl et al., 1988). Double selection eliminates the occurance of nonspecific cell clones. Two selectable markers instead of one are co-introduced into the cells in this procedure. A simultaneous selection is applied: The application of basal concentration for the first selective drug results in the removal of nonspecifically resistant clones. Increased concentrations of the second drug screens for high-level expressing clones. Several marker gene combinations have been tested and combinations like neo^R/DHFR, neo^R/$puromycin^R$ and neo^R/$hygromycin^R$ work efficiently. Transfectants were derived within 3 weeks which synthesized 5 to 100-fold more of the desired protein if compared to clone pools derived from single selection (Tab. 1.5.1). On the molecular level it was shown that increased expression in the double selected pools was due to increased transcription. Expression in these pools correlated well with the increased copy number of the transfected

Table 1.5.1 The combined or double selection is a tool to enrich transfected cells for pools exhibiting high-level expression without amplification.

Selection with	Number of clones[a)]	IFN-β expression[b)]
1 mg/ml G418	420	250
G418 + 3 μg/ml Puromyin	360	1000
G418 + 10 μg/ml Puromycin	40	25000

[a)] per T25 flask
[b)] productivity per million CHO cells and day

genes and/or their favorable site of integration into the genome. Several reports describe the use of the double selection procedure for the rapid access to high-level expressing cells. Page and Sydenham (1991) applied a double selection strategy with neo^R/DHFR and CHO cells to isolate recombinant cell lines secreting a humanized monoclonal antibody which is directed against the Campath-1 antigen (CDw52), an abundant glycoprotein present on lymphocytes, monocytes and cells of lymphoid malignancies. Upon double selection they received cell pools within two weeks exhibiting a 60-fold increase in antibody yield if compared to cell populations selected for basal DHFR levels. The double selection method was also applied to the transfer of retroviral vectors into packaging cell lines in order to produce stocks of recombinant retroviruses. For certain applications, e.g. gene therapy, high-titer stocks are essential. Upon double selection such stocks can be created rapidly. Tenfold increases in viral titers have been reported using double selection (Wirth et al., 1994) but 100-fold increases are possible if cotransfection is carried out with an expression plasmid carrying the neomycin resistance gene and a retroviral vector harboring the puromycin resistance gene (Morgenstern and Land, 1990) and screening is performed on puromyin gene expression (Wirth et al., 1994). Supernatants containing recombinant retrovirus, free of unwanted replication competent retroviruses, were harvested which exhibited viral titers up to 10^8 cfu/10^6 cells in 24 hours.

1.5.3.2 Screening with the Help of Bicistronic Expression Vectors

The coupling of genes at the DNA level is achieved in the double selection procedure, when cotransfected plasmids become physically linked as concatamers in a 'transgenome'. Alternatively, genes may be linked before transfection by cloning the selectable marker into the expression vector together with the gene of interest. In both cases the genes are independently transcribed as each gene is driven by its own promoter. Recently, a new series of expression plasmids were described, in which a linkage of a superior quality is accomplished by creation of a bicistronic transcription unit. In these bicistronic expression vectors a single promoter mediates transcription of one mRNA which contains both genes (see also 1.1.2.3). In such systems coupling of the genes is performed at the mRNA level. Bi- and polycistronic mRNAs are very common in bacterial systems, but are rare in higher eukaryotes.

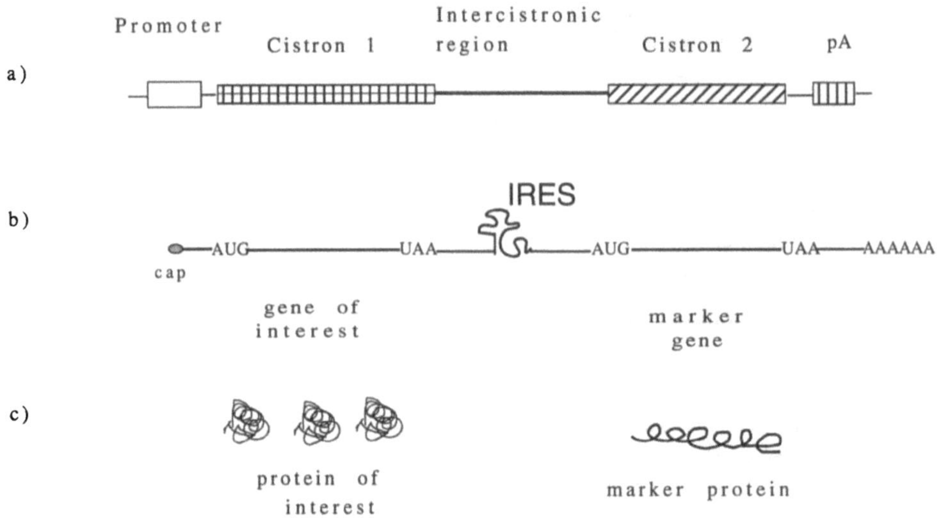

Fig. 1.5.4: Bicistronic expression vectors for screening purposes.
The second cistron consists of a selectable marker or a gene which codes for a reporter protein like the enzymes luciferase, green fluorescent protein (GFP), the secreted form of an alkaline phosphatase (SEAP) or a cell surface protein. In contemporary expression plasmids the intercistronic region consists of elements mediating an efficient internal initiation of translation. Such elements are derived from the 5'UTRs of poliovirus or EMC-virus. Both viruses are members of the picornavirus family.

In mammals, translation from monocistronic mRNAs is the rule. Translation efficiencies from a second cistron in bicistronic mRNAs usually are much lower than from the first cistron, ranging between 1/10 and 1/200. However, if a selectable marker is positioned as cistron 2, this low translational efficiency can be of advantage for two reasons: first, low-level expression of the selectable marker gene should not be harmful to the cell and should not lead, for example, to a retardation of the cell growth. Such negative effects on cell growth can sometimes be correlated to high-level expression of certain genes. Second, a strong selective pressure is applied if screening is performed with common concentrations of a selective drug for an underexpressed, selectable marker gene. This leads to the survival of cell clones with high-level expression of the nonselectable gene in cistron 1. There are several investigations which escribe the use of such bicistronic and even polycistronic expression vectors for screening, where the gene of interest is placed on cistron 1 position and the underexpressed selectable marker is placed on cistron 2 position (see Kaufman et al., 1987; Boel et al., 1987; Balland et al., 1988).

Kaufman et al. (1987) reported expression in CHO cells and COS cells mediated by bicistronic vectors carrying the adenosine deaminase gene (ADA) or a colony-stimulating factor (GM-CSF) gene and a DHFR-gene. Boel et al. (1987) used expression vectors for CHO cells based on the adenovirus major late promoter and a bicistronic unit with the cDNA for human pancreatic polypeptide precursor followed by a DHFR gene as selectable marker. Balland et al. (1988) investigated bicistronic Factor IX/Ecogpt constructs in CHO cells. In all these experiments, selection for the underexpressed marker resulted in a dramatically reduced number of cell clones. A 100–1000-fold increase in expression of the nonselected gene was compared to cell lines generated with a respective monocistronic plasmid reported by Balland et al. However, Boel et al. did not observe such a positive effect on the expression of the nonselected gene in their experiments. These differences need to be explained, and a critical evaluation is advised concerning these early experiments.

There are several reasons for the limited utility of the systems described above: Adequate levels of selection marker product must be produced to survive selection, necessitating very efficient transfection protocols (Kaufman et al., 1987; Kaufman et al., 1991). The expression level of the second cistron is non-predictable as it seems to be influenced by the length and type of the intercistronic region as well as by the first cistron itself. The strong selective pressure on the second gene was also suspected to favor undesired rearrangements and/or deletions of the first open reading frame in the transcription unit presumably allowing more efficient translation of the selectable gene. Later, these rearrangements were found to happen rather often in these expression systems (Kaufman et al., 1991; Adam et al., 1991). The problems encountered with these early vectors can be circumvented by two other approaches, which are described below.

In one strategy the selectable marker is replaced by a screening marker that does not put such strong pressure on the expression of a second cistron. Such markers code for gene products which are easily detectable in immunological assays or enzymatic reactions. These types of bicistronic vectors are first cotransfected with an ordinary selection plasmid. Thereafter, the clones arising from selection are screened for high expression of the screening marker in the bicistronic unit. Wirth et al., (1991) monitored firefly luciferase expression of a bicistronic expression vector to screen transfected BHK cells for high-level expression of IFN-β. Luciferase expression can be determined in an extremely sensitive, reliable and fast luminometric assay. Also *in situ* detection methods are available which allow the processing of a huge number of clones on a microtiter plate, tissue culture plate or a replica derived thereof (Wirth et al., 1991).

Wirth et al. (1990) reported the use of dicistronic expression vectors, containing the cDNA for human antithrombin III and the human IL-2 receptor (Tac-subunit) as screening marker, for the isolation of overexpressing CHO cells. Multiple rounds of sorting for the receptor expression yielded subpopulations of cells overexpressing the receptor itself *and* the antithrombin III molecule. The coupled coexpression of the two genes provided a general method for indirect screening of cells for expression of a gene of interest.

Recently, a new reporter gene became available, which can be monitored noninvasively in living cells. The green fluorescent protein (GFP) from the jellyfish *Aequorea victoria* emits green light when excited with UV or blue light. The chromophore in GFP is intrinsic to the primary structure of the protein and unlike other marker systems does not require substrate or cofactor addition. GFP spectra are similar to those of fluorescein and similar conditions for visualization of the chromophore can be used. Gene expression and protein localization experiments using GFP have been performed *in vivo, in situ* and in real time experiments (Chalfie et al., 1994; Inouye et al., 1994; Wang and Hazelrigg, 1994). GFP and derivatives have been functionally expressed in various organisms including *C. elegans, E. coli, Drosophila*, yeast, plant and mammalian cells (Chalfie et al., 1994; Inouye et al., 1994; Prasher, 1995; Pines, 1995; Brand 1995; Haseloff and Amos, 1995). In addition, suitable mutants with shifts in the wavelength of excitation and emission have been created, too (Heim et al., 1994). Monitoring gene expression with GFP as a second cistron in bicistronic vectors will facilitate screening procedures extraordinarily. Interestingly, FACS sorting and enrichment of high-level expression cells is possible (Wirth et al., unpublished observations). GFP monitoring therefore would substantially simplify and accelerate expression screening.

Several investigations have been performed which were aimed at improving translation efficiency from the second cistron eliminating thereby the undesired rearrangements upon selection. In a very special case it has been shown that the length of the intercistronic region influences translation from cistron 2 (Kozak, 1987). The efficiency of initiation from cistron 2 increased with the length of the intercistronic region. Efficiencies up to 50 % have been achieved with intercistronic regions between 80 and 140 bp in transient expression experiments. The vectors contained a bicistronic transcription unit with a minicistron followed by the preproinsulin gene. However, more general data with stable clones and different 5′ genes need to be obtained to support the initial and often cited observations of Kozak (1987). Translational enhancers like the adenoviral late leader have been tested in dicistronic expression vectors and also unusual translation mechanisms like the 'reach back' mechanism have been exploited (Peabody and Berg, 1986). Nevertheless, a real breakthrough for polycistronic expression vectors came with the discovery of the internal initiation mechanism in eukaryotic systems (see also chapter 1.1.2.3.1). This unusual mechanism was first found when the translation strategy of picornaviruses was investigated. Picornaviruses are positive-stranded RNA viruses with an uncapped RNA that codes for a long polyprotein. The mRNA has an unusually long 5′-UTR, which is rich in non-initiator AUGs. It was shown that translation is mediated by internal binding of ribosomes to a now well defined structure within the viral 5′-UTR allowing efficient translation (Jang et al., 1989; Pelletier and Sonenberg, 1988; Jackson et al., 1990; Nicholson et al., 1991). Several reports describe the use of picornaviral 5′-UTRs as the intercistronic region to increase translation of the selectable marker. Kaufman et al., (1991) used internal initiation mediated by the leader region of the EMC virus to guarantee sufficient expression of the DHFR gene in a bicistronic expression unit. Screening for amplification with methotrexate and high-level expression of the gene of interest could be achieved in CHO cell rearrangements or deletions as observed in the previous type of vectors.

Retroviral vectors have also been created harboring bicistronic and even polycistronic gene units (Adam et al., 1991, Ghattas et al., 1991; Koo et al., 1992). Adam et al. (1991) and Ghattas et al. (1991) constructed retroviral vectors in which the selectable marker gene expression was linked to a gene via elements mediating internal initiation of translation. Compared to existing retroviral vectors which realized marker gene expression through splicing or a second promoter, the bicistronic vector family was superior. Infectants exhibited a strongly correlated coexpression of both genes and did not show rearrangements or deletions upon selection.

The number of new viral elements mediating cap-independent initiation of translation is growing steadily. Even cellular elements have been indentified which enable cap-independent, internal initiation of translation (Macejak and Sarnow, 1991; Oh et al., 1992). Comparitive studies on the efficiency of translation of the different IRES elements have been performed recently *in vitro* (Bormann et al., 1995). The best IRESes were those from cardio-and aphtoviruses followed by those of enteroviruses, which exhibited up to 70 % of the efficiency of the IRES of the encephalomyocarditis virus (EMCV) in directing internal initiation of translation. The EMCV leader turned out to be one ot the most effective leaders and has been used in many expression experiments until now. Attempts to further optimize the EMCV-mediated initiation of a downstream cistron have been published (Davies and Kaufman, 1992). Interestingly, mutants in the initiator AUG region were shown to modulate the efficiency. The polioviral leader has also been extensively studied and is used to mediate internal initiation in artificial bicistronic mRNAs. Dirks et al. (1993) showed that in several cell lines second cistron initiation from this leader in bicistronic mRNAs is approximately threefold lower than cap-dependent initiation of the first cistron. Surprisingly, the internal initiation can be enhanced threefold by insertion of the 5'UTR of Xenopus β-globin, a leader that is known to promote translation.

1.5.4 Prospects

A variety of simple methods exist which allow the facilitated access to high-level producing cell clones. All methods have in common that gene transfer is rather random and a favorable integration or amplification is determined more or less by chance. Clearly, the tedious work is to screen for cells where this favorable event has happened. A more sophisticated way of achieving high-level expression is the 'directed transfer' of the gene to a locus which mediates a high-level expression. This is advantageous as the screening can be avoided (for a characterized locus) or can be carried out once (to identify a new locus). Methods which a allow a targeting of genes have been developed in the last years (Mansour et al., 1989; Jasin and Berg, 1988; Huang et al., 1991). They are based on homologous recombination or elements mediating a site specific recombination. The 'positive-negative selection' method for homologous recombination, for example, is currently widely used to target defined loci on the chromosome and to knock-out certain genes in ES cells for investigating their function in animals. This is done by creating transgenic mice with a 'loss of function'. Experiments are now performed which use these or similar methods to target genes to high expression loci (compare chapter 1.6). The nature of such a locus can be variable. It can consist of an element like a known strong cellular enhancer/promoter. An example is the targeting into the recombined immunoglobulin locus of hybridomas which presumably mediates the extraordinarily high expression of antibody genes in hybridomas. High-level expression of the replaced, non-essential immunoglobulin gene is overtaken from the foreign gene.

Another route is promising as well. It involves certain chromosomal elements which have been described in the last few years and which are responsible for mediating high-level expression in the ordered and highly organized chromatin structure. Such elements are divided into three groups according to their function and properties, the scaffold or matrix attachement regions (SAR or MAR), the locus control regions (LCR) and the dominant control regions (DCR). Targeting into such regions or yet identifiable elements is indeed tantalizing and can replace the screening procedures for high production. Finally, it is interesting that certain chromosomal elements, if supplied on the expression plasmid, render expression of a gene independent from the chromosomal surrounding after the integration step and result in a position-independent expression of the transfected gene.

It is obvious that current and future developments aim at further simplification and even replacement of screening for high-level expressing cells.

References

Adam, M.A., Ramesh, N., Miller, A.D. and Osbome, W.R.A. (1991) Internal initiation of translation in retroviral vectors carrying picornavirus 5' nontranslated regions. J. Virol. 65, 4985–4990.

Balland, A., Faure, T., Carvallo, D., Cordier, P., Ulrich, P., Foumet, B., la Salle, H.D. and Lecocq, J.-P. (1988) Characterization of two differently processed forms of human recombinant factor IX synthesized in CHO cells transformed with a polycistronic vector. Eur. J. Biochem. 172, 565–572.

Beckers, T., Hauser, H., Hüsken, D. and Engels, J.W. (1988) Expression and characterization of a des-methionine mutant interleukin-2 receptor (tac protein) with interleukin-2 binding affinity. J. Biol. Chem. 263, 8359–8365.

Boel, E., Berkner, K.L., Nexo, B.A. and Schwartz, T.W. (1987) Expression of human pancreatic polypeptide precursors from a dicistronic mRNA in mammalian cells. FEBS Lett. 219, 181–188.

Borman, A.M., Bailly, J.L., Girard, M. and Kean, K.M. (1995) Picornavirus internal ribosome entry segments: Comparison of translation effciency and the requirements for optimal internal initiation *in vitro*. Nucl. Acids. Res. 23, 3656–3663.

Brand, A. (1995) GFP in Drosophila. Trends. Genet. 11, 324–325.

Byrn, R.A., Mordenti, J., Lucas, C., Smith, D., Marsters, S.A., Johnson, J.S., Cossum, P., Chamow, S.M., Wurm, F.M., Gregory, T., Groopman, J.E. and Capon, D.J. (1990) Biological properties of a CD4 immunoadhesin. Nature 344, 667–670.

Capon, D.J., Chamow, S.M., Mordenti, J., Marsters, S.A., Gregory, T., Mitsuya, H., Byrn, R.A., Lucas, C., Wurm, F.M., Groopman, J.E., Broder, S. and Smith, D.H. (1989) Designing CD4 immunoadhesins for AIDS therapy. Nature 337, 525–531.

Chalfie, M., Tu, Y., Euskirchen, G., Ward, W.W. and Prasher, D.C. (1994) Green fluorescent protein as a marker of gene expression. Science 263, 802–805.

Davies, M.V. and Kaufman, R.J. (1992) The sequence context of the initiation codon in the encephalomyocarditis virus leader modulates efficiency of internal translation initiation. J. Virol 66, 1924–1932.

Dirks, W., Wirth, M., Hauser, H. et al. (1993) Multicistronische Expressionseinheiten und deren Verwendung. Patent. PCT EP9302294.

Ghattas, I.R., Sanes, J.R. and Majors, J.E. (1991). The encephalomyocarditis virus internal ribosome entry site allows efficient coexpression of two genes from a recombinant provirus in cultured cells and in embryos. Mol. Cell. Biol. 11, 5848–5859.

Gregersen, J.P., Mehdi, S., Gelderblom, H. and Zettlmeissl, G. (1990) A CD4 immunoglobulin fusion protein with antiviral effects against HIV. Arch. Virol. 111, 29–43.

Haseloff, J. and Amos, B. (1995) GFP in plants. Trends Genet. 11, 328–329.

Heim, R., Prasher, D.C., Tsien, R.Y. (1994) Wavelength mutation and posttranslational autoxidation of green fluorescent protein. Proc. Natl. Acad. Sci. USA 91, 12501–12504.

Huang, L.-C., Wood, B.A. and Cox, M.M. (1991) A bacterial model system for chromosomal targeting. Nucl. Acids Res. 19, 443–448.

Inouye, S. and Tsuji, F.I. (1994) Aequorea green fluorescent protein. Expression of the gene and fluorescence characteristics of the recombinant protein. FEBS Lett. 341, 277–280.

Jackson, R.J., Howell, M.T. and Kaminski, A. (1990) The novel mechanism of initiation of picornavirus RNA translation. TIBS 15, 477–483.

Jang, S.K., Davies, M.V., Kaufman, R.J. and Wimmer, B. (1989) Initiation of protein synthesis by internal entry of ribosomes into the 5'nontranslated region of encephalomyocarditis virus RNA *in vivo*. J. Virol. 63, 1651–1660.

Jasin, M. and Berg, P. (1988) Homologous integration in mammalian cells without target selection. Genes Dev. 2, 1353–1363.

Johnston, R.N., Beverley, S.M. and Schimke, R.T. (1983) Rapid spontaneous dihydrofolate reductase gene amplification shown by fluorescence-activated cell sorting. Proc. Natl. Acad. Sci. USA 80, 3711–3715.

Kaufman, R.J., Davies, M.V., Wasley, L.C. and Michnick, D. (1991) Improved vectors for stable expression of foreign genes in mammalian cells by use of the untranslated leader sequence from EMC virus. Nucl. Acids Res. 19, 4485–4490.

Kaufman, R.J., Murtha, P. and Davies, M.V. (1987) Translational efficiency of polycistronic mRNAs and their utilization to express heterologous genes in mammalian cells. EMBO J. 6, 187–193.

Kaufmann, R.J. and Murtha, P. (1987) Translational control mediated by eucaryotic initiation factor-2 is restricted to specific mRNAs in transfected cells. Mol. Cell. Biol. 2, 1568–1571.

Kavathas, P. (1990) Screening for expression by fluorescence activated cell sorting. DNA. Prot. Eng. Tech. 2, 10–14.

Kavathas, P. and Herzenberg, L.A. (1983) Amplification of a gene coding for human T-cell differentiation antigen. Nature 306, 385–387.

Koo, H.-M., Brown, A.M.C., Kaufman, R.J., Prorock, C.M., Ron, Y. and Dougherty, J.P. (1992) A spleen necrosis virus-base retroviral vector which expresses two genes from a dicistronic mRNA. Virology 186, 669–675.

Kozak, M. (1987) Effects of intercistronic length on the efficiency of reinitiation by eukaryotic ribosomes. Mol. Cell. Biol. 7, 3438–3445.

Lemay, G., Waksman, G., Roques, B.P., Crine, P. and Boileau, G. (1989) Fusion of a cleavable signal peptide to the ectodomain of neutral endopeptidase (E.C.3.4.24.11) results in the secretion of an active enzyme. J. Biol. Chem. 264, 15620–15623.

Macejak, D.G. and Sarnow, P. (1991) Internal initiation of translation mediated by the 5' leader of a cellular mRNA. Nature 353, 90–94.

Mansour, S.L., Thomas, K.R. and Capecchi, M. (1989) Disruption of the proto-oncogene int–2 in mouse embryo-derived stem cells: a general strategy for targeting mutations to non-selectable genes. Nature 336, 348–353.

McCracken, A.A. and Brown, J.L. (1984) A filter immunoassay for detection of protein secreting cell colonies. BioTechniques 2, 82–87.

Morgenstern, J.P. and Land, H. (1990) Advanced mammalian gene transfer: High titre retroviral vectors with multiple drug selection markers and a complementary helper-free packaging cell line. Nucl. Acids Res. 18, 3587–3596.

Nicholson, R., Pelletier, J., Le, S.-Y. and Sonenberg, N. (1991) Structural and functional analysis of the ribosome landing pad of poliovirus type 2: *in vivo* translation studies. J. Virol. 65, 5886–5894.

Oh, S.-K., Scott, M.P. and Sarnow, P. (1992) Homeotic gene *Antennapedia* mRNA contains 5'-noncoding sequences that confer translational initiation by internal ribosome binding. Genes Dev. 6, 1643–1653.

Page, M.J. and Sydenham, M.A. (1991) High level expression of the humanized monodonal antibody campath–1H in Chinese hamster ovary cells. Bio/Technology 9, 64–68.

Parkinson, J., Grinnell, B.W., Moore, R.E., Hoskins, J., Vlahos, C.J. and Bang, N.U. (1990) Stable expression of a secretable deletion mutant of recombinant human thrombomodulin in mammalian cells. J. Biol. Chem. 265, 12602–12610.

Peabody, D.S. and Berg, P. (1986) Termination-reinitiation occurs in the translation of mammalian cell mRNAs. Mol. Cell. Biol. 6, 2695–2703.

Pelletier, J. and Sonenberg, N. (1988) Internal initiation of translation of eukaryotic mRNA directed by a sequence derived from poliovirus RNA. Nature 334, 320.

Pines, J. (1995) GFP in mammalian cells. Trends Genet. 11, 326–327.

Prasher, D.C. (1995) Using GFP to see the light. Trends Genet. 11, 320–323.

Reil, H., Höxter, M., Moosmayer, D., Pauli, G. and Hauser, H. (1994) CD4 expressing human 293 cells as tools for studies in HIV-1 replication: The efficiency of translational frameshifting is not altered by HIV-1 infection. Virology 205, 371–375.

Seed, B. and Aruffo, A. (1987) Molecular cloning of the CD2 antigen, the T-cell erythrocyte receptor, by a rapid immunoselection procedure. Proc. Natl. Acad. Sci. USA 84, 3365–3369.

Strair, R.K., Towle, M.J. and Smith, B.R. (1988) Recombinant retroviruses encoding cell surface antigens as selectable markers. J. Virol. 62, 4756–4759.

Todd, R.D., Khurana, T.S., Sajovic, P., Stone, K.R. and O'Malley, K.L. (1989) Cloning of ligand-specific cell lines via gene transfer. Identification of a D2 dopamine receptor subtype. Proc. Natl. Acad. Sci. USA 86, 10134–10138.

Walls, J.D. and Grinnell, B.W. (1990) A rapid and versatile method for the detection and isolation of mammalian cell lines secreting recombinant proteins. BioTechniques 8, 138–142.

Walls, J.D., Berg, D.T., Yan, S.B. and Grinnell, B.W. (1989) Amplification of multicistronic plasmids in the human 293 cell line and secretion of correctly processed recombinant human protein C. Gene 81, 139–149.

Wang, S. and Hazelrigg, T. (1994) Implications for bcd mRNA localisation from spatial distribution of exu protein in Drosophila oogenesis. Nature 369, 400–403.

Wirth, M., Bode, J., Zettlmeissl, G. and Hauser, H. (1988) Isolation of overproducing recombinant mammalian cell lines by a fast and simple selection procedure. Gene 73, 419–426.

Wirth, M., Grannemann, R., Klehr, D. and Hauser, H. (1994) Screening retroviral helper cells for high efficient virus production using a simple selection procedure. J. Virol. 68, 566–569.

Wirth, M., Hoexter, M., Morelle, C. and Hauser, H. (1990) Bicistronic expression vectors facilitate screening for overexpressing mammalian cells, pp. 69–73. VCH publishers, Weinheim.

Wirth, M., Li, S.-Y., Schumacher, L., Lehmann, J., Zettlmeissl, G. and Hauser, H. (1990) Screening for and fermentation of high producer cell clones from recombinant BHK cells. In: Spier, R.E., Griffiths, J.B., Stephenne, J. and Croy, P.J. (eds.), pp. 44–50, Butterworth, London.

Wirth, M., Rehm, H.H., Rehm, H.-J., Reed, G., Pühler, A. and Stadler P. (1993) Genetic engineering of animal cells. VCH Weinheim, pp. 663–744.

Wirth, M., Schumacher, L. and Hauser, H. (1991) Use of dicistronic transcription units for the correlated expression of two genes in mammalian cells. In: Spier, R.E., Griffiths, J.B. and Meignier, B. (eds.), pp. 328–343. Butterworth, London.

Wysocki, L.J. and Sato, V.L. (1978) "Panning" for lymphocytes: A method for cell selection. Proc. Natl. Acad. Sci. USA 75, 2844–2848.

Zettlmeissl, G., Gregersen, J.-P., Duport, J.M., Mehdi, S., Reiner, G. and Seed, B. (1990) Expression and characterization of human CD4- Immunoglobulin fusion proteins. DNA Cell. Biol. 9, 347–353.

Zettlmeißl, G., Langner, K.D., Wirth, M. and Hauser, H. (1988) Erzeugung hochexprimierender Zellklone. Patent DE3806617.

1.6 Genetic Engineering of Antibodies and Derivatives from Mammalian Cells

Michaela Schäffner, Brigitte Kaluza and Ulrich H. Weidle

Abbreviations: Ab, antibody; mAb, monoclonal antibody; V, variable region; C, constant region; D, diversity region; J, joining region; Ig, immunoglobulin; CDR, complementarity determining region(s); FR, framework region of variable part of Ab; UT, untranslated region; IL, interleukin; IL2R, interleukin 2 receptor; Fab fragments, antigen binding fragments of Abs which are derived by papain digestion and contain the light chain and part of the heavy chain (variable region and first constant domain); ADCC, Ab-dependent cell-mediated cytotoxicity; CDC, complement-dependent cytotoxicity; VJ, variable region of rearranged L chain, VDJ, variable region of rearranged H chain; PCR, polymerase chain reaction; nt, nucleotide(s); cDNA, DNA complementary to mRNA; mRNA, messenger RNA.

1.6.1 Introduction

Antibodies are of great interest because they react with a variety of ligands and exhibit a plethora of effector functions. Therefore this class of molecules represents important tools for numerous diagnostic and therapeutic applications. The advent of recombinant DNA technology enabled expression of mAbs and several restructured versions thereof (Fig. 1.6.1) in prokaryotic and eukaryotic systems. However, appropriate modifications of the protein as a prerequisite for maintaining desired effector functions are mediated by mammalian cells only (Morrison, 1992). Therefore, in this article only mammalian cells are considered as expression systems; prokaryotic and lower eukaryotic systems are excellently reviewed in (Winter and Milstein, 1991; Plückthun, 1991).

In many cases hybridoma technology will lead to mAbs of the desired specificity and affinity, but of unsuitable isotype. The V regions encoding L and H chains of these mAbs can be combined with any C region by recombinant DNA techniques thus creating the desired isotype (Fig. 1.6.1).

Establishment of hybridomas is presently easily performed only with murine or rat B cells by conventional fusion techniques (Köhler and Milstein, 1974). Generation of human mAbs is more difficult and only a limited spectrum of specificities is accessible. Making use of recombinant DNA technology, C regions of any isotype and species can be combined with the V regions of the L and H chains of the cognate mAb, resulting in chimeric Abs (Morrison et al., 1984). This approach enables the design of mAbs with desired effector functions (Brüggemann et al., 1987; Steplewski et al., 1988) and if human C regions are fused to the rodent V regions, chimeric mAbs with reduced immunogenicity after therapeutic application in humans (Lo Buglio et al., 1989), Fig. 1.6.1, are generated.

Ab fragments, such as Fab, $F(ab')_2$ or Fv fragments can be generated by genetic engineering. Expression of these versions is preferentially accomplished in microbial

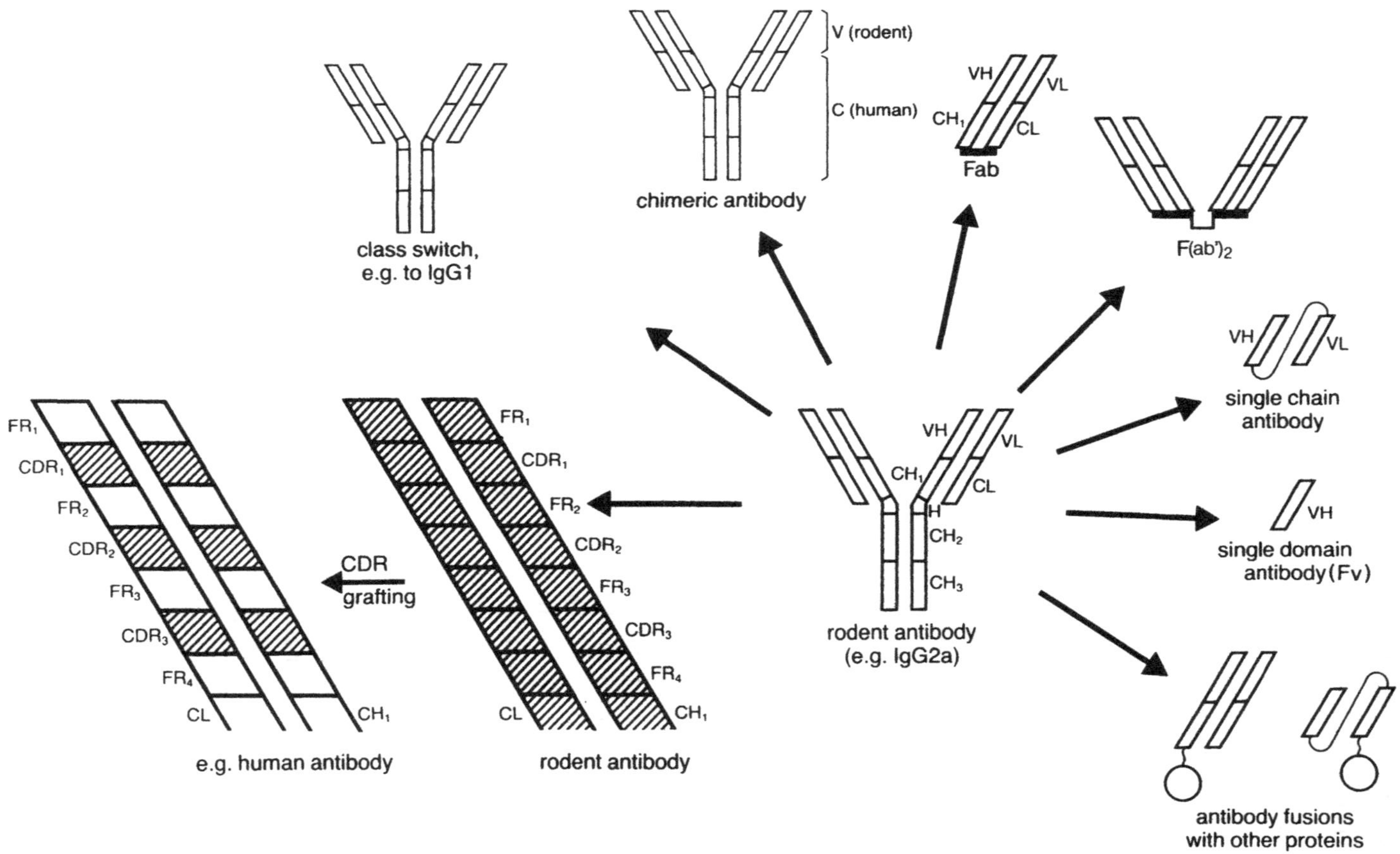

Fig. 1.6.1: Antibodies, restructured versions of antibodies, antibody fragments and fusion proteins available due to recombinant DNA techniques.

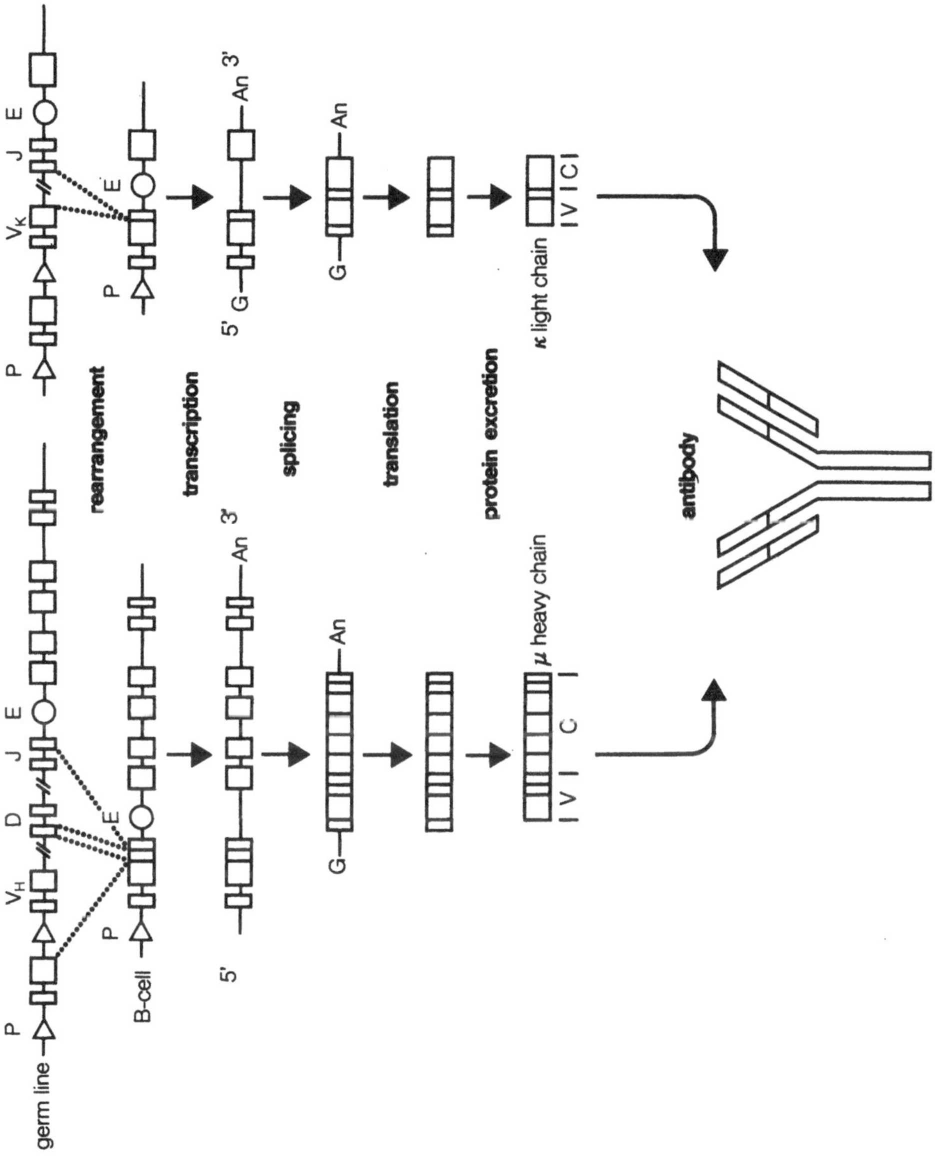

Fig. 1.6.2: Events leading from genomic rearrangement of the genes encoding L and H chains to secretion of functionally assembled antibody.
P = promoter; V_L and V_H, variable regions of L and H chains; L = light chain; H = heavy chain; D = diversity region; J = joining region; E = enhancer; A_n = polyadenylation site.

systems (Better at el., 1988; Skerra and Plückthun, 1988). In some cases, the H chain has been found to bind the antigen with little contribution of the corresponding L chain, resulting in a so-called "single-domain antibody" (Ward et al., 1989). Furthermore, V regions of L and H chains can be connected with short polypeptide linkers, resulting in "single-chain antibodies" (Bird et al., 1988; Bedzyk et al., 1990).

Alternatively, mAbs can be genetically fused with heterologous proteins. Antibody-toxin fusions have been generated for therapeutic applications (Chaudhary et al., 1989; Chaudhary et al., 1990). Soluble immuno-receptors were fused with mAbs with the aim of therapeutic application as immunmodulators (Waldmann, 1992). For analytical applications *in vitro*, fusions of Ab domains with other proteins are possible (Neuberger et al., 1984) for usage as signal transducers or for surface attachment. Alternatively, post-translational coupling can be performed, making use of thiol groups introduced by genetic engineering (Lyons et al., 1990).

Antigen specificity of mAbs is determined by structural modules called complementarity-determining regions (CDR). The CDR loops are located on each L and H chain of mAbs. Mutagenesis of amino acids located in the CDRs leads to mAbs with altered antigen binding specificities. Humanized mAbs result by transplanting rodent CDRs into appropriate positions of the framework regions of human L and H chains by mutagenesis techniques (Jones et al., 1986; Riechmann et al., 1988), resulting in molecules with reduced immunogenicity after administration into humans (Fig. 1.6.1).

The organization of mammalian Ig genes reflects the domain structure of Abs (Fig. 1.6.2), each protein domain being encoded by separate exons. The exon encoding the V domain is unique insofar as it is separated in germ-line and somatic cells into two (V and J) in L chains or three (V, D and J) in H chains separate gene segments, which are combined during B cell maturation in a process called rearrangement (Honjo, 1983). Thus, only mature B cells and their descendents contain functional V exons. The gene structure of Ig genes is very similar in mice and rodents, thus greatly facilitating the construction of chimeric Abs by "exon shuffling". During gene expression, the intervening sequences are removed from the primary transcript, yielding mRNA, which is translated subsequently into L and H chains (Fig. 1.6.2).

▷

Fig. 1.6.3: Cloning of VJ and VDJ regions of mAbs from DNA of hybridoma cells by PCR and insertion into compatible expression vectors. A and B: Promoterless VJ (L chain) and VDJ (H chain) fragments of rearranged immunoglobulin genes are cloned by PCR as fragments with SalI and NotI restriction endonuclease cleavage sites. For expression of chimeric mAbs in lymphoid cells, VJ and VDJ fragments are cloned directly into expression vectors. pUHWk contains the constant region of a human κ gene, pUHWγ1 contains the constant region of a human γ1 gene. In both vectors expression is driven by a heavy chain promoter/enhancer combination.

CAP, initiation site for mRNA; C, constant region; D, diversity region; E_k and E_H, enhancer for light and heavy chain; H, hinge region, J, joining region; S, signal sequence; V_L and V_H, variable region of L and H Ig genes; μPr, promotor of a mouse μ gene; 3′ UT, 3′ untranslated region; An (SV40), polyadenylation signal derived from simian virus 40; CH1, CH2, CH3, exons of the human γ1 gene; Ap^R, ampicillin resistence gene; neo, phosphotransferase conferring resistance to G418; arrows indicate the direction of transcription.

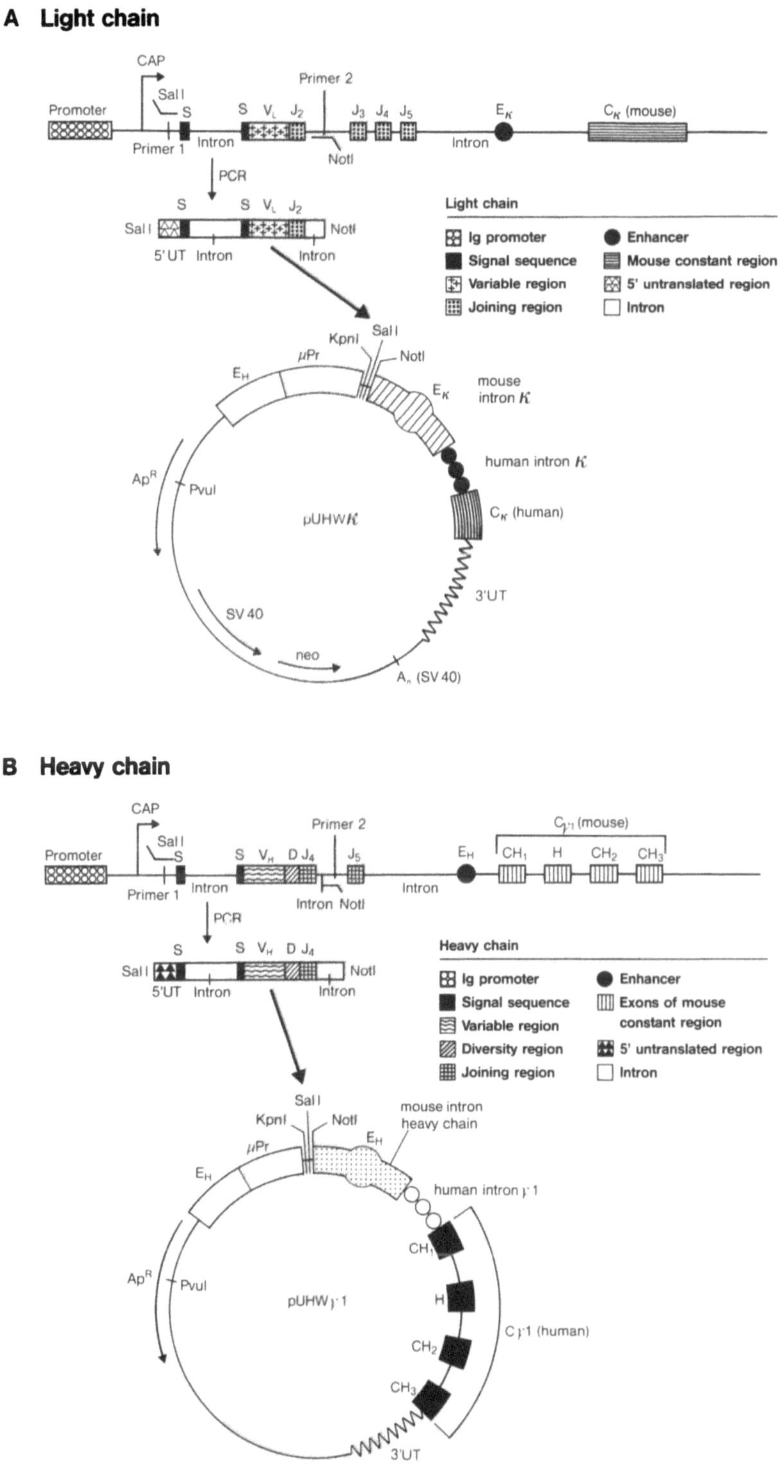
A Light chain
CAP
Primer 2
Sal I
S
Promoter
S V_L J_2
J_3 J_4 J_5
E_κ
C_κ (mouse)
Primer 1
Intron
NotI
Intron
PCR
S
S V_L J_2
Sal I
NotI
5' UT
Intron
Intron
Light chain
Ig promoter
Signal sequence
Variable region
Joining region
Enhancer
Mouse constant region
5' untranslated region
Intron
Sal I
KpnI
NotI
μPr
E_H
E_κ
mouse intron κ
human intron κ
Ap^R
PvuI
pUHWκ
C_κ (human)
3'UT
SV 40
neo
A_n (SV 40)
B Heavy chain
CAP
Primer 2
C_γ1 (mouse)
Sal I
S
Promoter
S V_H D J_4
J_5
E_H
CH_1 H CH_2 CH_3
Primer 1
Intron
Intron NotI
Intron
PCR
S
S V_H D J_4
Sal I
NotI
5'UT
Intron
Intron
Heavy chain
Ig promoter
Signal sequence
Variable region
Diversity region
Joining region
Enhancer
Exons of mouse constant region
5' untranslated region
Intron
Sal I
KpnI
NotI
mouse intron heavy chain
μPr
E_H
E_H
human intron γ1
CH_1
Ap^R
PvuI
pUHWγ1
H
Cγ1 (human)
CH_2
CH_3
3'UT

1.6.2. Cloning of V Regions

A prerequisite for expression of Ab versions discussed above is the cloning of the V regions encoding L and H chains. As starting materials, genomic DNA or mRNA from the appropriate hybridoma cell line can be used. V regions originating from genomic DNA can be inserted directly into expression vectors, whereas cDNAs have to be extensively modified to fit into vectors with genomic Ig elements. Isolation of V regions from bacteriophage libraries is laborious and often associated with technical difficulties (Morrison et al, 1984; Steplewski et al., 1988). An alternative procedure for isolation of genomic V regions is illustrated in Fig. 1.6.3 (Weissenhorn et al., 1991). cDNA sequences for L and H chain are determined after cloning the appropriate V regions by PCR. This is performed by transcribing mRNA into cDNA with reverse transcriptase making use of primers matching to the 5′ region of the C regions of L and H chains, addition of a dG-tail to the 5′ terminus of the cDNA by terminal deoxynucleotidyl-transferase and amplification of the so generated cDNA by PCR. Nucleotides in the 5′UT regions of L and H chain mRNA are identified as well as the appropriate J regions used for rearrangement. This information allows the design of primer pairs matching to 5′UT regions and to UT genomic sequences located 3′ of the appropriate J segment (the genomic J regions of mouse and human are completely sequenced).

Genomic VJ and VDJ segments are subsequently amplified from DNA of the hybridoma cell line by PCR and due to appropriate restriction endonuclease cleavage sites encoded by the primers, ligated directly into expression vectors for lymphoid cells (Fig. 1.6.3).

Cloning of L and H chain cDNA starting from mRNA can be performed using classical methods for screening for cDNA libraries with probes specific for L and H chain C regions (Shin and Morrison, 1989; Sambrock et al., 1989). With the advent of PCR technology, powerful methods for rapid cloning of C regions of Ig genes have emerged. One approach (Larrick et al., 1989a, Larrick et al., 1989b) uses an oligonucleotide homologous to the leader sequence of L and H chains as a 5′ primer. The leader sequence is encoded mainly by hydrophobic amino acids and for this reason is more or less conserved on the DNA level as well. As a 3′-primer, an oligonucleotide homologous to C region (CH1) sequences is employed, which can be deduced from known sequences if the isotype of the Ab to be cloned is known. However, due to leader sequence variations, not every V region can be cloned by this method. Another approach (Orlandi et al., 1989; Sastry et al., 1989; Le Bouef et al., 1989) makes use of the similarity of FR regions for primer design. The 5′-primer is a mixture of primers corresponding to a consensus sequence of FR1, the 3′-primer is mixture of primers corresponding to a consensus sequence of FR4 (which corresponds to the J segment), thus amplifying the mature V entity only. However, the unbiased amplification of complete V repertoires requires very complex sets of degenerate primers (Marks et al., 1991). This method will as a disadvantage modify the V region sequence in FR1 and FR4 to assume the primer sequence, the original sequence information in these parts being lost. The most versatile method uses the principle of inverse PCR (Triglia et al., 1988) for V region cloning (Kaluza et al., 1992). The principle of the method is outlined in Figure 1.6.4. mRNA encoding Ab L and H chains is transcribed into double-stranded cDNA,

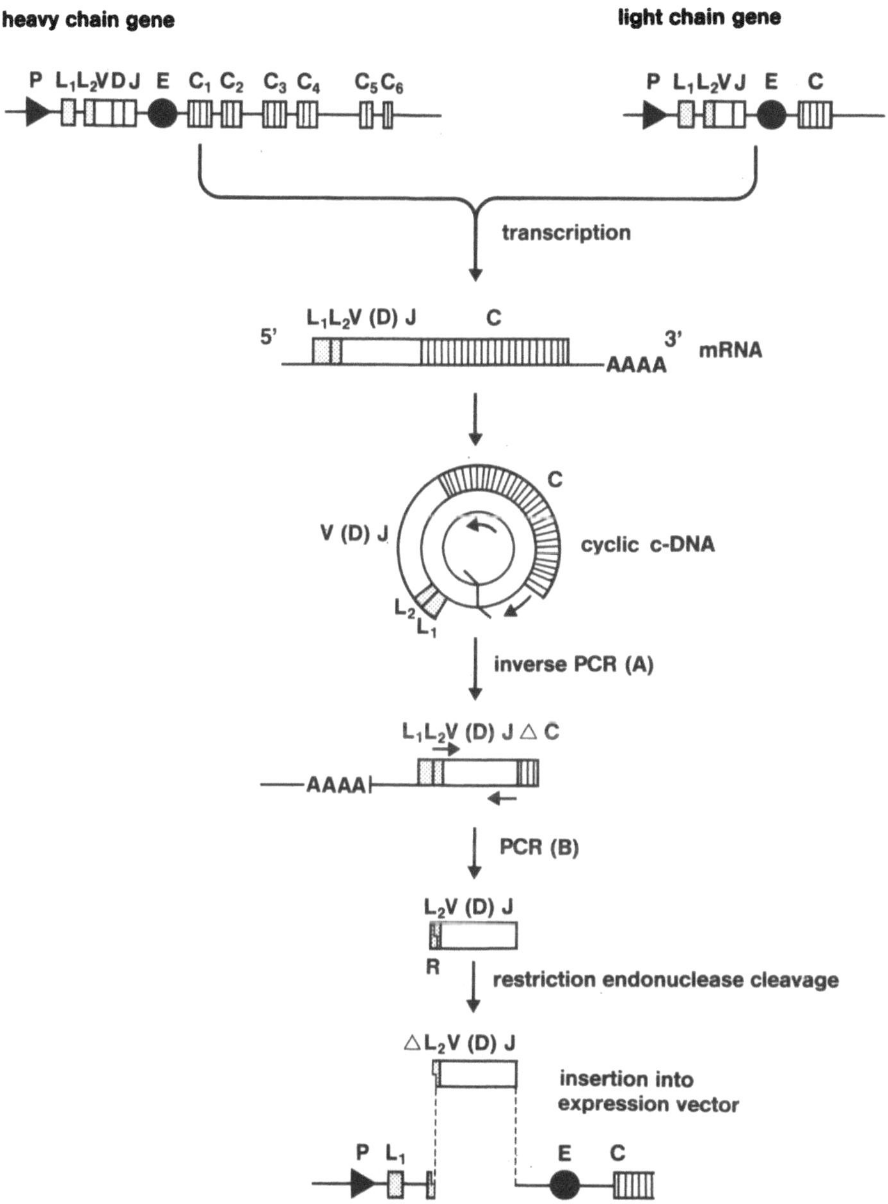

Fig. 1.6.4: Schematic outline of procedure for chimerization of mAbs conserving N-terminal amino acid sequences. Details are described in the text.
P, promoter; L_1, part of leader sequence encoded by the first exon of Ig genes; L_2, part of leader sequence encoded by VJ or VDJ exons of Ig genes; C, constant region; C_n, exons of Ig μ gene; E, enhancer; D, diversity region, J, joining region; V, variable region; R, restriction endonuclease cleavage site.

which is then circularized by ligation. Subsequently inverse PCR amplification (PCR(A)) is performed with primer pairs which anneal to the C and 3′-UT regions of the cDNA for L and H chains. The resulting amplification product contains the 5′UT region, the leader sequence, the complete V region and is flanked by 3′UT and C region sequences. The DNA sequence of the amplification product is determined and used for the design of primer pairs for PCR(B). The first primer matches to the L_2 region (the part of the signal sequence which is encoded within exon 2 of Ig L and H chain genes) and to the 5′ part of the V region. This primer introduces a restriction endonuclease site into L_2. The second primer matches to the 3′ part of the V(D)J region and ends exactly at the last nt of exon 2. The DNA fragment obtained by PCR(B) is cut with the appropriate restriction endonuclease to obtain a cohesive 5′ end and then ligated into especially designed expression vectors (Kaluza et al., 1992) to reconstitute chimeric Ig genes for L and H chains. This procedure allows the unbiased amplification of V regions of L and H chains of Ig genes with unaltered N-terminal amino acid sequences. Amino acids at the N-terminus seem to contribute to the correct folding of CDR regions (Chotia et al., 1989). For this reason, V region cloning by use of degenerate primers could lead to reduced Ab affinity.

A procedure for cloning genomic fragments encoding Ab V regions by inverse PCR has been described (Zwickl et al., 1990). It remains to be seen whether this procedure will be used as a standard method.

1.6.3 Expression Systems

Due to their ease of growth, bacteria are the system of choice for expression of Ab fragments (Better et al., 1988; Skerra and Plückthun, 1988). However, expression of functionally intact Ab was not feasible in microbial systems until now. Due to their inability to mediate glycosylation, bacteria are inappropriate expression systems for Abs with effector functions depending on glycosylation (Morrison, 1992). It has been shown that complete, glycosylated mAbs can be produced in yeast (Horwitz et al., 1988), albeit with a different glycosylation pattern and as a consequence with effector functions different from those of mammalian cell-derived mAb. For therapeutical applications, the potential immunogenicity of such mAbs has to be considered. For the production of complete Ab molecules with defined effector functions, mammalian cells are the system of choice, because they mediate correct post-translational modifications of the mAbs. Mammalian cells are transfected with expression vectors containing dominant marker genes for subsequent selection of stable transformants (Mulligan and Berg, 1981; Southern and Berg, 1982; Hartman and Mulligan, 1988). Most commonly used mammalian recipients are Chinese hamster ovary cells (CHO) and non-Ig producing myeloma cells such as Sp2/0 and P3X63.Ag8.653 (Weidle et al., 1987).

1.6.3.1 Baculovirus-System

L and H chains of a mouse mAb (IgG) directed against lipoprotein I of Pseudomonas Caeruginosa were introduced into the genome of baculovirus and Spodoptera frugiparda cells (fall armyworm) were infected with the recombinant virus (zu Putlitz et al., 1990). Infected insect cells stably secreted antigen-binding and glycosylated Ab at levels of 25–30 μg per ml containing 10^5 cells, corresponding to secretion levels of mAbs from hybridoma cell cultures (Harlow and Lane, 1988). The glycosylation pattern was different compared to hybridoma derived mAb. The baculo-derived mAb was able to bind complement component C1q, other effector functions, however, were not investigated. Using both double infection of insect cells with separate H- and L-chain-expressing viruses and infection with a double-recombinant virus containing both Ig L and H chain cDNA resulted in expression of a functional murine mAb (IgG1) with binding affinity to p-azophenylarsonate at levels of 5 μg/ml Sf9 monolayer culture supernatant (Hesemann and Capra, 1990).

1.6.3.2 Antibody Expression in Non-Lymphoid Cells

The first demonstration of functional expression of κ and $\gamma 1$ chains, as well as reconstituted Ab directed against creatine-kinase-MM in non-lymphoid cells in the transient COS cell system (Weidle et al., 1987), paved the way for the establishment of stable CHO cells secreting reconstituted Ab into the culture fluids (Weidle et al., 1987). In the absence of methotrexate selection, expression levels of about 100 ng Ab/10^6 cells/24 h/ml were described for confluent monolayers. The potential of engineering CHO cells for production of recombinant mAbs was demonstrated by expression of L and H chain cDNAs of the humanized mAb-Campath-1H in CHO cells with productivity of 100 μg/10^6 cells/24 h after methotrexate amplification (Page and Sydenham, 1991). Each cDNA was expressed under control of human β-actin promoter/polyadenylation signals (Page and Sydenham, 1991).

Using Glutamine-Synthetase as an alternative amplifiable marker gene (Bebbington et al., 1992; Cockett et al., 1990), a chimeric Ab directed against the tumor associated glycoprotein TAG72 of mamma and colon carcinomas (IgG4) used for tumor imaging was expressed up to 250 mg/l in suspension cell culture in serum-free medium after gene amplification of the CHO recipient cell line (Roberts et al., 1991; Field et al., 1992). In addition to engineering with respect to improvement of productivity, expression of reconstituted Ab in non-lymphoid cells paved the way for a number of scientific investigations. Glioma, phaeochromocytoma and other non-lymphoid cells transfected with expression constructs for genes of L and H chains of a hapten-specific IgM Ab yielded stable transformants secreting polymeric IgM composed of pentamers and hexamers in similiar yield obtained with a plasmocytoma (Cattaneo and Neuberger, 1987). Because these non-lymphoid cells do not express J chain, the results indicate that neither J chain nor other lymphoid-specific proteins are required for assembly and secretion of polymeric IgM.

Targeting of reconstituted Ab to the cytoplasm and the nucleus of transfected non-lymphoid cells was demonstrated recently (Biocca et al., 1990). This finding should initiate studies on the function of gene products by neutralizing their protein products with mAbs in the cytoplasm or the nucleus of appropriate cells.

1.6.3.3 Antibody Expression in Lymphoid Cells

Non-Ig producing cells such as Sp2/0 (Ochi et al., 1983) and X63Ag8.653 (Kearny et al., 1979) are excellent recipients for the production of Abs and heterologous proteins (Weidle and Buckel, 1987, Hendricks et al., 1989). Transfectomas generated with expression vectors, as shown in Fig. 1.6.3 C, D, secrete chimeric mAb in the range of 10–30 μg/10^6 cells/24 h/ml corresponding to the productivity of hybridoma cell lines (Weissenhorn et al., 1991; Weissenhorn et al., 1992). However, when Ig cDNAs are isolated, it is recommendable to reconstitute the exon-intron organization of Ig genes on the expression vectors. For this purpose V regions are joined with oligonucleotides providing both splice donor and unique restriction sites for insertion into an expression vector by conventional mutagenesis procedures giving rise to expression constructs mediating high-level secretion of chimeric mAbs in stably transfected myeloma cells (Gillies et al., 1989; Kaluza et al., 1991).

For expression of Ab cDNA, a strong constitutive promoter (SV40 early) or an H chain promoter/H chain enhancer combination is clearly preferable to an L chain promoter/L chain enhancer combination, corresponding roughly to the expression level obtained with Ig genes after deriving stably transfected myeloma cells (Weidle et al., 1987).

The question of intron requirement for Ig gene expression was addressed in (Neuberger and Williams, 1988). It was shown that an intron is required when transcription is driven by an Ig promoter/enhancer combination, although this requirement is not specific for a particular intron. The need of an intron seems to be dependent upon the promoter used. While an intron is required in the case of Ig or β-globin promoters, it is not required in the case of cytomegalovirus or heat-shock promoters, pointing to a connection between the promoter and RNA processing or export (see also chapter 1.1.2.1.3).

Improvement of expression of transfected Ig genes in lymphoid cells can be accomplished by gene amplification or making use of recently described new regulatory elements. Genes encoding a murine/human Ab were cloned adjacent to the gene coding for methotrexate-resistant dihydrofolate-reductase and introduced into myeloma cells and cell lines containing stably integrated genes were selected (Dorai and Moore, 1987). Co-introduction of the L and H chain constructs without amplification resulted in a 25-fold increase in secretion of intact Ab corresponding nearly to the productivity of hybridomas. Glutamine-Synthetase mediated amplification (Bebbington et al., 1992; Cockett et al., 1990) of the genes encoding a chimeric Ab directed against a tumor-associated antigen in mamma- and colon-carcinomas in the myeloma host cell line NSO resulted in Ab secretion up to 240 mg/l in simple batch culture (Field et al., 1992) (see also chapters 1.2 and 1.4).

A myeloma based expression system for production of large mammalian proteins was described in (Traunecker et al., 1991). Vectors include a promoter from an Ig κ V gene and a modified Ig H chain enhancer that lacks potential targets for negative regulatory proteins. As a recipient cell line, the mouse myeloma cell line J558L (Oi et al., 1983), which constitutively secretes Igλ1 L chains, is used. The gene of interest with its own initiation codon and leader sequences is engineered behind the Vκ promoter. Either genomic or cDNA forms of the gene can be used. To mimic the natural and stable Ig mRNA more closely, the engineered gene is forced to splice into the genomic Ig κ C region which also supplies polyadenylation signals. Two forms of recombinant proteins can be generated: either a chimeric protein containing a Cκ domain, or a non-chimeric protein depending on whether the translational stop codon is present in the engineered gene. The expression system was used to produce soluble forms of plasma-membrane molecules and receptors for structure/function studies and to generate Abs. Ig fusion proteins with CD2, CD4, CD8, TCRα, TCRβ, MHC class IIα and MHC class IIβ have been produced at levels up to 100 μg/ml. The system was also applied to produce mature interleukins, such as IL6 and IL7. In a further modification of the vector system, six His residues were engineered into the Cκ region to facilitate binding of proteins to Ni^{2+} chelate absorbents for elution under mild conditions, such as pH 5, for proteins which are prone to deleterious effects under drastic elution conditions, such as pH 2.8. Fusion of proteins, such as CD4 to the C regions of IgG or IgM also was demonstrated, giving rise to chimeric or pentameric receptor molecules, thus increasing the valence of the receptor, resulting in higher avidity.

Enhancer elements located in the 3'UT regions of L and H chain genes should prove useful tools for construction of improved Ig gene expression vectors. Omission of the 3′ κ gene enhancer leads to 20–40-fold lower expression of κ chains in transgenic animals (Meyer et al., 1990). Strong lymphoid-specific enhancers with homology to sequence motifs important with respect to the function of the intron enhancer have been discovered 3′ of the C regions of the H chain genes (Pettersson et al., 1990) and λ genes (Bich-Thuy and Queen, 1989; Hagmann et al., 1990).

Further improvement of expression of Ig genes might be mediated by locus control regions (LCR) (see also chapter 1.1.2.1.3). Such a region has been recently characterized for the β-globin locus (reviewed in Townes and Behringer, 1990). The LCR is a region of DNA at the 5′ end of the 50 kb β-globin locus, which mediates copy-dependent and integration-site independent expression of globin or heterologous genes when linked to them. Thus the LCR seems to be an element that can initiate an open chromatin structure of genes at great distances and activate heterologous genes placed within this domain (Forrester et al., 1990). Recent experimental evidence suggests the existence of LCR in the Ig gene locus (Plon and Groundine, 1991). Ig κ gene expression studies after stable integration of the transgenes have indicated that an intronic matrix association region (MAR) increases the expression of the transgene in S194 plasmacytoma cells (Blasquez et al., 1989) (see also chapter 1.1.2.1.3).

1.6.3.4 Transgenic Animals

Expression of transgenic Ig specific for a common pathogen could provide an animal with congenital immunity for that pathogen. Genes encoding mouse α H and κ L chains from Ab directed against phosphorylcholine (PC) were coinjected into ova to produce two transgenic pigs and three transgenic lambs (Lo et al., 1991). In the transgenic pigs, mouse IgA was detected in the serum, however, with little binding specificity to PC, presumably because secreted Ab included pig L chains. In transgenic sheep, mouse IgA was detectable in peripheral lymphocytes, but not in serum.

These studies need to be extended to obtain conclusive proof that the IgA transgene would be protective against pathogenic bacteria. Genes (κ, $\gamma 1$) encoding an anti-idiotypic mAb directed against a mAb with specificity towards the hapten 4-hydroxy-3-nitro-phenylacetate were introduced into the germ-line of mice, rabbits and pigs (Weidle et al., 1991). Serum Ab titers have been achieved in transgenic rabbits and pigs, respectively of 100 µg/ml and 1000 µg/ml. Unfortunately, only a few bands observed in isoelectric focusing gels were identical to those of the purified mouse mAb. One possible explanation may be that the level of κ chain expression was insufficient for complete allelic exclusion, resulting in Abs with L chains derived from rabbit and pig Ig L chain genes. The incorporation of additional sequences into the corresponding gene constructs, e.g. downstream L chain enhancer elements (Meyer et al., 1990), might be helpful in overcoming this low expression level. In principle it should be possible in the future to develop gene constructs allowing expression in transgenics sufficient to protect animals against the severe attack by viruses or bacteria in a manner equivalent to immunization. This strategy would be particularly useful for protection against diseases where vaccination is not allowed, difficult or impossible.

1.6.4 Manipulation of Antibodies by Gene Targeting

Homologous recombination between DNA sequences residing on the chromosome and newly introduced, cloned DNA sequences (gene targeting) allow the transfer of any modification of the cloned gene into the genome of a living cell (Capecchi, 1989a). Although mammalian cells can efficiently mediate recombination between homologous DNA sequences, they demonstrate an even greater propensity for mediating non-homologous recombination (Capecchi, 1989b). The frequency of recombination is roughly proportional to the extent of homology between introduced and chromosomal sequences (Capecchi, 1989b). Targeting frequency is neither increased by increasing the number of copies of incoming DNA nor by increasing the number of target sequences residing in the host genome (Capecchi, 1989a; 1989b).

The use of homologous recombination to introduce predetermined genetic changes in hybridoma Ig genes may have many applications for the production of Ig which is specifically optimized for therapeutic and diagnostic use to the detection and analysis of the molecular events which determine Ig expression. Homologous recombination between transferred and chromosomal Ig κ genes is reported in (Baker and Shulman, 1988). The test system is based on a hybridoma which synthesizes IgM(k) specific for the hapten 2,4,6-trinitrophenyl (TNP). A mutant hybridoma cell line with a deletion in the κ anti-TNP gene was transfected with a vector bearing a functionally rearranged TNP-specific Vκ segment. Rare transformants (frequency 4×10^{-3}) produced normal amounts of the functional κ chain. Analysis of the DNA indicated that the transformants were generated by homologous recombination (Baker and Shulman, 1988).

A chromosomal Ig gene mutation was corrected by transferring a pSV2neo vector encoding the C region of the Igµ H chain to mutant hybridoma cells bearing a 2 base pair deletion in the third C region of their chromosomal β gene (Baker et al., 1988).

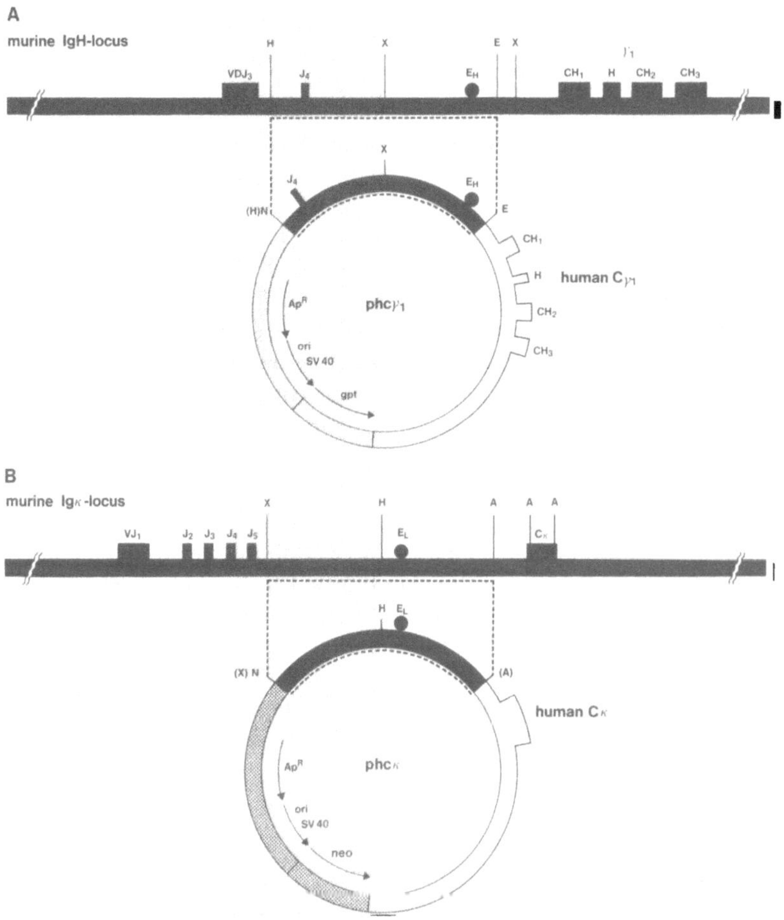

Fig. 1.6.5: Chimerization of monoclonal antibodies by gene targeting.
A and B: Targeting vectors for chimerization of H (A) and L (B) chains of murine mAbs.
Regions of homology between expression vectors and murine IgH and IgL loci are indicated.
A = Asp 700, B = Bam HI, E = Eco RI, H = Hind III, N = Not I, X = XbaI, overhang filled in by Klenow.
E_H and E_L, enhancers of H and L chains; V, variable region; D = diversity region; J, joining region, C, constant region; H = hinge region; CH1, CH2, CH3, exons of constant region of human γ1 gene.

- murine Ig sequences
- human Ig sequences
- sequences derived from pSV2 Eco gpt
- sequences derived from pSV2 neo
- exons of murine Ig genes
- exons of human Ig genes
- enhancer

Chimerization of Ab by gene targeting would obviate the necessity of cloning H and L chain V region segments. Production of antigen-specific H chain was achieved by targeting the human IgG1 H chain C region (Cγ1) exons to the genomic H chain locus of a hybridoma cell line secreting Ab specific for a tumor-associated antigen (Fell et al., 1989). Frequency of productive recombinations was 1 in 200 transfectants with accumulation of the chimeric protein up to 20 μg/ml in culture supernatants. Homologous recombination was mediated by a 2.3 kb region of the murine IgH locus located upstream of the switch region incorporated into the transfected plasmid carrying human Cλ1 exons and a selection marker. Clonal cell lines were established that expressed the chimeric protein, the murine Ig or both, suggesting that the parental cells retained more than one copy of the productive H chain gene. However, replacement of murine L chain C regions by human C regions was not performed in these studies. It should be possible to construct replacement vectors that would direct the generation of Fab fragments or modifications of mAbs conferring improved effector functions, such as biologic modifiers or enzymes.

We have undertaken an attempt to chimerize H and L chains of a murine mAb (IgG1) directed against the α chain of the human IL2R by gene targeting by replacing C regions of murine κ and γ1 chains by the corresponding human κ and γ1 C regions. The replacement vectors are outlined in Figure 1.6.5a and b. They carry 2 kb regions homologous to H and L locus murine intron sequences, the corresponding human C κ and γ1 regions and marker genes conferring resistance to mycophenolic acid and G418 after incorporation into the genome of the hybridoma cell line. Before transfection, they were linearized at their unique NotI sites. Chimerization of the H chain resulted in 14 clones secreting chimerized H chain (frequency 1 in 1000).

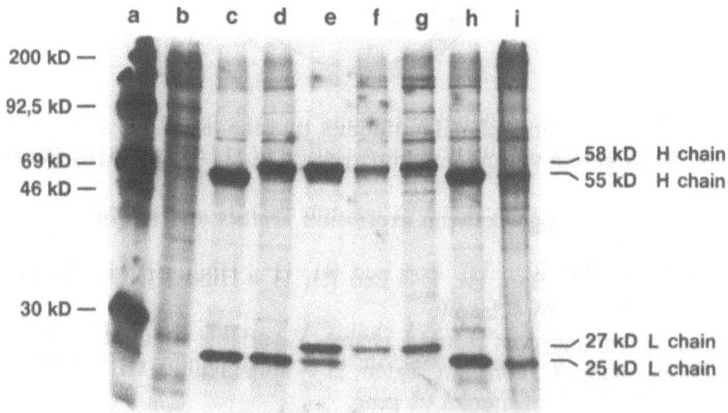

Fig. 1.6.5: Chimerization of monoclonal antibodies by gene targeting.
C: Analysis of secreted antibodies by hybridomas and cell lines derived by gene targeting.
Cells were labeled with [S^{35}]-methionine, supernatants adsorbed to Sepharose A, proteins sized on an 8 % SDS-polyacrylamide gel and labeled proteins visualized by fluorography.
lane a: marker proteins; line b: Ag8X653; lanes c and h: hybridoma cell line; lanes d–g: transfectomas derived by gene targeting.

Three of these clones expressed the murine H chain in dramatic excess compared to the chimerized H chain. This heterogeneous expression pattern was also maintained after subcloning experiments. Five of the clones expressed only chimeric H chain in addition to murine κ chains. One of these cell lines was transfected with the replacement vector for human κ genes, and cell lines resistant to mycophenolic acid and G418 were selected. Three cell lines (frequency 1 in 900) were selected that express exclusively H and L chimeric chains. Antigen binding specificity was maintained. Gel electrophoretic analysis revealed that chimeric H and L chains can be distinguished from their murine counterparts due to different mobility in polyacrylamide gels (Figure 1.6.5c). Expression studies revealed chimeric mAb up to 2 µg/10^6 cells/24 h/ml) in the culture supernatant with some heterogeneity between different clones and definitely below the productivity of the murine hybridoma cell line (10 µg/ml). The surprising finding of chimeric and murine H chain expression in some of the clones after the first step of the gene targeting experiment was further investigated by Southern Blotting experiments. Some of the deduced configurations after integration of the H chain replacement vector are schematically depicted in Figure 1.6.6. Illegitimate recombination should result in clones sterile with respect to chimeric H chain expression (Figure 1.6.6.d), however, assuming that the construct recombines in the vicinity of the cognate Ig H chain locus, chimeric H chain may be expressed due to differential splicing. Homologous recombination (Figure 1.6.6e) should result in clones secreting chimeric H chain, although minor quantities of murine H chain might be expressed due to differential splicing. Branch migration (Adair et al., 1989) with subsequent integration of the construct into random chromosomal locations results in clones with expression of murine H chains and depending on the site of integration varying amounts of chimeric H chain (Figure 1.6.6f).

In summary, gene targeting in hybridoma cells is accompanied by difficulties due to the potential secretion of murine chains in derived cell lines. Putative existence of several active murine loci for H and L chain in hybridomas as well as potential chromosomal heterogeneity contribute to complication of this experimental approach.

Expression of a third chain in addition to those of the cognate Ab may complicate the isolation of the cognate Ab secreted from hybridoma cells due to heterologous chain associations. Since wrong chains do not contribute to antigen specificity, they cause a reduction of the affinity of the Ab and therefore there is a need to eliminate these chains. We have tackled this problem by making use of the positive-negative selection system (Hartman and Mulligan, 1988) based on Gancyclovir-sensitivity of cells expressing herpes simplex virus thymidine kinase. This system imposes a positive selection for recombinants due to homologous recombination (Gancyclovir resistance) and a negative selection for recombinants due to illegitimate recombination (Gancyclovir sensitivity). Incorporating the neo gene sandwiched between sequences corresponding to murine κ chain intron, C region and 3′ flanking sequences followed by a herpes simplex virus thymidine kinase gene on a gene disruption vector, we were able to inactivate a second productively rearranged κ gene in a hybridoma cell line secreting an Ab directed against carcinoembryonic antigen (B. Rammer, personal communication).

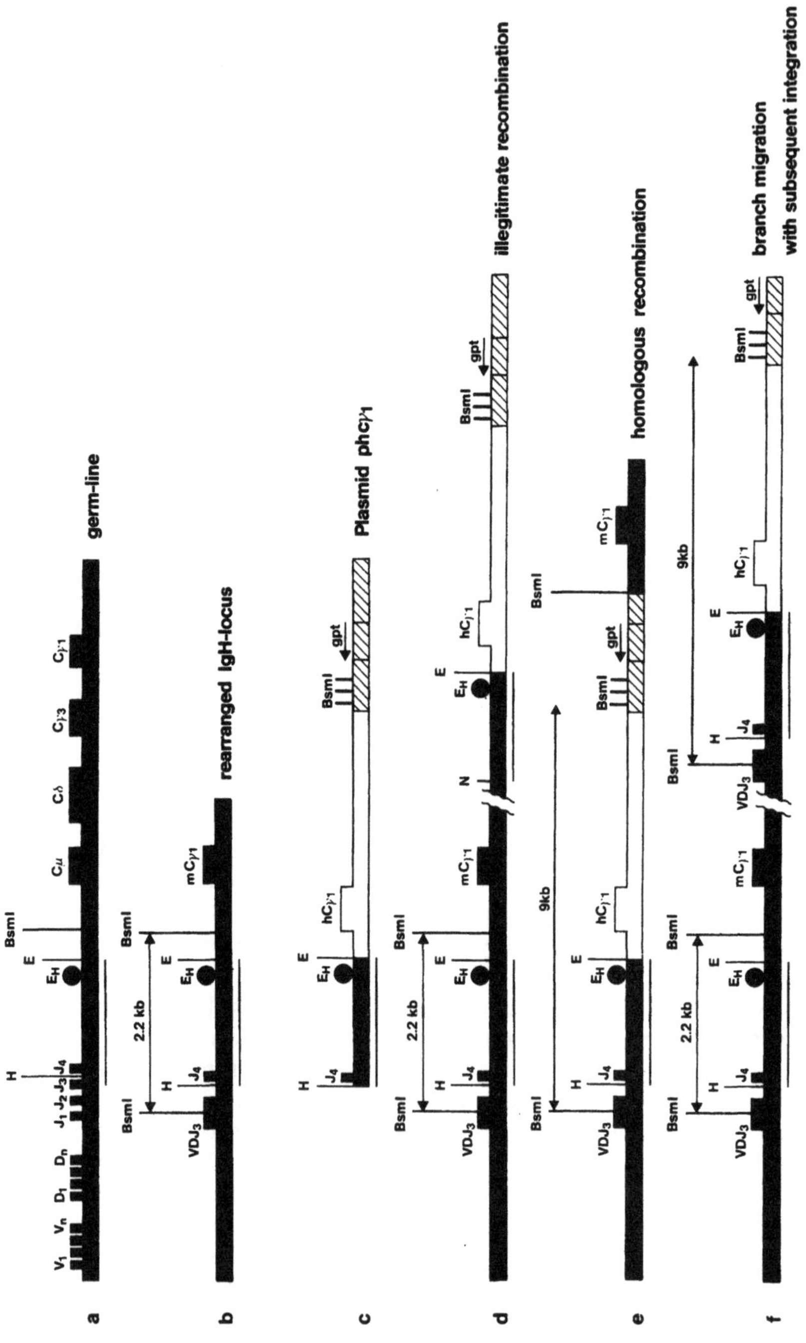
a
germ-line
b
rearranged IgH-locus
c
Plasmid phcγ1
d
illegitimate recombination
e
homologous recombination
f
branch migration
with subsequent integration
BsmI
2.2 kb
9kb
gpt
VDJ3
mCγ1
hCγ1
Cμ
Cδ
Cγ3
Cγ1

1.6.5 Chimeric Antibodies

Murine C regions are exchanged by any human C region of choice, thus transferring defined effector functions to the chimerized molecule. Effector functions to be considered are complement fixation and antibody-dependent cellular cytotoxicity (ADCC). With respect to ADCC, chimeric mAbs seem to be more potent than their murine counterparts, IgG1 being superior to other subclasses (Brüggemann et al., 1987; Steplewski et al., 1988; Shaw et al., 1988; Junghans et al., 1990). The murine version (IgG1) of a mAb directed against the α chain of the human IL2R was scored negative with respect to ADCC, while the chimeric version (IgG1) effectively mediated ADCC. Generally much fewer effector cells (up to 100-fold) seem to be required by chimeric (IgG1) mAbs to mediate ADCC. This phenomenon is mediated by the interaction of the mAb with FcγRI binding sites on effector cells, the CH2 region being required for interaction with the receptors (Duncan et al., 1988; Canfield and Morrison, 1991).

Complement activation is the prerequisite for lysis of target cells by a mechanism called complement-dependent cytotoxicity (CDC). Complement can activate phagocytes including macrophages and neutrophils due to interaction of activation fragments with receptors on these cells and induce cytolysis due to formation of membrane-attack complexes (MAC) by complement activation fragments. Fixation of human complement is isotype dependent. The human IgG subtypes differ in their ability to activate human complement. IgG1 and IgG3 being most effective, IgG2 much less active and IgG4 unable to activate complement (Brüggemann et al., 1987; Tao and Morrison, 1989). Aggregation mediates complement activation by IgG leading to multiple C1q binding sites and therefore IgM is frequently more efficient than other isotypes with respect to complement activation (Shulman et al., 1987).

◁

Fig. 1.6.6: Modes of integration of heavy chain replacement vector into the genome of a hybridoma cell line.
a) Germ-line configuration of murine IgH locus;
b) rearranged murine IgH locus;
c) replacement vector phcγ1;
d) illegitimate recombination;
e) homologous recombination;
f) branch migration with subsequent integration.
V, variable region; J, joining region; D, diversity region, C, constant region; gpt, E coli gene encoding xanthine-guanine phosphoribosyl-transferase; E_H, enhancer of H chain; E = EcoRI, H = HindIII, N = NotI. Underlinded are murine sequences present on the replacement vector.
■ murine IgH chain locus sequences
□ human constant region γ1 sequences
▨ Eco gpt sequences

We did not observe any significant CDC of an anti-IL2R mAb (directed against the α-chain) mediated by human complement when we compared the murine (IgG1) with the chimeric (IgG1) mAb with respect to lysis of a human neoplastic bone marrow cell line expressing the Tac protein of the IL2R (Kaluza et al., 1991; Diehl et al., 1982). Although there have been cases reported in which murine (Hale et al., 1983) and chimeric Ab (Riechmann, et al., 1988) mediated lysis of human nucleated cells in the presence of human complement, in most of the cases reported, CDC is inhibited by homologous restriction factors. These factors inhibit different steps of the homologous, but not of the heterologous complement activation cascade (Davies et al., 1989). The properties of antigen recognized, rather than the isotype of the Ab used, seem to be important for the ability to lyse cells with homologous complement: rat Abs directed against the pan-lymphocyte antigen Campath 1 and against MHC class I molecules are able to lyse human lymphocytes in the presence of human complement, while Abs of the same isotype directed against lymphocyte antigens CD45, CD3, CD7 or CD8 did not mediate lysis (Bindon et al., 1985). Campath 1 showed this property also as a chimeric Ab (Riechmann et al., 1988) while chimeric and humanized Ab of the same isotype (human IgG1) directed against Tac were CDC-negative with human complement (Junghans et al., 1990). Combining two chimeric Abs directed against different epitopes on the human CD4 molecule which are negative with respect to complement-activation individually, results in pronounced complement activation (Weissenhorn et al., 1992), easily explained by facilitated C1q binding due to aggregation.

Chimeric mAbs have great potential in the treatment of cancer, autoimmune diseases and prevention of allograft rejection. Chimerized Abs directed against a variety of tumor antigens are currently being evaluated in clinical trials: Abs directed against ganglioside GD2 present on tumors of neuroectodermal origin (Mueller et al., 1990; Barker et al., 1991); 17-1A, an antigen associated with gastrointestinal cancer (Shaw et al., 1987; Sun et al., 1987); human P glycoprotein (Hamada et al., 1990); B72.3, an antigen expressed on human carcinomas (Colcher et al., 1989; Hutzell et al., 1991); B6.2, also an antigen present on human carcinoma cells (Brown et al., 1987; Sahagan et al., 1986) and carcinoembryonic antigen (Neumaier et al., 1990; Koga et al., 1990).

Antibodies directed against the T cell surface antigen CD4 have the potential to induce tolerance in animal models (Waldmann, 1989a). Administration of murine and chimeric mAbs directed against CD4 have led to significant, albeit transient, improvement of pathological conditions such as rheumatoid arthritis, psoriasis and systematic vaculitis (Herzog et al., 1989; Reiter et al., 1991; Mathieson et al., 1990).

Another target of specific immune intervention is the α-chain of the IL2R (Tac protein) whose expression is induced in activated T cells forming the high-affinity IL2R (Waldmann, 1989b). Furthermore, certain malignancies of lymphatic origin display high levels of Tac antigen (Waldmann, 1989b). In animal models, mAbs directed against the α-chain of the IL2R have already been proven to be powerful and selective immunosuppressants to combat autoimmune diseases and allograft rejections (Diamantstein et al., 1989). In larger clinical trials IL2R Tac-targeted therapy using murine Abs has been shown to be useful in reducing rejection episodes after kidney transplantation (Carpenter et al., 1989; Soulillou et al., 1990) and there is evidence for their efficacy in graft-vs-host disease (Hervé et al., 1990). However, the clinical results are still disappointing compared to animal models and may be improved by making use of possibly more powerful engineered Abs. Combinations of mAbs directed against the α and the β-chain of the IL2R result in a high synergistic blocking effect on IL2-high-affinity binding and IL2-induced cell proliferation (Soulillou et al., 1990) and therefore may represent new tools for immuno-suppression.

In the first clinical trial, a chimeric mAb directed against a gastrointestinal antigen, an Ab response against the chimeric mAb was registered in only one of ten patients with metastatic colon cancer (Lo Buglio et al., 1989). Further possibilities are immunogenicity of the idiotypic element and of allotypes present on IgG molecules. Chimerization also may lead to improved pharmacokinetic properties of mAbs. Longer survival in humans was noted for chimeric mAbs compared to their murine equivalents (Lo Buglio et al., 1989). These properties may lead to enhanced biological activity and may be advantageous when the mAb is conjugated with a radionuclide for imaging or therapy of cancer (Goldenberg, 1989; 1991).

1.6.6 Humanized Antibodies

One approach to minimize antigenicity of the V region is humanization or CDR grafting of mAbs. This approach is based on the conserved structure of L and H chains of mAbs, with the FR regions acting as scaffolds for three loops on each chain (CDRs) which determine antigen specificity. In humanized mAbs, CDR regions of human L and H chains are exchanged against those of a rodent mAb. In order to maintain the affinity of the original mAb, choice and possibly modification of the FR is of primary importance. The first clinically relevant mAb to be humanized was CAMPATH 1 (Riechmann et al., 1988). CAMPATH 1 is an antigen present on almost all human lymphocytes and monocytes. The CDRs of a rat IgG2a mAb directed against this antigen were transplated into human acceptor sites. For restoration of affinity, an amino acid substitution was necessary (Riechmann et al., 1988). The humanized mAb induced remission of non-Hodgkin's lymphoma (Hale et al., 1988) and may be useful for bone marrow purging (Cobbold et al., 1990).

Generation of humanized mAbs can be performed by using site-directed mutagenesis on single-stranded DNA (Jones et al., 1986); or by constructing the whole V region using overlapping oligos incorporating the rodent CDR into a human framework (Queen et al., 1989). Both of those techniques yield a lower percentage of correct product and require the synthesis of long oligonucleotides. However, the advent of PCR technology has lead to approaches for rapid generation of humanized mAbs (Lewis and Crowe, 1991).

In the meantime a number of mAbs have been humanized and their properties evaluated in various *in vitro* and *in vivo* systems. Reshaping of the anti-Tac mAb, which is known to bind to the α-chain (p55) of human IL2R, resulted in a humanized mAb with about one-third of the affinity of murine anti-Tac (Queen et al., 1989). This humanized mAb exhibits new features for immunotherapy of malignant and immune disorders (Junghans et al.,1990) and prolongs cardiac survival in cynomologus monkeys (Brown et al., 1991) without toxic side effects and therefore may be of value as an adjunct to standard immunosuppressive therapy in humans. Reshaping of a murine mAb directed against human CD4 by transplanting the CDR on human H chains derived from myeloma proteins KOL or NEW resulted in two humanized mAbs, whereby the avidity of the KOL-based reshaped mAb was only slightly reduced, whereas that of the NEW-based reshaped mAb was very poor (Gorman et al., 1991).

Another field of application for humanized mAbs is neutralization of pathogenic viruses. Respiratory syncytial virus (RSV) is the major cause of respiratory illness in young children admitted to hospitals and therefore causes one of the major childhood diseases, giving rise to bronchiolitis and pneumonia in children throughout the world (Brandt et al., 1973). A murine mAb directed to RSV was reshaped (Tempest et al., 1991), resulting in loss of affinity for RSV, but additional alterations to one of the FR restored binding affinity and specificity. This reshaped mAb cross-reacted with all clinical isolates of RSV tested and both prevented disease and cured mice even when administered four days after infection. This mAb may prove useful in the management of this major childhood disease.

Herpes simplex virus (HSV) infections range from asymptotic to life threatening. Many of the mAbs directed against major cell surface glycoproteins gB and gD have shown high neutralizing activities *in vitro* and *in vivo*. Two murine mAbs directed against gB and gD glycoproteins were humanized with the aid of computer modeling (Co et al., 1991). The binding, virus neutralization and cell protection results all indicate that both humanized mAbs have retained the binding activities and biological properties of the murine mAbs (Co et al., 1991).

Treatment of human cancer is a major challenge for the therapeutic application of human mAbs. The epidermal growth factor receptor (EGFR) is overexpressed in a variety of malignant tissues, especially in gliomas. Therefore, mAb A425 which binds to the EGFR was reshaped (Kettleborough et al., 1991). FR which might be critical for antigen binding were identified using a molecular model of the V regions of mAb A425. To test the importance of these residues, nine versions of the reshaped human 425 H-chain and two versions of the human 425 L-chain were designed and constructed. The different versions of 425-reshaped human Ab showed a wide range of activities for antigen, indicating that substitutions at certain positions in the FR significantly influenced binding to antigen. One version of the reshaped 425 Ab bound to antigen with an avidity approaching that of the mouse 425Ab. Amplification and overexpression of the tyrosine kinase receptor HER2 is associated with multiple human malignancies and appears to be involved in progression of 25–30 % of human breast and ovarian cancers (Slamon et al., 1987; Slamon et al., 1989). The medium patient survival time is inversely correlated with the extent of amplification (Slamon et al., 1989). Recently, several stimulating ligands of the HER2 protein tyrosine kinase have been identified (Peles et al., 1992; Holmes et al., 1992). A murine mAb, Ab4D5, directed against the extracellular domain of HER2 was developed, which specifically inhibits the growth of tumor cells overexpressing HER2 in monolayers and soft agar (Lupu et al., 1990). A humanized version, hum Ab40-1 was derived which does not block proliferation of the human breast carcinoma cell line SK-BR-3, which overexpresses HER2, despite tight antigen binding (Carter et al., 1992). One of seven additional versions designed by molecular modeling binds HER2 250-fold more tightly (Carter et al., 1992). This variant has the potency to block SK-BR-3 cell proliferation comparable to the murine mAb. This variant also exhibits improved ADCC, but it does not efficiently kill WI-38 cells, which express HER2 at lower levels. Antigen binding thermodynamics and antiproliferative effects of chimeric and humanized anti-$p185^{HER2}$ Ab Fab fragments were extensively investigated by circular dichroism, radioimmunoassay, scanning calorimetry and isothermal titration calorimetry (Kelley et al., 1992).

1.6.7 Bifunctional Antibodies

Antibodies with dual Ag specificities are potentially valuable molecules in immunotherapy, because they can cross-link cytotoxic T cells to effector cells thus triggering lethal destruction of the unwanted cells (Segal and Snider, 1989). Targeted antigens include the α, β, γ and δ chains of the T cell receptor on cytotoxic T cells and the FcγRIII receptor (CD16) on NK cells (Gililand et al., 1988; Lanzevecchia and Scheidegger, 1987; Staerz and Bevan, 1986; Berg et al., 1991). To be effective, the antigen on the effector cells must be able to trigger signal transduction for cytotoxicity after binding of the Ab. Recently bispecific Abs were used to treat ovarian carcinoma in the nude mouse (Garrido et al., 1990). Peripheral blood lymphocytes from patients were incubated overnight with ILII, treated with heteroconjugates containing anti-CD3 cross-linked to an antitumor Ab and then injected intraperitoneally into tumor-bearing mice with the result of dramatic tumor regression. Bifunctional Abs have been prepared by chemical cross-linking, by disulfide exchanges, by establishment of fused hybridomas (quadromas), or by transfecting cloned Ab genes into a hybridoma cell line (Lenz and Weidle, 1990). Heterogeneous products and purification of the bispecific Ab from many other Ab products due to random chain combinations are major problems to be solved. Leucine zipper sequences derived from the transcription factors *fos* and *jun* were used recently to promote heterodimer formation to ensure efficient production of $F(ab')_2$, bispecific Ab directed against p55 of the IL2 receptor and CD3 on cytotoxic T cells (Kostelny et al., 1992). The Ab was highly effective in recruiting cytotoxic T cells by promoting lysis of IL2R bearing cells.

1.6.8 Antibody Fusion Proteins

Construction of fusion proteins with Ig represents an approach to endow Ig with novel properties. Fusion can be performed within the C region thus maintaining the antigen binding properties or within the V region, eliminating the antigen binding function, but leaving the effector functions of the C region of the Ab.

Fusion proteins between mAbs and DNA polymerase I Klenow activity, part of myc and Staphylococcus aureus nuclease, were the first fusion proteins to be functionally investigated (Neuberger et al., 1984; Williams and Neuberger, 1986). IL2 is a lymphokine with the potential to activate immune reative cells, to increase vascular permeability and to induce local inflammation at the site of a tumor and is therefore of interest in cancer therapy. The V as well as the C regions of mAbs have been replaced with IL2. An IL2-Ab C region fusion protein was shown to induce proliferation of an IL2-dependent cell line and the lysis of a IL2R positive cell line in the presence of complement (Landolfi, 1991). Ab V region IL2 fusion proteins were shown to bind to both tumor cells and IL2R on activated T cells. Tumor cells coated with the fusion protein were shown to induce T cell proliferation and the fusion protein significantly enhanced destruction of tumor cells by cell-mediated cytotoxicity (Fell et al., 1991).

A major problem in using soluble receptors as therapeutic agents is their short half-life in the circulation. This problem can be solved by fusing the extracellular domain of receptors with C regions of Ig molecules. A fusion protein consisting of TNF receptor and the C region of an Ig H chain was shown to be 100–100,000-fold more active as a TNF receptor antagonist than a mAb directed against the TNF receptor (Peppel et al., 1991). The chimeric protein is very stable and circulates with a half-life of 48 h when injected into mice. IL4 was shown to be responsible for prodution of IgE, thus neutralization of IL4 by IL4 receptor-IgG chimeric proteins might ameliorate IgE-mediated allergic conditions (Paul, 1991). IL6 is a multifunctional interleukin molecule and it is suggested to play a role in several diseases, such as B cell lymphoma, multiple myeloma, Kaposi's sarcoma and polyclonal T cell activation associated with autoimmune diseases, such as rheumatoid arthritis (Van Snick, 1990; Hirano et al., 1990). IL6 receptor-Ig fusion proteins might be evaluated clinically in the near future as receptor antagonists in the indicated disease conditions. *In vivo* administration of soluble IL1 receptor was shown to delay transplant rejection in mice receiving allografts (Fanslow et al., 1990). Furthermore, experimental allergic encephalomyelitis was ameliorated by administration of soluble IL1 receptor (Hannum et al., 1990). Fusion proteins with improved half-life might exhibit improved therapeutic chacteristics.

Soluble CD4 is an antagonist of gp120 mediated infection of T lymphocytes, macrophages and other cells by HIV (Smith et al., 1987). In order to improve the very short half-life of soluble CD4 and to endow it with improved biological functions, CD4 has been joined to Ig C regions (Capon et al., 1989). A fusion protein obtained by linkage of four extracellular Ig-like domains of CD4 to a human H chain IgG1 gene at the hinge region was able to bind to HIV particles and mediated the lysis to HIV infected cells by a complement-dependent mechanism (Gregersen et al., 1990).

CD44H is the principle cell surface receptor for hyaluronate, which is a major glycosaminoglycan of the extracellular matrix. CD44H expression is enhanced in a variety of malignant tumors and correlates with tumor aggressiveness, supporting the notion that interaction between CD44H and hyaluronate may play an important role in tumor growth and dissemination. It was shown that *in vivo* tumor formation of human Namalva cells, stably transfected with CD44H, can be suppressed by a soluble human CD44-Ig fusion protein, indicating that disruption of the interaction between CD44H and its physiologic ligand may provide a novel strategy for controlling tumor growth in vivo (Sy et al., 1992).

References

Adair, G.M., Nairn, R.S., Wilson, J.H., Seidman, M.M., Brotherman, K.A., McKinnon, C. and Scheerer, J.B. (1989) Targeted homologous recombination at the endogenous adenine phosphoribosyltransferase locus in Chinese hamser cells. Proc. Natl. Acad. Sci. USA 86, 4574–4578.

Audrain, M., Boeffard, F., Soulillou, J.P. and Jacques, Y. (1991) Synergistic action of monoclonal antibodies directed against p55 and p75 chains of the human IL2-receptor. J. Immunol. 146, 884–892.

Baker, M.D. and Shulman, M.J. (1988) Homologous recombination between transferred and chromosomal immunoglobulin k genes. Mol. Cell. Biol. 8, 4041–4047.

Baker, M.D., Pennell, N., Bosnoyan, L. and Shulman, M.J. (1988) Homologous recombination can restore normal immunoglobulin production in a mutant hybridoma cell line. Proc. Natl. Acad. Sci. USA 85, 6432–6436.

Barker, E., Mueller, B.M., Handgretinger, R., Herter, M., Yu, A.L. and Reisfeld, R.A. (1991) Effect of a chimeric anti-ganglioside GD2 antibody on cell-mediated lysis of human neuroblastoma cells. Cancer Res. 51, 144–149.

Bebbington, C.R., Renner, G., Thomson, S., King, D., Abrams, D. and Yarranton, G.T. (1992) High level expression of a recombinant antibody from myeloma cells using a glutamine-synthetase gene as an amplifiable marker. Bio/Technology 10, 169–175.

Bedzyk, W.D., Weidner, K.M., Denzin, L.K., Johnson, L.S., Hardman, K.D., Pantoliano, M.W., Asel, E.D. and Voss, E.W. jr. (1990) Immunological and structural characterization of a high affinity anti-fluorescein single-chain antibody. J. Biol. Chem. 265, 18615–18620.

Berg, J., Lötscher, E., Steimer, K.S., Baesinger, J., Jäck, H.M. and Wabl, M. (1991) Bispecific antibodies that mediate killing of a cell infected with immunodeficiency virus of any strain. Proc. Natl. Acad. Sci. USA 88, 4273–4277.

Better, M., Chang, C.P., Robinson, R.R. and Horwitz, A.H. (1988) Escherichia coli secretion of an active chimeric antibody fragment. Science 240, 1041–1043.

Bich-Thuy, L. and Queen, C. (1989) An enhancer associated with the mouse immunoglobulin $\lambda 1$ gene is specific for λ light chain producing cells. Nucl. Acid Res. 17, 5307–5322.

Bindon, C.I., Hale, G., Clark, M. and Waldmann, H. (1985) Therapeutic potential of monoclonal antibodies to the leucocyte-common antigen. Transplantation 40, 538–544.

Biocca, S., Neuberger, M.S. and Cattaneo, A. (1990) Expression and targeting of intracellular antibodies in mammalian cells. EMBO J. 9, 101–108.

Bird, R.E., Hardman, K.D., Jacobsen, J.W., Johnson, S., Kaufman, B.M., Lee, S.M., Lee, T., Pope, S.H., Riordan, G.S. and Whitlow, M. (1988) Single-chain antigen binding proteins. Science 242, 423–426.

Blasquez, V.C., Xu, M., Moses, S.C. and Garrard, W. (1989) Immunoglobulin k gene expression after stable integration. I. Role of the intronic MAR and enhancer in plasmacytoma cells. J. Biol. Chem. 264, 21183–21189.

Brandt, C.D., Kim, H.W., Orrobio, J.O., Jeffries, B.C., Wood, S.C., Chanock, R.M. and Parrott, R.H. (1973) Epidemiology of respiratory syncytical virus infection in Washington DC III. Composite analysis of eleven consecutive yearly epidemies. Am. J. Epidemiol. 98, 355–364.

Brown, B.A., Davis, G.L., Saltzgaber-Muller, J., Simon, P., Ho, M.K., Shaw, P.S., Stone, B.A., Sands, H. and Moore, G.P. (1987) Tumor-specific genetically engineered murine-human chimeric monoclonal antibody. Cancer Res. 47, 3577–3583.

Brown, P.S. jr., Parenteau, G.L., Dirbas, F.M., Garsia, R.J., Goldman, C.K., Bukowski, M.A., Junghans, R.P., Queen, C., Hakimi, J., Benjamin, W.R., Clark, R.E. and Waldmann, T.A. (1991) Anti-Tac-H, a humanized antibody to the interleukin 2 receptor, prolongs primate cardiac allograft survival. Proc. Natl. Acad. Sci. USA 88, 2663–2667.

Brüggemann, M., Williams, G.T., Bindon, C.T., Clark, M.R., Walker, M.R., Jefferis, R., Waldmann, H. and Neuberger, M.S. (1987) Comparison of the effector functions of human immunoglobulins using a matched set of chimeric antibodies. J. Exp. Med. 166, 1351–1361.

Canfield, S.M. and Morrison, S.L. (1991) The binding affinity of human IgG for its high affinity Fc receptor is determined by multiple amino acids in the CH2 domain and is modulated by the hinge region. J. Exp. Med. 173, 1483–1491.

Capecchi, M.R. (1989a) Altering the genome by homologous recombination. Science 244, 1288–1292.

Capecchi, M.R. (1989b) The new mouse genetics: altering the genome by gene targeting. Trends in Genet. 5, 70–76.

Capon, D.J., Chamow, S.M., Mordenti, J., Marsters, S.A., Gregory, T., Matsuya, H., Byrn, R.A., Lucas, C., Wurm, F.M. and Groopman, J.E. (1989) Designing CD4 immunoadhesins for AIDS therapy. Nature 337, 525–531.

Carpenter, C.B., Kirkman, R.L., Shapiro, M.E., Mildford, E.L., Tilney, N.L., Waldmann, T.A., Zimmermann. C.E., Ramos, E.L. and Storm, T.B. (1989) Prophylactic use of monoclonal anti-IL2-receptor antibody in cadaveric renal transplantation. Am. J. Kidney Dis. 14, Suppl. 2, 54–57.

Carter, P., Presta, L., Gorman, C.L., Ridgway, J.B.B., Henner, D., Wong, W.L.T., Rowland, A.M., Kotts, C., Carver, M.E. and Shepard, H.M. (1992) Humanization of an anti-p185HER2 antibody for human cancer therapy. Proc. Natl. Acad. Sci. USA 89, 4285–4289.

Cattaneo, A. and Neuberger, M. (1987) Polymeric immunoglobulin M is secreted by transfectants of non-lymphoid cells in the absence of immunoglobulin J chain. EMBO J. 9, 2753–2758.

Chaudhary, V.K., Batra, J.K., Gaib, M.G., Willingham, M.C., Fitzgerald, D.J. and Pastan, I. (1990) A rapid method of cloning functional variable regions of antibody genes in Escherichia coli as single chain immunotoxins. Proc. Natl. Acad. Sci. USA 87, 1066–1070.

Chaudhary, V.K., Queen, C., Junghans, R.P., Waldmann, T.A., Fitzgerald, D. J. and Pastan, I. (1989) A recombinant immunotoxin consisting of two antibody variable domains fused to Pseudomonas exotoxin. Nature 339, 393–397.

Chotia, C., Lesk, A.M., Tramontano, A., Levitt, M., Smith-Gill, S.J., Air, G., Sheriff, S., Padlan, E.A., Davies, D., Tulip, W.R., Colman, P.M., Spinelli, S., Alzan, P.M. and Poljak, R.J. (1989) Conformations of immunoglobulin hypervariable regions. Nature 342, 877–883.

Co, M.S., Deschamps, M., Whitley, R.J. and Queen, C. (1991) Humanized antibodies for antiviral therapy. Proc. Natl. Acad. Sci. USA 88, 2869–2873.

Cobbold, S.P., Hale, G., Clark, M.R. and Waldmann, H., (1990) Purging in auto-and allografts: monoclonal antibodies which use human complement and other natural effector mechanisms. Prog. Clin. Biol. Res. 333, 139–151.

Cockett, M.I., Bebbington, C.R. and Yarranton, G.T. (1990) High level expression of tissue inhibitor of metalloproteinase in Chinese hamster ovary cells using glutamine synthetase gene amplification. Biotechnology 8, 662–667.

Colcher, D., Milenic, D., Roselli, M., Raubitschek, A., Yarranton, G., King, D., Adair, J., Whittle, N., Bodmer, M. and Schlom, J. (1989) Characterization and biodistribution of recombinant and recombinant/chimeric constructs of monoclonal antibody B72.3. Cancer Res. 49, 1738–1745.

Davies, A., Simmons, D.L., Hale, G., Harrison, R.A., Tighe, H., Lachmann, P.J. and Waldmann, H. (1989) CD 59, an LY 6-like protein expressed in human lymphoid cells, regulates the action of the complement membrane attack complex on homologous cells. J. Exp. Med. 170, 637–654.

Diamantstein, T., Volk, H.-D., Hahn, H.J., Tilney, N.L. and Kupiec-Weglinski, J.W., (1989) Interleukin 2 receptor targeted immunotherapy-update and attempt at synthesis. Prog. in Immunol. 1, 619–626.

Diehl, V., Kirchner, H.H., Burrichter, H., Stein, H., Fonatsch, C., Gerdes, J., Schaadt, M., Heit, W., Vehanska-Ziegler, B., Ziegler, A., Heinz, F. and Sueno, K. (1982) Characteristics of Hodgkin's disease-derived cell lines. Cancer Treatm. Rep. 66, 615–632.
Dorai, H. and Moore, G.P. (1987) The effect of dihydrofolate reductase-mediated gene amplification on the expression of transfected immunoglobulin genes. J. Immunol. 139, 4232–4341.
Duncan, A.R., Wood, J.M., Partridge, L.J., Burton, D.R. and Winter, G. (1988) Localization of the binding site for the human high-affinity Fc receptor in IgG. Nature 332, 563–564.
Fanslow, W.C., Sims, J.E., Sassenfeld, H., Morrisey, P.J., Gillies, S., Dower, S.K. and Widmer, M.B. (1990) Regulation of alloreactivity in vivo by a soluble form of the interleukin-1 receptor. Science 248, 739–742.
Fell, H.P., Gayle, M.A., Grosmaire, L. and Ledbetter, J.A. (1991) Genetic construction and characterization of a fusion protein consisting of a chimeric F(ab') with specificity for carcinomas and human IL2. J. Immunol. 146, 2446–2452.
Fell, H.S., Yarnold, S., Hellström, I., Hellström, K.E. and Folger, K.R. (1989) Homologous recombination in hybridoma cells: heavy chain chimeric antibody produced by gene targeting. Proc. Natl. Acad. Sci. USA 86, 8507–8511.
Field, R.P., Brand, H., Renner, G.L., Robertson, H.A. and Boraston, R. (1992) Production of a chimeric antibody for tumor imaging and therapy from Chinese hamster ovary (CHO) and myeloma cells. Prod. Biol. Anim. Cells Cult (ESCAT 10 Meet., 742–744).
Forrester, W.C., Epner, E., Driscoll, M.C., Enver, T., Brice, M., Papayannopoulou, T. and Groudine, M. (1990) A deletion of the human β-globin locus activation domain causes a major alteration in chromatin structure and replication across the entire β-globin locus. Genes Dev. 4, 1637–1649.
Garrido, M.A., Valdayo, M.J., Winkler, D.F., Titus, J.A., Hecht, T.T., Perez, P., Segal, D. M. and Wunderlich, J.R. (1990) Targeting human T-lymphocytes with bispecific antibodies to react against human ovarian carcinoma cells growing in nu/nu mice. Cancer Res. 50, 4227–4232.
Gililand, L.K., Clark, M.R. and Waldmann, H. (1988) Universal bispecific antibody for targeting tumor cells for destruction by cytotoxic T cells. Proc. Natl. Acad. Sci. USA 85, 7719–7723.
Gillies, S.D., Lo, K.-M. and Wesolowski, J. (1989) High-level expression of chimeric antibodies using adapted cDNA variable region cassettes. J. Immunol. Meth. 125, 191–202.
Goldenberg, D.M. (1989) Future role of radiolabeled monoclonal antibodies in oncological diagnosis and therapy. Sem. Nucl. Med. 19, 332–339.
Goldenberg, M.D. (1991) Challenges to the therapy of cancer with monoclonal antibodies. J. Natl. Cancer Inst. 83, 78–79.
Gorman, S.D., Clark, M.R., Routledge, E.G., Cobbold, S.P. and Waldmann, H. (1991) Reshaping a therapeutic CD4 antibody. Proc. Natl. Acad. Sci. USA 88, 4181–4185.
Gregersen, J.P., Mehdi, S., Gelderblom, H. and Zettlmeissl, G. (1990) A CD4-immunoglobulin fusion protein with antiviral effects against HIV. Arch. Virol. 111, 29–43.
Hagmann, J., Rudin, C.M., Harsch, D., Chaplin, D. and Storb, U. (1990) A novel enhancer in the immunoglobulin l locus is duplicated and functionally independent for NFkB. Genes Dev. 4, 978–992.
Hale, G., Bright, S., Chumbley, G., Hoang, T., Metcalf, D., Munro, A.J. and Waldmann, H. (1983) Removal of T cells from bone marrow for transplantation: a monoclonal antilymphocyte antibody that fixes human complement. Blood 62, 873–882.
Hale, G., Dyer, H.J., Clark, M.R., Philipps, J.M., Marcus, R., Riechmann, L., Winter, G. and Waldmann, H. (1988) Remission induction in non-Hodgkin lymphoma with reshaped human monoclonal antibody CAMPATH-1H. Lancet 2, 1394–1399.

Hamada, H., Miura, K., Ariyoshi, K., Heika, Y., Sato, S., Kameyama, K., Kurosawa, Y. and Tsuruo, T. (1990) Mouse-human chimeric antibody against the multidrug transporter P-glycoprotein. Cancer Res. 50, 3167–3171.
Hannum, C. H., Wilcox, C. J., Arend, W. P., Joslin, F. G., Dripps, D. J., Heimdal, P. L., Armes, L. G., Sommer, A., Eisenberg, S. P. and Thompson, R. C. (1990) Interleukin-1 receptor antagonist activity of a human interleukin-1 inhibitor. Nature 343, 336–340.
Harlow, E. and Lane, D. (1988) Antibodies: a laboratory manual, p 271. Cold Spring Harbor Laboratory, Cold Spring Harbor, NY.
Hartman, S. C. and Mulligan, R. C. (1988) Two dominant-acting selectable markers for gene transfer studies in mammalian cells. Proc. Natl. Acad. Sci. USA 85, 8047–8051.
Hendricks, M. B., Luchette, C. A. and Banker, M. J. (1989) Enhanced expression of an immunoglobulin-based vector in myeloma cells mediated by coamplification with a mutant dihydrofolate reductase gene. Bio/Technol. 7, 1271–1274.
Hervé, P., Wijdens, J., Bergerat, J. P., Bordigoni, P., Milpied, N., Cahn, J. Y., Clément, C., Béliard, R., Morel-Fourier, B., Racodot, E., Troussard, X., Benz-Lemoine, E., Gaud, C., Legros, M., Attal, M., Kloft, M. and Peters, A. (1990) Treatment of corticosteroid resistant acute graft-versus-host disease by in vivo administration of anti-Interleukin 2 receptor monoclonal antibody (B-B10). Blood 75, 1017–1023.
Herzog, C., Walker, C., Müller, W., Rieber, P., Reiter, C., Riethmüller, G., Wassmer, P., Stockinger, H., Majdic, O. and Pichler, W. J. (1989) Anti-CD4 antibody treatment of patients with rheumatoid arthritis: effect on clinical course and circulating T-cells. J. Autoimm. 2, 625–642.
Hesemann, C. A. and Capra, D. J. (1990) High level production of a functional immunoglobulin heterodimer in a baculovirus expression system. Proc. Natl. Acad. Sci. USA 87, 3942–3946.
Hirano, T., Akira, S., Taga, T. and Kishimoto, T. (1990) Biological and clinical aspects of interleukin-6. Immunol. Today 11, 443–449.
Holmes, W. E., Sliwkowski, M. X., Aktia, R. W., Henzel, W. J., Lee, J., Park, J. W., Yansura, D., Abadi, N., Raab, H., Lewis, G. D., Shepard, H. M., Kuang, W.-J., Wood, W. I., Goeddel, D. and Vandlen, R. L. (1992) Identification of Heregulin, a specific activator of pl85erbB2. Science 256, 1205–1210.
Honjo, T. (1983) Immunoglobulin genes. Ann. Rev. Immunol. 1, 499–528.
Horwitz, A. H., Chang, C. P., Better, M. and Hellström, A. H. (1988) Secretion of functional antibody and Fab fragment from yeast cells. Proc. Natl. Acad. Sci. USA 85, 8678–8682.
Hutzell, P., Kashmiri, S., Colcher, D., Primus, F. J., Hand, P. H., Roselli, M., Finch, M., Yarranton, G., Bodmer, M., Whittle, N., King, D., Loullis, C. C., McCoy, D. W., Callahan, R. and Schlom, J. (1991) Generation and characterization of a recombinant/chimeric B72.3 (human gamma 1 antibody). Cancer Res. 51, 181–189.
Jones, P. T., Dear, P. H., Foote, J., Neuberger, M. S. and Winter, G. (1986) Replacing the complementarity determining regions in a human antibody with those of a mouse. Nature 321, 522–525.
Junghans, R. P., Waldmann, T. A., Landolfi, N. F., Avdalovic, N. M., Schneider, W. R. and Queen, C. (1990) Anti-Tac-H, a humanized antibody to the interleukin-2 receptor with new features for immunotherapy in malignant and immune disorders. Cancer Res. 50, 1495–1502.
Kaluza, B., Betzl, G., Shao, H., Diamantstein, T. and Weidle, U. H. (1992) A general method for chimerization of monoclonal antibodies by inverse polymerase chain reaction conserving authenic N-terminal sequences. Gene 122, 321–328.
Kaluza, B., Lenz, H., Russmann, E., Hock, H., Rentrop, O., Majdic, O., Knapp, W. and Weidle, U. H. (1991) Synthesis and functional characterization of a recombinant monoclonal antibody directed against the α-chain of the human interleukin-2 receptor. Gene 107, 297–305.

Kearny, J.F., Radbruch, A., Liesegang, B. and Rajewsky, K. (1979) A new mouse myeloma cell line that has lost immunoglobulin expression but permits the construction of antibody-secreting hybrid cell lines. J. Immunol. 123, 1548–1550.

Kelley, R.F., O'Connell, M.P., Presta, L., Eigenbrot, C., Covarrubias, M., Snedecor, B., Bourell, J.H. and Vetterlein, D. (1992) Antigen binding thermodynamics and antiproliferative effects of chimeric and humanized anti-p185HER2 antibody Fab fragments. Biochemistry 31, 5434–5441.

Kettleborough, C.A., Saldhana, H., Heath, V.J., Morrison, C.J. and Bendig, M.M. (1991) Humanization of a mouse monoclonal antibody by CDR-grafting: the importance of framework residues on loop conformation. Prot. Eng. 4, 773–783.

Koga, H., Kanda, H., Nakashima, M., Watanabe, Y., Endo, K. and Watanabe, T. (1990) Mouse-human chimeric monoclonal antibody to carcinoembryonic antigen (CEA): in vitro and in vivo activities. Hybridoma 9, 43–56.

Köhler, G. and Milstein, C.C. (1974) Continuous culture of fused cells secreting antibody of predefined specificity. Nature 256, 495–497.

Kostelny, S.A., Cole, M.S. and Tso, J.Y. (1992) Formation of a bispecific antibody by the use of leucine zippers. J. Immunol. 148, 1547–1553.

Landolfi, N.F. (1991) A chimeric IL2/Ig molecule possesses the functional activity of both proteins. J. Immunol. 146, 915–919.

Lanzevecchia, A. and Scheidegger, D. (1987) The use of hybrid hybridomas to target human cytotoxic T lymphocytes. Eur. J. Immunol. 17, 105–111.

Larrick, J.W., Danielson, L., Brenner C.A., Abrahamson, M., Fry, K.E. and Borrebaeck, C.A.K. (1989b) Rapid cloning of rearranged immunoglobulin genes from human hybridoma cells using mixed primers and the polymerase chain reaction. Biochem. Biophys. Res. Com. 160, 1250–1256.

Larrick, J.W., Danielson, L., Brenner C.A., Wallace, E.F., Abrahamson, M., Fry, K.E. and Borrebaeck, C.A.K. (1989a) Polymerase chain reaction using mixed primers: cloning of human monoclonal antibody variable region genes from single hybridoma cells. Bio/Technology 7, 934–938.

Le Bouef, R.D., Galin, F.S., Hollinger, S.K., Peiper, S.C. and Blalock, J.E. (1989) Cloning and sequencing of immunoglobulin variable region genes using degenerate oligodeoxyribonucleotides and polymerase chain reaction. Gene 82, 371–377.

Lenz, H. and Weidle, U.H. (1990) Expression of heterobispecific antibodies by genes transfected into producer hybridoma cells. Gene 87, 213–218.

Lewis, A.P. and Crowe, J.S. (1991) Immunoglobulin complementarity-determining region grafting by recombinant polymerase chain reaction to generate humanized monoclonal antibodies. Gene 101, 297–302.

Lo Buglio, A.F., Wheeler, R.H., Trang, J., Haynes, A., Rogers, K., Harvey, E.B., Sun, L., Grayeb, J. and Khazaeli, M.B. (1989) Mouse/human chimeric monoclonal antibody in man. Proc. Natl. Acad. Sci. USA 86, 4220–4224.

Lo, D., Pursel, V., Sandgren, E., Behringer, R. Rexroad, C., Palmiter, R.D. and Brinster, R. (1991) Expression of mouse IgA by transgenic mice, pigs and sheep. Eur. J. Immunol. 21, 1001–1006.

Lupu, R., Colomer, R., Zugmaier, G., Sarug, J., Shepard, M., Slamon, D. and Lippman, M. (1990) Direct interaction of a ligand for the erbB2 oncogene product with EGF receptor and pl85erbB2 . Science 249, 1552–1555.

Lyons, A., King, D.J., Owens, R.J., Yarranton, G.T., Millican, A., Whittle, N.R. and Adair, J.R. (1990) Site-specific attachment to recombinant antibodies via introduced surface residues. Protein Engineering 3, 703–708.

Mansour, S., Thomas K. and Capecchi, M. (1988) Disruption of the proto-oncogene int–2 in mouse embryo-derived stem cells: a strategy for targeting mutations to non-selectable genes. Nature 336, 348–352.

Marks, J.D., Hoogenborn, H.R., Bonnert, T.P., McCafferty, J., Griffith, A.D. and Winter, G. (1991) By-passing immunization: human antibodies from V gene libraries displayed on phage. J. Mol. Biol. 222, 581–597.

Mathieson, P.W., Cobbold, S.P., Hale, G., Clerk, M.R., Olivera, D.B.G., Lockwood, C.M. and Waldmann, H. (1990) Monoclonal antibody therapy in systemic vaculitis. N. Engl. J. Med. 323, 250–254.

Meyer, K.B., Sharpe, M.J., Surani, M.A. and Neuberger, M.S. (1990) The importance of the 3'-enhancer region in immunoglobulin k gene expression. Nucl. Acids Res. 18, 5609–5615.

Morrison, S.L. (1992) In Vitro Antibodies: Strategies for production and application. Ann. Rev. Immunol. 10, 239–265.

Morrison, S.L., Johnson, M.J., Herzenberg, L. and Oi, V.T. (1984) Chimeric human antibody molecules with human constant region domains. Proc. Natl. Acad. Sci. USA 81, 6851–6855.

Mueller, B.M., Romerdahl, C.A., Gillies, S.D. and Reisfeld, R.A. (1990) Enhancement of antibody-dependent cytotoxicity with a chimeric anti-CD2 antibody. J. Immunol. 144, 1382–1386.

Mulligan, R. and Berg, P. (1981) Selection for animal cells that express the Escherichia coli gene coding for xanthine-guanine phosphoribosyltransferase. Proc. Natl. Acad. Sci. USA 78, 2072–2076.

Neuberger, M.S. and Williams, G.T. (1988) The intron requirement for immunoglobulin gene expression is dependent upon the promoter. Nucl. Acids Res. 116, 6713–6724.

Neuberger, M.S., Williams, G.T. and Fox, R.O. (1984) Recombinant antibodies possessing novel effector functions. Nature 312, 604–608.

Neumaier, M., Shively, L., Chen, F.S., Gaida, F.J., Ilgen, C., Paxton, R.J., Shively, J.E. and Riggs, A.D. (1990) Cloning of the genes of T84.66, an antibody that has a high-specificity and affinity for carcinoembryonic antigen, and expression of chimeric human/mouse T84.66 genes in myeloma and Chinese hamster ovary cells. Cancer Res. 50, 2128–2134.

Ochi, A., Hawley, R.G., Hawley, T., Shulman, M.J., Traunecker, A., Köhler, G. and Hozumi, N. (1983) Functional immunoglobulin M production after transfection of cloned immunoglobulin heavy chains into lymphoid cells. Proc. Natl. Acad. Sci. USA 80, 6351–6355.

Oi, V.T., Morrison, S., Herzenberg, L.A. and Berg, P. (1983) Immunoglobulin gene expression in transformed lymphoid cells. Proc. Natl. Acad. Sci. USA 80, 825–829.

Orlandi, R., Güssow, D.H., Jones, P.T. and Winter, G. (1989) Cloning immunoglobulin variable domains for expression by the polymerase chain reaction. Proc. Natl. Acad. Sci. USA 86, 3833–3837.

Page, M.J. and Sydenham, M.K. (1991) High level expression of the humanized monoclonal antibody CAMPATH–1H in Chinese hamster ovary cells. Bio/Technology 9, 64–68.

Paul, W.E. (1991) Interleukin-4: A prototypic immuporegulatory lymphokine. Blood 77, 1859–1870.

Peles, E., Bacus, S.S., Koski, R.A., Lu, H.S., Wen, D., Odgen, S.G., Levy, R.B. and Yarden, Y. (1992) Isolation of the Neu/HER–2 stimulatory ligand: a 44 kd glycoprotein that induces differentiation of mammary tumor cells. Cell 69, 205–216.

Peppel, K., Crawford, D.G. and Beutler, B. (1991) A tumor necrosis factor (TNF) receptor-IgG heavy chain chimeric protein as a bivalent antagonist of TNF activity. J. Exp. Med. 174, 1483–1489.

Pettersson, S., Cook, G.P., Bruggemann, M., Williams, G.T. and Neuberger, M.S. (1990) A second B cell specific enhancer 3' of the immunoglobulin heavy-chain locus. Nature 344, 165–168.

Plon, S.E. and Groundine, M. (1991) Unrevalling immunoglobulin expression. Current Biol. 1, 13–14.

Plückthun, A. (1991) Antibody-engineering: advances from the use of Escherichia coli expression Systems. Bio/Technology 9, 545–551.

Queen, C., Schneider, W.P., Selick, H., Payne, P.W., Landolfi, N.F., Duncan, J.F., Avdalovic, N.M., Levitt, M., Junghans, R.P. and Waldmann, T.A. (1989) A humanized antibody that binds to the interleukin 2 receptor. Proc. Natl. Acad. Sci. USA 86, 10029–10033.

Reiter, C., Kakavand, B., Rieber, E.P., Schattenkirchner, M., Riethmüller, G. and Krüger, K. (1991) Treatment of rheumatoid arthritis with monoclonal CD4 antibody M-T151. Clinical results and immunopharmacological effects in an open study, including repeated application. Arthritis and Rheumatism 34, 525–536.

Riechmann, L., Clark, M., Waldman, H. and Winter, G. (1988) Reshaping human antibodies for therapy. Nature 332, 323–327.

Roberts, G., Bebbington, C.R., Renner, G., Gofton, C.M., Thomson, S. and McCormack, M. (1991) Efficient expression of recombinant antibodies in mammalian cells. J. Cell. Bioch., Suppl. 15 E, 122.

Sahagan, B.G., Dorai, H., Saltzgaber-Muller, J., Toneguzzo, F., Giundon, C.A., Lilly, S.P., McDonald, K.W., Morrissey, D.V., Stone, B.A., Davis, G.L., McIntosh, P.K. and Moore, G.P. (1986) A genetically engineered murine/human chimeric antibody retains specificity for human tumor-associated antigen. J. Immunol. 137, 1066–1074.

Sambrook, J., Fritsch, E.F. and Maniatis, T. (1989) Molecular cloning - a laboratory manual. Cold Spring Harbor Laboratory Press, Cold Spring Harbor, New York.

Sastry, L., Alting-Mees, M., Huse, W.D., Short, J.M., Sorge, J.A., Hay, B.N., Janda, K.D., Benkovic, S.J. and Lerner, R.A. (1989) Cloning of the immunological repertoire in Escherichia coli for generation of monoclonal catalytic antibodies: construction of a heavy chain variable region specific cDNA library. Proc. Natl. Acad. Sci USA 86, 5728–5732.

Segal, D.M. and Snider, D.P., (1989) Targeting and activation of cytotoxic lymphocytes. Chem. Immunol. 47, 179–213.

Shaw, D.R., Khazaeli, M.B. and Lo Buglio, A.F. (1988) Mouse/human chimeric antibodies to a tumor-associated antigen: biologic activity of the four human IgG subclasses. J. Natl. Cancer Inst. 80, 1353–1359.

Shaw, D.R., Khazaeli, M.B., Sun, L.K., Ghrayeb, J., Dadonna, P.E., McKinney, S. and Lo Buglio, A.F. (1987) Characterization of a mouse/human chimeric monoclonal antibody (17–1A) to a colon cancer tumor-associated antigen. J. Immunol. 138, 4534–4538.

Shin, S.-U. and Morrison, S.L. (1989) Production and properties of chimeric antibody molecules. Meth. in Enzymol. 178, 459–476.

Shulman, M.J., Collins, C., Pennell, N. and Hozumi, N. (1987) Complement activation by IgM: evidence for the importance of the third constant domain of the mu heavy chain. Eur. J. Immunol. 17, 549–554.

Skerra, A. and Plückthun, A. (1988) Assembly of a functional immunoglobulin Fv fragment in E. coli. Science 240, 1038–1041.

Slamon, D.J., Clark, G.M., Wong, S.G., Levin, W.J., Ullrich, A. and McGuire, R.L. (1987) Human breast cancer: correlation of relapse and survival with amplification of the HER–2/Neu oncogene. Science 235, 177–182.

Slamon, D.J., Godolphin, W., Jones, L.A., Hoff, J.A., Wong, S.C., Keith, D.E., Lewin, W.J., Stuart, S.G., Udore, J., Ullrich, A. and Press, M.F. (1989) Studies of the HER–2/neu protooncogene in human breast and ovarian cancer. Science 244, 707–712.

Smith, D.H., Byrn, R.A., Marsters, S.A., Gregory, T., Groopman, J.E. and Capon, D.J. (1987) Blocking of HIV–1 infectivity by a soluble, secreted form of the CD4 antigen. Science 238, 1704–1707.

Soulillou, J.P., Cantarovich, D., Le Mauff, B., Giral, M., Robillard, N., Hourmant, M., Hirn, M. and Jaques, Y. (1990) Randomized controlled trial of a monoclonal antibody against the interleukin-2 receptor (33B3.1) as compared with rabbit antilymphocyte globulin for prophylaxis against rejection of allografts. New Engl. J. Med. 322, 1175–1182.

Southern, P. and Berg, P. (1982) Transformation of mammalian cells to antibiotic resistance with a bacterial gene under the control of the SV 40 early region promoter. J. Mol. Appl. Genet. 1, 327–341.

Staerz, U.D. and Bevan M.J. (1986) Use of anti-receptor antibodies to focus T cell activity. Immunol. Today 7, 241–245.

Steplewski, Z., Sun, L.K., Shearmann, C.W., Grayeb, J., Dadonna, P.C. and Koprowski, H. (1988) Biological activity of human-mouse IgGl, IgG2, IgG3 and IgG4 chimeric monoclonal antibodies with antitumor specificity. Proc. Natl. Acad. Sci. USA 85, 4852–4856.

Sun, L.K., Curtis, P., Rakowicz-Szulczynska, E., Grayeb, J., Chang, N., Morrison, S.L. and Koprowski, H. (1987) Chimeric antibody with human constant regions and mouse variable regions directed against carcinoma-associated antigen 17–1A. Proc. Natl. Acad. Sci. USA 84, 214–218.

Sy, M.-S., Guo, Y.-J. and Stamenkovic, I. (1992) Inhibition of tumor growth in vivo with a soluble CD44-immunoglobulin fusion protein. J. Exp. Med. 176, 623–627.

Tao, M.H. and Morrison, S.L. (1989) Studies of aglycosylated chimeric mouse-human IgG. Role of carbohydrate in the structure and effector functions mediated by the human IgG constant region. J. Immunol. 143, 2595–2601.

Tempest, P.R., Bremner, P., Lambert, M., Taylor, G., Furze, J.M., Carr, F.J. and Harris, W.J. (1991) Reshaping a monoclonal antibody to inhibit human respiratory syncytial virus infection in vivo. Bio/Technology 9, 266–271.

Townes, T.M. and Behringer, R.R. (1990) Trends in Genet. 6, 219–233.

Traunecker, A., Oliveri, F. and Karjalainen, K. (1991) Myeloma based expression system for production of large mammalian proteins. TIBTECH 9, 109–113.

Triglia, T., Peterson, M.G. and Kemp, D.J. (1988) Procedure for in vitro amplification of DNA segments that lie outside the boundaries of known sequences. Nucl. Acids. Res. 16, 8186.

Van Snick, J. (1990) Interleukin-6: An overview. Ann. Rev. Immunol. 8, 253–278.

Waldmann, H. (1989a) Manipulation of T-cell responses with monoclonal antibodies. Ann. Rev. Immunol. 7, 407–444.

Waldmann, T.A. (1989b) Multichain interleukin 2 receptor: a target for immunotherapy in lymphoma. J. Natl. Cancer Inst. 81, 914–923.

Waldmann, T.A. (1992) Immune receptors: targets for therapy of leukemia/lymphoma, autoimmune diseases and for the prevention of allograft rejection. Ann. Rev. Immunol. 10, 675–704.

Ward, E.S., Güssow, D., Griffiths, A.D., Jones, R.T. and Winter, G. (1989) Binding activities of a repertoire of single immunoglobulin variable domains secreted from E. coli. Nature 341, 544–546.

Weidle, U.H. and Buckel, P. (1987) Establishment of stable mouse myeloma cells constitutively secreting human tissue-type plasminogen activator. Gene 57, 131–141.

Weidle, U.H., Borgya, A., Mattes, R., Lenz, H. and Buckel, P. (1987) Reconstitution of functionally active antibody directed against creatine-kinase from separately expressed heavy and light chains in non-lymphoid cells. Gene 51, 21–29.

Weidle, U.H., Koch, S. and Buckel, P. (1987) Expression of antibody c-DNA in murine myeloma cells: possible involvement of additional regulatory elements in transcription of immunoglobulin genes. Gene 60, 205–216.

Weidle, U.H., Lenz, H. and Brem, G. (1991) Genes encoding a mouse monoclonal antibody are expressed in transgenic mice, rabbits and pigs. Gene 28, 185–191.

Weissenhorn, W., Scheuer, W., Kaluza, B., Schwirzke, M., Reiter, C., Flieger, D., Lenz, H., Weiss, E.H., Rieber, E.P., Riethmüller, G. and Weidle, U.H. (1992) Combinatorial functions of two chimeric antibodies directed to human CD4 and of one antibody directed to the α-chain of the human interleukin-2 receptor. Gene 121, 271–278.

Weissenhorn, W., Weiss, E., Schwirzke, M., Kaluza, B. and Weidle, U.H. (1991) Chimerization of antibodies by isolation of rearranged genomic variable regions by the polymerase chain reaction. Gene 106, 273–277.

Williams, G.T. and Neuberger, M.S. (1986) Production of antibody-tagged enzymes by myeloma cells: application to DNA polymerase I Klenow fragment. Gene 43, 319–324.

Winter, G. and Milstein, C. (1991) Man-made antibodies. Nature 349, 293–299.

zu Putlitz, J., Kubasek, W.L., Duchêne, M., Marget, M., von Specht, B.U. and Domdey, H. (1990) Antibody production in baculovirus-infected insect cells. Bio/Technology 8, 651–654.

Zwickl, M., Zaninetta, D., McMaster, G.K. and Hardman, N. (1990) Selective cloning of B cell hybridoma specific rearranged immunoglobulin gene loci using the polymerase chain reaction. J. Immunol. Meth. 130, 49–55.

1.7 Safety Evaluation of Products Derived From Mammalian Cell Lines

David Onions

1.7.1 Introduction

The rapid development of recombinant DNA technology over the last decade has spawned a new industry that is likely to have as profound effect on our lives as the parallel revolution in information technology.It is estimated that by 2003 the pharmaceutical market will be worth $ 250 billion and at least 10% of these products will be directly produced by biotechnology (Drews 1993). New systems for expressing proteins including transgenic plants and animals will take their place alongside the established systems in *E. coli*, yeast and mammalian cells. Nevertheless, for the near future mammalian cells will be the principal system for the production of these proteins.

The production of safe products must be the overriding concern in the development of a new pharmaceuticals based on recombinant DNA technology; the release of an unsafe product could have major deleterious consequences for the development of this nascent industry. The regulatory procedures affecting products from cell lines are still evolving but they are founded on the principles developed for the production of vaccine viruses from cell lines. The rigour of testing required in safety evaluation is sometimes seen as an irritation by scientists involved in the development of products but the history of vaccine development indicates its importance. Contamination of both human and veterinary vaccines has been recorded. In the early phase of poliovirus vaccine production, primary Rhesus monkey kidney cells contaminated with large amounts SV40 were used. New diseases have been created from the trans-species transmission of viruses from contaminated cells as occurred when a duck adenovirus was introduced into a susceptible chicken population through a vaccine. The resulting epidemic resulted in the deaths of many thousands of birds.

The experience of these incidents has led to a broad approach to safety testing in which the cell substrates are rigorously screened during the establishment of the cell bank and the product derived from the cells are screened at several stages during manufacture. These tests are supplemented by a validation of the manufacturing process to remove or inactivate abroad range of viruses (Lees and Onions, 1991; CPMP, 1991c). This review concentrates on the first of these phases of testing, the evaluation of the cell substrate.

1.7.2 The Regulatory Framework

A broadly similar regulatory framework has evolved in Europe the USA and Japan although there are important differences in procedures. As markets become international it becomes cost-effective to devise a testing programme that the meets requirements of all of these major markets.

In the European Community legislation initiated in 1984 began the process of regulating high technology biotechnology products, including monoclonal antibodies and products derived by recombinant DNA methods. The competent authorities of member states were obliged to consult with each other and the Committee for Proprietary Medicinal Products (CPMP) concerning the use of products derived by biotechnology. With the onset of the single market a three tier system of regulation operates. National procedures are limited to applications to a single member state but a decentralised system, covering the majority of medicinal products; will reflect mutual recognition of approval between certain member states. For high technology medicinal products derived by biotechnology, a centralised procedure is compulsory. It is proposed that applications for authorisation will be eventually be submitted directly to a European Agency for the Evaluation of Medicines which will encompass the functions of the Committee for Proprietary Medicinal Products (CPMP) and the Committee for Veterinary Medicinal Products (CVMP).

An important function of the CPMP has been to provide specific guidelines on the quality and safety of biotechnology products. Several such guidelines have now been published covering both recombinant DNA products and monoclonal antibodies (CPMP 1989; 1991; 1992; 1992b; 1994; 1996). Similar guidelines have been published by the Food and Drug Administration, Centre for Biologics Evaluation (CBER) in the United states of America (CBER 1985; 1992; 1993; 1994). In Japan more general guidelines on production of products from cell lines have been produced by the Ministry of Health and Welfare (1988) and by the International Conference on Harmonisation (ICH 1996).

1.7.3 Characterisation of the Cell Banks

An essential feature of the development of cell based technologies has been the establishment of a seed lot system based on characterised cell banks. A *Master Cell Bank* (MCB) is generated under good manufacturing practice (GMP) conditions and consists of a series of uniform vials of the cells laid down at the same time and stored immersed in liquid nitrogen or in its vapour phase. The *Manufacturer's Working Cell Bank* (MWCB), is derived from a vial of the MCB and cells revived from this bank are used as the source of production cells. Until recently there was a difference in practice between the United States of America and Europe with the USA favouring the majority of testing on the MWCB while, in Europe, the emphasis has been on fully characterising the MCB. In general both the FDA and the CPMP now stress the importance of rigorous testing of the MCB with some repeat testing on the MWCB.

In order to ensure the stability of the production system and to exclude the appearance of latent contaminants or major changes in the pattern of known contaminants, it is necessary to generate an *Extended Cell Bank* (ECB). The extended cell bank is generated from the MWCB and is maintained for at least 10 generations beyond those used in normal production. In continuous production systems it is difficult to determine the number of generations of the production cells. Under these circumstances *post production cells* are removed from the fermenter at the end of the production run, and may be passaged for at least 10 further cell generations and then tested.

Characterisation of the MCB consists of three major elements:
- Determining the identity of the cells.
- Validating the sequence of the gene insert in the case of recombinant DNA products.
- Determining the freedom from contaminating adventitious agents including, bacteria, yeast, fungi, mycoplasma and viruses.

1.7.3.1 Identity of the Cells

The need to ensure the identity of cells is revealed by disturbing statistics gathered from cell repositories. In one survey, 466 lines from 62 different laboratories were examined and 75 (16%) were found to be mistakenly identified (Nelson-Rees 1978). In a more recent survey 35% of 96 lines were found to be incorrectly identified, 25% of the human lines were from a different species and 11% from a different human source (Hakku et al. 1984).

Traditionally identity is determined by a combination of cytogenetics and isoenzyme studies. Karotypic analysis on human cell lines using giemsa or G-banding can be very precise, yielding differences not detected by DNA fingerprinting, (Chen 1988; Gilbert et al., 1990). However on rodent cells it is generally a less satisfactory technique and DNA fingerprinting using minisatellite probes for hypervariable loci, is increasingly coming into use and may eventually replace more traditional techniques (Gilbert et al. 1990). Refinements to genetic fingerprinting techniques using random primers or sequencing of hypervariable regions will steadily improve the acceptability and accuracy of these techniques. The pattern of antigen receptor gene rearrangements in lymphoid cells and the pattern of retrovirus proviral integrations in cells can also form useful adjuncts in DNA fingerprinting.

1.7.3.2 Validating the Gene Insert

Increasing emphasis is being placed on the genetic stability of recombinant DNA products (CBER 1992; Corran et al., 1993). Data on nucleic acid sequence and protein structure are required to give a complete picture of the identity and purity of the product. The origin of the DNA insert, the method of its preparation and the DNA sequence of the whole construct including the regulatory sequences must be specified.

The sequence of the cloned gene encompassing all the expressed elements should be determined at the MCB stage and at least once in post production cells. Where multiple copies of a gene are introduced or produced by amplification, analysis of copy number and gene pattern should be undertaken by Southern blotting and the mRNA sequence determined.

1.7.4 Tests for Adventitious Agents

The sources of microbiological contamination in cell systems arises from several sources (Fig. 1.7.1). Viruses can persist from the original animal from which the cells were derived, an extreme case of this is exhibited by the endogenous retroviruses which are present in the cells as part of their normal genetic makeup. A second and important source of contamination arises from products of animal origin used to grow or store the cells. Although there is an encouraging trend towards the use of defined tissue culture media which do not utilise such materials, most cells have been stored in foetal bovine serum and possibly removed from tissue culture flasks with trypsin of animal origin. A final source of contamination is from the operators or the environment under which the cells are handled. Good manufacturing practice should limit the possibility of contamination from operators but it does occur.

Appropriate tests for bacterial and mycotic sterility must be undertaken. However, most problems arise from mycoplasma and viral contamination of the cell lines. The use of antibiotic free medium imposes a strict discipline on the operators and enables bacterial contamination to be detected easily.

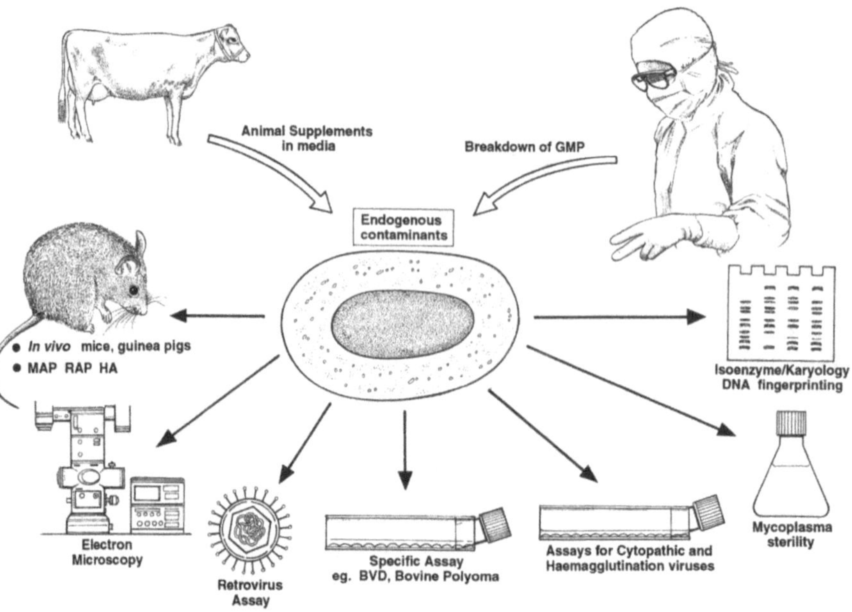

Fig. 1.7.1: Assays for characterising cells for cell banking and source of contaminants.

1.7.4.1 Mycoplasma Screening

Mycoplasma are insidious and common contaminants of cells derived from academic laboratories and were a source of considerable concern in the foundation of hybridoma master cell banks from such sources. Human operators are a major source of mycoplasmal contamination. In a survey of 34,697 cell cultures 3955 were contaminated by mycoplasma and 41 % were of human origin usually *M. orale or M. hominus and salvarium.* (Del Giudice and Gardella 1984). *M. hyorhinus* was another frequent contaminant and possibly arose from animal material used in the cell cultures.

Mycoplasma may go unnoticed in that they do not necessarily produce an observable change in the appearance of the cells although they can have major effects on their metabolism and responses to external signals. They can also dramatically alter the susceptibility of cells to virus infection.

Simple screening systems for mycoplasma involve the use of direct detection of mycoplasma on the cells using the bisbenzimidazole fluochrome stain for DNA (Hoechst 33258). The sensitivity of this system can be significantly improved by co-culturing the test article cells with a cell line like Vero that is mycoplasma free. However, screening by agar and broth culture should also be carried out and detailed testing procedures are given in the FDA point to consider document (CBER 1993).

1.7.4.2 *In Vivo* Assays the MAP, and HAP Tests

Amongst the most widely used cells in biotechnology are murine hybridoma cells and CHO cells derived from the Chinese hamster. A standard method of detecting a wide range of possible indigenous contaminants in these cells are the mouse and hamster antibody production tests (MAP and HAP) (Fig. 1.7.2). In this procedure the test article cells are inoculated into SPF mice or hamsters and the animals followed to determine if an antibody response develops, indicating the presence of the virus. Twenty-two viruses are listed as potential contaminants of murine cell lines in appendix 1 of CPMP's guidance notes for the production and marketing of murine monoclonal antibodies (CPMP 1994) (Table 1.7.1) and of these 16 are normally detected in standard MAP tests. The remaining viruses, lymphocytic choriomeningitis virus (LCMV), mouse cytomegalovirus, mouse rotavirus (EDIM), thymic virus and lactic dehydrogenase viruses and retroviruses require specific assay systems. Retrovirus tests must always be conducted and it is usual to ensure freedom from LCMV by an appropriate *in vivo* test.

The significance of LCMV virus as a contaminant has been reinforced by the recent description of infection of animal house workers by LCMV. (Mahy et al., 1991). LCMV is an arenavirus that establishes a silent, persistent infection in mice following perinatal transmission whereas transmission to adults causes severe, often fatal lymphocytic choriomeningitis. In the outbreak described by Mahy, the infection was first uncovered when one of the workers in the facility developed meningitis. Subsequently a total of 8 of 82 workers were shown to have serological evidence of infection and cell lines passaged in nude mice were found to be infected with the virus, one line having been infected since 1975.

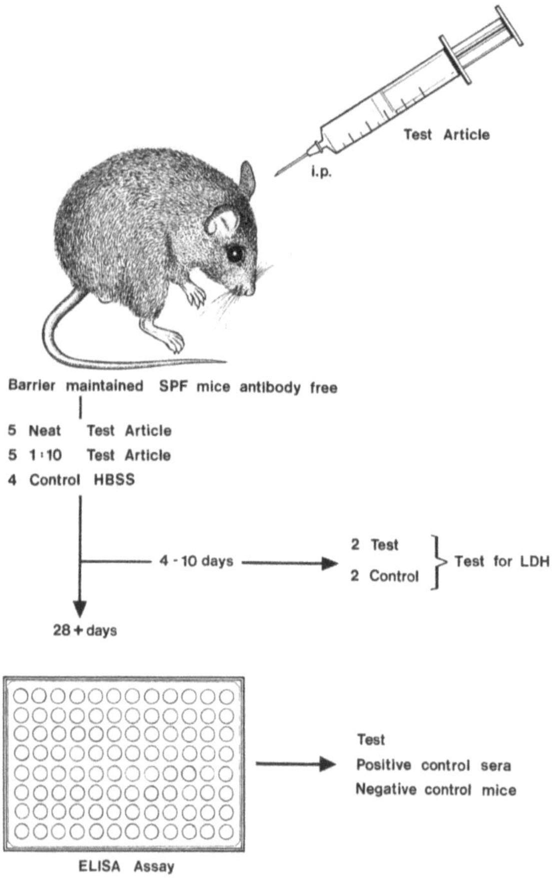

Fig. 1.7.2: MAP/RAP/HAP test.

In general, with the exception of retroviruses, there should be no evidence of virus contamination by any virus. Under no circumstances can cells containing the viruses in group 1, which can cause disease in man, or polyoma virus from group 2, be used for the production of biologicals. (see Table 1.7.1).

As a supplement to the MAP or HAP tests, *in vivo* challenge of animals and inoculation of embryonated eggs is used to reveal the presence of adventitious agents. At least 5 guinea pigs, 10 adult mice and 10 suckling mice from two litters, are inoculated intramuscularly with test article material. A further 10 adult mice receive an intracranial inoculation of the test article. These procedures are valuable in demonstrating the presence of certain viruses that are difficult to detect in *vitro* assays. Morbidity and mortality are taken as indicators of infection in the animals and the egg cultures are inspected regularly for gross changes and the contents screened by haemagglutination at the end of the assay.

Table 1. Potential Contaminants of Mouse and Rat Cells

Group	Virus	Species Affected
I	Hantavirus (Haemorrhagic fever with renal syndrome)*	M, R
	Lymphocytic choriomeningitis virus (LCMV)*	M
	Rat rotavirus*	R
	Reovirus type 3 (reo 3)*	M, R
	Sendai virus*	M, R
II	Ectromelia virus*	M
	K virus (K)	M
	Kilham rat virus (KRV)	R
	Lactic dehydrogenase virus (LDH)	M
	Minute virus of mice (MVM)	M, R
	Mouse adenovirus (MAV)*	M
	Mouse cytomegalovirus (MCMV)	M
	Mouse encephalomyelitis virus (MEV, Theiler's or GCVII)	M
	Mouse hepatitis virus (MHV)	M
	Mouse rotavirus (EDIM)	M
	Pneumonia virus of mice (PVM)*	M, R
	Polyoma virus	M
	Rat coronavirus (RCV)	R
	Retroviruses*	M, R
	Sialodacryoadenitis virus (SDA)	R
	Thymic virus	M
	Toolan virus (HI)*	R

M: mouse; R: rat

* Knowm to be capable of replicating in vitro in cells of human and monkey origin.
From CPMP (1989b)

1.7.4.3 Assays for Bovine Viruses and Other Viruses of Animal Origin

A significant cause for concern has been the presence of viruses of animal origin as contaminants of cell lines. Most emphasis has been placed on bovine viruses that may be present in serum used to grow or store cells but it should not be forgotten that a range of other animal products, like porcine trypsin, may be used in tissue culture. A complete testing schedule should take account of these non-bovine viruses.

To detect bovine viruses at least two bovine indicator cell lines, susceptible to the major viruses of concern are inoculated with a test article. The detection of viruses depends on two principal properties. First the viruses may be cytopathic, producing a recognisable change in the appearance of the cell sheet. Often particular virus groups produce a characteristic change. For instance herpesviruses often produce syncitia by fusing the membranes of adjacent cells while adenoviruses block the Na^{++} pump so that the cells become characteristically swollen. Some viruses also haemagglutinate erythrocytes or the infected cells may bind *i.e.haemadsorb* red blood cells. This provides a sensitive indication of the presence of some viruses like parainfluenza viruses that may be minimally cytopathic. However certain viruses require very specific systems to reveal their presence and amongst these are BVD virus, bovine polyomavirus and the uncharacterised agent of bovine spongiform encephalopathy.

1.7.4.3.1 BVD Virus

The BVD virus is a member of the pestivirus group related to swine-fever virus of pigs and border disease virus of sheep. It is frequently transmitted in utero so that calves are born persistently infected with the virus. Usually these strains are non-cytopathic and they will be present in the foetal serum at high titre as, the calf or foetus is tolerant to the virus. (Brownlie et al., 1984). BVD is capable of infecting cells of a range of different species (King and Harkness, 1975) and as it is not often cytopathic it is not easily revealed. In order to reliably detect the virus it is necessary to infect a permissive cell system and, after repeated passage of the cells over at least 3 weeks, it may be revealed by immunofluoresence with specific anti-sera. More recently the polymerase chain reaction (PCR) has been used to detect both infected serum batches and infected cell cultures.

1.7.4.3.2 The Bovine Polyomavirus

The bovine polyomavirus is a good example of a virus that has been known to science for some time but has received little attention until recently as it has not been important in veterinary medicine. However, it is an extremely common contaminant of bovine serum, it is possibly a zoonotic virus and it belongs to a family of potentially oncogenic viruses. It is therefore of little surprise to find that it has become the focus of some concern in the biotechnology industry.

Bovine polyomavirus was first isolated from kidneys of newborn calves. Subsequently Parry and his colleagues (Parry et al., 1983) detected a polyomavirus in a rhesus monkey kidney cell line used to grow hepatitis A virus. Other primate cell lines including vero cells were shown to harbour a related virus. The high prevalence of antibodies to bovine polyoma virus in the UK cattle population, 62% being positive, indicated that bovine serum was the likely source of these viruses in cell culture. A surprising and worrying feature of the sero-epidemiology of this virus is that while antibody to it was generally absent from the human population, 71% of veterinarians, 50% of cattle farmers and 40% of abattoir workers had detectable antibody that was not cross reactive with human polyomaviruses (Parry and Gardner, 1986). This virus must therefore be considered potentially zoonotic and ought to be screened for in serum and in cell lines. Fortunately the complete sequence of the virus has been determined which enables PCR based assays to be used. These have confirmed the prevalence of the virus with 70% of commercial foetal bovine serum batches being positive (Schuurman et al., 1990).

1.7.4.3.3 BSE and Scrapie

The development of a new disease in cattle, bovine spongiform encephalopathy, serves as a salutary warning of how technology can influence the development of disease, a lesson that should not be lost on the biotechnology industry. It is now believed that the origin of the disease, first recognised in cattle in the United Kingdom in 1986 arose from the consumption of feedstuffs containing sheep protein contaminated with the scrapie agent or from a pre-existing agent of cattle spongiform encephalopathy. In turn it was a change in the treatment of animal carcases in the rendering process that allowed the extremely resistant agent of BSE to get into the food chain (Wilesmith et al., 1988). Indeed the scrapie agent is highly resistant to many physical treatments and even the recommended treatments by autoclaving at 134–138 °C for 18 minutes, treating with sodium hydroxide 1N for 1h at 20 °C or treatment with sodium hypochlorite (2% available chlorine) for 1h at 20 °C are not guaranteed to abolish infectivity (CPMP 1992b). The nature of the agent remains unresolved but is the subject of intense scientific interest. Among the competing hypotheses are an infectious protein, or prion and an unconventional virus with free nucleic acid protected by a large molar excess of the abnormal cellular protein PrP^{sc}, that is deposited as fibrils in the brains of affected animals (Hope et al., 1988).

A further factor of concern is that animals including mice and pigs can be experimentally infected with the agent by intracerebral inoculation. Moreover in this University we have recently recorded the presence of spongiform encephalopathy in cats. We have not observed this pathology in brain sections in the previous ten years and it must remain a possibility that the agent has been transmitted to cats through the consumption of infected food (Leggett et al., 1990).

BSE must clearly be a matter of concern in evaluating products directly derived from bovine material. However it is not of general concern in products derived from cell culture. First, it is now required to use serum from non-endemic areas for BSE, like New Zealand or the United States of America. Secondly in infected animals the titre of the agent is very high in brain tissue 10^9 whereas in serum it is undetectable (i.e. $< 10^2$ infectious units/g) in the *in vitro* assay used to detect the agent. Moreover, while high titres of the scrapie agent can infect rat neuronal cells in culture, most cells are probably resistant to infection. In general because the assay for scrapie requires a year or more it is not practicable to assay for the agent in cell lines or serum batches. Where necessary a validation of the downstream purification process can be conducted to test its capacity to remove the scrapie agent.

1.7.4.4 *In Vitro* Assays for Viruses

In addition to the specific assays for bovine assays a general in vitro assay is required to detect viruses particularly those capable of infecting human cells. Usually a test article prepared from the cell line is used to inoculate between 2 to 6 cell lines and the test article cell line itself. The FDA recommend a minimum of 2 cell lines including an a human diploid cell line like MRC-5 and a monkey cell line like vero. the CPMP recommend that in addition a cell line from the species of origin of the test cells be included and a bovine cell line. However, as described above bovine viruses are examined in specific assays and it is more usual to widen the range of cells to include a human epithelial cell line like HeLa, that is permissive for adenoviruses, and the RK13 cell line permissive for rubella.

The principal of the assay is that the cells are examined for a total of 28 days with a blind passage of negative cells at 14 days. In addition to observing for cytopathic effect the cell system is examined for haemadsorbing and haemagglutinating viruses with human O, guinea-pig and chicken erythrocytes. It is of course essential to have both negative controls and positive control cultures infected with viruses that will display haemagglutination, haemadsorption and a cytopathic effect. One cautionary point needs to be emphasised. While many tissue culture adapted strains grow well in these systems this is not always true for primary isolates. Experience is required to detect the subtle changes that might be associated with such wild type viruses.

1.7.4.5 Retrovirus Assays

Retroviruses pose one of the most important challenges in developing products from cell systems as these viruses are ubiquitous and associated with life-threatening conditions including cancer and immunosuppression. However, by conducting an integrated approach to testing, at each stage of the production cycle, the problem of retrovirus contamination can be contained.

Many of the pathogenic properties of these viruses are associated with their replication cycle (Jarrett and Onions, 1992). The viruses are enclosed in an outer envelope containing viral glycoproteins. These glycoproteins bind to receptors on cells and, in part, define the host range of the viruses. Once bound the genomic RNA of the virus is liberated into the cytoplasm where it is transcribed into double stranded DNA by a virion encoded enzyme, reverse transcriptase. The DNA copy or provirus migrates to the nucleus where it is covalently integrated into chromosomal DNA. At this stage the virus may remain as a latent infection or it may be transcribed into new mRNA and genomic RNA. New virus particles are assembled at the cell surface and are budded from the cytoplasm. This process is not cytopathic

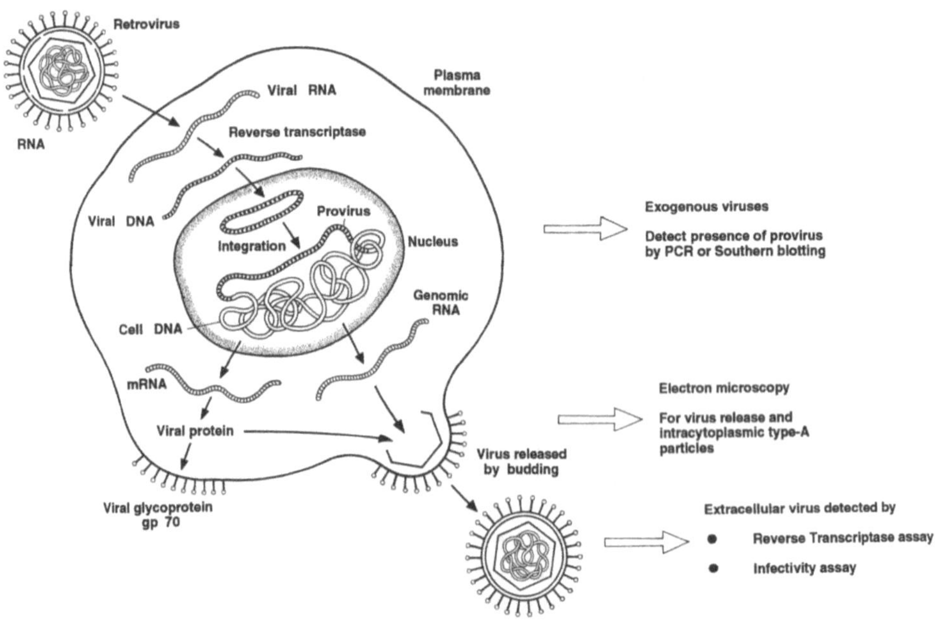

Fig. 1.7.3: Replication and Assay Points for Retrovirus.

for the oncovirus sub-group of retroviruses so they continue to shed virus while the cells replicate and function normally (Fig. 1.7.3).

In probably all vertebrate species some retroviruses have succeeded in infecting the germ line so that provirus are passed from generation to generation as conventional genes (Fig. 1.7.4). Usually the expression of these viruses is tightly repressed in the individual but tissue culture cells may spontaneously begin to produce them. In addition to these endogenous viruses, many species can be infected by exogenous retroviruses transmitted from animal to animal like conventional viruses. In some cases these can undergo recombination with endogenous retrovirus elements to generate new subgroups with altered host cell ranges. Some members of the oncovirus group, including the murine leukaemia viruses (MuLV) can recombine with cellular transforming genes (oncogenes) so that the progeny virions carry these within their genome. This forms one of the principal methods by which these viruses transform cells.

Retroviruses are sub-divided into the *oncoviruses* like MuLV the *lentiviruses* which includes HIV and the *spumaviruses* that do not have any clear disease associations.

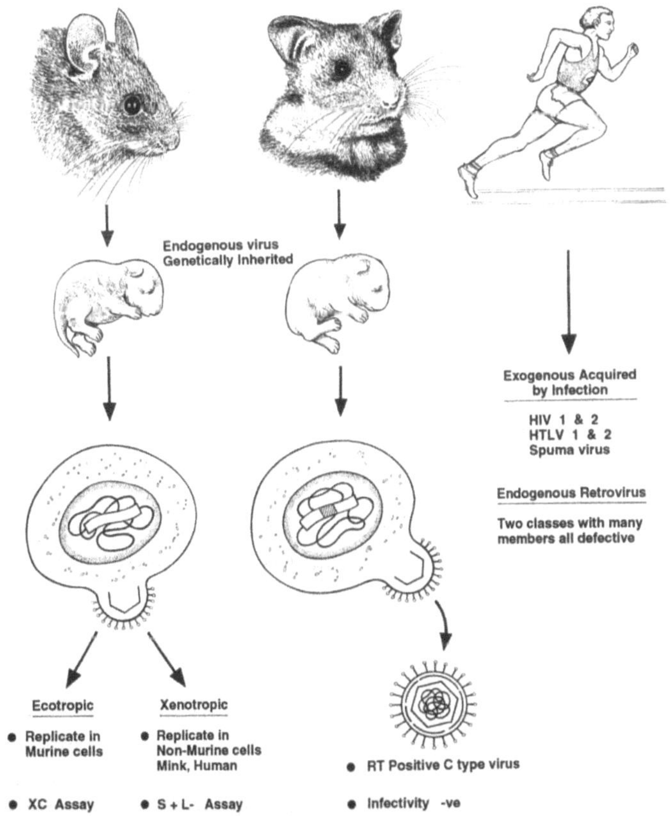

Fig. 1.7.4: Retroviruses endogenous and exogenous.

The oncoviruses are further divided by the organisation of their genomes and by their appearance in the electron microscope so that MuLV is described as a type-C virus because its core is formed during the budding process.

The murine leukaemia viruses exemplify the complexity of these viruses. Cells may release a number of different viruses defined by their host range. Ecotropic viruses infect only murine cells whereas xenotropic viruses released by most murine hybridomas infect cells of many species but not murine cells. Murine xenotropic viruses replicate efficiently in mink cells but less efficiently in human cells and indeed the xenotropic viruses from murine hybridomas do not usually replicate in human cells. However, the situation can be more complex as ecotropic viruses can undergo recombination with other endogenous sequences to generate polytropic viruses capable of replicating in murine and non-murine cells. These polytropic viruses are often the viruses that initiate leukaemia in laboratory mice. Amphotropic murine viruses also replicate in murine and non-murine cells but their origin is from wild mice. However, they have been used as the basis of retrovirus packaging lines now being used in gene therapy protocols.

A screening system for retroviruses must take account of this complexity of host ranges and different assays are required to detect these different virus groups (Onions and Lees, 1991). For murine cells the ability of ecotropic viruses to produce syncitia on rat XC cells forms the basis of a quantitative assay, one XC plaque being formed by one infectious virus particle. Similarly xenotropic and amphotropic viruses can be detected by their ability to rescue defective oncogene containing viruses from so called S+L- cells. The oncogene containing virus rescued by these defective viruses transforms adjacent cells producing a focus of rounded cells; again one focus is induced by one infectious virus so that the assays are quantitative. Polytropic viruses may be positive in the XC assay but they also produce a degenerative cytopathic effect on mink cells, so that by a combination of XC and S + L-assays one can detect the retroviruses of concern from murine cells.

In these assays it is usual to plate the supernatant from the test cell line directly onto the XC or S + L-, which constitutes a direct assay. However if the virus is present at low titre it may be missed for two reasons. First, if the titre is 1000 infectious units per litre and if 1 ml is assayed the virus will be missed on 37% of occasions simply because of the distribution of the virus in the sample as given by the Poisson distribution. Moreover while one virus particle produces one plaque or focus not every virus "hit" does this. Consequently direct assays become inaccurate at titres below 10^1 ml. One method of circumventing this problem is to conduct extended assays in which the indicator cells are passaged 5 times over a 3 week period, which allows the virus infection to spread throughout the culture system. These assays are quantal i.e. positive or negative, rather than quantitative assays, but they are more sensitive than their direct equivalents. A further level of sensitivity can be achieved by co-culturing the test cells with the indicator cell lines (Fig. 1.7.5).

In devising a retrovirus screening assay at least three independent methods are used:

- Transmission electron microscopy is used a general screening technique on cells for adventitious agents and it is a valuable method of detecting type-C retrovirus particles and non-infectious retrovirus like elements called intracisternal type-A particles. (Other type A particles can be the pre-formed cores of type B or D viruses).

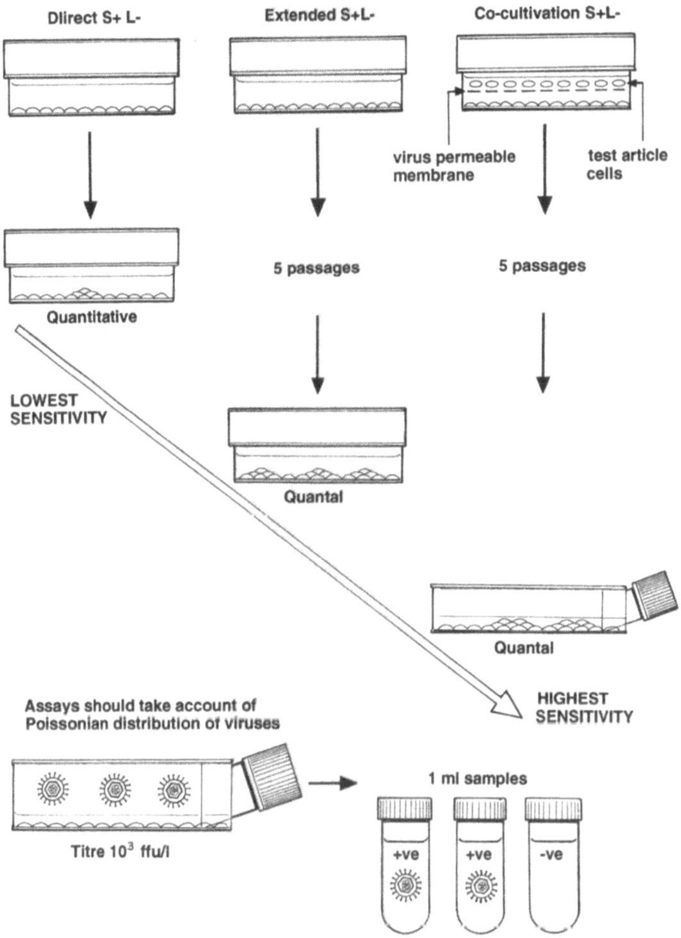

Fig. 1.7.5: Sensitivity of retrovirus assays.

- When testing a number of clones for cell banking purposes it may be appropriate to use direct assays as an initial screen but extended or preferably co-cultivation assays should be used on apparently negative clones to be sure of their status.
- The reverse transcriptase assay provides a broad screening system for many different retroviruses. It does not detect infectious virus but detects the presence of extra-cellular virus particles containing this enzyme. The distinction is important as, for instance, CHO cells may release reverse transcriptase containing particles but these are defective containing deletions within their genomes. However one should be cautious in always assuming that CHO viruses will be non-infectious as recombination is frequent in retroviruses and repair of such defects can occur. While reverse transcriptase assays are simple to conduct their control and interpretation requires care. Contaminating cellular polymerases, particularly γ DNA

polymerase, can use the synthetic RNA templates used to monitor reverse transcriptase activity. For this reason it is essential to include a synthetic DNA template to measure DNA polymerase contamination. Retroviruses also differ in their cation requirements for reverse transcriptase activity and the assay should be conducted separately with Mn^{++} and Mg^{++} as the divalent cation.

Whenever evidence for retrovirus production is produced by any of these systems the cell line should be co-cultured with human cells to determine if the virus is capable of infecting human cells.

Although some cell lines might initially be negative for retrovirus production they may, at subsequent time points become positive. The stability of non-producer cells can be tested by treating them with thymidine analogues like iododeoxyuridine.

In the case of cell lines other than murine and CHO cells it is usually necessary to devise specific retrovirus assays. For instance, in human cells it is essential to screen for the human leukaemia retroviruses, HTLV-I & II, and the lentiviruses HIV I & II although these are not likely to be contaminants of non-lymphoid cells (Onions and Lees, 1991b). These viruses are exogenous so that it is possible to use DNA hybridisation and PCR approaches as a supplement to culture techniques.

Retrovirus contaminants can sometimes go unnoticed for long periods. For instance many stocks of the marmoset cell line B598 used to produce EBV for preparing human hybridomas are also contaminated with a squirrel monkey type-D retrovirus (SMRV) (Popovic et al., 1982).

1.7.5 Integrating Testing with the Production Cycle

Safety testing should form an integral part of the overall quality management process and should not be regarded as an irritating requirement to satisfy regulatory authorities. Much valuable information on the stability and reproducibility of production processes can be gained from good safety testing. All the procedures for safety evaluation should be conducted under the formal conditions of Good Laboratory Practice by an experimental testing house. In addition to testing the cell banks, tests are also recommended or required for the bulk harvest, the purified bulk and final vialed product. These tests are given in Table 1.7.2.

1.7.5.1 The Bulk Harvest

Conventional sterility and mycoplasma tests should be conducted on the bulk harvest but in addition a limited version of the in vitro assay for virus contamination is recommended. This involves two cell lines usually, vero and MRC-5 although this may be varied according to the cell system used. The principle point of this assay is to detect viruses that may have been introduced due to a breakdown in GMP. If bovine material is used in the production media it is also advisable to undertake tests for bovine viruses.

Table 2. Typical Testing Strategy for a Product Derived from Rodent Cells

	Fermenter/Bulk Harvest	Purified Bulk	Final Product
Sterility	R	R	R
Mycoplasma	R		
Virus assays:			
In vitro (limited)	R		
Bovine viruses	O		
Retrvirus assays:			
RT	R		
XT and/or SL	R		
DNA contamination test		R	
Rabbit Pyrogen test			R
Abnormal Toxicity (general safety)			R

R recommended
O may be required

Table 3. Example of a Testing Strategy for a Rodent Cell Bank

	Cell Bank	End of Production Cells
Sterility	R	R
Mycoplasma	R	R
Virus assays:		
In vitro	R	R
In vivo	R	R
MAP/RAP/HAP	R	
Bovine viruses	R	R
Retrovirus assays:		
TEM	R	R
RT	R	R
XC	R	R
S^+L^-	R	R
Co-cult	O	O
Isoenzyme/analysis/ Karyology/DNA fingerprinting	R	R

R recommended
O may be required

An important feature of the testing protocol for retroviruses is that it should provide data that can be related to the validation of downstream clearance. The validation determines the capacity of the downstream processes to inactivate or clear retroviruses and other viruses. If there is already evidence that retroviruses are being released from the MCB the tests should determine if the titre has increased or the host range has changed. For instance if the virus could only be detected by extended assayin the MCB a direct assay on the bulk harvest would indicate if the titre is increasing. For Lines like CHO it is necessary to rely on assays like reverse transcriptase or negative slam electron microscopy to determine the level of virus particles. CHO clones vary in their reverse transcriptase (RT) status and the levels of reverse transcriptase activity can change during the growth of the cells in a fermenter. When taking aliquots of the bulk harvest for analysis it is important that these are clarified in a low speed centrifuge immediately after removal from the fermenter and are stored at $-70\,°C$ or below. This reduces the problems of DNA polymerase contamination and ensures that the retroviruss remain viable for infectious assays (Fig. 1.7.6).

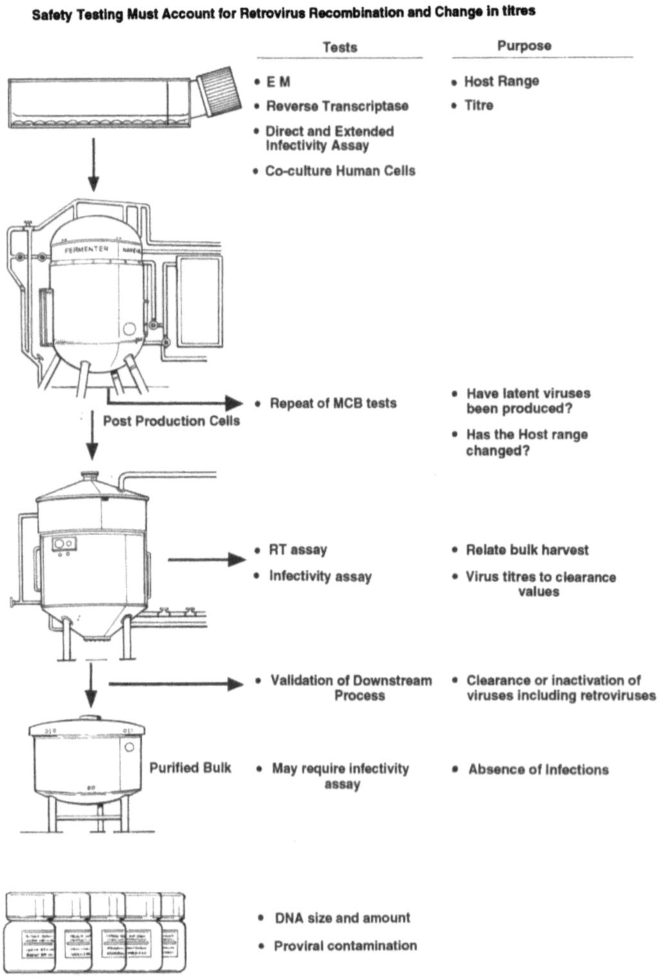

Fig. 1.7.6: Integrated testing for retroviruses.

The FDA often recommend that a sample of the bulk harvest is taken for examination by negative staining for virus particles. In my view this is an unsatisfactory procedure of limited value. Even an experienced microscopist has difficulty in distinguishing retroviral particles in the large amounts of microsomal debris usually present in fermenter samples. If this assay is required it can be improved by a combination of rate zonal and isopycnic density ultracentrifugation to partially purify particles of retroviral size and density.

1.7.5.2 The Purified Bulk/Final Lot Bulk

At this stage the mycoplasma, sterility and residual DNA contamination tests must be conducted. Conventionally, DNA testing uses random probes from the cell line of interest although other technologies that are quantitative down to the sensitivity demanded may be used. The WHO and CPMP recommend that products should not contain greater than 100pg/dose whereas the FDA's recommendation is more rigorous at 10pg/dose. In practice, however the acceptable level of DNA contamination is open to discussion with the relevant regulatory authority.

In the case of systems known to be producing particular viruses like EBV or certain retroviruses, it may be necessary to confirm that the downstream process has removed these viruses. This batch test on the product is in addition to the Validation of Downstream Process which is also required (Lees and Onions, 1991).

1.7.5.3 The Final Filled Product or Dosage Preparation

Biosafety tests for pyrogenicity and abnormal toxicity are carried on the final product. The abnormal toxicity or general safety test, involves the inoculation of the final product into test animals and serves as a general screen for abnormal properties. Traditionally the pyrogen test is an *in vivo* test in rabbits but in the U.S.A. a validated Limulus Amoebocyte Lysate (LAL) test may be used.

Increasing emphasis is being placed on validating the potency and stability of the product. In particular it is now recognised that post-translational and processing changes can affect the final product. For those proteins administered chronically, like factor VIII non-self epitopes generated by abnormal glycosylation could be of importance. Cells forced to produce high levels of glycoprotein can result in the production of abnormal glycans including the G0 agalactosyl element associated with the initiation autoimmune diseases. Similarly enzymatic alterations associated with production or processing can lead to modifications like cyclization of N-terminal glutamine or internal asparagine residues (Corran et al., 1993). At present there are no firm guidelines on methodologies to be used in determining purity and stability of the product. but the manufacturer in association with a testing agency is expected to develop appropriate testing protocols. These may include combinations of peptide mapping, N-terminal sequencing, tests for conformational epitopes and assays based on mass spectrometry and NMR. In addition functional assays provide an important verification of the specificity of the product.

1.7.6 Conclusion

Safety evaluation is a crucial element in the development of recombinant DNA products and monoclonal antibodies. Public confidence in these technologies depends on ensuring that safe products are produced. Moreover a carefully constructed safety testing scheme contributes to the total quality management of the development and production systems.

References

Brownlie, J., Clarke, M. C. and Howard C. J. (1984) Experimental production of fatal mucosal disease in cattle. Vet. Rec. 114, 535–536.

Centre for Biologics and Research (CBER) U.S. (1985) Points to consider in the production and testing of new drugs and biologicals produced by recombinant DNA technology.

Centre for Biologics and Research (CBER) U.S. (1993) Points to consider in the characterization of cell lines used to produce biologicals (1993).

Centre for Biologics and Research (CBER) U.S. (1994) Points to consider in the manufacture of monoclonal antibody products for human use (Draft 1994).

Centre for Biologics and Research (CBER) U.S. (1992) Supplement to the points to consider in the production and testing of new drugs and biologicals produced by recombinant DNA technology: nucleic acid characterization and genetic stability (1992).

Chen, T. R. (1988) Re-evaluation of HeLa, HeLa S3 and HEp-2 karyotypes. Cytogenetics and Cell Genetics 48, 19–24 (1988).

Committee for Proprietary Medicinal Products (1989): Ad Hoc working Party on Biotechnology/Pharmacy and working Party on Safety of Medicines. Notes to applicants for marketing authorizations on the pre-clinical biological safety testing of medicinal products derived from biotechnology (and comparable products derived from chemical synthesis). J. Biological Standardization 17, 203–212 (1989).

Committee for Proprietary Medicinal Products (Revision 1995): Production and Quality control of Monoclonal Antibodies (111/527/94) (1994).

Committee for Proprietary Medicinal Products (Revision 1994): Production and Quality Control of Medicinal Products Derived by Recombinant DNA Technology (111/3417/92) (1992).

Committee for Proprietary Medicinal Products (1991): Ad Hoc working Party on Biotechnology/Pharmacy and working Party on Safety of Medicines. Production and Quality Control of Cytokine Products Derived by Biotechnological Processes. Biologicals 19, 125–131 (1991).

Committee for Proprietary Medicinal Products (1991b): Ad Hoc working Party on Biotechnology/Pharmacy and working Party on Safety of Medicines. Production and Quality control of Human monoclonal Antibodies. Biologicals 19, 133–138 (1991b).

Committee for Proprietary Medicinal Products (1996): Guidance on Virus Validation Studies: The Design, Contribution and Interpretation of Studies Validating the Inactivation and Removal of Viruses (CPMP/BWB/268/95).

Committee for Proprietary Medicinal Products (1992): Ad Hoc working Party on Biotechnology/Pharmacy and working Party on Safety of Medicines. Guidelines for minimizing the risk of transmitting agents causing spongiform encephalopathy via medicinal products. Biologicals 20, 155–158 (1992b).

Corran, P. H., Griffiths, E., Robertson, J. S., and Geisow, M. J. (1993) Harmonizing viewpoints on ensuring consistency and stability of rDNA-derived biologicals: Report of meeting held at National Institute of Biological standards and control, 23–24 November 1992. Trends in Biotechnology 11, 77–80 (1993).

Del Giudice, R. A., and Gardella, R. S. (1984) Mycolasma infection of cell culture: effects, incidence and detection. In: In Vitro Monograph 5. Uuses and Standardization of Vertebrate Cell Culture , pp 104–115, Tissue Culture association, Gaithersburg, M. D. (1984).

Drews J. (1993) Into the 21st Century. Bio/technology 11, S16–S20 (1993).

Gilbert, D. A., Reid, Y. A., Gail, H. M., Pee, D., White, C., Hay, R. J. and O'Brien, S. J. (1990) Application of DNA fingerprints for cell line individualization. American Journal of Human Genetics 47, 499–514 (1990).

Hakku, B., Halton, D.M., Mally, M., and Peterson, W.D. Jr. (1984) Cell Characterization by Use of Multiple Genetic Markers. In: Eukaryotic Cell Cultures. Acton, R.T. and Lynn, J.D., (eds.) pp. 13–31. Plenum Press (1984).

Hope, J., Reekie, L.J.D., Hunter, N., Multhaup, G., Beyreuther, K., White, H., Scott, A. C., Stack, M.J., Dawson, M. and Wells, G.A.H. (1988) Fibrils from brains of cows with a new cattle disease contains scrapie associated protein. Nature 336, 390–392 (1988).

International Conference on Harmonisation: Q5D: Quality of Biotechnology/Biological Products Derivation an Characterization of Cell Substrates used for Production of Biotechnological/Biological Products. Draft 5 (1996).

Jarrett, O. and Onions, D. (1992) Leukaemogenic Viruses. In: Leukaemia. Ed JA Whittaker, J.A., (ed.) pp 34–63. Blackwell Scientific Publications (1992).

King, A.A. and Harkness, J.W. (1975) Viral contamination of foetal bovine serum. Vet. Rec. 97, 16 (1975).

Lees, G. and Onions, D.E. (1991) Validation of Downstream Processing. In: Production of biologicals from animal cells in culture. Spier, R.E., Griffiths, J.B. and Meignier, (eds.), pp 783–788 Butterworth–Heinemann (1991).

Leggett, M.M., Dukes, J. and Pirie, H.M. (1990) A spongiform encephalopathy in a cat. Vet. Rec. 127, 586–588 (1990).

Mahy, B.W.J., Dykewicz, C., Fisher-Hoch, S., Ostroff, S., Tipple, M. and Sanchez, A. (1991) Virus zoonoses and their potenial for contamination of cell cultures. In: Developments in Biological Standardization 75. Horaud, F. and Brown, F. (eds.), pp 183–189, S. Karger Basel (1991).

Ministry of Health and Welfare Japan. (1988) Documents necessary for approval of drugs manuafactured using cell culture technology (1988).

Nelson-Rees, W.A. (1978) The identification and monitoring of cell line specificity. In: Origin and natural history of Cell Lines. Progress in Clinical and Biological Research 20, 25–79 (1978).

Onions, D.E. and Lees, G. (1991) Evaluating the safety of murine and human hybridomas: New problems and new techniques. In: Production of biologicals from animal cells in culture. Spier, R.E., Griffiths, J.B. and Meignier, (eds.), pp 34–38, Butterworth-Heinemann (1991).

Onions, D.E. and Lees, G. (1991) Human retroviruses and herpesviruses: problems and solutions in safety testing biologicals. In: Developments in Biological Standardization 75. Horaud, F. and Brown, F., (eds.), pp 145–158, S. Karger Basel (1991).

Parry, J.V., Lucas, M.H., Richmond, J.E. and gGardner, S.D. (1983) Evidence for ab ovine origin of the polyomavirus detected in foetal rhesus monkey kidney cells, FRHk-4 and -6. Archives of Virology 78, 151–165 (1983).

Parry, J.V. and Gardner, S.D. (1986) Human exposure to bovine polyomavirus: A zoonosis? Archives of Virology 87, 287–296 (1986).

Popovic, M., Kalyanaraman, V.S., Reitz, M.S. and Sarngadharan, M.G. (1982) Identification of the RPMI 8226 retrovirus and its dissemination as a significant contaminant of some widely used human and marmoset cell lines. Int. J. Cancer. 30, 93–99 (1982).

Schuurman, R., Sol, C. and van der Noorda, J. (1990) The complete nucleotide sequence of bovine polyomavirus. J. Gen. Virol. 71, 1723–1735 (1990).

Wilesmith, J.W., Wells, G.A.H., Cranwell, M.P. and Ryan, J.B.M. (1988) Bovine spongiform encephalopathy: epidemiological studies. Vet. Rec. 123, 638–644 (1988).

2 Biological Aspects of Animal Cells

2.1 Metabolic Control of Animal Cell Culture Processes

Roland Wagner

Abbreviations

AcCoA = acetyl coenzyme A, ADP = adenosine diphosphate, AEC = adenylate energy charge, Ala = alanine, AlaAT = alanine aminotransferase, AMP = adenosine monophosphate, ASA = aspartate semialdehyde, Asp = aspartate, AspAT = aspartate aminotransferase, ATP = adenosine triphosphate, BHK = (Syrian) baby hamster kidney, cAMP = cyclic adenosine 3',5'-monophosphate, CER = carbon dioxide evolution rate, cGMP = cyclic guanosine 3',5'-monophosphate, CHO = chinese hamster ovary, CP = carbamoylphosphate, CPS = carbamoylphosphate synthetase, CTP = cytidine triphosphate, DAPI = 4',6-diamidino-2-phenylindole, DHAP = dihydroxyacetone-3-phosphate, DHFR = dihydrofolate reductase, DHP = dihydroxyacetone phosphate, DIF = defined Iscove's-Ham's F-12 serum-free medium, DNA = deoxyribonucleic acid, 1,3-DPG = 1,3-diphosphoglycerate, DO = dissolved oxygen, *E. coli* = *Escherichia coli*, eq. = equation, FACScan = fluorescence activated cell scanner, FAD = flavine adenine dinucleotide, Frc = fructose, Frc-1-P = fructose-1-phosphate, Frc-1,6-DP = fructose 1,6-diphosphate, Frc-6-P = fructose-6-phosphate, GA = glycerine aldehyde, Gal-1-P = galactose-1-phosphate, GAP = glycerinaldehyde-3-phosphate, GAPDH = glycerinaldehyde-3-phospate dehydrogenase, GDH = glutamate dehydrogenase, Glc = glucose, Glc-1-P = glucose-1-phosphate, Glc-6-P = glucose-6-phosphate, Gln = glutamine, Glnase = glutaminase, Glu = glutamate, GlcN = glucosamine, GlcN-6-P = glucosamine-6-phosphate, GlcNAc-1-P = N-acety-glucosamine-1-phosphate, GlcNAc-6-P = N-acetyl-glucosamine-6-phosphate, GPDH = glucose-6-phosphate dehydrogenase, GS = glutamine synthetase, GTP = guanosine triphosphate, HeLa = human cervix tumor cell line (Henrietta Lachs), HEPES = N-[hydroxyethyl]piperazine-N'-[2-ethanesulfonic acid], HK = hexokinase, HPLC = high performance liquid chromotography, HSA = human serum albumin, IL-2 = interleukin-2, IsoCit = isocitrate, KGA = α-ketoglutaric acid, Lac = lactate, LDH = lactate dehydrogenase, Mal = malate, MAS = malate-aspartate-shuttle, MDH = malate dehydrogenase, ME = malic enzyme, 2-ME = 2-mercaptoethanol, NAD = nicotine adenine dinucleotide (oxidized), NADP = nicotine adenine dinucleotide phosphate (oxidized), NMR = nuclear magnetic resonance, NTP = nucleotide triphosphate, OAA = oxalacetic acid, OMP = orotidine monophosphate, OUR = oxygen uptake rate, PCA = perchloric acid, PDH = pyruvate dehydrogenase, PEP = phosphoenol pyruvate, PFK = phosphofructokinase, 3-PG = 3-phosphoglycerate, PGL = 6-phosphoglucono-δ-lactone, PK = pyruvate kinase, PLP = pyridoxal phosphate, PPC = pentose phosphate cycle, PRPP = 5 phosphoriboxy-1-diphosphate, PurTP = purine triphosphates, PyrTP = pyrimidine triphosphates, Rib-5-P = ribose-5-phosphate, RNA = ribonucleic acid, RQ = respiratory quotient, Rub-5-P = ribulose-5-phosphate, SF = *Spodoptera frugiperda*, Suc = succinate, SucCoA =

succinyl coenzyme A, TCC = tricarbonic acid cycle (citric acid cycle), t-PA = tissue plasminogen activator, *T. thermophila* = *Tetrahymena thermophila*, UDP = uridine diphosphate, UDP-Gal = UDP-galactose, UDP-GalNAc = UDP-N-acetyl-galactosamine, UDP-Glc = UDP-glucose, UDP-GlcNAc = UDP-N-acetyl-glucosamine, UDP-GNAc = sum of UDP-GalNAc and UDP-GlcNAc, UMP = uridine monophosphate, UTP = uridine triphosphate

2.1.1 Introduction

In contrast to plant cells which normally get their cellular energy from photosynthesis, animal cells need a carbohydrate source and the amino acid glutamine for their metabolic ATP and NADPH producing processes. Then this energy can be used for the maintenance metabolism and/or for the synthesis of particular desired products. If this is a recombinant product and is synthesized from amplified genes the consumed cellular energy can amount to a substantial part of the total energy of the cell. Therefore, a central part of the product obtaining process should be the optimization of the cellular energy household which might be possible by the generation of well-balanced physical, chemical and physiological conditions of the cell-environment system or the design of higher metabolic efficiency. However, optimal knowledge of the biochemical constitution of the biological producer is an absolute prerequisite for a successful advance and should always be the first step of a product oriented process design. Additionally, the exact control and maintenance of growth phases where highest productivity is achieved should be aspired to, since it is very beneficial for a safe and low cost production process promising constant product integrity.

2.1.2 Carbon Sources

Galactose, fructose, and mainly glucose are possible carbon sources used for the *in vitro* cultivation of mammalian cells (Fig. 2.1.1). Especially glucose passes the glycolysis at very high rates in continuous mammalian cells. Cell specific flux rates were measured to 0.13 nmol s^{-1} 10^{-6} in comparison to 0.07 nmol s^{-1} 10^{-6} for insect cells (Neermann and Wagner, 1996). In contrast, the entrance enzyme hexokinase has only a low affinity to galactose (Delhotal et al., 1984) which enters the metabolism at the galactose-1-phosphate level. Gal-1-P can be transformed to Glc-6-P, a metabolite of the glycolysis which marks the cross point of the anabolic PPC and the catabolic oxidation pathway. Fructose can also be phosphorylated to Frc-6-P by the hexokinase. Additionally, it uses the fructose-1-phosphate-pathway and enters the glycolysis on the level of GAP. One can assume that the second possibility is always realized if glucose is also present. Similar to galactose the affinity of the hexokinase for fructose is 20-fold lower than for glucose resulting in a reduced metabolism. Finally, different amounts of energy are provided by the three substrates (Tab. 2.1.1). In particular, galactose oxidation only yielded 6 moles ATP per mole substrate in comparison to 8 for glucose and fructose.

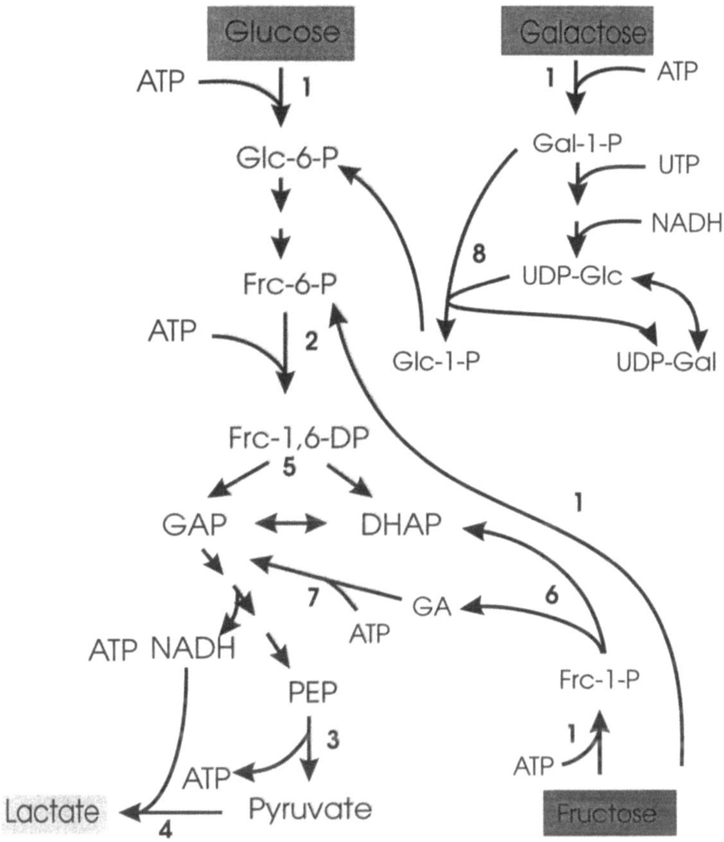

Fig. 2.1.1: Incorporation of the three major carbon sources glucose, galactose and fructose into the metabolism of *in vitro* cultivated mammalian cells. 1 = HK, 2 = PFK, 3 = PK, 4 = LDH, 5 = aldolase, 6 = Frc-1-P aldolase, 7 = triose kinase, 8 = Gal-1-P uridyltransferase.

Table 2.1.1. Comparison of the energy yield during glycolysis for the three main carbon sources in animal cell culture. A synthesis of two ATP from one cytoplasmatically produced NADH has been assumed.

Carbon Source	ATP Demand	ATP Profit	ATP Yield	NADH Demand	NADH Profit	NADH Yield	Total ATP
Glucose	2	4	2	0	2	2	**6**
Fructose	2	4	2	0	2	2	**6**
Galactose	3	4	1	1	2	1	**3**

2.1.2.1 Aerobic Glycolysis is the Result of a Misfunction Cascade

The high glycolytic flux has been found as a general phemomenon in continous cell lines (Donelly and Scheffler, 1976). As demonstrated in Figure 2.1.2 the lactate production depends strictly on the glucose consumption such that most of the available energy is wasted by this metabolism. Due to the fact that theoretically 2 mole lactate are produced from 1 mole glucose, a molar lactate/glucose ratio of only 1 principally demonstrates a well optimized process. Only a small part of total lactate is derived from glutamine. Probably most of this amino acid is oxidized aerobically indicating that this is the main energy source in continuous mammalian cells (Reitzer et al., 1979). Fitzpatrick et al., (1993) found that glucose contributes up to 45% to the energy metabolism during exponential growth and only 3 % and 0.2 % are metabolized in the PPC and the TCC, respectively. Hexokinase, phosphofructokinase, and pyruvate kinase (Fig. 2.1.1) are believed to be the key enzymes responsible for the abnormal metabolic behavior considering that this is the result of a reversible phosphorylation process caused by oncogene encoded protein kinases. Hexokinase is reduced to 11 % of its original level and could be the rate limiting step of the glycolysis (Fitzpatrick et al., 1993; Neermann and Wagner, 1996). Assuming an apparent pyruvate kinase which leads to an accumulation of many glycolytic

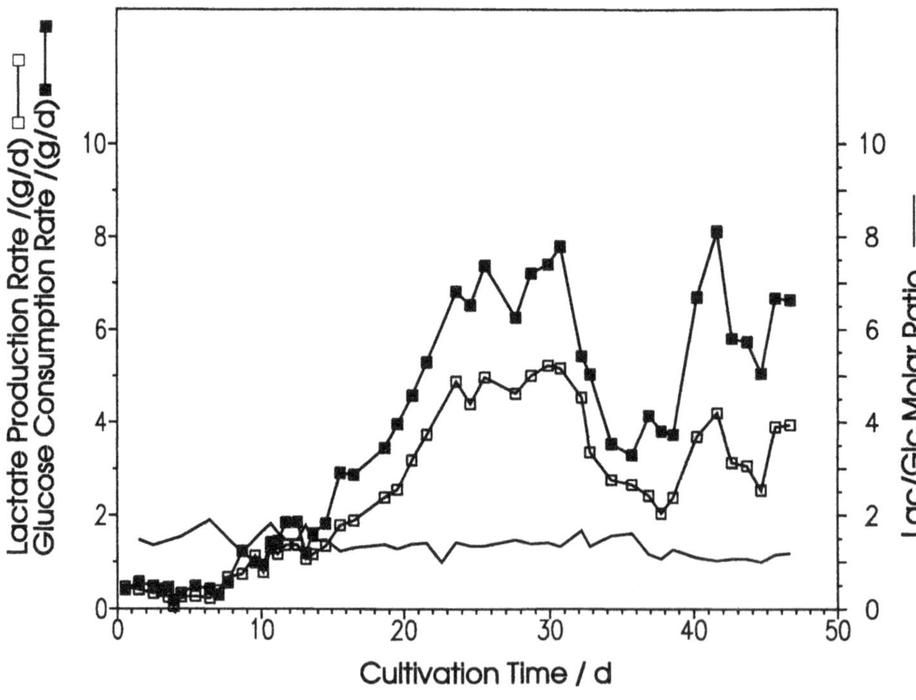

Fig. 2.1.2: Progress of glucose consumption and lactate production of a BHK 21 cell line cultivated in a continuously perfused 1-l-bioreactor. A little more than one mol of lactate is produced for one mol glucose consumed. The lactate production is directly dependent on the glucose uptake rate indicating that this is the main source for lactate.

intermediates including Frc-6-P it presumably resulted in a breakdown of the regulatory control of the phosphofructokinase. This enzyme is normally inhibited by its educt ATP and consequently, in balance with the pyruvate flow used for the TCC and the resulting ATP formation. However, very high levels of Frc-6-P overcome the inhibition effect of ATP, and the regulation of glycolysis by phosphofructokinase is lost. Due to the high rates of glycolytic ATP production, the cytosolic ATP/ADP ratio increases and causes a decrease of the mitochondrial ATP production as the concentration of ADP in this compartment is reduced. As a result, the transmembrane potential increases and the rate of oxidative phosphorylation decreases whereas the mitochondrial $NADH/NAD^+$ ratio is elevated (Williamson and Cooper, 1980). An additional theory relies on a defect malate-aspartate-shuttle in tumor cells which is responsible for the proton import into the mitochondria. This metabolic problem together with a reduced number of mitochondria in tumors would be responsible for a dramatic increase of the cytosolic NADH concentration. However, the demand for energy-supplying ATP in the cell via glycolysis is dependent on the presence of NAD^+ which has to be regenerated by the reduction of pyruvate to lactate. As a result, high amounts of lactate are secreted into the medium. This metabolic defect could explain the high glucose oxidation rate.

2.1.2.2 Mammalian and Insect Cells Differ in their Metabolism

Mammalian cell types are harvested from differentiated tissues. They are primarily mortal and have only a limited capability for cell division (20–50 cell divisions). Therefore, they are generally used for processes after immortalization which can be obtained by harvesting tumor cells, spontaneous events or by a directed introduction or activation of oncogenes or tumor viruses. In contrast, cells derived from insects are harvested from the actively dividing tissues of the larvae which retain the potential of recreating new tissues during metamorphosis of the animal. These cells do not show the atypical biochemical misfunction of secondary derived and degenerated permanent cells from mammalian tissues. Therefore, the lactate production and secretion in insect cell cultures is normally very poor and does not exceed a concentration of $0.5\,g\,l^{-1}$ (Deutschmann and Jäger, 1994). Higher lactate concentrations only appear when oxygen limitation occurs. Additionally, lactate can be reused by these cells when glucose is nearly eliminated in the medium (below $0.2\,g\,l^{-1}$ for SF9 cells).

2.1.2.3 The Pentose Phosphate Cycle Competes with the High Glycolytic Rate for Glucose

As shown in Figure 2.1.3 the pentose phosphate cycle starts from Glc-6-P of the glycolytic pathway. The major role of the PPC is the biosynthesis of NADPH used as reduction equivalents for anabolic processes and ribose-5-phosphate units for the synthesis of nucleic acids. The use of ribose units for DNA duplication or RNA transcription should affect the glycolytic flux and might influence the productivity. Reduction of protein production is assumed during S-phase dependent DNA syn-

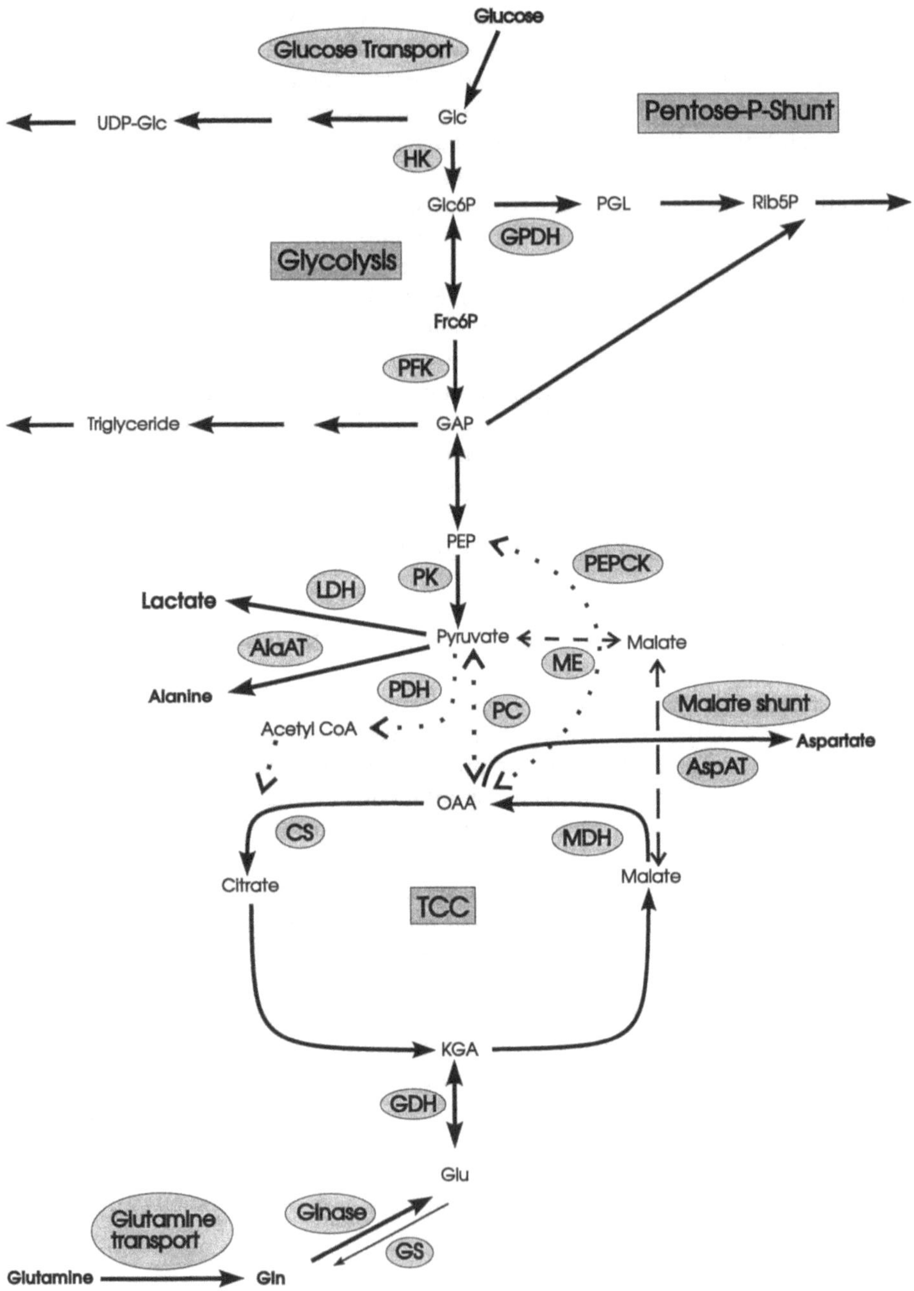

Fig. 2.1.3: Metabolic network of glucose and glutamine degradation. Arrows with dotted line indicate that the respective metabolic way does not exist in continuous mammalian cell lines or is active at very low rates.

Table 2.1.2. Energy used (expressed as ATP) for the formation of Rib-5-P from different carbon sources.

Carbohydrate	ATP Cons.	NADPH Prod.	CO_2 Prod.
Glucose	1	2	1
Fructose (Frc-1-P)	2	2	1
Fructose (Frc-6-P)	1	2	1
Galactose	2	2	1

thesis. Probably the PPC is also controlled by the cell cycle. Glucose-6-phosphate dehydrogenase as the key enzyme also catalyzes the release of CO_2 (Fig. 2.1.1). This entrance step seems to be the most probable candidate for a regulation of this pathway. Reitzer et al. (1979) demonstrated that an increase in glucose concentration from 0.2 mmol l^{-1} to 10 mmol l^{-1} reduces the flux of glucose carbon through the PPC by 50 % while simultaneously increasing the flux of glucose carbon through glycolysis to lactic acid. In contrast, when galactose is used as a substrate instead of glucose the amout of Glc-6-P is too low to provide stationary concentrations high enough to supply sufficient Rib-5-P for the DNA synthesis and the NADPH in exponentially growing cultures. Fructose is also a bad candidate to perform the main task of the PPC because it enters the glycolysis on the level of Frc-6-P and GAP (Fig. 2.1.1). Furthermore, the enzymatic transformation of galactose and fructose to Rib-5-P needs more energy than the use of glucose (Tab. 2.1.2).

2.1.3 Glutamine Metabolism

2.1.3.1 Glutamine is the Main Energy Source in *in Vitro* Cultivated Continuous Mammalian Cells

Typically, 30 to 98 % of the cellular energy in continuous mammalian cell lines is derived from glutamine which is a much higher portion than in normal quiescent cells considering that this is the main energy source (Fitzpatrick et al., 1993; McKeehan, 1986; Reitzer et al., 1979; Stoner and Merchant, 1972; Zielke et al., 1984).

2.1.3.2 Glycolysis and Glutaminolysis are Reciprocally Synchronized

It has been demonstrated that up to 13 % of the glutamine is transformed into lactate (Zielke et al., 1980). The regulation of glucose utilization has been found to be dependent on the glutamine energy metabolism. Decreasing levels of glucose induce an increased glutamine utilization (Donnelly and Scheffler, 1978; Reitzer et al., 1979; Sumbilla et al., 1981; Zielke et al., 1978). When medium was deprived of glutamine below a critical concentration, the metabolism changed from alanine secretion to alanine consumption (Wagner et al., 1988).

The reciprocal regulation of the glycolysis and glutaminolysis of normal and transformed cells has been investigated by feeding them with radioactive substrates. Reitzer et al., (1979) used HeLa cells and showed that a higher glucose concentration (0.2–10 mmol l^{-1}) increased the rate of glucose metabolism to lactate in relation to the PPC. In contrast, the addition of 2 mmol l^{-1} glutamine to cultured normal human fibroblasts stimulates the glucose to pass with more than 77 % of its total amount the PPC. The oxidation of glutamine, however, is inhibited at higher medium concentrations of 0.2–2 mmol l^{-1} (Kucha et al., 1981, Zielke et al., 1978). Fitzpatrick et al. (1993) reported a very low flux of glucose via TCC in transformed cells. No activity of the key enzymes connecting glycolysis with the tricarboxylic acid cycle, such as pyruvate dehydrogenase, pyruvate carboxylase and phosphoenolpyruvate carboxykinase, could be detected in BHK and CHO cells (Neermann and Wagner, 1996). Flux distribution showed glycolytic rates very similar to lactate formation rates. No glucose- or pyruvate-derived carbon entered the tricarboxylic acid cycle, indicating that glucose is mainly metabolized via glycolysis and lactate formation. About 8 % of utilized glucose was metabolized via the pentose phosphate shunt, while 20 % to 30 % of utilized glucose followed pathways other than glycolysis, the tricarboxylic acid cycle or the pentose phosphate shunt. About 18 % of utilized glutamine was oxidized, consistent with the notion that glutamine is the major energy source for mammalian cell lines.

However, in general, an active TCC is a prerequisite for efficient growth, since only this pathway supplies sufficient reduction equivalents and ATP. Certainly, *in vitro* cultivated continuous mammalian cells are restricted in using C_3 metabolites. They passively lose the membrane permeating pyruvate and its transaminase product alanine. The latter is found in the supernatant whereas the loss of pyruvate is prevented by the adjustment of 5 mmol l^{-1} of this metabolite in the medium. Despite its membrane diffusion capability, alanine is also thought to be extruded actively because it is accumulated up to more than 8 mmol l^{-1} (Wagner et al., 1988). In addition, a general defect is also found for anaplerotic activities (pyruvate carboxylase, amino transferases) such that the membrane permeable glutamine should be added in concentrations of about 4 mmol l^{-1} to feed the TCC.

Table 2.1.3. ATP productivity from different glutamine degradation pathways (Glacken et al. (1988), Häggström (1991), modified)

End Product	Gln Specific ATP Production/mol mol^{-1}	Gln Specific NH_4 Production/mol mol^{-1}	ATP/NH_4 Ratio
CO_2	27 (24 + NADPH)	2	13.5
Lactate + CO_2	9	2	4.5
Alanine + CO_2	9 (6 + NADPH)	1	9
Aspartate + CO_2	9	1	9

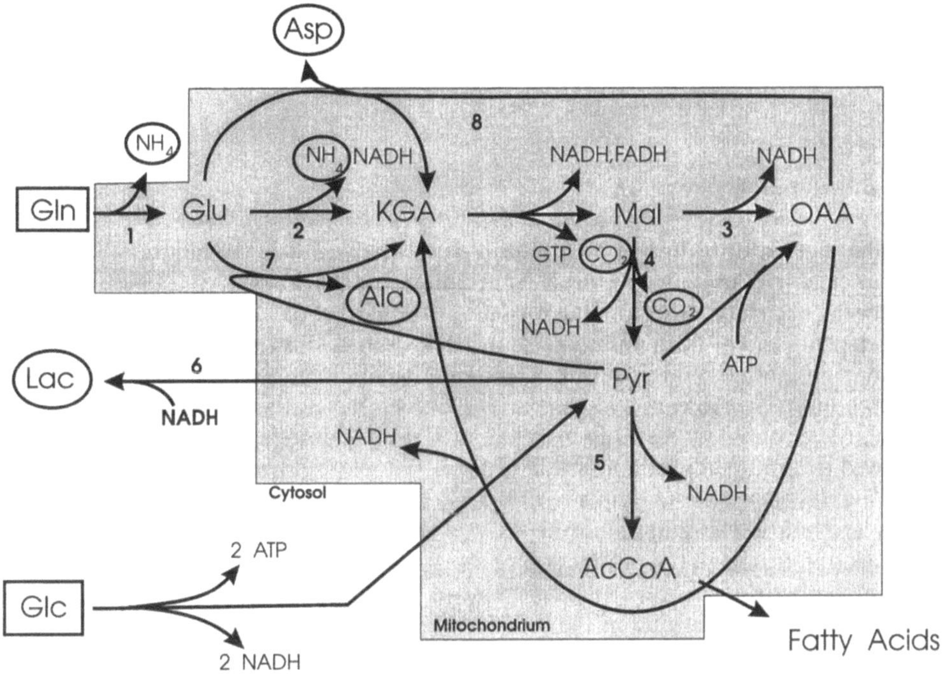

Fig. 2.1.4: The oxidative metabolism of glutamine. The critical reaction steps are indicated by the specific enzyme [according to Glacken (1988), modified]. 1 = Glnase, 2 = GDH, 3 = MDH, 4 = ME, 5 = PDH, 6 = LDH, 6 = AlaAT, 8 = AspAT, □ = nutrients, ○ = end products.

2.1.3.3 ATP Productivity

The carbon of glutamine passes many branched catabolic pathways (Fig. 2.1.4) and every route results in different amounts of ATP (Tab. 2.1.3). The degradation to lactate produces 6–9 mole ATP provided that every mol NADH gives 3 moles ATP. In contrast, the complete oxidation results in a glutamine specific ATP production of 24–27 mol mol^{-1}. Incomplete oxidation to the amino acids alanine and aspartate yields 9 moles ATP. This catabolic pathway is known as glutaminolysis.

Table 2.1.4. Relative portion of glutamine oxidation to ATP production

Cell Type	ATP from Gln Oxid. %	Reference
CHO	40	Donnelly and Scheffler, 1976
Human Fibroblasts	30	Zielke et al., 1978
HeLa (Carbon Source: Glc)	70	Zielke et al., 1980
HeLa (Carbon Source: Frc)	98	Zielke et al., 1980

The portion of ATP derived from glutamine oxidation in relation to glucose metabolism varies from one cell type to the other and is dependent on the environmental conditions. In Table 2.1.14 the relative part of the glutamine oxidation to the ATP production is listed for some cell types.

2.1.3.4 Metabolic Network

Alanine, one end product of glutaminolysis, is secreted into the medium. In mammalian cells it is mainly produced from glutamine (Wagner et al., 1988) whereas insect cells use glucose for alanine formation (Bédard et al., 1993). The direct metabolic dependence of both amino acids in BHK cells is shown in Figure 2.1.5. If the medium glutamine concentration is lowered below 1.5 mmol l^{-1} alanine can be remetabolized by the cells (Wagner et al., 1988). One can assume that this change in metabolic behavior can result in a change of the physiological conditions probably followed by modified productivity. The degradation of glutamine offers several considerable biochemical set points (Fig. 2.1.4). The glutaminase can be considered as a key enzyme of the glutaminolysis as it is an irreversible reaction. Additionally, this reaction results in the production of the growth inhibitor ammonium. However, it is in competition to some transaminations where the ammonia is transferred to other compounds resulting in the biosynthesis of diverse amino acids. Several isoenzymes of the glutaminase are known (Blitz et al., 1982; Horowitz and Knox, 1976; McGivan et al., 1980; Pinkus and Windmueller, 1977; Sevdalian et al., 1980). Liver and kidney type are lovated within the mitochondria. However, at present, no information is available about the type of isoenzyme which is responsible in continuous cell lines. This question may attract attention in the future.

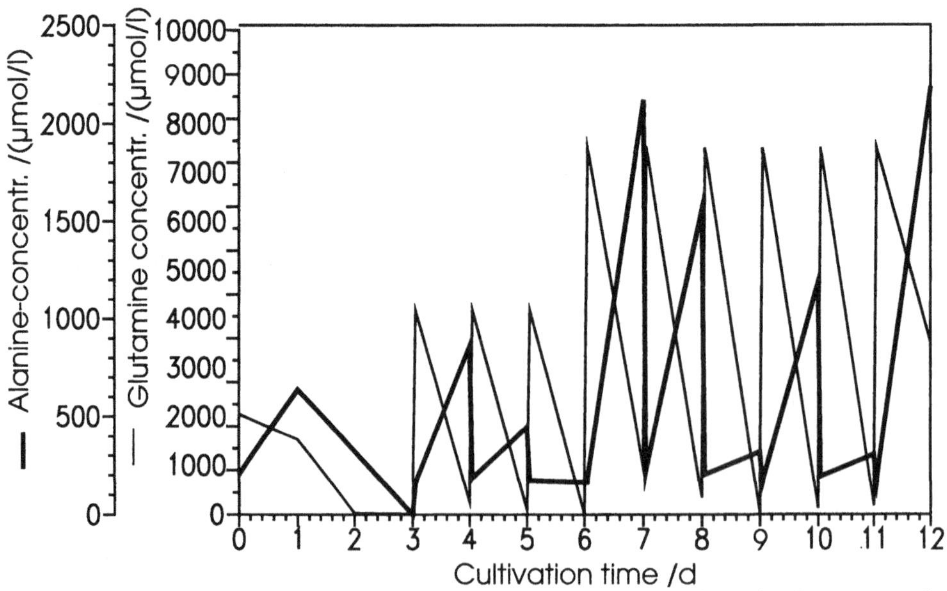

Fig. 2.1.5: Reciprocal consumption of glutamine and production of alanine in a repeated batch culture with recombinant BHK cells grown on microcarriers.

Moredith and Lehninger (1984a; 1984b) found that glutamine is not transformed via malic enzyme in ascites tumor but the glutamate from glutamine in mitochondria is degradated to aspartate by the action of the AspAT. The cosynthesized α-ketoglutarate is stoichiometrically transformed to oxalic acetic acid which participates in the transaminase reaction. In conclusion, the total malate is converted to OAA by the MDH. Reitzer et al., (1980) showed that more than 80 % of the NADPH formed is derived from the activity of the malic enzyme. In contrast, they found radioactive lactate derived from radioactive glutamine (Reitzer et al., 1979). This contradiction, together with an NADPH supply from a PPC fed by glucose, has not yet been clarified (McKeehan, 1986). An inhibition of the malic enzyme leads to the conversion of the total glutamate from glutamine to aspartate depending on the acitivty of the PEP-carboxykinase and the oxalacetate-decarboxylase (Moredith and Lehninger, 1984a; 1984b). This might be advantageous because the secreted end product, aspartate, does not inhibit cell growth in comparison to the alternative lactate (Glacken, 1988).

2.1.4 Oxygen and Carbon Dioxide

2.1.4.1 Carbon Dioxide

Normally, CO_2 is the end product of aerobic metabolism and it is secreted into the medium in *in vitro* cultures. However, in normal cells carbon dioxide is needed as a substrate for the carboxylation of pyruvate to oxalacetic acid catalyzed by the pyruvate carboxylase (Fig. 2.1.6), an important anaplerotic reaction which feeds the TCC where the 2 carbon atoms derived from AcCoA are released again via CO_2. However, no activity of the key enzymes connecting glycolysis with the tricarboxylic acid cycle, such as pyruvate dehydrogenase, pyruvate carboxylase and phosphoenolpyruvate carboxykinase, could be detected in continuous mammalian cell lines (Neermann and Wagner, 1996). Furthermore, CO_2 is the control parameter for the bicarbonate buffer system in media for mammalian cells (in contrast, for insect cell culture the histidine buffer system is used). Therefore, bioreactors must perform the removal of CO_2 by the gas out stream and the control of the CO_2 portion in the gas mixture to maintain a pH value which is optimal for growth and production. In addition, it has been estimated that generally, alkaline conditions promote the formation of lactate and that an acid pH supports the shift from a glycolytic to a more oxidative metabolism.

2.1.4.2 Oxygen

With respect to its function as the end acceptor of the electrons during mitochondrial respiration, oxygen is an important nutrient. However, its solubility in aqueous solutions is very low, in fact, O_2 is often the first limiting compound. According to Henry's distribution coefficient the oxygen solubility is dependent on temperature and other medium compounds. Pure oxygen has a solubility (c_L) in water of

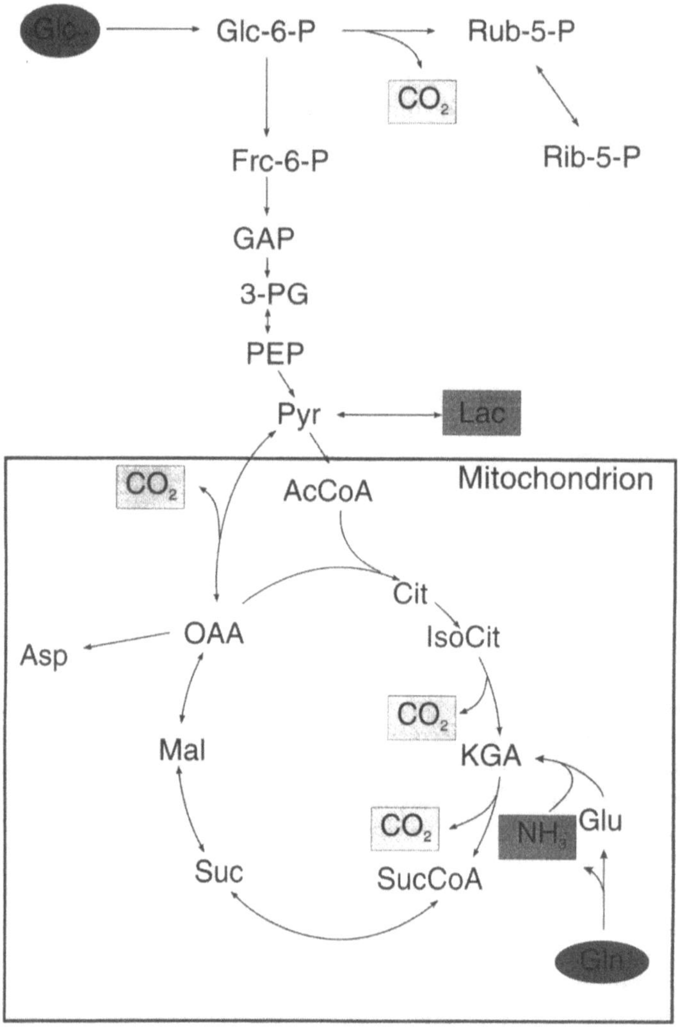

Fig. 2.1.6: The role of CO_2 in glycolysis, TCC and PPC.C_1 carbon is released during PPC whereas during PC and TCC C_1 and C_6 carbon atoms are released.

2.18 mmol l^{-1} at a temperature of 273.15 K (0 °C) and a pressure of 1015 hPa (1 bar). It reduces to 0.43 mmol l^{-1} (air saturation) at the physiological temperature of mammalian cells (37 °C) when the aeration is performed with air. Additionally, $c_L(O_2)$ is influenced by the type and concentration of other dissolved compounds. Carbohydrates reduce c_L with increasing concentrations whereas proteins and peptides show an additional dependence from the molar mass. In particular, a sodium chloride concentration of 2–4 mol l^{-1} reduces the O_2 solubility to 60 % (at 25 °C) of its normal value. Table 2.1.5 shows different temperature-dependent oxygen concentrations for air saturated water according to equation (1).

Table 2.1.5. Oxygen concentrations in air saturated water for different temperatures used in animal cell cultures.

Temperature [°C]	**Oxygen Concentration in Air Saturated Water [$mg\,l^{-1}$]**	**Oxygen Concentration in Air Saturated Water [$mol\,l^{-1}$]**
20	8.85	0.55
27	7.87	0.49
30	7.54	0.47
37	6.87	0.43
41	6.52	0.41

$$c_L(O_2) = 14.16 - 0.394 \cdot T + 7.714 \cdot 10^{-3} \cdot T^2 - 6.46 \cdot 10^{-5} \cdot T^{-3} \quad (1)$$

$c_L(O_2)$ = oxygen concentration in air saturated water [$mg\,l^{-1}$]

T = water temperature [°C]

As pointed out before (2.1.2.1) glycolysis is not repressed by high oxygen concentrations in *in vitro* cultivated mammalian cells. The optimal dissolved oxygen concentration varies with the cell type but is generally about 50% of air saturation. Cells are often oxygen tolerant and accept concentrations between 10 and 100% of saturation with air (Fleischaker and Sinskey, 1981; Miller et al., 1987). However, higher oxygen concentrations can lead to cell death as the result of oxidative damage of cellular constituents (Brosemer and Rutter, 1961). In contrast, at low oxygen concentrations (< 10% of air saturation) glucose and glutamine consumption rates decrease suggesting that growth is oxygen limited (Tab. 2.1.6). Oxygen concentrations below 0.5%, however, promote an increase of glucose and glutamine consumption and press the metabolism to the formation of abnormally high amounts of lactate.

Table 2.1.6. Physiological reaction to different oxygen concentrations in the medium.

Sol. Oxygen Concentr. [% of air saturation]	**Cell Sp. Glc and Gln Cons. Rate**	**Metabolism**	**Growth**
> 100	low	oxidative damage	inhibition
10–100	regular	regular	exponential
0.5–10	low	oxygen limitation	reduced
0.1–0.5	very high	Inhibition of TCC and oxid. phosphorylation intensive Lac prod.	static

2.1.4.3 Respiratory Quotient

The ratio of carbon dioxide evolution rate (CER) and oxygen uptake rate (OUR) is expressed as the respiratory quotient ($RQ = q_{CO2}/q_{O2}$). It can be measured on-line and reflects the physiological state of cells and varies inevitably with the nature of substrates and products. However, CER is difficult to determine in bicarbonate-buffered cell cultures, but can be estimated according Bonarius et al. (1995). The complete oxidation of glucose and glutamine has a mean RQ of 1 whereas amino acids such as alanine or aspartate show RQs of 0.83 and 1.17, respectively (Tab. 2.1.7). If alanine or aspartate are end products of the glutamine metabolism the RQ is calculated to values of 0.67 and 0.33. However, incomplete oxidation of glutamine to lactate also results in a respiratory quotient of 1. The release of the growth inhibitor ammonium is observed for the oxidation of glutamine to the secreted end products lactate, alanine, and aspartic acid. But only the transformation to lactic acid results in 2 molecules NH_3 per molecule glutamine. Amino acids such as alanine or aspartic acid that are also considered to be metabolized to lactate can only be fermented without oxygen consumption, a respiratory quotient cannot be estimated.

In addition, the metabolism used from cells in culture can be very roughly estimated by quantifying the gas-in and gas-out amounts of O_2 and CO_2. The relative amount of oxidative and fermentative metabolism can provide evidence for the metabolic efficiency when RQ is compared to the amino acid uptake rates and the formation of lactate. In addition, a fine analysis of the metabolic situation is possible by use of ^{14}C radioactive substrates and the release of $^{14}CO_2$ (Fitzpatrick et al., 1993; Neermann and Wagner, 1995; Petch and Butler, 1994) (see Figure 2.1.6).

2.1.5 End Products

2.1.5.1 Lactate

In contrast to insect cell cultures, mammalian cells produce large amounts of lactic acid and secrete it into the medium. If a critical concentration is exceeded (Tab. 2.1.8) lactate can become inhibitory to cells grown in culture (Reuveny et al., 1986a). This acid can dramatically reduce the medium pH. This, however, can be avoided by controlling it within an optimum range (pH 6.7–7.4) for growth and product formation. Generally, lactate concentrations below 20 mmol l^{-1} (1.8 g l^{-1}) are considered not to show any negative effects. Glacken et al. (1986) recommended keeping the glucose concentration at low levels in order to reduce lactic acid accumulation and simultaneously increase the respiration.

2.1.5.2 Ammonia

Free ammonium can inhibit growth and production of cell cultures (Butler and Spier, 1984) (Tab. 2.1.9). *In vitro* cultured mammalian cells obtain ammonium from two major sources. It is either derived directly from the medium or results from

Table 2.1.7. Balance sheets of some catabolic pathways for possible substrates of animal cells.

subst. [mol]	**chemical structure**	**O_2 cons. [mol]**	**H_2O cons. [mol]**	**CO_2 prod. [mol]**	**H_2O prod. [mol]**	**NH_3 prod. [mol]**	**endpd. [mol]**	**chemical structure**	**RQ**
1 Glc	$C_6H_{12}O_6$	6	–	6	6	–	–	–	1.00
1 Gln	$C_5H_{10}O_3N_2$	2	–	2	–	2	1 Lac	$C_3H_6O_3$	1.00
1 Gln	$C_5H_{10}O_3N_2$	3	–	2	–	1	1 Ala	$C_3H_6O_2N$	0.67
1 Gln	$C_5H_{10}O_3N_2$	3	–	1	–	1	1 Asp	$C_4H_7O_4N$	0.33
1 Ala	$C_3H_7O_2N$	–	1	1	–	1	1 Lac	$C_3H_6O_3$	–
1 Asp	$C_4H_7O_4N$	–	1	1	–	1	1 Lac	$C_3H_6O_3$	–

Table 2.1.8. Influence of different lactate concentrations on cell growth at constant pH.

Lactate Conc./mmol l^{-1}	Growth	Productivity	Reference
< 20	no effect	no effect	Wagner et al., 1988 Miller et al., 1988b
20–40	no effect	inhibition	Glacken et al., 1988
> 40–60	slight inhibition	inhibition	Glacken et al., 1988
> 60	inhibition	inhibition	Glacken, 1988

Table 2.1.9. Effect of different medium concentrations of ammonium on growth and productivity

Ammon. Conc./mmol l^{-1}	Growth	Productivity	Reference
< 2	no effect	no effect	Glacken, 1988
2–5	inhibition	no effect	Glacken et al., 1988 McQueen and Bailey, 1990b Newland et al., 1990
> 5	inhibition	inhibition	Glacken, 1988

cellular metabolism. In both cases glutamine is involved. The first ammonium generating process is the normal temperature dependent degradation of glutamine to the five-member ring structure pyrrolidone carboxylic acid under the formation of NH_3 according to following reaction (Tritsch and Moore, 1962) (Fig. 2.1.7).

This may yield a significant amount of ammonium as calculated by Negrotti et al. (1989) based on a half-life for glutamine of only 7 days at 37 °C and pH 7.2.

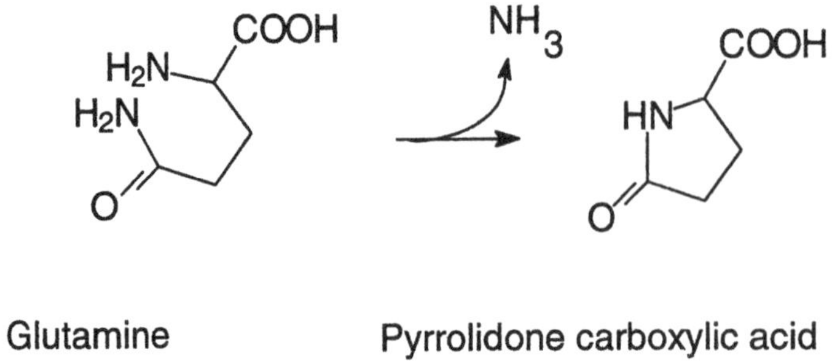

Fig. 2.1.7: Thermic cycling of glutamine to pyrrolidone carboxylic acid under the formation of NH_3.

However, ammonium ions are mainly produced by two steps of the glutaminolysis (GLNase, GDH). Then, it is secreted into the medium or used for the formation on carbamoylphosphate. This metabolic pathway consumes 3 moles ATP per mole NH_4^+ and additionally, fixes one mole CO_2. A heat shock has been reported to increase ammonia accumulation in the growth phase, but reduce its production in the production phase in CHO cells (Tsao et al., 1992).

2.1.5.2.1 Uptake of Ammonia and its Toxicity

It has been shown that the infusion of NH_4Cl stimulated the uracil synthesis in mice liver (Zaharevitz et al., 1988). The responsible reaction is catalyzed by the mitochondrial carbamoylphosphate synthetase (CPSI) which produces large quantites of carbamoylphosphate after application with high amounts of ammonium. Since ornithine is the acceptor for carbamoylphosphate in the mitochondria, it can be detected directly after addition of ammonia to the cell cultures. Hence, the higher concentration of ornithine in the culture supernatant could be an indicator for an elevated activity of the enzymes of the urea cycle and an increased release of metabolites from the mitochondria.

Recently, Martinelle and Häggström (Martinelle and Häggström, 1993) suggested that ammonium ions compete with potassium ions for inward transport, over the cytoplasmic membrane, via potassium transport proteins like the Na^+/K^+-ATPase and the $Na^+K^+2CL^-$-cotransporter. This model is based on the observation that the ammonium induced intracellular pH change was inhibited by the addition of potassium ions. As a result they postulated that one important toxic effect of ammonia/ammonium is an increased demand for maintenance energy, caused by the need to maintain ion gradients over the cytoplasmic membrane and that potassium ions can be used to detoxify ammonia.

Doyle and Butler (1990) described a pH dependency of the cytotoxic effect of ammonium chloride. They assumed a defect in the energy metabolism caused by a decrease of the metabolite pools of the TCC as the mechanism of the toxicity. The formation of α-ketoglutarate by the action of the GDH is considered to be the main target. This reaction is highly influenced by its product ammonia and can be stopped or even returned by the presence of high ammonium concentrations (Figure 2.1.4). Then, the metabolites of the TCC would be consumed followed by a reduced formation of ATP. The lower energy supply could influence the membrane integrity which could explain the appearance of mitochondrial metabolites such as CP and ornithine in the cytoplasm.

The pH effect of ammonia has been confirmed by others (Goochee and Monica, 1990; McQueen and Bailey, 1990a; 1990b) who assumed an effect of ammonium by a reduction of the intracellular pH value. They described that the key enzyme PFK is already inactivated by a decrease of 0.2 units of the intracellular pH. However, they found out that the cell specific yield coefficient for glucose under the influence of added ammonium differed from that influenced by a reduced intracellular pH. Since the glutamine has not been totally consumed by the cells they assumed that a toxic metabolite appeared during these experiments (see 2.1.5.2.2 for the toxic metabolite). Moreover, cells can be adapted to ammonium concentrations (Schumpp and Schlaeger, 1992). Cells were tolerant up to maximum values of about $30 \, mmol \, l^{-1}$ (Ryll et al., 1994).

2.1.5.2.2 Ammonia Increases the Intracellular Concentrations of Activated Sugars
The addition of NH_4Cl to *in vitro* cultivated mammalian cells has been shown to increase the intracellular amount of UDP-GlcNAc and UDP-GalNAc (Ryll et al., 1994). This effect could be simulated by applying glucosamine and galactosamine which are known to be effective inhibitors for the cell growth of mammalian cells, especially tumor cells (Bekesi et al., 1969; Bekesi and Winzler, 1970); Keppler et al., 1970). This effect had been used in the development of anti-cancer strategies (Bekesi et al., 1969; Bekesi and Winzler, 1970; Friedman and Skehan, 1980; Keppler, 1977; Keppler et al., 1970). Additionally, Scholtissek et al., (1975) reported the inhibition of the multiplicity of different viruses. The inhibitory effect of glucosamine is explained by the elevation of the UDP-N-acetylaminohexose (UDP-GNAc) concentration and the simultaneous reduction of the UTP concentration (Bekesi et al., 1969; Bekesi and Winzler 1970; Chou et al., 1984; Friedman and Skehan, 1980; Keppler et al., 1982; Krug et al., 1984; Plagemann and Erbe, 1973; Wice et al. 1979), suggesting UDP-GlcNAc and UDP-GalNAc as key metabolites of the growth control. The formation of activated sugars reduced the UTP pool which resulted in the *de novo* synthesis of UMP due to the disconnected feed back inhibition from UTP. Glucosamine is then transformed to UDP-GlcNAc (Figure 2.1.8) which is in equilibrium with UDP-GalNAc. This reaction had been shown by using ^{14}C pulsed glucosamine and galactosamine (Keppler et al., 1982; Krug et al., 1984; Plagemann and Erbe, 1973) and by step change experiments with added NH_4Cl and $^{15}NH_4Cl$ to an exponentially growing culture with simultaneous HPLC analysis of the intracellularly-appearing UDP-GNAc (Ryll et al., 1994). Interestingly, an expansion of the UDP-GNAc pool had only been observed when glucose was used as carbon source. However, the replacement of glucose by galactose and mannose did not lead to an increase of the intracellular UDP-GNAc concentration (Wice et al., 1979) resulting in ammonium-tolerant cells (Ryll et al., 1994). This is in accordance with the fact that glucose is needed for the formation of fructose-6-phosphate which is the NH_3 acceptor (Fig. 2.1.8).

2.1.5.2.3 Ammonia Can Effect Protein Glycosylation
A dramatic change of the intracellular concentration of activated sugars could have an influence on the product through modulation of the glycosylation system (Goochee and Monica, 1990; Maiorella, 1992; Ryll et al.; 1994, Wice et al., 1979). UDP-GlcNAc had been shown to act as a physiological regulator of the UDP-glucoronyltransferase activity (Burchell et al., 1983). Additionally, Morin et al., (1983) described the effect of tunicamycin and D-glucosamine on L1210 leukemia cells. Both drugs enlarge the UDP-GlcNAc pool and lead to the expansion of the endoplasmic reticulum and the nuclear membrane. Tunicamycin acts as an inhibitor of glycosylation processes and provokes drastic changes of the plasma membrane. In fact, application of NH_4Cl and glucosamine have been observed to change *N*-glycosylation by promoting the formation of higher antennary structures (Gawlitzek et al., 1996) (see also Chapter 2.2 Glycosylation: A Post-Translational Modification).

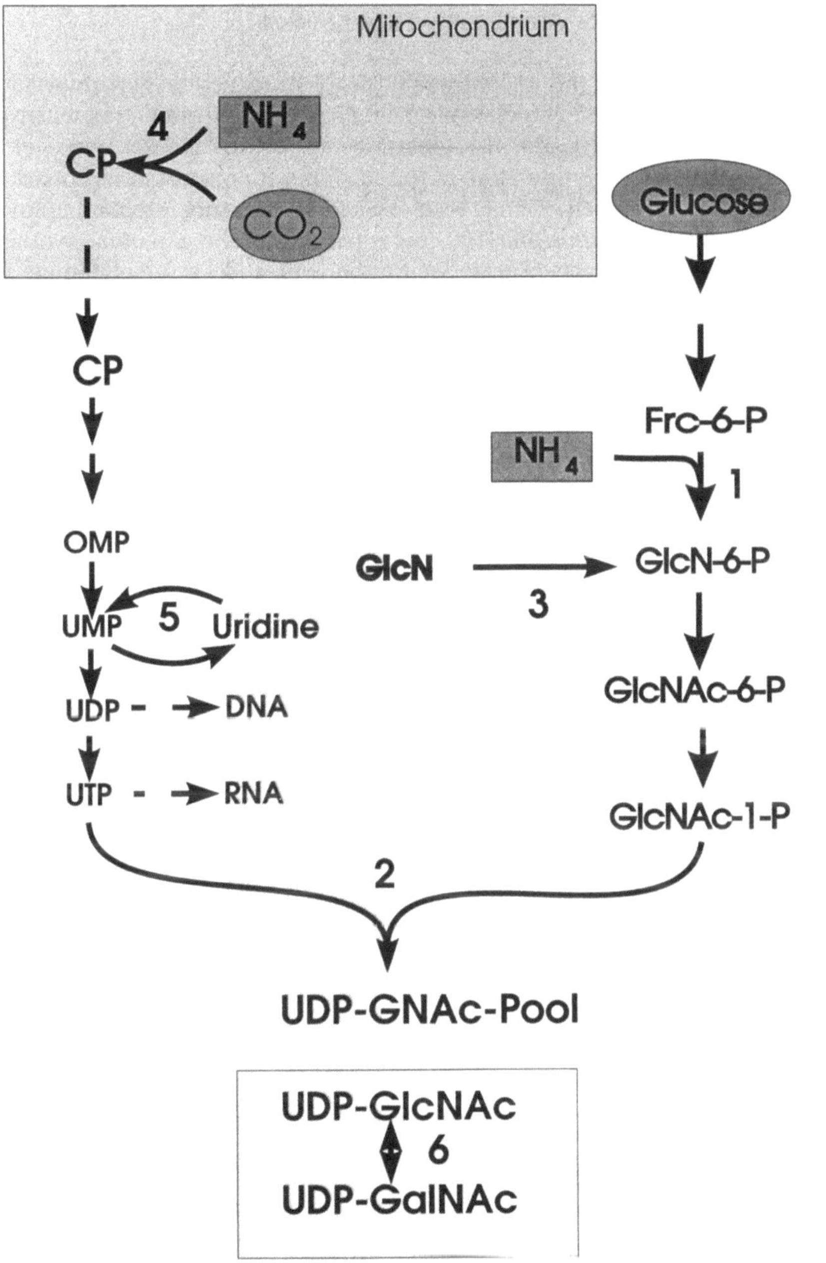

Fig. 2.1.8: Metabolic uptake of ammonia. UDP-GlcNAc is intracellularly accumulated and to toxic concentrations resulting in a growth inhibition. 1 = glucosamine phosphate isomerase, 2 = UDP-glucosamine pyrophosphorylase, 3 = HK, 4 = CPS I, 5 = uridine kinase, 6 = UDP-N-acetyl-D-glucosamine-4-epimerase.

2.1.5.3 Reduction of Ammonium Concentration

In order to overcome this uncontrollable problem of thermal glutamine degradation, several heat stable and autoclavable media based on a formulation where glutamine is replaced by acetyl-L-alanyl-L-glutamine are on the market. Another approach involves the direct manipulation of the cell line with respect to its nutrient demand. In one case a CHO cell line has been selected by exposing it to methionine sulphoximine for an overexpression (1000 copies per cell) of the glutamine synthetase gene. Another strategy involved introducing the gene coding for glutamine synthetase directly into the cells, resulting in both cases in a glutamine autotrophic cell line, and thus eliminating the demand for an exogenous supply of glutamine (Bebbington et al., 1992; Sanders et al.,1987).

A further critical ammonium producing process is the transformation of glutamate to α-ketoglutarate (Figure 2.1.4). If the total glutamate is converted via GDH one mole ammonia is synthesized from one mole glutamate. In contrast, if glutamate is transformed via AspAT and AlaAT then aspartate and alanine are the end products. Hence the GDH represents a key enzyme and a possible regulatory step. GDH is a mitochondrial as well as a cytosolic enzyme (Kovacevic, 1972). In general, alanine, not aspartate, is the main secreted amino acid in most cell lines (BHK 21, hybridoma; Fig. 2.1.5). A metabolic control could be achieved by addition of specific GDH inhibitors such as fumarate and PLP (Smith et al., 1976). Fumarate can enter the cells and pyridoxal is a component of standard media formulations (Nakano et al., 1982). However, it is questionable if pyridoxal can enter the cells (Glacken, 1988). High medium concentrations are reported to possibly inhibit GDH, since then pyridoxal is phosphorylated intracellularly. In addition, PLP is also a cofactor for transaminases and probably able to act as a reciprocal regulator. Furthermore, leucine is critical since it activates the GDH (Smith et al., 1976). Therefore, a reduced medium concentration of this amino acid should be advantageous in some cases.

Normally, arginine is present in standard cell cultures. However, the amount of secreted ornithine produced from arginine is very low in optimally growing cell cultures. In contrast, if ammonium chloride is added the rate of secreted ornithine derived from arginine increases, supporting the assumption that the respective enzyme arginase is induced resembling a detoxification process for ammonia. If some enzymes of the urea cycle are expressed in cells, a genetic completion of this pathway should be evaluated.

A third approach could utilize glucosamine-6-phosphate isomerase, the enzyme that accepts ammonium ions to form UDP-GNAc (see 2.1.2.3). A change of the characteristics of the enzyme should reduce its ammonium utilization capacity and result in ammonium-tolerant cells. Mannose-6-phosphate has been shown to inhibit the enzyme.

2.1.5.4 Maintenance Energy

For process optimization, it is advantageous to know the minimum essential energy necessary for survival of the cell. This maintenance energy is defined as the amount of substrate mobilized at a growth rate of $\mu = 0$. Usually, the substrate is used for the formation of ATP. According to equation (2) the ATP consumption rate is expressed as:

$$q_{ATP} = \frac{\mu}{Y_{ATP}} + m_{ATP} \tag{2}$$

m_{ATP} = maintenance energy (cell specific consumption rate expressed in mol g^{-1} s^{-1})
Y_{ATP} = ATP specific cell yield expressed in g mol^{-1}

The maintenance energy can be calculated from the steady state values of q_{ATP} or the substrate against the steady state growth rate in a chemostate (Glacken, 1988). Miller et al., (1987) have estimated that the maintenance energy for hybridoma cells can amount to up to 60% of the total energy expenditure. This means that the cells must raise most of the energy produced for their maintenance. In contrast, microbial organisms usually have a lower maintenance energy demand (Wang et al., 1979).

There are two ATP-consuming processes under maintenance conditions. The maintenance of ion gradients and the so-called ⟨futile⟩ cycles.

2.1.5.4.1 Ion Gradicnts

Ion gradients across semi-permeable membranes are a fundamental characteristic of all living cells. The Na^+-K^+-ATPase actively pumps ions across a membrane against a concentration gradient by consuming ATP. In addition to the maintenance of a K^+-rich intracellular environment, the ion gradient generated by this enzyme is additionally connected with the intracellular amount of the mediator Ca^{2+}, glycolysis, and oxidative phosphorylation. Additionally, the Na^+-K^+-ATPase is complexly associated with the growth cycle, the cell cycle, and DNA synthesis (Kennedy and Lever, 1984; Leister et al., 1985; Rozengurt and Heppel, 1975). The energy demand for this enzyme is high. 85% of the maintenance energy is used for ion gradients. It was estimated that for rabbit renal cells 50% of the total energy production is required for ion gradients (Harris et al., 1981). There is some evidence that the energy for the ATPase is compartmentalized, e.g. glycolysis provides all the ATP for the ion gradients in muscle (Lynch and Paul, 1983), the heart (Paul et al., 1979), in kidney cells (Lynch, 1987; Lynch and Balaban, 1987) and in ascites tumor cells (Scholnick et al., 1973), while in some other cell systems the total energy is derived from oxidative phosphorylation. However, information about biotechnologically important standard cell lines is scarce. However, glutamine metabolism is the most probable candidate for continuous mammalian cell lines whereas in insect cells glucose oxidation should be the energy supplier.

2.1.5.4.2 ⟨Futile⟩ Cycles

Glacken (1988) discussed these metabolic reactions as cycles in which hydrolysis of ATP is the only net result. Two systems are thought to be active in cultured cells. First, the glutamate dehydrogenase/transhydrogenase system (Fig. 2.1.9a), whose reaction is probably driven by ammonia and located in the mitochondria (Tagler et al., 1975). Second, the glutaminase/glutamine synthetase system (Fig. 2.1.9b), whose cycle may also be driven by ammonia. A high flux of ⟨futile⟩ cycles can be calculated to be responsible for wasting large amounts of energy. Glacken (1988) proposed some inhibitors of these reactions to forestall the reactions.

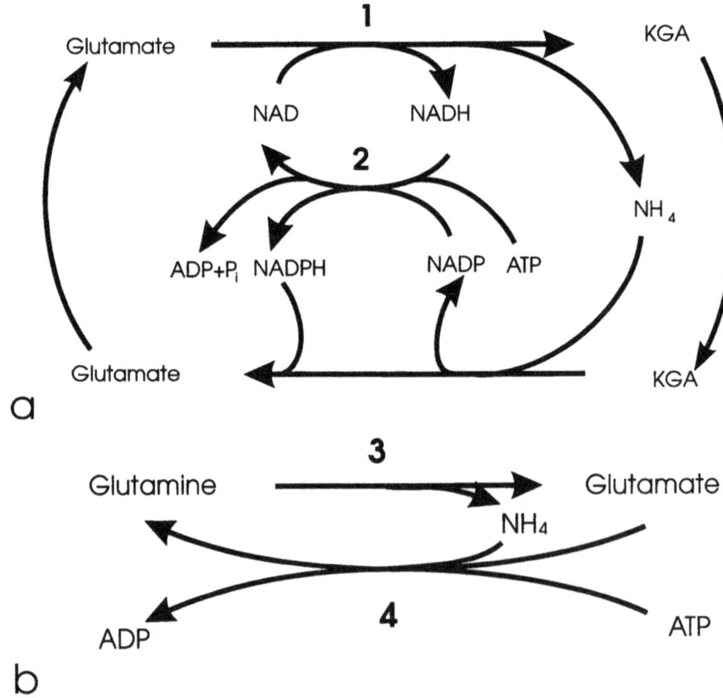

Fig. 2.1.9: The two ⟨futile⟩ cycles thought to be present in higher cells. GDH/transhydrogenase (a) and glutaminase/glutamine synthetase (b). Both cycles are driven by ammonia. 1 = GDH, 2 = transhydrogenase, 3 = Glnase, 4 = GS.

Phosphinate 9 is a very specific inhibitor of glutamine synthetase (Farrington et al., 1987) whereas transhydrogenases can be efficiently inhibited by the addition of rhein (Kean et al., 1971).

2.1.6 Control of Growth Dependent Production Processes

The development and optimization of bioprocesses and the control of cellular growth is generally performed by the determination of specific nutrients such as glucose, glutamine and some essential amino acids as well as secreted waste products, especially lactate and ammonia that are easy to monitor in the medium. All these parameters are involved in a complex network as shown above (2.1.3.4). The specific rates are regulated such that the specific growth phase is not truly reflected by these indirect metabolic parameters of the culture supernatant. Therefore, several approaches have been used for the direct control of the cell reactor by the observation of more general metabolites that are involved in central growth control (see also Chapter 3.4 for On-line and Off-Line Process Analysis).

2.1.6.1 Environmental Control

Generally, long-term cultivation of animal cell cultures has been monitored by the control of the environmental medium conditions particularly glucose, lactate and amino acid concentrations. However, media are considered to be optimally developed only for a limited range of cell lines such that often most of the non-consumed nutrients are wasted by high perfusion rates due to the early limitation of only one compound. One solution to overcome this problem is the exact knowledge of the cell specific consumption rates for many, if not all, compounds, and thus to supplement the medium for complete consumption. This requirement is performed in the medium-recycle-fermentor where only the exhausted medium components are supplemented by a high concentrated formulation until the rest of the nutrients are also depleted, or accumulated inhibitory products become a handicap (Kempken et al., 1991). This control system requires a high analytic expenditure but does not consider the change in nutrient uptake rates if medium concentrations decrease and the accumulation of inhibitory compounds occur during fermentation, which can induce modified physiological conditions.

2.1.6.2 Intracellular Control

2.1.6.2.1 NMR Offers the Accessibility to Special Metabolites

Several processes using immobilizing techniques (e. g. hollow fiber and fluidized bed bioreactors) prevent direct or representative access to the cells such that indirect methods for the control of the procedure become necessary. The development and optimization of these dense cell systems is supported by the use of NMR spectroscopy. In a few cases, immobilized and suspended cells have been compared (Brodelius and Vogel, 1984). Intracellular pH, the levels of phosphorylated metabolites, and approximate P/O ratios were investigated. ^{31}P-NMR offers information about nucleosides and inorganic phosphate whereas ^{13}C-NMR *in vivo* spectroscopy is used for monitoring the intracellular glucose and lactate content as well as other metabolites of the primary metabolism (Mancuso et al., 1994). Additionally, ^{15}N-NMR shows the metabolism of the important nitrogen but the relatively high costs of ^{15}N isotopes limit this application to selected questions. Fernandez et al., (1988) used NMR for monitoring a perfused membrane reactor with hybridoma cells. Step changes by the addition of high amounts of glucose (10 mmol l^{-1}) resulted in a sharp increase of intracellular lactate and nucleoside triphosphates (NTP). Further, Mancuso et al. (1990) determined the cell concentration in hollow fiber bioreactors with ^{23}Na-NMR and ^{31}P-NMR. However, due to the special reactor equipment necessary for the NMR spectrometer and the specific NMR devices as well as the isotopic labeled precursors application for process development is limited and use for routine processes is not practical.

2.1.6.2.2 Flow Cytometry Distinguishes Between the Proliferative and the Non-Proliferative State of Growth

In rapidly growing cell cultures the portion of dividing cells is higher than in quiescent systems. Due to the fact that the DNA content is duplicated an efficient control of growth dependent processes is possible by monitoring the cell cycle status (G_1, S,

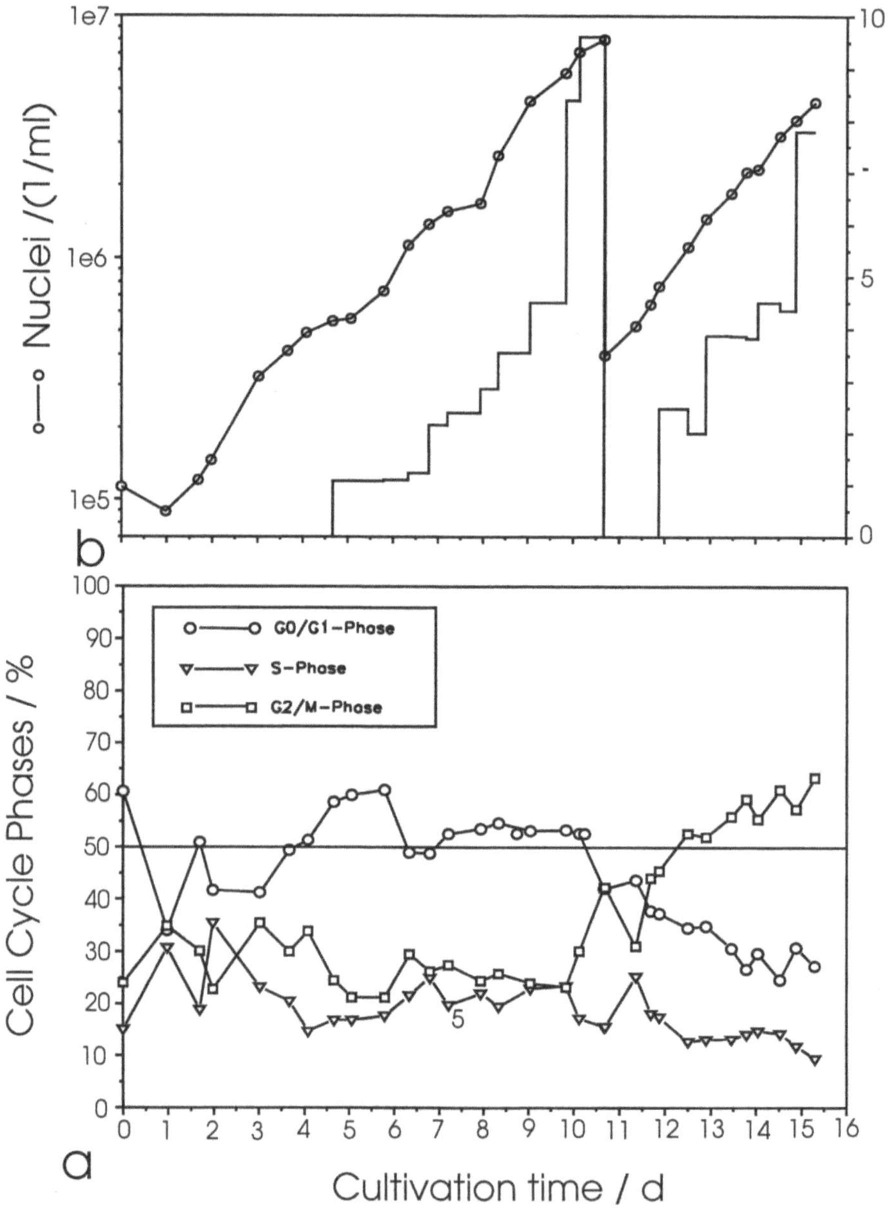

Fig. 2.1.10: Progress of cell cycle phases for a continuously perfused cultivation of a recombinant BHK 21 cell line (a) and the respective cell growth (b). During the second phase of the culture which started at day 10.6 optimized perfusion conditions have been adjusted resulting in an increased portion of G_2+M phase cells.

G_2+M) of single cells. By using suitable DNA staining methods (eg. DAPI or propidium iodide) a FACScan is able to distinguish between these three cell cycle phases. A high portion of S-phase cells that can increase to about 60 % of the total cell amount demonstrates exponential growth state. Deutschmann et al., (1995) could control the perfusion rate of a cultivation of recombinant BHK cells (Fig. 2.1.10b). Optimal conditions with respect to an exponential growth rate were adjusted by maintaining the portion of cells in the G_2+M phase of the cell cycle to about 60 % (Fig. 2.1.10a).

2.1.6.2.3 The Intracellular Ribonucleotide Pool Generates Robust Control Parameters

Nucleotides have been shown to be key parameters for growth dependent processes (Ryll and Wagner, 1992; Wylegalla et al., 1985). They are involved in a number of cellular processes and have widespread regulatory potential (Atkinson, 1977). They participate as substrates, products, effectors or energy donators in many cellular reactions. Especially, their pool sizes are considered to trigger essential life functions by circulating them with the growth cycle.

Due to their fundamental regulatory potential, fluctuations of nucleotide pool bulks have been considered as affecting alterations in transport processes, macromolecular synthesis and cell growth. Some evidence has been found, that the pool sizes of ATP or of UTP or the adenylate energy charges are influence or dependent upon the cell cycle (Rapaport et al., 1979), the stimulation of cells by serum (Grummt et al., 1977) or colchicine (Chou et al., 1984), and growth control (Murphree et al., 1974). Additionally, in plant cells a specific variation of the intracellular concentrations of nucleotides has been shown to be dependent on the growth cycle (Wylegalla et al., 1985).

Recently, high performance detection systems have been developed to quantify intracellular nucleotide and activated sugar nucleotide concentrations in animal cells (Ryll and Wagner, 1991). This method has replaced time consuming enzymatic methods such as the luciferin-luciferase system which has often been used for calculation of the AEC (El Hag et al., 1990; Redegeld et al., 1989) and which possesses the disadvantage of detecting only one single compound.

The intracellular amount of single nucleotides shows no significant change with respect to altered growth behavior except in the case of UDP-GlcNAc, which increases as a result of ammonium enrichment (Ryll and Wagner, 1992). In addition, the combintion of nucleotides expressed as energy charge (eq. 3) is not relevant in *in vitro* cultivated mammalian cells as it is maintained at a very high level, above 0.95 even if cells pass the starvation phase (Ryll et al., 1991a; 1991b).

$$AEC = \frac{(ATP + 0.5\ ADP)}{(ATP + ADP + AMP)} \tag{3}$$

However, the specific combination of intracellular pools of specific nucleotides within an equation results in growth specific functions. In particular, two combinations give a characteristic response: the quotient of the sum of purine triphosphates and pyrimidine triphosphates (NTP ratio, eq. 4) and the uridine ratio (U ratio) expressed as the quotient of UTP and UDP-GNAc (eq. 5).

$$NTP = \frac{PurTP}{PyrTP} = \frac{(ATP + GTP)}{(UTP + CTP)} \tag{4}$$

$$U = \frac{UTP}{UDP - GNAc} \tag{5}$$

It has been shown that the growth cycle dependent variation of nucleotide pools is a fundamental and universal attribute of almost all *in vitro* cultivated systems. It is characteristic in mammalian and insect cells and in complete cellular organisms such as ciliates (Ryll and Wagner, 1992; Ryll et al., 1995) as well as in plant cells (Meyer and Wagner, 1985; Wylegalla et al., 1985) and bacteria (Ryll, 1992).

2.1.6.2.4. NTP and U Ratio During Growth Cycle

During the lag-phase a constant intracellular NTP ratio is adjusted as a prerequisite for the following exponential cell growth (Fig. 2.1.11). This ratio remains constant during the exponential growth phase and increases linearly from the beginning of reduced growth to the end of this phase. Generally, this growth phase is not easily recognized if only the cell number is determined. In contrast, the NTP ratio drastically increases. A relatively small alteration of the culture conditions results in a sharp change in this parameter. During the stationary and following starvation stage, where nutrient limitation or inhibition by a specific compound occurres, the NTP ratio decreases again as a result of a reduced PurTP pool.

The U ratio shows the opposite behavior (Fig. 2.1.11). The value increases up to the early exponential phase and also remains relatively constant during this phase. When comparing the NTP and the U ratio it is remarkable that the calculated end

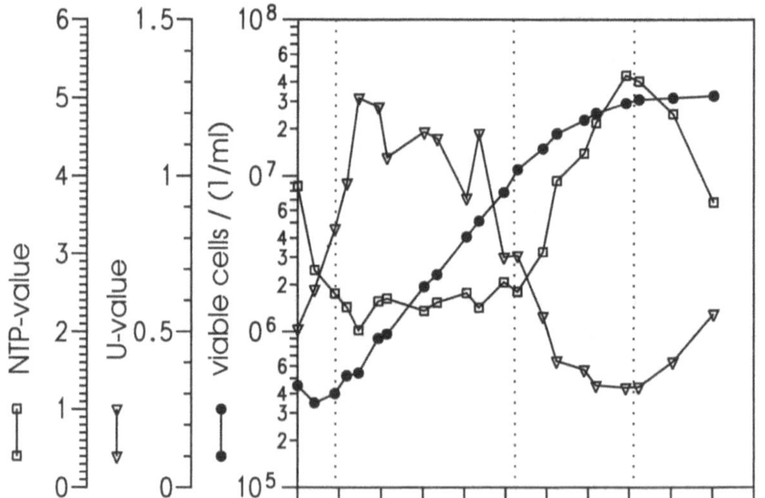

Fig. 2.1.11: Progress of NTP and U value (a) during a continuously perfused cultivation of a mouse mouse hybridoma cell line in a 1-l stirred bioreactor. The vertical dotted lines indicate the phase borders for the adaptation (lag), exponential (log), reduced exponential and starvation phase.

of lag-phase is not identical to the end of lag-phase as estimated by cell counting. Here, one advantage of the cell analysis was obvious in recognizing the precise end of a growth phase which was not exactly detectable by the estimated phase boundaries from the cell numbers (dotted vertical lines). Determination of cell physiology allows a fine tuning of cell growth that has never been possible with the standard methods. This is identical with the U-ratio, particularly at the end of the exponential growth where the value decreases drastically before the real end of the log-phase can be determined by the cell number.

2.1.6.3 NTP to U-Plot

The results obtained from the nucleotide calculation can be applied for culture control by using the following NTP versus U plot (Fig. 2.1.12). Every change of the cells during the growth cycle is observed within a range of a cell specific function. A particular area of this function which is framed by the rectangle which characterizes an upper and a lower limit of a permitted region corresponds for example to the exponential growth phase. Every calculated parameter of a culture which is located within this permitted area must be positioned in an exponential state of growth. All the other values on the left and the right side of the exponential region belong to lag phase conditions, to the reduced exponential growth or to the starvation phase. The general behavior of the cells is as follows: If the cells reduce their growth the values move to the left upper region (increased NTP but decreasing U value). This movement can be countered by increasing the perfusion rate or a batchwise

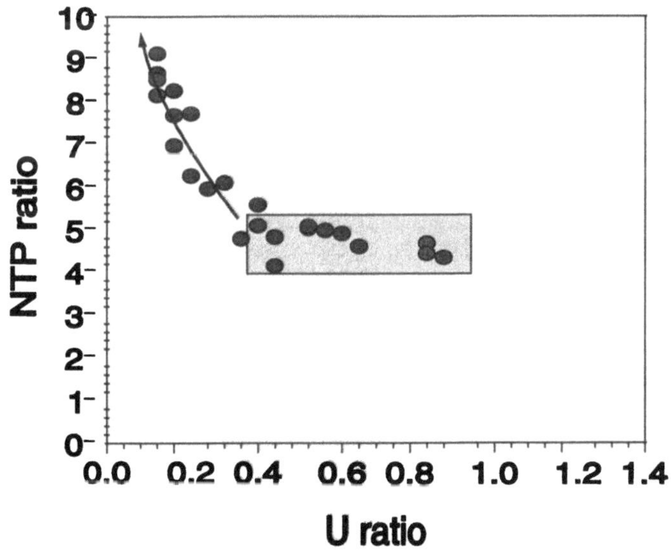

Fig. 2.1.12: The NTP/U-plot for the control of cell cultivation processes for a batch cultivation of HeLa cells. Points within the rectangl represents the exponential growth phase and have been proposed to designate as "log-box" whereas values outside this region point to the lag phase, phase of reduced growth or the starvation phase (left upper side).

medium exchange. When arriving at maximum point of the left upper side the values move down again (reduced NTP value). During this phase where the values move from the upper region either to the left lower or the outer right region a bleeding rate or a batchwise cell harvest is recommended. As a result of its simple operation this function is proposed for the observation, control and regulation of bioprocesses based on *in vitro* cell cultures. This direct control of mammalian cell culture has been shown to be very efficient and has lead to optimal conditions as has only been possible with classical methods.

2.1.6.4. Extracellular Nucleotides Act as Effectors of Cell Proliferation and Productivity

Nucleotides are involved in the control of central growth processes. Natural adenosine derivates such as NADH, NAD, ATP, ADP, AMP and adenosine except inosine inhibit cell proliferation of several normal and tumor cells *in vitro* (Di Virgilo et al., 1990; Ghose et al., 1989; Green and Chan, 1973; Henderson and Scott, 1981; Lohmann and Hugo, 1989; Parsons and Hayward, 1983; Rapaport, 1983; 1988; Rapaport et al., 1983; Snyder et al., 1978). The key enzyme for inhibition of cell proliferation is the intracellular adenosine kinase (Ghose et al., 1989; Green and Chan, 1973; Henderson and Scott, 1981). Extracellularly applied adenosine is taken up and intracellularly phosphorylated by the adenosine kinase to AMP which primarily inhibits the PRPP-Synthetase. As a result, the complete nucleotide pool except AMP, ADP and ATP, decreases (Henderson and Scott, 1981). The reduced NADP pool interferes with oxidative PPC and may reduce cholesterol, fatty acid synthesis and cell growth (Eigenbrodt et al., 1985; Snyder et al., 1978). It has been shown that the intracellular accumulation of AMP and the inhibition of PRPP represent the main effects of natural adenosine derivates. As a result of reduced pyridine nucleotides which act as coenzymes in many important enzymatic reactions several key reactions were inhibited (eg. dehydrogenases such as LDH, MDH and GAPDH). Consequently, Frc-1,6-DP accumulates analogous to the "aerobic glycolysis effect" as already mentioned under 2.1.2.1 (Chiaretti et al., 1979; Lopez-Alarcon and Eboli, 1986; Lopez-Alarcon et al., 1979). In AMP-arrested cells, intracellular pyruvate accumulates and glutaminolysis is stimulated suggesting that AMP predominantly inhibits LDH followed by a drastic reduction of glycolytic flux and energy production. In consequence, higher amounts of lactate are produced from glutamine (Hugo et al., 1992). The controlled addition of adenosine to production processes remains to be investigated but the relatively high cost of the substances does not promise technologically realizable practice. However, the application of these metabolites to cell cultures may give a better access to the understanding of questions about the metabolic control of cells.

Cyclic nucleotides such as cAMP and cGMP are known to be regulatory mediators and they are also involved in the secretion of biologically active proteins (Goldberg et al., 1975; Hadden et al., 1974; Perkins, 1973). Dalili and Ollis (1988) described a 41 % higher antibody secretion rate after treatment with cGMP (Ca^{2+}/dcGMP). Further, the addition of 1 mmol l^{-1} cAMP (as dcAMP) enhanced the specific production rate by 37 % but inhibited cellular growth. Thus the regulation of cell growth by cAMP may offer the advantage of maintaining cells in a viable non-proliferating state in high cell density fermentors with increased productivity.

2.1.6.5 Turn Over Rates

When controlling the cell specific oxygen uptake rates and the intracellular nucleotide pool additional important informations are available about the turn over rates for ATP which give indications about the metabolic activity of an organism. Calculation is performed on the assumption that theoretically, the amount of formed phosphate bonds per oxygen consumed (P/O-quotient) reached the value 3 because maximally 3 molecules ATP are produced when 2 electrons from NADH were transferred to one oxygen atom. Due to the fact that the oxidative phosphorylation is a proton linked force and not a chemically linked reaction this theoretical value is never completely obtained. This is caused by leakages for the protons, connected transport processes and the influence of the concentrations of ATP, ADP, P_i, O_2 and H_3O^+. Assuming that a P/O quotient of 2 is reached the turn over rate for ATP can be estimated for intracellular steady state concentrations (Ryll, 1992) (Tab. 2.1.10).

Mammalian and insect cells have an ATP turn over rate of nearly 1 minute. The protozoon organism *T. thermophila* is clearly faster with 10 seconds and *E. coli* can convert its total ATP pool within 0.1 seconds. The turnover rates are connected to the growth rates of the cells. BHK 21, HeLa and Sf 21 have a growth rate μ of up to $0.8\,d^{-1}$ whereas, for this experiment, the two eucaryotic and procaryotic organisms show values of 0.26 and $0.316\,h^{-1}$, respectively[116].

Table 2.1.10. Estimated turn over rates for the ATP pool (according to Ryll, 1992). Steady state rates are calculated for $\mu = 0.38$.
[1] = $\mu mol\ g^{-1}\ s^{-1}$; [2] = $\mu mol\ g^{-1}$

Cell Line		Cell Spec. O_2 Uptake [amol s^{-1}]	Cell Spec. ATP Prod. [amol s^{-1}]	Cell Spec. ATP Conc. [amol]	Turn Over Rate of the ATP Pool [s]
Hybridoma		20–30	80–120	4000–5500	46–50
BHK 21	**Batch Perfusion**	30–70	120–280	8000–14000	50–67
	Chemostate Steady State	38	152	6500	43
HeLa		30–50	120–200	9000–10000	50–75
Sf21		25–50	100–200	9000–11000	55–90
Tetrahymena thermoph.		1000–1700	4000–6800	30000–65000	7.9–9.6
E. coli		1.75–3.5[1]	7–14[1]	0.6–1.4[2]	0.09–0.1

2.1.7 Cells Secrete Proteases

During cultivation of animal cells, proteases are released into the medium. This process happens during the proliferation phase of the cells as well as after their death and following cell lysis (Lind et al., 1991a; 1991b). In particular, cultures of transformed cells often show a higher protease activity than those of normal cells (Goldberg and Dice, 1974; Hatcher et al., 1977). However proteases had been of lower interest when only medium, supplemented with serum, was used because serum consists of up to 10 % protease inhibitors (Cock and Chen, 1988), particularly serine proteases such as α-antitrypsin and antithrombin III (Highsmith and Rosenberg, 1974; McGilligan and Thomas 1991; Valeri et al., 1980; Venneröd et al., 1976; Wong et al., 1982). Additionally, the total protein content in relation to the amount of product is very high, such that proteases have considerable substrates besides the recombinant protein or monoclonal antibody. In comparison, in cultures with serum-free or particularly protein-free medium proteases can have a negative influence upon cell proliferation as well as causing protein substrate loss (Murakami, 1989). Furthermore, a low protein content may cause higher protease activity (Schlaeger et al., 1987).

2.1.7.1 Proteolytic Activities Derived from Cell Culture Media

Supplementation of culture media with HEPES buffer, which is often routinely used in standard cell culture, can cause problems because HEPES was shown to have a significant proteolytic effect on chromogenic substrate – suggesting that it is probably also active on proteins synthesized by cells (Kratje et al., 1994). Despite the presence of protease inhibitors a specific proteolytic activity in serum albumin-containing media (especially for HSA) has been observed which was not detected when serum-free medium conditions existed (Kratje et al., 1994). This result showed, that in addition to several other advantages, such as the absence of mycoplasmas and viruses and the presence of defined medium conditions, the serum-free state may also provide the benefit of reducing the proteolytic potential on the products which results in a better stability during the subsequent downstream process operations.

2.1.7.2 Proteolytic Activities Derived from the Cells

CHO cells are considered to secrete a cysteine endopeptidase continuously (Satoh et al., 1990) and hybridoma cells seem to release an acid proteinase (Karl et al., 1990). In culture supernatants of recombinant BHK cells a specific protease which specifically cleaves behind the amino acids arginine and lysine has been observed only when drastical changes of the growth conditions occured (e.g. from serum-containing to serum-free medium) (Kratje and Wagner, 1994). This protease only shows a short transient activity while the cells adapt to the new environment of reduced protein concentration.

The protease equipment has been shown to be cell specific by comparing culture supernatants from BHK and hybridoma cells. Characterization by specific protease inhibitors and activators discloses a significant difference between the two cell lines. As already proposed by Murakami (1989), the addition of the serine protease inhibitor aprotinin showed a total inhibition of the proteolytic activities in culture supernatants of a mouse-mouse hybridoma cell line. No effect was observed for a BHK cell line with serine protease inhibitors. Here, the plasminogen activator specific inhibitor, α-aminocaproic acid, could reduce the total activity below 50% of the nontreated supernatant. Additionally, an activation was found for 2-ME which probably stabilized the proteolytic enzymes, but it could not activate proteases derived from the hybridoma cells (Kratje et al., 1994). This might be explained by the fact that BHK cells are originally adherent cells and a lot of proteolytic operations are involved in the complex process of adherance. However, BHK cells can also grow in suspension. This tremendous physiological difference can also have an effect on the protease composition and on the quality and quantity of the proteolytic attack on recombinant products as shown for recombinant interleukin-2 produced from this host cell line in serum-free suspension cultures (Sugimoto et al., 1992).

2.1.8 Productivity and Proliferation

Some cells produce better when they are in a proliferative state of growth, and others secrete better when they are quiescent. In particular, the production of monoclonal antibodies from hybridoma cells has been found to be often independent from the cell division but they also continue to produce during the stationary and the starvation phases of the culture (Birch et al., 1985; Merten et al., 1985; Miller et al., 1988b; Renard et al., 1988; Reuveny et al., 1986b; Velez et al., 1986). Whereas in some cases the antibody production rate appears to be constant (Renard et al., 1988; Velez et al., 1986) in others the specific rate of antibody production is higher during the stationary and starvation phases than during exponential growth (Birch et al., 1987, Emery et al., 1987). In contrast to growth kinetics, little information is available about the relation between productivity and the cell propagation. There is however, some evidence from dense cell cultivation systems that generally production of monoclonal antibodes is independent from proliferation in hybridoma cells whereas recombinant cell lines often seem to reduce their productivity when they are cultivated in dense-cell immobilizing systems (Ryll et al., 1990a; Ryll et al., 1990b). However, many factors which affect hybridoma growth also increase the cell specific rate of antibody production (Dalili and Ollis, 1988; Glossy et al., 1988; Miller et al., 1988a; Philips et al., 1987; Reuveny et al., 1986a) so it is not clear whether the increased secretion is a stress-related effect, as has been suggested (Miller et al., 1988a; Øyaas et al., 1989) or whether it is the result of an enhanced antibody production at low growth rates. Studies using continuous cultures indicate that the specific antibody production rate is higher at reduced growth rates (Birch et al., 1985; Miller et al., 1988b) and that these fluctuations are explained to be under cell cycle control. Several hybridoma cell lines have been shown to present the highest antibody production during G_1 and G_2 (Hayter et al., 1992) and G_1 and G_0 phase (Deutschmann et al., 1996) followed by a decline during M phase. The production is not interrupted during S phase of the cell cycle. These results confirm that antibody production from hybridoma cells seems not to be restricted to specific cell cycle phases but the rate is synchronized with the phases. Duration of the total cell cycle is dominantly determined by the length of the G_1 phase, since S, G_2, and M phases are relatively constant. Hence, the proportion of cells passing the G_1, phase is dependent on the cell growth rate. Assuming that the antibody productivity has its maximum during the G_1 phase in hybridoma cells, an increased production rate at low growth rates or during the stationary phase when the majority of the cells are in G_1 and G_0 phase is understandable.

A chemostate culture provides a useful tool for the determination of kinetic parameters, particularly with regard to growth rates. By using this method and different other cultivation modes for myeloma cells it has been shown that the productivity of a recombinant chimeric antibody is growth-dependent (Robinson et al., 1992). Additionally, Kubbis and Stockinger (1990) were the first who showed a cell cycle dependent productivity for CHO cells carrying recombinant DHFR and t-PA genes. A maximum increase up to 157 % of the intracellular DHFR content has been detected during G_1 and/or at G_1/S transition followed by a continuous increase to 200 % during S and G_2/M. However, the coamplified t-PA gene has been shown not to be cotranslated with the respective DHFR gene. The amount continuously increased from relative 100 % in G_1 to 127 % in the early S and it reached its maximum of 159 % during late S, whereas it decreased to 118 % during G_2/M-phase. In contrast, investigations with BHK cells using asynchronous and synchronized cultures (serum depletion, aphidicolin and colcemid treatment) suggest that the maximum productivitiy for recombinant IL-2 occurs during the G_2/M-phase of the cell cycle (Deutschmann et al., 1996) (see also Chapter 3.3.2.1.4).

2.1.9 Conclusion

The control of animal cell cultures by monitoring the metabolic state of the cells has been shown to be an advantageous tool for further development of growth dependent production processes. It offers the access to a defined characterization of the specific state of a cell culture, which has not yet been possible with standard environmental control of nutrient parameters such as glucose and amino acid concentrations and waste products such as lactate and other secreted end metabolites as well as other extracellular compounds in the culture supernatant. Variations of pool sizes as demonstrated for the key metabolites UDP-GlcNAc and UDP-GalNAc can also influence the growth rate and the product quality. Exact analysis of cells offers, in addition to the development (e.g. serum-free media) and control of *in vitro* cell culture processes, a multitude of applications in the field of medicine and drug research by determining precise cell physiology. Cells can be exactly characterized by the NTP and U values. The knowledge of the correct cell physiology is an important prerequisite for further directed metabolic optimization, for the investigation of cell phenomena and possibly for the precise timing of successful genetic experiments. Major impact is expected from application of the knowledge about cell specific rates and fluxes of the metabolic network to recognize potential bottlenecks as targets for a genetic manipulation within a directed metabolic engineering and design.

References

Atkinson, D.A. (1977) Cellular energy metabolism and its regulation. Academic Press New York.

Bebbington, C.R., Renner, G., Thomson, S. King, D., Abrams, D. and Yarranton, G.T. (1992) High-Level Expression of a Recombinant Antibody from Myeloma Cells Using a Glutamine Synthetase Gene as an Amplifiable Selectable Marker. Bio/Technology, *10*, 169–175.

Bédard, C., Tom, R. and Kamen, A. (1993) Growth, Nutrient Consumption, and End-Product Accumulation in Sf-9 and BTI-EAA Insect Cell Cultures: Insight into Growth Limitation and Metabolism. Biotechnol. Progr. *9*, 615–624.

Bekesi, J.G., Bekesi, E. and Einzler, R.J. (1969) Inhibitory effect of D-glucosamine and other sugars on the biosynthesis of protein, ribonucleic acid, and deoxyribonucleic acid in normal and neoplastic tissues. J. Biol. Chem., *244*, 3766–3772.

Bekesi, J.G. and Winzler, R.J. (1970) Inhibitory effects of D-glucosamine on growth walker 256 carcinosarcoma and on protein, RNA, and DNA synthesis. Cancer Res. *30*, 2905–2912.

Birch, J.R., Thompson, P.W., Boraston, R., Oliver, S. and Lambert, K. (1987) In: Plant and Animal Cells: process Possibilities (Webb, C. and Mavituna, F., eds.) Ellis Horwood Ltd., Chistester, 162–171.

Birch, J.R., Thompson, P.W., Lambert, K. and Boraston, R. (1985) The Large Scale Cultivation of Hybridoma Cells Producing Monoclonal Antibodies. In: Large Scale Mammalian Cell Culture (Feder, J. and Tolbert, W. R., eds.) Academic Press, New York, 1–18.

Blitz, R.M., Letteri, J.M., Pellegrino, E.D. and Pinkus, L. (1982) Glutamine: A new metabolic substrate. Adv. Exp. Med. Biol. *157*, 423–429.

Bonarius, H.P.J., De Gooijer, C.D, Tramper, J. and Schmid, G. (1995). Determination of the Respiratory Quotient in Mammalian Cell Culture in Bicarbonate Buffered Media. Biotechnol. Bioeng. *45*, 524–535.

Brodelius, P. and Vogel, H.J. (1984) Ann. N. Y. Acad. Sci. *434*, 496.

Brosemer, R.W. and Rutter, W.J. (1961) The Effect of Oxygen Tension on the Growth and Metabolism of a Mammalian Cell. Exp. Cell Res. *25*, 101–113.

Burchell, B., Weatherill, P.J. and Berry, C. (1983) Evidence indicating that UDP-N-acetylglucosamine does not appear to simulate hepatic microsomal UDP-glucuronosyltransferase by interaction with the catalytic unit of the enzyme. Biochem. Biophys. Acta *735*, 309–313.

Butler, M. and Spier, R.E. (1984) The effects of glutamine utilisation and ammonia production on the growth of BHK cells in microcarrier cultures. J. Biotechnol. *1*, 187–196.

Chiaretti, B., Casciaro, A., Minotti, G., Eboli, M.L. and Galeotti, T. (1979) Quantitative evaluation of the activity of the malate-aspartate shuttle in Ehrlich Ascites tumor cells. Cancer Res. *39*, 2195–2199.

Chou, I.-N., Zeigher, J. and Rapaport, E. (1984) Imbalance of total cellular nucleotide pools and mechanism of the colchicine-induced cell activation. Proc. Natl. Acad. Sci. USA *81*, 2401–2405.

Cook, J.R. and Chen, J.-K. (1988) Enhancement of Transformed Cell Growth in Agar by Serine Protease Inhibitors. J. Cell Phys. *136*, 188–193.

Dalili, M. and Ollis, D.F. (1988) The Influence of Cyclic Nucleotides on Hybridoma Growth and Monocloinal Antibody production. Biotechnol. Lett. *10*, 781–786.

Delhotal, B. Lemonnier, F., Couturier, M. Wolfram, C. Gautier, M. and Lemonnier, A. (1984) Comparative use of fructose and glucose in human liver and fibroblastic cell cultures. In Vitro Cellul. Dev. Biol. *20*, 699–706.

Deutschmann, S.M. and Jäger, V. (1994) Optimization of the Growth Conditions of Sf21 Insect Cells for High Density Perfusion Culture in Stirred Tank Bioreactors. Enzyme Microb. Technol., *16*, 506–512.

Deutschmann, S. M., Vogel, J., Jäger, V. and Wagner, R. (1996) Comparison of the dependency of monoclonal antibody and recombinant protein production with the cell cycle. Cytotechnology, in preparation.

DiVirgilo, F., Pizzo, P., Zanovello, P., Bronte, V. and Covallo, D. (1990) Extracellular ATP as a possible mediator of cell mediated cytotoxicity. Immunology Today *11*, 274–277.

Donnelly, M. and Scheffler, I. E. (1976) Energy metabolism in respiration deficient and wild-type Chinese hamster fibroblasts in culture. J. Cell Physiol. *18*, 39–52.

Doyle, D. and Butler, M. (1990) The effect of pH on the toxicity of ammonia to a murine hybridoma. J. Biotechnol. *15*, 91–100.

Eigenbrodt, E., Fister, P. and Reinacher, M. (1985) New perspectives on carbohydrate metabolism in tumor cells. In: Regulation of Carbohydrate Metabolism, Vol. II (Beitner, R. ed.) CRC Press, Boca Reaton, FL., 141–179.

El Hag, I. A., Ross, G. and Stenram, U. (1990) Decrease of liver energy charge, ATP and glutathione at concomitant intraarterial administration of adriamycin and degradable starch microspheres in rat. Selective Cancer Therapeutics *6*, 135–144.

Emery, A. N., Lavery, M., Williams, B. and Handa, A. (1987) In: Plant and Animal Cells: Process possibilities (Webb, C. and Mavituna, F., eds.) Ellis Horwood Ltd., Chichester, 137–148.

Farrington, G. K., Kumar, A. and Wedler, F. C. (1987) Design and synthesis of phosphonate inhibitors of glutamine synthetase. J. Med. Chem. *30*, 2062–2067.

Fernandez, E. J., Mancuso, A. and Clark, D. S. (1988) NMR Spectroscopy Studies of Hybridoma Metabolism in a Simple Membrane Reactor. Biotechnol. Progr. *4*, 173–183.

Fitzpatrick, L., Jenkins, H. A. and Butler, M. (1993) Glucose and Glutamine Metabolism of a Murine B-Lymphocyte Hybridoma Grown in Batch Culture. Appl. Biochem. Biotechnol. *43*, 93–116.

Fleischaker, R. J. and Sinskey, A. J. (1981) European J. Appl. Microbiol. Biotechnol. *12*, 193–197.

Friedman, S. J. and Skehan, P. (1980) Membrane-active drugs potentiate the killing of tumor cells by D-glucosamine. Proc. Natl. Acad. Sci. USA *77*, 1172–1176.

Gawlitzek, M., Nimtz, M., Valley, U., Wagner, R., Conradt, H. S. (1996) Ammonium ion/glucosamine dependent increase of oligosaccaride complexity in recombinant glycoproteins secreted from cultivated BHK-21 cells. Biotechnol. Bioleng., in press.

Ghose, A. K., Viswanadhan, V. N., Sanghvi, Y. S. Nord, L. D., Willis, R. C. Revanker, G. R. and Robins, R. K. (1989) Structural mimicry of adenosine by the antitumor agents 4-methoxy- and 4-amino-8-(β-D-ribofuranosylamino)pyrimidino-[5,4 d]pyrimidine as viewed by a molecular modeling method. Proc. Natl. Acad. Sci. USA, *86*, 8242–8246.

Glacken, M. W. (1988) Catabolic Control of Mammalian Cell Culture. Bio/Technology *6* 1941–1050.

Glacken, M. W. Adema, E. and Sinskey, A. J. (1988) Mathematical descriptions of hybridoma culture kinetics I. Initial metabolic rates. Biotechnol. Bioeng. *32*, 491–506.

Glacken, M. W., Fleischaker, R. J. and Sinskey, A. J. (1986) Reduction of Waste Product Excreteion via Nutrient Control: Possible Strategies for Maximizing Product and Cell Yields on Serum in Cultures of Mammalian Cells. Biotechnol. Bioeng., *28*, 1376–1389.

Glassy, M. C., Tharakan, J. P. and Chou, P. C. (1988) Serum free media in hybridoma culture and monoclonal antibody production Biotechnol. Bioeng. *32*, 1015–1028.

Goldberg, A. L. and Dice, J. F. (1974) Intracellular Protein Degradation in Mammalian and Bacterial Cells. Ann. Rev. Biochem., *43*, 835–869.

Goldberg, N. D., Haddox, M. K., Nicol., S. E. Glass, D. B., Sanford, C. H., Kuehl, F. A. and Estersen, R. (1975) Biological Regulation Through Opposing Influences of cyclic GMP and Cyclic AMP. In: Advances in Cyclic Nucleotide Research (Drummond, G. I., Greengard, P. and Robison, G. A., eds.) Raven Press, New York, 307–330.

Goochee, C.F. and Monica, T. (1990) Environmental Effects on Protein Glycosylation. Bio/Technology *8*, 421–427.
Green, H. and Chan, T.-S. (1973) Pyrimidine starvation induced by adenosine in fibroblasts and lymphoid cells: Role of adenosine deaminase. Science *182*, 836–837.
Grummt, F., Paul, D. and Grummt, I. (1977) Regulation of ATP pools, rRNA and DNA synthesis in 3T3 cells in response to serum of Hypoxanthine. Eur. J. Biochem. *76*, 7–12.
Häggström, L. (1991) Energetics of Glutaminolysis – A Theroretical Evaluation. In: Production of Biologicals from Animal Cells in Culture (Spier, R., Griffiths, J. B., Meignier, B., eds.) Butterworths-Heinemann, Oxford, pp. 79–81.
Hadden, J.W., Hadden, E. and Goldberg, N.D. (1974) Cyclic GMP and Cyclic AMP in Lymphocyte Metabolism and Proliferation. In: Cyclic AMP, Cell Growth, and Immune Response (Braun, W., Lichtenstein, L.M. and Parker, C.W. eds.), 237–246.
Harris, S.I., Balban, R.S., Barrett, L. and Mandel, L.J. (1981) Mitochondrial respiratory capacity and Na^+-dependent and K^+-dependent adenosine triphosphatase-mediated ion transport in the intact renal cell. J. Biol. Chem., *256*, 319–328.
Hatcher, V.B., Obermann, M.S., Wertheim, M.S., Rhee, C.Y., Tsien, G. and Buck, P.G. (1977) The Relationship between Surface Protease Activity and the Rate of Cell Proliferation in Normal and Transformed Cells. Biochem. Biophys. Res. Comm. *76*, 602–608.
Hayter, P.M., Kirkby, N.F. and Spier, R.E. (1992) Relationship between hybridoma growth and monoclonal antibody production. Enzyme Microb. Technol. *14*, 454–461.
Henderson, J.F. and Scott, F.W. (1981) Inhibition of animal and invertebrate cell growth by naturally occurring purine bass and ribonucleosides. Pharmac. Ther. *8*, 539–571.
Highsmith, R.F. and Rosenberg, R.D. (1974) The inhibition of human plasmin by human antithrombin III-heparin cofactor. J. Biol. Chem., *249*, 4335–4338.
Horowitz, M.L. and Knox, W.E. (1976) A phosphate-activated glutaminase in rat liver different from that in kidney and other tissues. Enzymol. Clin. Chin. *9*, 241–255
Hugo, F., Mazurek, S., Zander, U. and Eigenbrodt, E. (1992) *In vitro* Effect of Extracellular AMP on MCF-7 Breast Cancer Cells: Inhibition of Glycolysis and Cell Proloferation. J. Cellul. Physiol. *153*, 539–549.
Karl, D.W., Donovan, M. and Flickinger, M.C. (1990) A novel acid proteinase released by hybridoma cells. Cytotechnology, *3*, 157–169.
Kean, E.M., Gutman, M. and Singer, T.O. (1971) Studies on respiratory chainlinked nicotinamide adenine-dinucleotide dehydrogense. 22 Rhein, a competitive inhibitor of dehydrogenase. J. Biol. Chem. *246*, 2346–2351.
Kempken, R., Büntemeyer, H. and Lehmann, J. (1991) Antikörper-Produktion im Medium-Kreislauf-Fermenter (MKF). BioEnineering *1*, 16–25.
Kennedy, B.G. and Lever, J.B. (1984) Regulation of Na^+K^+-ATPase activity in MDCK kidney epithelial-cell cultures – Role of growth state, cyclic AMP, and chemical inducers of dome formation and differentiation. J. Cell. Physiol. *121*, 51–63.
Keppler, D.O.R. (1977) Uridine triphosphate deficiency, growth inhibition, and death in ascites Hepatoma cells induced by a combination of pyrimidine biosynthesis inhibition with uridylate trapping. Cancer Res. *37*, 911–917.
Keppler, D.O.R., Rudiger, J.F.M., Bischoff, M. and Decker, K.F.A. (1970) The trapping of uridine phosphates by D-galactosamine, D-glucosamine, and 2-deoxy-D-galactose. Eur. J. Biochem. *17*, 246–253.
Keppler, D.O.R., Schulz-Holstege, C. and Fauler, J. (1982) Uridylate trapping, induction of UTP deficiency, and stimulation of pyrimidine synthesis *de novo* by D-galactosone. Biochem. J. *206*, 139–146.
Kovacevic, Z. (1972) Possibility for the transport of reducing equivalents from the cytosol to the mitochondrial compartment in Ehrlich ascites tumor cells by the malata-aspartate shuttle. Eur. J. Biochem., *25*, 372–378.

Kratje, R. B., Lind, W. and Wagner, R. (1994) Evaluation of the Proteolytic Potential of *In Vitro*-Cultivated Hybridoma and Recombinant Mammalian Cells. J. Biotechnol., *32*, 107–125.

Krug, E. Zweibaum, A., Schulz-Holstege, C. and Keppler, D. (1984) D-glucosamine-induced changes in nucleotide metabolism and growth of colon carcinoma cells in culture. Biochem. J. *217*, 701–708.

Kubbis, M. and Stockinger, H. (1990) Cell Cycle-Dependent DHFR and t-PA Production in Cotransfected, MTX-Amplified CHO Cells Revelaed by Dual-Laser Flow Cytomtery. Exp. Cell. Res. *188*, 267–271.

Kuchka, M., Markus, H. B. and Mellham, W. J. (1981) Influence of hexose conditions on glutamine oxidation of SV-40 Transformed diploid fibroblast human cell sines. Biochem. Med. *26*, 356–365.

Leister, K. J., Wenner, C. E. and Tomei, L. D. (1985) Correlation of quabain-sensitive ion movements with cell cycle activation. Proc. Natl. Acad. Sci. *82*, 1599–1603.

Lind, W., Lucki-Lange, M., Jäger, V. and Wagner, R. (1991a) Characterization of protease activity in protein-free culture supernatants of recombinant mammalian cells. In: Production of Biologicals from Animal Cells in Culture (Spier, R. E., Griffiths, J. B. and Meignier, B., eds.) Butterworths, UK, pp. 196–202.

Lind, W., Lietz, M., Jäger, V. and Wagner, R. (1991b) Proteolytic activities in serum-free supernatants of mammalian cell lines. In: Animal Cell Culture and Production of Biologicals (Sasaki R., Ikura, K., eds.), Kluwer, Dordrecht, 319–327.

Lohmann, W. and Hugo, F. (1989) The effect of NADH on different human and mouse cell lines. Naturwissenschaften, *76*, 72–74.

Lopez-Alarcon, L. and Eboli, M. L. (1986) Oxidation of reduced cytosolic nicotinamide adenine dinucleotide by the malate-aspartate shuttle in the K-562 human Leukemia cell line. Cancer Res., *46*, 5589–5591.

Lopez-Alarcon, L., Eboli, M., De Liverali, E., Palombini, G. and Galeotti, T. (1979) Evidence for the oxidation of glycolytic NADH by the malate-aspartate shuttle in Ehrlich Ascites Tumor cells. Arch. Biocgem. Biophys. *192*, 391–396.

Lynch, R. M. (1987) Energy metabolism of renal cell lines A6 and MDCK: regulation by Na^+-K^+-ATPase. Am J. Physiol. *252*, C225–C231.

Lynch, R. M. and Balaban, R. S. (1987) Coupling of aerobic glycolysis and Na^+-K^+ ATPase in renal cell line MDCK. Am. J. Physio. *253*, C269–C276.

Lynch, R. M. and Paul, R. J. (1983) Compartmentation of glycolytic and glycogenolytic metabolism in vascular smooth muscle. Science *222*, 1344–1346.

Maiorella, B. L. (1992) *In Vitro* Management of Ammonia's Effect on Glycosylation of Cell Products Through pH Control. US Patent Number 5,096,816.

Mancuso, A., Fernandez, E. J., Blanch, H. W. and Clark, D. S. (1990) A Nuclear Magnetic Resonance Technique for Determining Hybridoma Cell Concentration in Hollow Fiber Bioreactors. Bio/Technology *8*, 1282–1285.

Mancuso, A., Sharfstein, S. T., Tucker, S. N., Clark, D. S. and Blanch, H. W. (1994) Examination of Primary Metabolic Pathways in a Murine Hybridoma with Carbon-13 Nuclear Magnetic Resonance Spectroscopy. Biotechnol. Bioeng. *44*, 563–585.

Martinelle, K. and Häggström, L. (1993) Mechanisms of ammonia and ammonium ion toxicity in animal cells: Transport across cell membranes. J. Biotechnol. *30*, 339–350.

McGilligan, K. and Thomas, D. W. (1991) Evaluation of Assays for Detecting α-1-Protease Inhibitor during Purification from Rat Serum. Analyt. Biochem. *193*, 260–265.

McGivan, J. D., Lacey, J. H. and Joseph, S. K. (1980) Localization and some properties of phosphatedependent glutaminase in disrupted liver mitochondria. Biochem. J. *192*, 537–542.

McKeehan, W. L. (1986) In: Carbohydrate Metabolism in Cells (Morgan, J., ed.) Plenum Press, New York, 111–150.

McQueen, A. and Bailey, E. (1990a) Mathematical modelling of the effect of ammonium ion on the intracellular pH of hybridoma cells. Biotechnol. Bioeng. *35*, 897–907.
McQueen, A. and Bailey, E. (1990b) Effect of ammonium ion and extracellular pH on hybridoma cell metabolism and antibody production. Biotechnol. Bioeng. *35*, 1067–1077.
Merten, O.-W., Reiter, S., Himmler, G., Scheirer, W. and Katinger, H. (1985) Production Kinetics of Monoclonal Antibodies. Dev. Biol. Standard. *60*, 219–227.
Meyer, R. and Wagner, K.G. (1985) Nucleotide pools in suspension cultured cells of *Datura innoxia* I. Changes during growth of the batch culture. Planta *166*, 439–445.
Miller, W.M., Blanch, H.W. and Wilke, C.R. (1988a) A Kinetic Analysis of Hybridoma Growth and Metabolism in Batch and Continuous Suspension Culture: Effect of Nutrient Concentration, Dilution Rate, and pH. Biotechnol. Bioeng., *32*, 947–965.
Miller, W.M, Wilke, C.R. and Blanch, H.W. (1987) Effects of Dissolved Oxygen Concentration on Hybridoma Growth and Metabolism in Continuous Culture. J. Cellul. Physiol. *132*, 524–530.
Miller, W.M., Wilke, C.R. and Blanch, H.W. (1988b) Transient Response of Hybridoma Cells to Lactate and Ammonia Puls and Step Changes in Continuous Culture. Bioprocess Eng. *3*, 113–122.
Moredith, R.W. and Lehninger, A.L. (1984a) The pathway of glutamate and glutamine oxidation by tumor cell mitochondria: Role of mitochondrial $NAD(P)^+$-dependent malic enzyme. J. Biol. Chem. *259*, 6215–6221.
Moredith, R.W. and Lehninger, A.L. (1984b) Purification, kinetic behavior and regulation of $NAD(P)^+$ malic enzyme of tumor mitochondria. J. Biol. Chem. *259*, 6222–6227.
Morin, M.J., Porter, C.W., McKernan, P. and Bernacki, R.J. (1983) The biochemical and ultrastructural effects of tunicamycin and D-glucosamine in L 1210 leukemic cells. Agric. Biol. Chem. *51*, 253–255.
Murakami, H. (1989) Serum-free media used for cultivation of hybridomas. In: Monoclonal Antibodies: Production and Application, Alan R. Liss, Inc. pp. 107–141.
Murphree, S. Moore, E.C. and Peterson, D. (1974) Temporal variation of adenine ribonucleotides during the cell cycle of chinese hamster fibroblasts in culture. Exp. Cell Res. *83*, 189–190.
Nakano, E.T., Cianpi, N.A. and Young, D.V. (1982) The identification of a serum viability factor for SV3T3 cells as biotin and its possible relationship to Krebs cycle activity. Arch. Biochem. Biophys. *215*, 556–563.
Neermann, J. and Wagner, R. (1995) Comparative Analysis of Glucose and Glutamine Metabolism in Transformed Mammalian Cell Lines, Insect and Primary Liver Cells. J. Cellul. Physiol. submitted.
Negrotti, M., Murray, K., Reidy, T.J., Krok, K. and Seaver, S.S. (1989) Effects of glutamine and ammonia on hybridoma cell proliferation and antibody production (Abstract). Proc. Second Engineering Foundation Conference on Animal Cell Culture, Santa Barbara CA, USA, p.65.
Newland, M. Greenfield, P.F. and Reid, S. (1990) Hybridoma growth limitations: The roles of energy metabolism and ammonia production. Cytotechnology *3*, 215–229.
Øyaas, K., Berg, T.M., Brakke, O. and Levine, D.W. (1989) Hybridoma Growth and Antibody Production under Conditions of Hyperosmotic Stress. In: Advances in Animal Cell Biology and Technology or Bioprocesses (Spier, R. E., Griffiths, J. B., Stephenne, J. and Crooy, P. J., eds.) Butterworths, London, 212–220.
Parsons, P.G. and Hayward, I.P. (1983) Differential effects of NAD, Nicotinamide and related compounds upon growth and nucleoside incorporation ion human cells. Biochem. Pharmacol. *32*, 871–876.
Paul, R.J., Bauer, M. and Pease, W. (1979) Vascular smooth-muscle aerobic glycolysis linked to sodium and potassium transport processes. Science, *206*, 1414–1416.

Perkins, J. P. (1973) Adenyl Cyclase. In: Advances in Cyclic Nucleotide Research (Greengard, P. and Robison, G. A. eds.) Raven Press, New York, 155–223.
Petch, D., and Butler, M. A. (1994) Profile of energy metabolism in a murine hybridoma: Glucose and glutamine utilization. J. Cell. Physiol. *161*, 71–76.
Phillips, H. A., Scharer, J. M., Bols, N. C. and Moo-Young, M. (1987) Effect of Oxygen on Antibody Productivity in Hybridoma Culture. Biotechnol. Lett. *9*, 745–750.
Pinkus, L. M. and Windmueller, H. G. (1977) Phospate-dependent glutaminase of smallintestine: localization and role in intestinal metabolism. Arch. Biochem. Biophys. *182*, 506–517.
Plagemann, P. G. W., and Erbe, J. (1973) Transport and metabolism of glucosamine by cultured novikoff rat hepatoma cells and effects on nucleotide pools. Cancer Res. *33*, 482–492.
Rapaport, E. (1983) Treatment of human tumor cells with ADP or ATP yields arrest of growth in the S phase of the cell cycle. J. Cell. Physiol. *114*, 279–283.
Rapaport, E. (1988) Experimental cancer therapy in mice by adenine nucleotides. Eur. J. Cancer Clin. Oncol., *24*, 1491–1497.
Rapaport, E., Fishman, R. F. and Gercel, C. (1983) Growth inhibition of human tumor cells in softagar cultures by treatment with low levels of adenosine 5′-triphosphate. Cancer Res. *43*, 4402–4406.
Rapaport, E., Garcia-Blanco, M. A. and Zamecnik, P. C. (1979) Regulation of DNA replication in S phase nuclei by ATP and ADP pools. Proc. Natl. Acad. Sci USA *76*, 1643–1647.
Redegeld, F. A. M., Moison, R. M. W., Koster, A. S. J. and Noordhoek, J. (1989) Alterations in energy status by menadione metabolism in hepatocytes from fasted and fed rats. Archives of Biochemistry and Biophysics, *273*, 215–222.
Reitzer, L. J., Wice, B. M. and Kennell, D. (1979) Evidence That Glutamine, Not Sugar, Is the Major Energy Source for Cultured HeLa Cells. J. Biol. Chem. *254*, 2669–2676.
Reitzer, L. J., Wice, B. M. and Kernell, D. (1980) The pentose cycle: Control and essential function in HeLa cell nucleic acid synthesis. J. Biol. Chem. *255*, 5616–5626.
Renard, J. M. Spagnoli, R., Mazier, C. Salles, M. F. and Mandine, E. (1988) Evidence that Monoclonal Antibody production Kinetics is Related to the Integral of the Viable Cells Curve in Batch Systems. Biotechnol. Lett. *10*, 91–96.
Reuveny, S., Velez, D., Macmillan, J. D. and Miller, L. (1986a) Factors affecting cell growth and monoclonal antibody production in stirred reactors. J. Immunol. Methods *86*, 53–59.
Reuveny, S., Velez, D., Macmillan, J. D. and Miller, L. (1986b) Comparison of cell propagation methods for their effect on monoclonal antibody yield in fermentors. J. Immunol. Methods *86*, 61–69.
Robinson, D. K., Widmer, J. and Memmert, K. (1992) Effect of specific growth rates on productivity in continuous open and partial cell retention animal bioreactors. J. Biotechnol. *22*, 41–50.
Rozengurt, E. and Heppel, L. A. (1975) Serum rapidly stimulates quabain-sensitive Rb^{+}-86 influx in quiescent 3T3 cells. Proc. Natl. Acad. Sci. *72*, 4492–4495.
Ryll, T. (1992) Untersuchungen intrazellulärer Ribonucleotidpools tierischer Zellkulturen für die Kontrolle von Produktionsprozessen. Thesis, Technical University Braunschweig, Germany.
Ryll, T., Lucki-Lange, M., Jäger, V. and Wagner, R. (1990a) Production of recombinant human interleukin 2 with BHK cells in a hollow fibre and a stirred tank reactor with protein-free medium. J. Biotechnology, *14*, 377–392.
Ryll, T., Lucki-Lange, M., Jäger, V. and Wagner, R. (1990b) Comparison of the production of recombinant IL-2 in stirred tank and hollow fibre bioreactors under protein-free medium conditions. In: Trends in animal cell culture technology (Murakami, H. ed.), Kodansha Pub., Tokyo, Japan, pp. 231–233.
Ryll, T., Jäger, V. and Wagner, R. (1991a) Intracellular concentration of ATP and other nucleotides during continuous cultivation of hybridoma cells. In: Production of Biologicals

from Animal Culture (Spier, R.E., Griffiths, J.B. and Meignier, B., eds.) Butterworths, UK, pp. 236–242.

Ryll, T., Jäger, V. and Wagner, R. (1991b) Variation in the ratios and concentrations of nucleotide triphosphates and UDP-sugars during a perfused batch cultivation of hybridoma cells. In: Animal Cell Culture and Production of Biologicals (Sasaki, R. and Ikura, K., eds.), Kluwer, Dordrecht, 307–317.

Ryll, T. and Wagner, R. (1991) An improved ion-pair HPLC method for the quantification of a wide variety of nucleotides and sugar-nucleotides in animal cells. J. Chromatogr. *570*, 77–88.

Ryll, T. and Wagner, R. (1992) Intracellular ribonucleotide pools as a tool for monitoring the physiological state of in vitro cultivated mammalian cells during production processes. Biotechnol. Bioeng. *40*, 934–946.

Ryll, T., Valley, U. and Wagner, R. (1994) Biochemistry of Growth Inhibition by Ammonium Ions in Mammalian Cells. Biotechnol. Bioeng. *44*, 184–193.

Ryll, T., Valley, U. and Wagner, R. (1995) The control of cellular growth by intracellular ribonucleotide pools as a fundamental system in *in vitro* cultivated microorganisms. J. Cell. Physiol. in preparation.

Sanders, P.G., Hussein, L., Coggins, L. and Wilson, R. (1987) Gene Amplification: The Chinese Hamster Glutamine Synthetase Gene. Develop. biol. Standard. *66*, 55 53.

Satoh, M., Hosoi, S. and Sato, S. (1990) Chinese Hamster Ovary Cells Continuously Secrete a Cysteine Endopeptidase. In Vitro Cell. Dev. Biol. *26*, 1101–1104.

Schlaeger, E.J., Eggimann, B. and Gast, A. (1987) Proteolytic Activity in the Culture Supernatants of Mouse Hybridoma Cells. Develop. Biol. Standard. *66*, 403–408.

Scholnick, P., Lang, D. and Racker, E. (1973) Regulatory mechanisms in carbohydrate metabolism: timulation of aerobic glycolysis by energy-linked ion transport and inhibition by dextran sulfate. J. Biol. Chem. *248*, 5175–5182.

Scholtissek, C., Rott, R. and Klenk, H.-D. (1975) Two different mechanisms of the inhibition of the multiplication of enveloped viruses by glucosamine. Virology *63*, 191–200.

Schumpp, B. and Schlaeger, E.-J. (1992) Growth study of lactate and ammonia double-resistant clones of HL-60 cells. In: R. E. Spier, J. B. Griffiths and C. MacDonald (eds.), Animal Cell Technology: Developments, Processes and Products, Butterworth-Heinemann Ltd., Oxford, p. 183–185.

Sevdalian, D.A., Ozand, P.T. and Zielke, H.R. (1980) Increase in glutaminase activity during the growth cycle of cultured human diploid fibroblasts. Enzyme *252*, 142–144.

Smith, E.L., Austin, B.M., Blumenthal, K.M. and Nye, J.F. (1976) Glutamate dehydrogenases. In: The Enzymes (Boyer, P.B., ed.) Academic Press *11*, 294–366.

Snyder, F.F., Hershfield, M.S. and Seegmiller, J.E. (1978) Cytotoxic and metabolic effects of adenosine and adenine on human lymphoblasts. Cancer Res. *38*, 2357–2362.

Stoner, G.D., and Merchant, D.J. (1972) Amino acid utilization by L-M strain mouse cells in a chemically defined medium. In Vitro *5*, 70–72.

Sugimoto, S., Lind, W. and Wagner, R. (1992) Activation of a specific proteolytic activity in suspension cultures of recombinant adherent cells. In: Animal Cell Technology: Developments and Products (Spier, R.E., Griffiths, J.B. and MacDonald, C., eds.) Butterworths-Heinemann, Kent, UK, pp. 547–553.

Sumbilla, C.M., Zielke, C.L., Reich, W.D., Ozand, P.T. and Zielke, H.R. (1981) Comparison of the oxidation of glutamine, glucose, ketine bodies and fatty acids by human diploid fibroblasts. Biochem. Biophys. Acta *675*, 301–304.

Tagler, J.M., Akerboom, T.P.M., Hoek, J.B., Meijer, A.J., Vaartjes, W., Ernster, L. and Williamson, J.R. (1975) Ammonia and energy metabolism in isolated mitochondria and intact liver cells. In: Normal and Pathological Development of Energy Metabolism (Hommes, F.A. and Van der Berg, C.J., eds.) Academic Press, 63–75.

Tritsch, G. L. and Moore, G. E. (1962) Spontaneous decomposition of glutamine in cell culture media. Exp. Cell. Res. *28*, 360–364.
Tsao, E. I., Bohn, M. A., Numsuwan, V., Omstread, D. R. and Munster, M. J. (1992) Feects of heat shock on the production of human erythropoietin from recombinant CHO cells. Biotechnol. Bioeng. *40*, 1190–1196.
Valeri, A. M., Wilson, S. M. and Feinmann, R. D. (1980) Reaction of antithrombin with proteases. Evidence for a specific reaction with papain. Biochim. Biophys. Acta *614*, 526–533.
Velez, D., Reuveny, S., Miller, L. and Macmillam, J. D. (1986) Kinetics of monoclonal antibody production in low serum growth medium. J. Immunol. Methods. *86*, 45–52.
Venneröd, A. M., Laake, K., Solberg, A. K. and Strömland, S. (1976) Inactivation and binding of human plasma kallikrein by antithrombin III and heparin. Thromb. Res. *9*, 457–462.
Wagner, R., Ryll, T., Krafft, H. and Lehmann, J. (1988) Variation of Amino Acid Concentrations in the Medium of HU β-IFN and HU IL-2 Producing Cell Lines. Cytotechnology *1*, 145–150.
Wang, D. I. C., Cooney, C. L., Demain, A. L., Dunnill, P., Humphery, A. L. and Lilley, A. I. (1979) Fermentation and Enzyme Technology; John Wiley and Sons.
Wice, B. M., Trugnan, G., Pinto, M., Rousset, M., Chevalier, G., Dussaulx, E., Lacroix, B. and Zweibaum, A. (1979) The intracellular accumulation of UDP-N-acetylhexosamines is concomitant with the inability of human colon cancer cells to differentiate. J. Biol. Chem. *260*, 139–146.
Williamson, J. R., Cooper, R. H. (1980) Regulation of Citric Acid Cycle in Mammalian Systems. FEBS Letters *117*, K73–K85.
Wong, R. F., Chang, T. L. and Feinman, R. D. (1982) Reaction of antithrombin with proteases. Nature of the reaction with trypsin. Biochemistry, *21*, 6–12.
Wylegalla, C., Meyer, R., Wagner, K. G. (1985) Nucleotide pools in suspension-cultured cells of *Datura innoxia*. Planta *166*, 446–451.
Zaharevitz, D. A., Napier, E. A., Anderson, L. W., Strong, J. M. and Cysyk, R. L. (1988) Stimulation of uracil nucleotide synthesis in mouse liver, intestine and kidney by ammonium chloride infusion. Eur. J. Biochem. *175*, 193–198.
Zielke, H. R., Ozand, P. T., Tildon, J. T., Sevdalian, D. A. and Cornblath, M. (1978) Reciprocal Regulation of Glucose and Glutamine Utilisation by Cultured Human Diploid Fibroblasts. J. Cell Physiol. *95*, 41–48.
Zielke, H. R., Sumbilla, C. M. and Sevdalian, D. S. (1980) Lactate: A Major Product of Glutamine Metabolism by Human Diploid Fibroblastes. J. Cell Physiol. *104*, 433–441.
Zielke, H. R., Zielke, C. L. and Ozand, P. T. (1984) Glutamine: A Major Energy Source for Cultured Mammalian Cells. Fed. Proc. *43*, 121–125.

2.2 Glycosylation: A Post-Translational Modification

Angela Savage

2.2.1 Introduction

One of the major goals in the application of biotechnology towards human therapeutics is the discovery and development of protein-based pharmaceuticals using recombinant techniques. These recombinant proteins are often used for the augmentation or replacement of naturally occurring proteins in the treatment of numerous pathological conditions. The majority of proteins secreted by mammalian cells, including many proteins of pharmacological importance, are glycoproteins with carbohydrate content varying from about 3 % by weight (as in immunoglobulin G) to more than 40 % (as in erythropoietin). It is becoming increasingly clear that polypeptides which are glycosylated in their native state require glycosylation for full *in vivo* efficacy. Furthermore, the therapeutic profile of an administered polypeptide may be profoundly influenced by its glycosylation. Therefore, it is essential in designing a human therapeutic protein that the structure is as close as possible to the naturally occurring product, and so it is imperative that correct glycosylation is achieved. To date, cDNAs encoding a number of therapeutically significant glycoproteins have been cloned and expressed in heterologous cells and many products are now commercially available.

Because of their ability to perform complex post-translational modifications, cultured mammalian cells are becoming more frequently used to produce proteins for therapeutic and diagnostic use. Post-translational modifications include a series of reactions, involving the making or breaking (formation or cleavage) of covalent

Table 2.2.1. Post-translational modifications; glycosylation is the one of the most important modifications found in recombinant human therapeutics.

Post-translational modifications			
N-Terminus αNH2	**C-Terminus αCOOH**	**Side chains (amino acid)**	**Peptide bonds**
Acylation acetyl pyroglutamyl formyl myristoyl palmitoyl methyl	Amidation unsubstituted substituted ADP-ribosylation Prenylation farnesyl geranylgeranyl	N ε Acetylation (Lys) Methylation γ Carboxylation (Glu) Hydroxylation (Pro/Lys) Hydroxylation (Asn/Asp) Phosphorylation (Ser/Thr/Tyr) Sulphation (Tyr) Adenylation (Tyr) Ubiquitination Glycosylation (Asn/Ser/Thr)	Proteolysis Cross linking Proline isomerisation

bonds, whereby the newly-synthesized polypeptide chain is converted to a functional protein (Han and Martinage, 1991; 1992; 1993). Although a large array of such post-translational modifications is now recognised (see Tab. 2.2.1), one of the major modifications of interest to biotechnologists using mammalian cells as expression system is glycosylation.
(See Spellman, 1990; Rasmussen, 1991; Cumming, 1991; Jenkins and Curling, 1994).

2.2.1.1 Glycosylation and Glycoforms

Glycoproteins, rather than being viewed as single entities are now considered as a collection of glycosylation variants containing different oligosaccharide structures (microheterogeneity) that arise as a consequence of the oligosaccharide biosynthetic pathways. A glycoprotein with only one glycosylation site may be associated with many different oligosaccharide structures at that site so that it is, in fact, a set of glycosylated variants; the family of glycoproteins resulting from this variation in glycosylation are known as glycoforms. When a glycoprotein has two different glycosylation sites and three different oligosaccharides at each site, then nine possible glycoforms are present. Frequently, proteins contain multiple sites with each site being associated with many different oligosaccharide structures. This results in a spectrum of glycoforms which have different structures and thereby the potential to modulate the function of the proteins. It is now clear that the oligosaccharide moieties of glycoproteins can play an important role in defining several biological properties of glycoproteins, including clearance rate, immunogenicity, and biological activity as well as affecting the physicochemical characteristics.

There are several reasons why oligosaccharides are ideal to carry out the many roles attributed to them. The first is that in terms of primary structure, oligosaccharides can display staggering diversity when compared to peptides or nucleic acids. This diversity arises from differences in the constituent monosaccharides, their various linkage positions, anomeric configurations, the possibility of linear or branched sequences as well as substitution by non-carbohydrate groups. Secondly, because protein-linked oligosaccharides are often located at the molecular surface they are optimally situated for modulating bioactivities or serving as recognition determinants. Thirdly, oligosaccharides are hydrodynamically quite large; α_1-acid glycoprotein, with five glycans and a MW of 37kD, could essentially be completely enshrouded by oligosaccharide.
(See Rademacher et al. 1988; Allen and Kisailus, 1992; Opdenakker, 1993; Fukuda and Hindsgaul, 1994; Lennarz and Hart, 1994).

2.2.2 Structures and Conformations of Oligosaccharides

Although a large number of monosaccharides and their derivatives is theoretically available, only a limited number is found in nature. However, in glycoproteins expressed in animal cells only those shown in Figure 2.2.1 are expected and it is reasonable to assume that all residues are in the pyranose (six-membered) ring form and, apart from Fuc, which has the L configuration, have the D absolute configur-

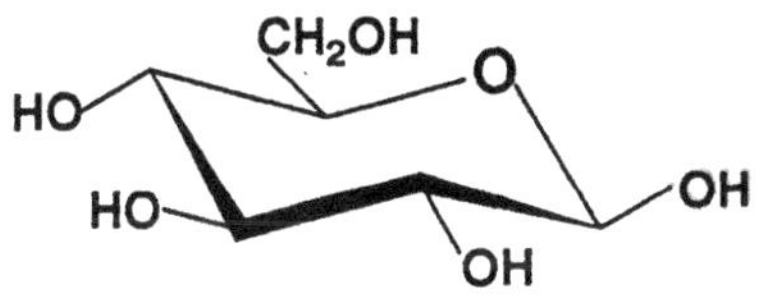

β-D-Glucose

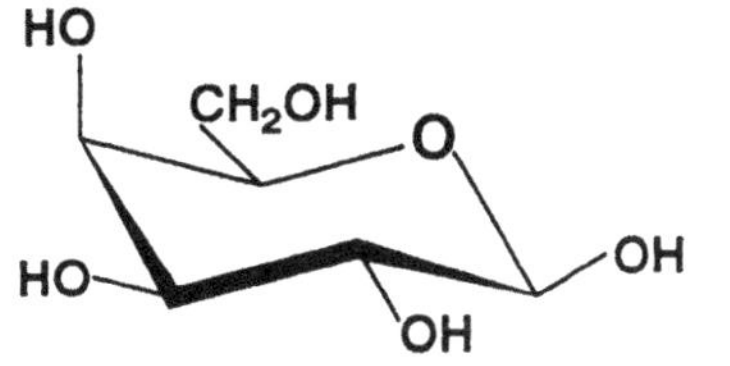

β-D-Galactose

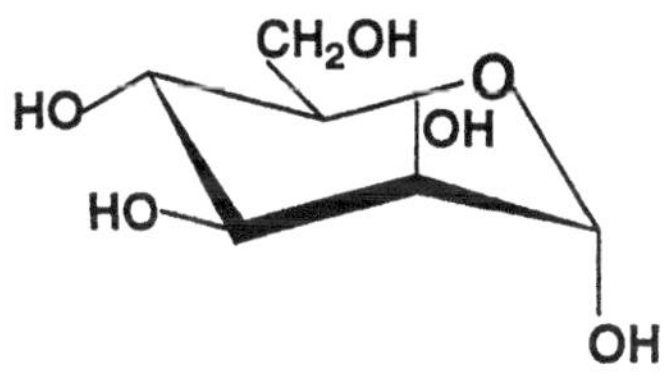

α-D-Mannose

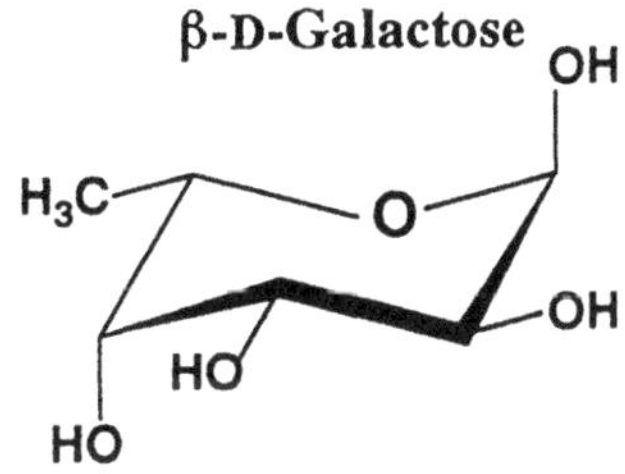

α-L-Fucose

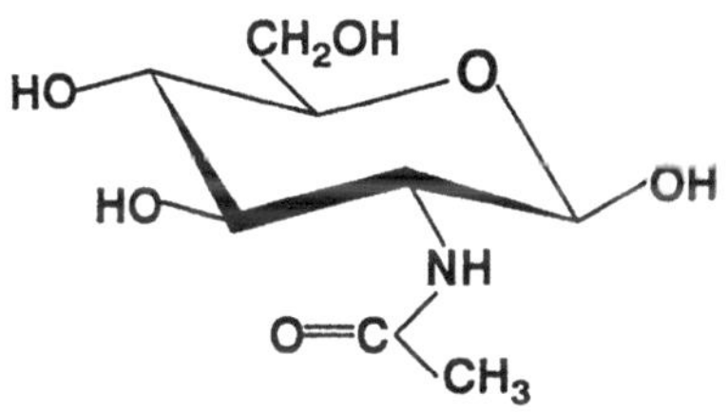

N-Acetyl-β-D-glucosamine

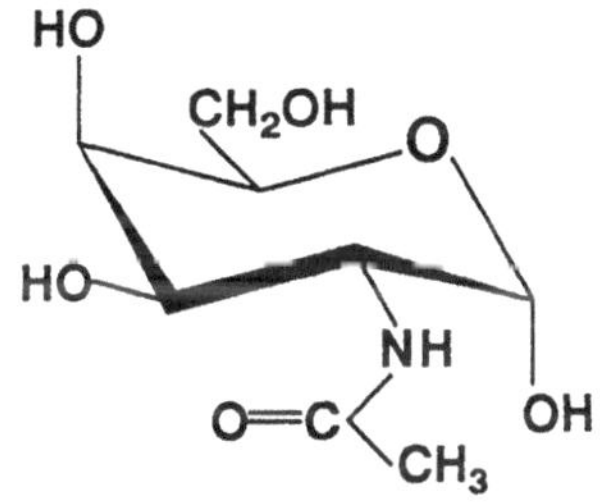

N-Acetyl-β-D-galactosamine

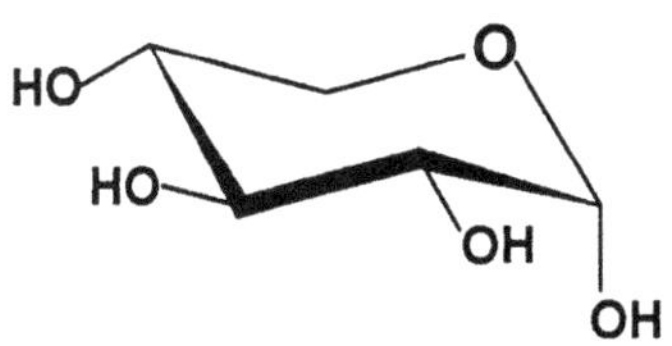

α-D-Xylose

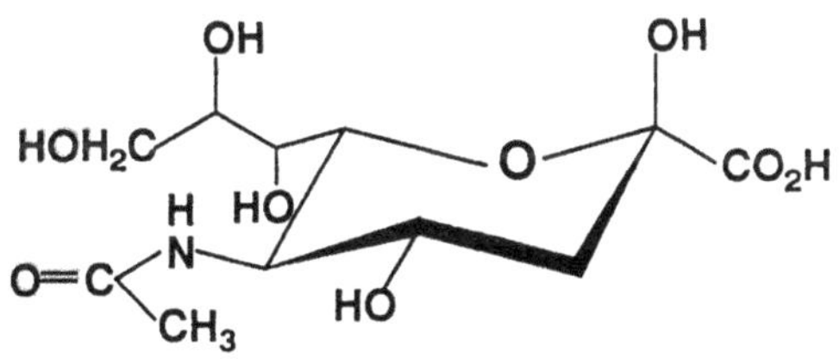

α-*N*-Acetylneuraminic Acid

Fig. 2.2.1: Structures of monosaccharides found in mammalian glycoproteins.

(i) GlcNAc-Asn

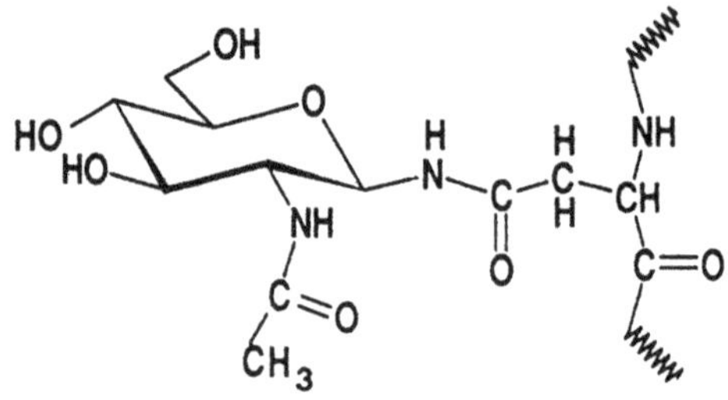

(ii) GalNAc-Ser/Thr

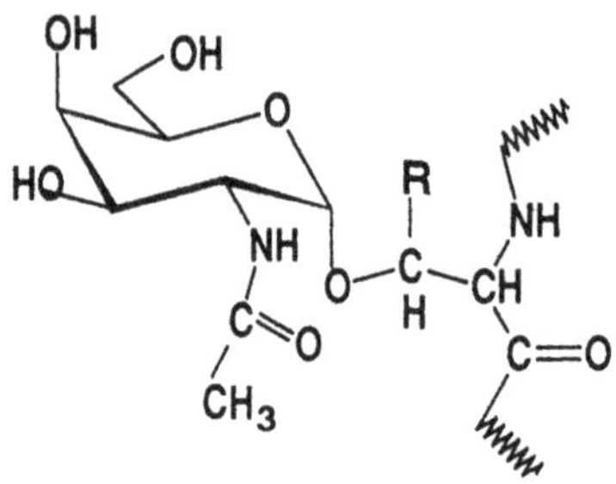

Fig. 2.2.2: Examples of (i) *N*- and (ii) *O*-carbohydrate-peptide linkages found in various glycoproteins. R = H Ser; R = CH_3 Thr.
A given glycoprotein expressed in animal cells may contain only *N*-linked oligosaccharides or *O*-linked oligosaccharides or both.

ation (at C-5)[1]. In general, Glc is not found in glycoproteins; rather its *N*-acetyl derivative, GlcNAc, is widely distributed. However, Glc and Xyl have recently been found in epidermal growth factor (EGF) modules. Oligosaccharides containing *N*-acetylneuraminic acid (sialic acid) are termed acidic while those devoid of NeuAc are termed neutral.
(See Lis and Sharon, 1993; Kobata, 1992; Montreuil et al., 1995).
Oligosaccharides in mammalian systems are covalently bound to protein *via* two types of linkages (Fig. 2.2.2), (i) the *N*-glycosidic linkage from GlcNAc to Asn and (ii) the *O*-glycosidic linkages such as that between GalNAc/GlcNAc/Fuc/Xyl and Ser/Thr. A given glycoprotein expressed in animal cells may contain only *N*-linked oligosaccharides or *O*-linked oligosaccharides or both.

2.2.2.1 Structures of *N*-Linked Oligosaccharides

Although a dazzling array of *N*-linked oligosaccharides has been reported, they all contain a pentasaccharide core (Fig. 2.2.3) and may be classified according to the number and types of substituents attached to the pentasaccharide core (Fig. 2.2.4). In oligomannose type structures (Fig. 2.2.4i), two to six additional mannose units are attached at positions 3 or 6 on the mannose of the 1–6 arm or at position 2 of mannose of the 1–3 arm of the core. Complex-type oligosaccharides (Fig. 2.2.4iia)

[1] For a discussion of carbohydrate nomenclature see Allen and Kisailus, 1992.

Fig. 2.2.3: The pentasaccharide core of *N*-linked oligosaccharides.

are formed by the addition of Galβ1-4GlcNAc (*N*-acetyllactosamine) units at positions 6, 4 or 2 on the mannose of the 1–6 arm or at positions 4 or 2 on mannose of the 1–3 arm of the core to give di-, tri-, tetra-, or penta-antennary structures (Fig. 2.2.4iib). A combination of one or two mannose residues on the 1–6 arm of the core along with at least one *N*-acetyllactosamine unit on the 1–3 arm gives rise to a hybrid type structure (Fig. 2.2.4iii).

Enormous diversity, particularly of complex-type structures arises, not only from the number of antennae present, but also from the variation in the carbohydrate moieties which constitute the non-reducing ends. These terminal residues, where much of the functionality lies, may be additional *N*-acetyllactosamine units (Fig. 2.2.4iib), or one or more of the following, NeuAc/NeuGcα2-3/6, Fucα1-2/3/4, Galα1-3, GlcNAcα1-4 or GalNAcβ1-4. Finally, terminal sugars may be substituted by, for example, a sulphate group.

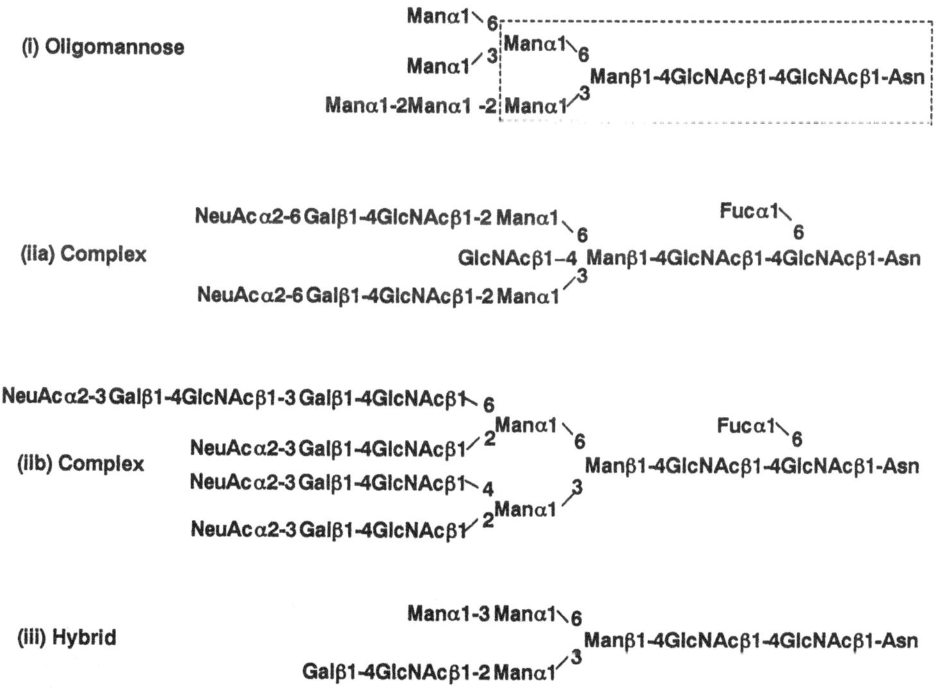

Fig. 2.2.4: (i) Oligomannose structure found in recombinant tissue plasminogen activator: The common pentasaccharide core is outlined by a dotted line, (iia) Complex biantennary structure with proximal fucose and bisecting GlcNAc found in melanoma tissue plasminogen activator, (iib) Complex tetra-antennary structure with proximal fucose and repeating lactosamine unit found in recombinant erythropoietin and (iii) Hybrid oligosaccharide found in recombinant tissue plasminogen activator.

Diversity may also arise at the core since the GlcNAc linked to Asn may be substituted by Fucα1-6 and/or Fucα1-3 (the latter has been found in product from insect cell lines), while the core Man linked β1-4 may be substituted by GlcNAc at position 4 to give a "bisecting" GlcNAc (Fig. 2.2.4iia).

2.2.2.2 Structures of *O*-Linked Oligosaccharides

The linkage between GalNAc and Ser/Thr is the most common *O*-glycosidic link and the core regions associated with this linkage are much less ordered than the cores of oligosaccharides which are *N*-linked *via* GlcNAc to Asn. However, the core type most frequently found in recombinant animal cells is the core-type 1; Galβ1-3GalNAc, which may be further substituted by NeuAc in α2-3 linkage to the Gal and/or α2-6 linkage to the GalNAc (Fig. 2.2.5).
Two types of unusual glycosylation have been found in the epidermal growth factor homology regions (EGF modules) of some multidomain proteins. The first, found

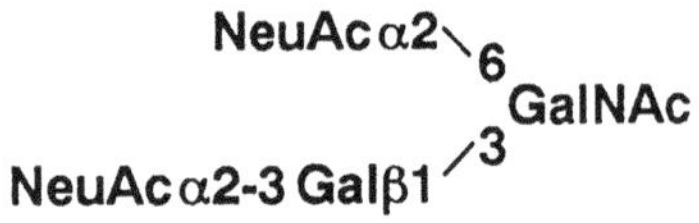

Fig. 2.2.5: Structure of *O*-linked tetrasaccharide found in many recombinant glycoproteins.

in tissue plasminogen activator (t-PA), is *O*-Fuc which is usually unsubstituted. However, the novel tetrasaccharide NeuAcα2-6Galβ1-4GlcNAcβ1-3Fucα1-Ser/Thr has recently been reported in the EGF module of human factor IX. The second is the sequence ± (Xylα1-3)Xylα1-3Glcβ1-Ser/Thr, which was first found in blood coagulation factors VII and IX.

GlcNAc (unsubstituted) is found linked to Ser/Thr in nucleoplasmic and cytoplasmic proteins. However, it is anticipated that such glycosylation will not be found in secreted (recombinant) products.

2.2.2.3 Conformation of Oligosaccharides

Considerable progress has been made in the past decade in understanding the tertiary structure of free oligosaccharides in solution. The overall oligosaccharide solution conformation at a given moment is determined by the orientation of the constituent monosaccharides to one another, expressed in terms of the interglycosidic torsion angles, ϕ, ψ and ω (Fig. 2.2.6). NMR spectroscopy, in conjunction with conformational energy calculations and X-ray crystallography studies, provides a powerful method for experimentally determining these angles. Results indicate that the Manα1-6 linkage, which has three torsional angles (ϕ, ψ and ω) of the core region of *N*-linked oligosaccharides, exhibits considerable conformational flexibility.
(See Homans, 1993; Montreuil et al., 1995).

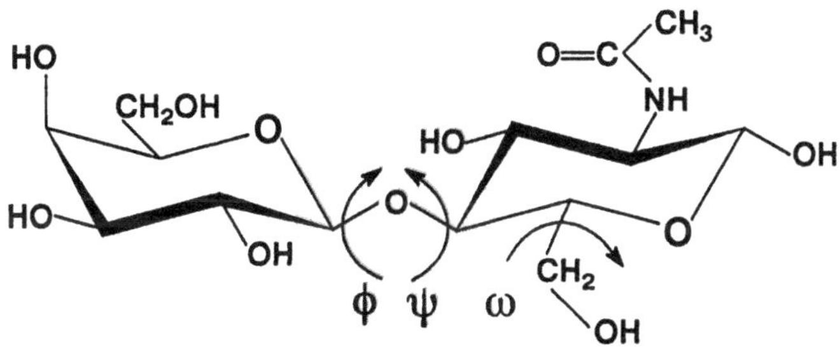

Fig. 2.2.6: Definition of the torsion angles ϕ(H1-C1-O1-CX), ψ (C1-O1-CX-HX) and ω (H5-C5-C6-O6). The example illustrates a 1–4 linkage in Galβ1–4GlcNAc. Note the possibility of an extra degree of rotational freedom for a 1–6 linkage.

2.2.3 Intracellular Biosynthesis

Oligosaccharide biosynthesis does not take place on a template in the manner of DNA, RNA and protein biosynthesis. Rather, glycosylation occurs as a post-translational event and so can be viewed as being cell-controlled, as the information for the final oligosaccharide structure(s) found are not coded in the DNA of that protein. Engineering the protein portion of glycoproteins is straightforward. However, dictating their covalently attached carbohydrates is more complicated because they are the products of transferases and processing glycosidases encoded in the cellular genome. More than 25 glycosyltransferase enzymes have been cloned to date, and the structural details of many of these genes have been reviewed recently.
(See Cummings, 1992; Schacter and Brockhausen, 1992; Joziasse, 1992; Montreuil et al., 1995).

2.2.3.1 Glycosylation Sites

Potential sites for glycosylation occur at the amino acids Asn or Ser/Thr in the consensus sequences shown in Table 2.2.2. However, the consensus sequence, or

Table 2.2.2. Consensus sequences for glycosylation in mammalian cells.

Glycosylation Type	Consensus Sequence	Comment
N-Asn *via* GlcNAc	Asn-X-Ser/Thr-Y	Pro at X inhibits glycosylation Pro at Y decreases likelihood
O-Ser/Thr *via* GalNAc	Ser/Thr in Pro-rich region	Pro − 1 and + 3 relative to Ser/Thr increases likelihood
O-Ser/Thr *via* Fuc	Cys-X-X-Gly-Ser/Thr-Cys	
O-Ser/Thr *via* Glc-Xyl	Cys-X-Ser-X-Pro-Cys	
O-Ser/Thr *via* GlcNAc	None to date	Not found in secreted product

sequon, does not guarantee glycosylation; fewer than half of the potential sites in secreted glycoproteins are actually glycosylated. Therefore, it should be noted that the cDNA sequence reveals little information about actual glycosylation; such information can only be obtained experimentally. Variations at glycosylation sites can exist not only in the presence or absence of oligosaccharides but also in the type of oligosaccharide (oligomannose or complex) found at a particular site as well as the exact oligosaccharide present (e.g. biantennary/triantennary or asialo/sialylated) and the degree of processing of each.
(See Spellman, 1990; Dwek, 1995).
Many biosynthetic steps involve the addition of sugars, one at a time, through the action of highly-specific enzymes known as glycosyltransferases. The monosaccharide is transferred from a nucleotide sugar donor (e.g. UDP-Gal/Glc/GalNAc/GlcNAc/Xyl, GDP-Fuc/Man or CMP-NeuAc) to the oligosaccharide acceptor as follows,

$$\text{Nucleotide} - \text{sugar} + \text{acceptor} \rightarrow \text{sugar} - \text{acceptor} + \text{nucleotide}$$

One of the interesting features of glycosyltransferases is their ability to discriminate between different oligosaccharide structures. This substrate specificity directs the synthetic pathway along various routes and is a major factor in the control of oligosaccharide synthesis. The study of highly purified glycosyltransferases has provided evidence for the hypothesis that every sugar-sugar linkage requires a separate transferase for its synthesis so that well over 100 different enzymes must exist to account for all documented oligosaccharide structures.

2.2.3.2 Biosynthesis of *N*-Glycans

The biosynthesis of *N*-glycans begins in the rough endoplasmic reticulum (Scheme 2.2.1) with the co-translational transfer of a large oligosaccharide ($Glc_3Man_9GlcNAc_2$) from dolichol pyrophosphate oligosaccharide to an Asn residue in the polypeptide. (This step is highly inhibited by the antibiotic tunicamycin resulting in the cessation of Asn-linked biosynthesis). The next step is "trimming"; the removal of three glucose and four mannose residues within the lumen of the endoplasmic reticulum and *cis* Golgi apparatus due to the processing actions of specific α-glucosidases and α-mannosidases. The product of this processing is the structure $Man_5GlcNAc_2$-Asn which is the starting point for the synthesis of all complex and hybrid *N*-glycans.

"Elongation"; the addition, in the medial Golgi, of GlcNAc in a β1-2 linkage to the Manα1-3 arm of the core by GlcNAc-transferase I, is a "go" signal for the subsequent removal of two of the Man units, the addition of Fuc in the core by a fucosyltransferase and the addition of one, two, three or four more GlcNAc residues to produce bi-, tri-, tetra- and penta-antennary structures. Movement of these structures to the trans Golgi elicits "terminal glycosylation"; the addition of Gal, GalNAc, NeuAc and Fuc. The addition of a "bisecting" GlcNAc serves as a "stop" signal for further branching and for the addition of core Fuc.

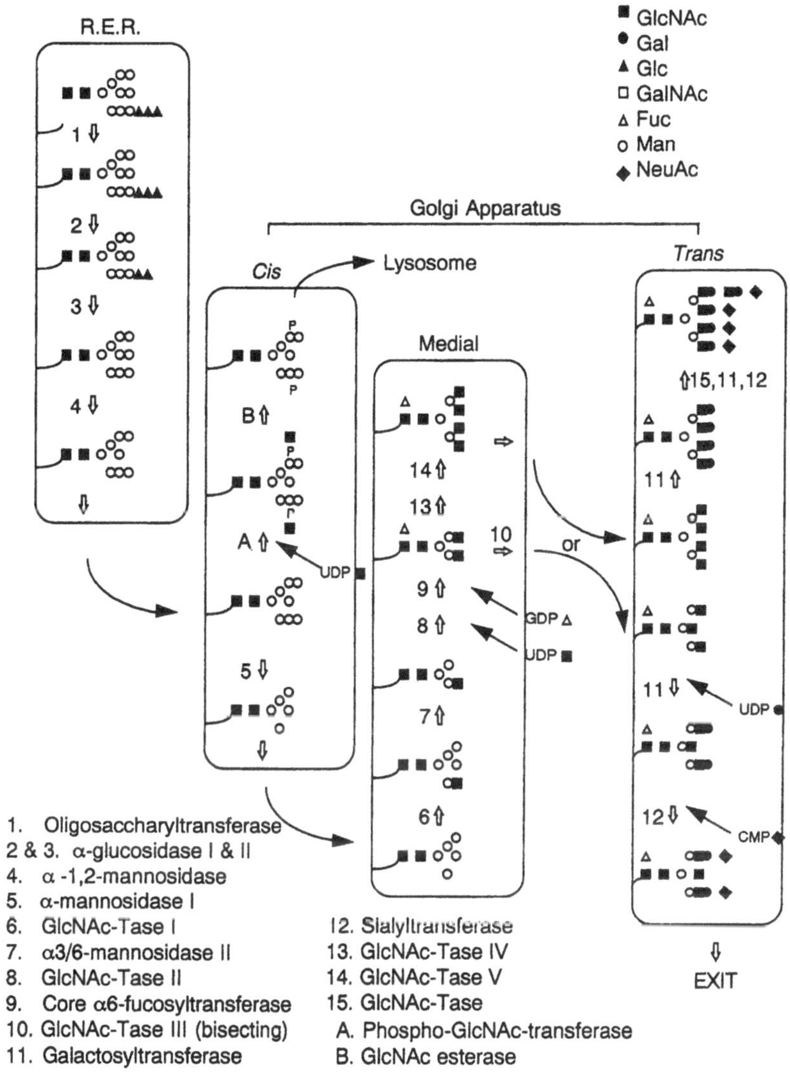

Scheme 2.2.1: Biosynthesis of *N*-linked oligosaccharides (adapted from Kornfeld and Kornfeld, 1985).

2.2.3.3 Biosynthesis of *O*-Glycans

The synthetic pathway for *O*-linked oligosaccharides (Scheme 2.2.2) differs from the *N*-glycan pathway in that there is no transfer of a large oligosaccharide from lipid carrier to protein, nor is there processing by the action of glycosidases. Initiation of *O*-glycosylation occurs with the action of polypeptide αGalNAc-transferase. The intracellular location for this step in mammalian cells has been a source of controversy. Some reports suggest that initiation of *O*-glycosylation occurs in the

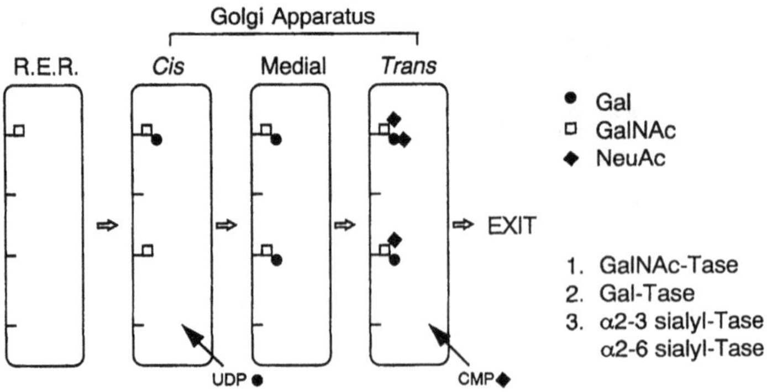

Scheme 2.2.2: Biosynthesis of *O*-linked oligosaccharides.

endoplasmic reticulum or pre-Golgi compartments while other reports indicate it occurs in the Golgi. It has been hypothesized that the intracellular location for initiation of *O*-glycosylation may be dependent upon cell type and target protein. The *O*-glycans are then built up by the successive addition of α- or β-Gal , β-GlcNAc or α-GalNAc at positions 3 or 6 of the GalNAc to give rise to at least 8 different core types, only one of which, Galβ1-3GalNAc, occurs frequently in recombinant mammalian cells. The action of sialyltransferases gives rise to two different tri-saccharides or the tetrasaccharide shown in Figure 2.2.5.

Each enzyme involved in the pathway in Scheme 2.2.2 has a substrate specificity unique for *O*-linked structures. For example, the α2,3 and α2,6 sialyltransferases are distinct from the sialyltransferases which act upon *N*-linked oligosaccharides.

2.2.3.4 Factors Affecting Glycosylation

Protein translation occurs on the basis of a mRNA template, assuring high fidelity of protein structure. In contrast, oligosaccharide processing occurs as a result of the sequential actions of several enzymes in different intracellular compartments. It should not be surprising that the outcome of this set of reactions varies with the environmental conditions. *In vivo*, this variability apparently serves an important function within the cell, permitting fine regulation of the clearance rate and biological activity for at least some proteins. *In vitro*, potential environmental effects on oligo-saccharide structure cannot be ignored, given the importance of oligosaccharide structure in defining clearance rate, biological specific activity, and immunogenicity. (See Jenkins and Curling, 1994; Goochee and Monica, 1990; Goochee et al., 1991; 1992; Andersen and Goochee, 1994; Montreuil et al., 1995).

2.2.3.4.1 *N*-Glycosylation

The protein exerts an influence on glycosylation through the position of the sequon in the polypeptide. Sequons close to the *N*- or *C*-termini are generally less efficiently glycosylated. Since the conformation of the protein influences the accessibility of the enzymes involved in biosynthesis, sequons presented at the exposed turns of β-pleated sheets, for example Asn-297 in the C_{H2} domain of immunoglobulin G

(IgG), are characteristically glycosylated with complex sugars. Another influential factor may be the local environment of amino acids that surrounds the sequon. In particular, folding or packaging of the protein that involves disulfide bridges may hinder glycosylation, as in the case with interleukin-6. It has been shown that eliminating the disulfide bond between Cys-45 and Cys-51 increases the efficiency with which the sequon at Asn-46 is glycosylated. The polypeptide sequence may determine the speed with which protein folding renders the sequon inaccessible, and this is believed to result in "competition" between the rate of folding and the addition of the dolichol-linked precursor. If the protein folds too rapidly upon entering the endoplasmic reticulum, an unglycosylated Asn-X-Ser/Thr sequon may become inaccessible as substrate for oligosaccharyltransferase.

That protein structure exerts control on the type of oligosaccharide biosynthesised is exemplified by the consistency of oligosaccharide structure of the same protein expressed in different cells. For example, biantennary complex-type oligosaccharides are found for interferon-β1 while tetraantennary complex-type oligosaccharides are found for erythropoietin expressed in the same cells.

Oligosaccharide processing can differ between *N*-glycosylation sites on the same protein, as demonstrated by recent studies of the *N*-linked oligosaccharides of t-PA (t-PA, section 2.2.5.3.1). The site at Asn-117 always carries oligomannose structures while Asn-448 is almost always fully occupied by complex-type oligosaccharides. In addition, it has been hypothesized that oligosaccharide processing at a given glycosylation site may be dependent upon events occurring at an adjacent glycosylation site, by, for example, steric interference or by the influence on protein tertiary structure. For example, variable occupancy at Asn-184 affects the fine structure of the glycan population at Asn-448. These data clearly demonstrate that oligosaccharide processing is influenced by the local molecular environment at each *N*-glycosylation site.

The internal and external environment of the cell and its developmental stage also influence the extent to which a sequon is glycosylated. The availability of the dolichol-linked precursor and the oligosaccharyl transferase, together with the primary, secondary, and tertiary structure of the protein, controls the kinetics of the catalytic transfer of the oligosaccharide precursor to the Asn nitrogen (see also Chapter 2.1.5.2.3).

2.2.3.4.2 *O*-Glycosylation

In plasma glycoproteins only a small number of Ser/Thr sites are *O*-glycosylated. For example, Ser 126 of erythropoietin (EPO) is *O*-glycosylated while 16 other EPO Ser/Thr sites are never occupied. None of the more than 60 Ser/Thr sites of t-PA are *O*-glycosylated. Variable site occupancy is evident at Thr-3 of interleukin-2 (IL-2) produced by human lymphocytes and recombinant CHO, mouse L and Ltk- cells, while no other IL-2 Ser/Thr sites are *O*-glycosylated.

In contrast, the extracellular domains of some mammalian cell surface glycoproteins possess large numbers of *O*-linked oligosaccharides. For example, more than 80 *O*-linked oligosaccharides are found in the extracellular domain of leukosialin, representing glycosylation at approximately 85 % of the possible Ser/Thr sites.

2.2.4 Glycosylation Potential

Prokaryotic cells do not possess a glycosylation capability analogous to that of eukaryotic cells. Therefore, human polypeptides of therapeutic value which are glycosylated in their native state are generally expressed in recombinant form in eukaryotic cells. The type of glycosylation associated with different expression systems is shown in Table 2.2.3 and glycosylation can be further manipulated by judicious selection of the tissue and cell-line used.
(See Rasmussen, 1991; Jenkins and Curling, 1994; Rademacher et al., 1988; Goochee et al., 1992; Conradt et al., 1990; Montreuil et al., 1995).

Table 2.2.3. Glycosylation potential of different expression systems.

Expression system	***N*-Oligosaccharide type**	***O*-Oligosaccharide type**
Bacteria	None	None
Yeast	Oligomannose/hypermannose	Man_{1-5}
Insect	Oligomannose/fucosylated small complex/paucimannose	Neutral disaccharide only Galβ1-3GalNAc
Mammalian	Oligomannose/complex/hybrid	Sialylated/neutral

2.2.4.1 Bacterial Cells

Bacteria such as *E. coli* lack the biosynthetic machinery to carry out either Asn- or Ser/Thr-linked glycosylation or to form proper disulphide bonds. It is likely that non-glycosylated protein from *E. coli* will have incorrect folding and be more hydrophobic than the native counterpart, thus presenting difficulties in handling and formulation. In addition, the absence of oligosaccharides leads to unmasking of protein epitopes which may be antigenic.

2.2.4.2 Yeast Cells

Yeast glycosylate the sequon for *N*-glycosylation with $Glc_3Man_9GlcNAc_2$ but are unable to process the oligomannose chains to complex oligosaccharides. Rather, the $Man_9GlcNAc_2$ structure is further mannosylated to yield elaborate mannan structures sometimes containing more than 50 mannose sugars. However, mutant strains permit synthesis of recombinant proteins with truncated *N*-linked, oligomannose oligosaccharides. *O*-Glycosylation, in which up to five Man units are linked to Ser/Thr is also accomplished, but not necessarily at the same sites as *O*-glycosylation in animal cells. Two problems may be encountered with glycoproteins produced in yeast. The first is the likelihood of decreased circulatory lifetime due to clearance *via* the Man/GlcNAc receptor of the reticulendothelial system and the second is the possibility of antigenicity of the oligomannose chains.

2.2.4.3 Insect Cells

Insect cells appear to use the same initial steps for *N*-glycosylation as mammalian cells and can process the oligomannose structure to yield eventually the core $Man_3GlcNAc_2$ moiety. While it was initially thought that insect cells lack the capacity to elongate *N*-glycan chains beyond this core pentasaccharide to generate complex or hybrid *N*-glycans, recent studies have shown the presence of unusual complex oligosaccharide structures such as the trifucosylated oligosaccharide with terminal GalNAc found in honeybee venom phospholipase A (Fig. 2.2.7) and structures containing Xyl linked β1-2 to the core β-linked Man.
(See März et al., 1995).

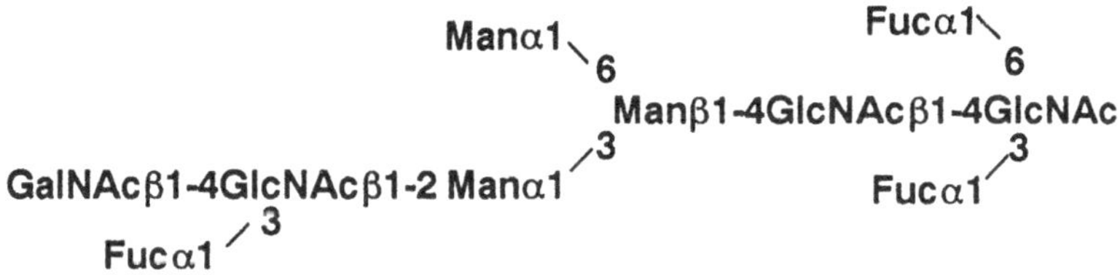

Fig. 2.2.7: Structure of oligosaccharide found in honeybee venom phospholipase A; the α-1,3 fucosylation of the Asn-bound GlcNAc constitutes both an antigenic and an allergenic determinant.

Recent reports indicate further significant differences between the glycosylation potentials of insect and mammalian cells. Analysis of the *N*-glycans of a number of recombinant products revealed partly α1-6-fucosylated paucimannose ($Man_3GlcNAc_2$ and $Man_2GlcNAc_2$) in addition to oligomannose (Man_9- to Man_5-) structures instead of the complex glycans of the native glycoprotein. Analysis of the *O*-glycans usually revealed the presence of only the neutral disaccharide Galβ1-3GalNAc, since sialylation is usually not accomplished in insect cells. These data conflict with other reports which suggested that insect cells possess the potential to synthesise sialylated complex-type *N*-glycans. However, it is now accepted that under certain conditions, insect cells are able to process nascent *N*-glycans to sialylated complex structures with various degrees of branching and that transfection-based, time-dependent activation, either at the gene or the protein level, of normally silent glycosyltransferases can take place, resulting in the production of oligosaccharide structures usually not found in these cells. GlcNAc transferases I and II have been identified in three lepidopteran cell lines which explains the presence of fucosylated cores (antennal GlcNAc is required for the action of fucosyltransferase). Surprisingly, α1-3-fucosylated oligosaccharides have never been found on recombinant glycoproteins.

2.2.4.4 Mammalian Cells

Animal cells are capable of processing oligomannose structures to complex-type structures as shown in scheme 2.2.1 and it is now well accepted that the oligosaccharide processing is species dependent and cell-type dependent within a given species. The influence of cell type on glycosylation appears to be related primarily to the presence, concentration, kinetic characteristics, and compartmentalisation of the individual glycosyltransferases and glycosidases. Differences in oligosaccharide structure among mammalian cells (microheterogeneity) are frequently attributable to differences in the presence of various glycosyltransferases. For example, difference in the specific *N*-acetylglucosaminyltransferase (GlcNAc-Tase) activities will dictate the degree of branching which will occur.

The species-related differences in the sugar chains of glycoproteins have been well documented. For example, fibronectin isolated from bovine plasma contains four different asparagine-linked sugar chains, while that isolated from human plasma contains only two types of biantennary complex-type sugar chains. The tissue-related differences of *N*-glycosylation have also been well documented.

To date there is no reported example of a heterologous cell expression system which glycosylates proteins in precisely the same way as the normal cell source for the protein of interest. Heterologous eukaryotic cells usually express additional saccharide determinants (which may be antigenic) or may not express certain required determinants. The potential for *N*-glycosylation of various animal host cells currently used for the production of recombinant glycoproteins is shown in Table 2.2.4.

Among the points to note from this table are;

- The most widely used animal host, the Chinese hamster ovary cell, is known to lack the functional enzyme α2,6-sialyltransferase, leading to exclusively α2,3-linked terminal sialic acid residues.
- Two baby hamster kidney-21 cell lines have been characterised; BHK-21A does not sialylate but instead may add on the GalNAcβ1-4Gal structure, leading to a shortened $t_{1/2}$. BHK-21B synthesises sialylated bi-, tri-, and tetraantennary structures
- The terminal Galα1-3GalNAc structure is present in New World monkeys and nonprimate mammals, but absent in humans and Old World primates, or anthropoid apes. It is also found on 50% of malignant breast specimens examined. Since humans normally produce antibodies that recognise this determinant, such antibodies may function in antitumour defence.
- The Fucα1-3Gal is gene present in CHO cells, but is not normally expressed (Tanigawara et al., 1990).

Differences in *O*-glycosylation site occupancy are also displayed among mammalian cells. For example, recombinant CHO cells secrete a ratio of glycosylated to nonglycosylated IL-2 on the order of 9 : 1, while recombinant BHK and Ltk-cells produce approximately equal proportions of glycosylated and nonglycosylated IL-2.

The glycosylation potential of animal cells may be further manipulated by the following,

- The use of glycosylation mutants of mammalian cells which are well characterised and which attach a unique set of oligosaccharides (Stanley, 1992).
- The introduction of "missing" biosynthetic enzymes using cloning techniques, for example the α2-6 sialyltransferase, which is absent in CHO cells (Rasmussen, 1991).

Table 2.2.4. Structural features of *N*-linked oligosaccharides from secretory recombinant glycoproteins expressed in various mammalian cell lines.

<table>
<tr><th rowspan="2">Host cell line:</th><th rowspan="2">CHO</th><th colspan="2">BHK</th><th rowspan="2">C127</th><th rowspan="2">Ltk$^-$</th></tr>
<tr><th>21A</th><th>21B</th></tr>
<tr><td>proximal Fucα1-6</td><td>+</td><td colspan="2">+</td><td>+</td><td>+</td></tr>
<tr><td>α2-6NeuAc</td><td>−</td><td colspan="2">−</td><td>+</td><td>+</td></tr>
<tr><td>α2-3NeuAc</td><td>+</td><td>−</td><td>+</td><td>+</td><td>+</td></tr>
<tr><td>α2-3/6NeuGc</td><td>+</td><td colspan="2">+/−</td><td>+</td><td>+</td></tr>
<tr><td>tri/tetra-antennarity</td><td>+</td><td colspan="2">+</td><td>+</td><td>+</td></tr>
<tr><td>Galβ1-4GlcNAc repeats</td><td>+</td><td colspan="2">+</td><td>+</td><td>+</td></tr>
<tr><td>sulphate</td><td>−</td><td colspan="2">−</td><td>+</td><td>+</td></tr>
<tr><td>Galα1-3Gal</td><td>+</td><td colspan="2">−</td><td>+</td><td>|</td></tr>
<tr><td>branched repeats</td><td>?</td><td colspan="2">−</td><td>+</td><td></td></tr>
<tr><td>bisecting GlcNAc</td><td>−</td><td colspan="2">−</td><td>+</td><td>+</td></tr>
<tr><td>GalNacβ1-4GlcNAc</td><td>−</td><td>+</td><td>−</td><td>−</td><td>−</td></tr>
<tr><td>Fucaα1-2Gal</td><td></td><td colspan="2">+/−</td><td></td><td></td></tr>
<tr><td>Fucα1-3Gal</td><td>+/−</td><td colspan="2"></td><td></td><td></td></tr>
<tr><td>NeuAcα2-3GalNAc</td><td></td><td colspan="2">+/−</td><td></td><td></td></tr>
<tr><td>phosphate</td><td></td><td colspan="2"></td><td></td><td></td></tr>
</table>

- Using site-directed mutagenesis, the opportunity exists to introduce or delete glycosylation sites. There have been many examples of deleting Asn- and Ser/Thr glycosylation sites. In some cases, the mutant protein is poorly secreted, but it is not yet possible to predict when this result will be obtained. A number of cases where new Asn- and Ser/Thr glycosylation sites have been introduced have also been reported by Pfeiffer et al. (1994).
- Another *in vivo* approach to controlling the structures of the carbohydrates of recombinant proteins is the use of processing inhibitors. These compounds include swainsonine, castanospermine, deoxynojirimycin and deoxymannojirimycin.
- Remodeling of oligosaccharide structures by treatment of purified glycoprotein with glycosidases or glycosyltransferases (Rasmussen, 1991). This has been achieved with the commercially available Ceredase®.
- Appropriate combinations of the above.

In addition to being species and host cell-type dependent, glycosylation in animal cells is also dependent on the three-dimensional (3-D) structure of the protein to be glycosylated and the protein environment of the glycosylation site (as discussed in section 2.2.3.4). The latter is not surprising because, unlike protein translation, oligosaccharide processing is not template driven. Rather, glycosylation occurs as a result of the sequential actions of several enzymes in different intracellular compartments, and these reactions can be affected by extracellular factors that influence enzyme activity or substrate availability. Furthermore, the oligosaccharides of secreted glycoproteins can be modified extracellularly by glycosidases released from cultured cells.

2.2.4.5 Influence of Cell Culture Conditions

Cell culture conditions can affect glycosylation either as a result of changes in the general cell-culture methodology or in specific culture variables.
(See Goochee and Monica, 1990; Goochee et al., 1991; 1992; Andersen and Goochee, 1994.) (see also Chapter 2.1)

2.2.4.5.1 Cell Culture Methodology

Numerous recent studies have demonstrated that the specific choice of method used to culture cells can influence the oligosaccharide structure of the protein produced (see chapter 2.1.5.2.3). Furthermore, for a particular choice of culture methodology, glycoprotein oligosaccharide structure can change as a function of time. For example, when recombinant human thyrotropin (TSH) is produced in CHO cells using a hollow fiber bioreactor, the product possesses significantly less sialic acid and galactose than TSH produced on microcarrier beads in a large-scale bioreactor. The sialic acid level of the TSH produced in the hollow fiber bioreactor also varies significantly over the different bioreactor operational phases. It has also been shown that for recombinant CHO cells, the occupancy of *N*-linked glycosylation sites decreases as a function of time.

2.2.4.5.2 Specific Culture Variables

A number of recent reviews (Jenkins and Curling, 1994; Goochee and Monica, 1990; Goochee et al., 1991; 1992; Andersen and Goochee, 1994) have reported on studies which have attempted to identify the specific factors in the cell-culture environment responsible for such variations.

- Ammonium ions can reduce the sialylation of both *N*- and *O*-linked oligosaccharides by directly altering the *trans*-Golgi environment.
- The presence of glycosidases, in particular sialidases, can lead to degradation of oligosaccharide structures.
- Changes in glycosyltransferase activities concomitant with cell growth-stage may affect the biosynthetic pathway leading to e.g. changes in the degree of branching.
- Glucose starvation, leading to depletion of the cellular energy state and key nucleotide sugars (e.g. UDP-GlcNAc), may affect the initial step of *N*-linked oligosaccharide biosynthesis. (see Chapter 2.1.5.2.2)

- Many *in vitro* studies have documented the role of hormones in regulating the oligosaccharide structures of glycoproteins.
- Increases in the partial pressure of dissolved carbon dioxide, ranging from 10 mmHg to 160 mmHg, were shown to increase the ratio of *N*-glycolylneuraminic acid to *N*-acetylneuraminic acid by approximately threefold.
- The growth factor interleukin-6, when added to the medium of a myeloma cell line, reduced *N*-acetylglucosaminyltransferase III (GlcNAc-Tase-III) activity, but increased GlcNAc-Tase-IV and GlcNAc-Tase-V activity, leading to changes in the degree of branching.
- Other factors in batch culture such as initial glutamine concentration, the concentration and purity of bovine serum albumin, the medium lipid composition and the common medium surfactant Pluronic F68, EDTA and HEPES buffer also appear to affect glycosylation.
- The addition of the antibiotic tunicamycin, which blocks synthesis of the oligosaccharide precursor, results in a nonglycosylated protein.
- The effects of changing the physical parameters of cell culture have also been investigated. Mild or even severe hypoxia has minimal effects on the glycosylation of t-PA produced by recombinant CHO cells. Similarly, pH changes within the range 6.9–8.2 in the cell culture medium do not have a dramatic effect on the glycosylation profile of a recombinant glycoprotein hormone expressed in CHO cells; however, there is some evidence for underglycosylation at pH levels outside this range.

2.2.5 Biological Function and Therapeutical Significance

Just as a biosynthesising cell would not expend the considerable energy in unnecessarily glycosylating a protein, the biotechnologist will not introduce the added expense and complexity of production and analysis needed for correct glycosylation unless there are sound therapeutic and commercial reasons for doing so. Erythropoietin, a circulating glycoprotein hormone that stimulates erythropoiesis, has the distinction of being the first recombinant glycoprotein produced industrially for clinical use. Other clinically important glycoproteins include the thrombolytic agent t-PA and recombinant factor VIII which can be used in patients with haemophilia A, replacing the defective or absent component of the blood coagulation cascade.

The enzymatic modification of the glycan of the enzyme glycocerebrosidase (glucosylceramidase), which is essential for its clinical use (under the trade name Ceredase®, allows the treatment of patients with Gaucher's disease. It is the first, and thus far probably the only, case of enzyme replacement therapy, a concept suggested some 30 years ago.

Glycans make an enormous contribution to the surface presented by a glycoprotein. Indeed, the molecular volume of a disialylated biantennary *N*-glycan is comparable with that of an immunoglobulin domain. It is therefore not surprising that for this reason, coupled with the fact that cells would not invest so much energy in glycosylating proteins in a complex biosynthetic pathway, were there not some good functional reasons for doing so, that the presence of *N*- and *O*-linked carbohydrates on mammalian glycoproteins has been shown to play an important role in many biological areas.

The numerous functions attributed to oligosaccharides may be divided into two overlapping categories; those which include a purely physicochemical role such as aiding the conformation and stability or protecting polypeptide chains from proteases or thermal inactivation and those with contributions involving molecular interaction, such as controlling secretion and the half-life of proteins and cells, the modulation of biological and immunological activity. Which, if any, of these properties are altered by a change in glycosylation depends upon the individual protein and the nature of the change.

Because carbohydrate chains on proteins contribute significantly to the biologically active conformation of polypeptides and effect their overall biological properties, they influence the overall pharmacokinetic properties of the glycoprotein. Improvement of any of these properties for a therapeutic glycoprotein, with consequent impact on dosage, mode of administration, frequency of administration, formulation or storage, is of both medical and economic benefit.

Studies on the effects of *N*- and *O*-linked oligosaccharides on the properties of glycoproteins have been conducted by comparing the properties of glycosylated and non-glycosylated proteins. The latter have been achieved by production in the presence of tunicamycin which inhibits *N*-glycosylation. Since no inhibitor of *O*-glycosylation is currently available, studies have been conducted with cell lines defective in *O*-glycosylation. Finally, site directed mutagenesis has been utilized to explore the importance of individual *N*- and *O*-linked oligosaccharide sites on glycoprotein properties.
(See Rasmussen, 1991; Cumming, 1991; Varki, 1993; Dwek, 1995; Goochee et al., 1992).

2.2.5.1 Physicochemical Roles

(See Rasmussen, 1991; Jenkins and Curling, 1994; Goochee et al., 1992).

Solubility. As a general rule, the sugar chains of a glycoprotein tend to have greater solubility in aqueous solution than does the protein portion due to the hydrophilicity of the hydroxyl groups. In addition, the hydrophobic faces of the sugars are able to align with the exposed hydrophobic side-chains of the protein, thus further increasing solubility. In most cases, solubility increases with NeuAc content, the number of glycan chains and their antennarity. However, removal of NeuAc with neuraminidase increases solubility in some cases. Glycosylation also inhibits protein aggregation.

Resistance to Thermal Inactivation. In some cases, partial or complete removal of glycoprotein oligosaccharides leads to increased susceptibility to thermal denaturation. For example, enzymatic removal of sialic acid from EPO results in a significant increase in the rate of thermal inactivation at 70 °C. Complete removal of *N*- and *O*-linked oligosaccharides results in a further increase in the rate of EPO thermal denaturation.

Resistance to Protease Digestion. *N*-linked oligosaccharides frequently play a role in protecting a glycoprotein from extracellular proteolytic attack. For example, aglycosyl hCG is more susceptible to proteolysis by chymotrypsin while removal of just terminal sialic acid from EPO results in increased susceptibility to proteolysis by trypsin.

2.2.5.2 Molecular Interaction Role

The accumulation of much information in recent years on the structures of oligosaccharides in glycoproteins has enabled investigations of their roles in molecular recognition (See Rasmussen, 1991; Jenkins and Curling, 1994; Goochee et al., 1992). Among these roles are:

Secretion. The importance of glycosylation for protein secretion is protein dependent. In some cases it is of no consequence, while in others active protein is not secreted unless glycosylation has occurred. For example, point mutations to remove individually the three *N*-linked and one *O*-linked glycosylation sites of EPO suggest that two *N*-glycosylation sites (Asn-38 and Asn-83) and the single *O*-glycosylation site (Ser-126) are required for efficient EPO secretion from recombinant CHO cells. It has been hypothesised that *N*- and *O*-linked oligosaccharides promote protein secretion for some proteins by assisting in the achievement of the proper tertiary or quaternary structure necessary for exit from the endoplasmic reticulum and by improving its physicochemical effects as outlined above.

Plasma Clearance Rate and Bio-Distribution. The physicochemical effects of oligosaccharides discussed above influence the turnover and half-life of glycoproteins. More significantly, however, interaction with endogenous soluble or cell-surface associated lectins is another important consideration. Examples include the asialoglycoprotein receptor of hepatocytes in the liver which binds glycoproteins with complex oligosaccharides containing terminal Gal, GalNAc or GlcNAc. Desialylation of EPO leads to rapid *in vivo* clearance and hence decreased *in vivo* activity, although the specific activity of EPO is increased on desialylation. In addition, the Man/GlcNAc receptor of endothelial cells of the liver and resident macrophages cells of the liver and spleen are involved in clearance of, for example, t-PA which contains an oligomannose structure at Asn-117.

Immunogenicity/Antigenicity. Oligosaccharides can affect the antigenicity of a protein either by constituting an antigenic determinant itself (such as the ABO blood-group determinants), or by masking antigenic sites on the protein. Many circulating antibodies are targeted against specific oligosaccharide determinants. For example, approximately 1% of circulating human IgG is specific for the terminal Galα(1,3)Galβ(1,4)GlcNAc epitope. It should be noted that the terminal Galα(1,3) moiety is produced by some host cell types (see Tab. 2.2.4) in lieu of sialylation. The immunogenicity (ability to elicit an immune response) of oligosaccharides, unless covalently linked to a protein, is less clear.

Effects on Biological Activity. For many glycoproteins, full glycosylation equates with full biological activity. For example, removal of sialic acid moieties from factor

IX results in complete loss of enzymatic activity, while tetraantennary complex-type oligosaccharides increase the *in vivo* biological activity of EPO. Removal of the complex oligosaccharide at Asn-184 of t-PA can affect its fibrin binding properties while the oligomannose-type oligosaccharides at Asn-117 increase the *in vitro* activity. However, there are several instances where glycosylation status has no major impact on the biological efficacy of a protein *in vitro*. For example, although desialylated human EPO shows a thousand-fold reduction in specific activity *in vivo* compared to its native form, there is little effect *in vitro*.

2.2.5.3 Glycosylation of Recombinant Therapeutics

Although over a dozen recombinant glycoproteins are currently either commercially available or in clinical trials, the following sections will be limited to discussion of just two of these, in order to illustrate the most salient aspects.

2.2.5.3.1 Tissue Plasminogen Activator (t-PA)

Human tissue plasminogen activator is a serine protease secreted by endothelial cells which is involved in the vascular thrombolysis system. It converts (by proteolytic cleavage) plasminogen into plasmin which in turn is able to degrade the fibrin network that holds blood clots intact. Recombinant t-PA (rt-PA) is used clinically as a thrombolytic agent for the treatment of myocardial thrombosis. t-PA has a specific affinity for fibrin and its enzymatic activity is markedly increased in the presence of fibrin. Therefore, t-PA may be clinically superior to streptokinase and urokinase, which do not have an affinity for fibrin and sometimes cause systemic hyperfibrinogenolysis.

t-PA contains five domains; a fibronectin "finger" domain, an epidermal growth factor (EGF) domain, two kringles, and the catalytic serine protease domain. Binding sites for fibrin are present within the „finger" and kringle domains. The amino acid sequence of t-PA contains three *N*-glycosylation sequons at Asn-117, 184, and 448 (Tab. 2.2.2). Asn-117 carries almost exclusively oligomannose oligosaccharides, whereas Asn-184 and 448 carry predominantly complex oligosaccharides in t-PA type I while Asn-184 is not glycosylated in t-PA type II. Each of these species has a subpopulation of glycoforms and can further exist in a single-chain and double-chain form. Variable occupancy at Asn-184 affects the fine structure of the glycan population at Asn-448 demonstrating that glycosylation at one site can influence the processing at another. In addition, recombinant t-PA expressed in CHO or human embryonic kidney cells, as well as in melanoma-derived t-PA, carries an *O*-glycosidically linked fucose residue in the EGF domain.

A great variety of complex oligosaccharide structures have been reported at Asn-184 and 448 for t-PA obtained from different sources. In a study of Bowes melanoma t-PA (mt-PA) the major oligosaccharide at Asn-184 was deduced to be a biantennary complex chain with fucose at the proximal GlcNAc and some chains appeared to have a bisecting GlcNAc residue. The oligosaccharides at Asn-448 consisted of a heterogeneous mixture of tri- and tetraantennary chains, a portion of which were missing terminal sialic acid and galactose residues but containing the proximal fucose and bisecting GlcNAc. NeuAc was found in both α2-3 and α2-6 linkages. Another study agreed, on the whole, with these results, but found in addition, sulphate groups on some complex chains and oligomannose oligosaccharides at Asn-184 and 448 in addition to Asn-117.

The oligosaccharide structures of t-PA derived from colon cells shows a very different pattern. There was a higher ratio of complex to oligomannose chains and the former did not contain a bisecting GlcNAc or terminal GlcNAc residues.

The glycan chains of recombinant t-PA produced in CHO cells have been studied in detail by a number of laboratories. While differences in results are evident, e.g. the relative amounts of tri-and tetraantennary structures present, the oligosaccharides show a pattern similar to that from colon-derived t-PA, with no terminal or bisecting GlcNAc residues found and sialic acid only in α2-3 linkage (as expected for CHO cells, see Tab. 2.2.4). As outlined in section 2.2.4.5, differences in oligosaccharide structures may be related to the cell culture conditions.

The oligosaccharide structures reported for rt-PA obtained from the murine C127 are qualitatively similar to those from CHO cells, with more highly-branched complex structures. One major difference, however, is the presence of the Galα1-3Gal epitope in the murine product.

The pharmocokinetics of recombinant, C127-derived human t-PA containing the terminal Galα1-3Gal epitope have been compared in chimpanzees, (who also have natural circulating antibodies specific for that determinant) with CHO-derived human t-PA which does not contain this epitope. The C-127-derived t-PA was cleared no faster than the CHO-derived counterpart. Since the clearance of t-PA is a complex phenomenon, the results of this study are difficult to interpret.

Human t-PA is cleared through the mannose receptor of liver endothelial cells and Kupffer cells *via* the oligomannose structure at Asn-117. Enzymatic removal of this oligosaccharide using Endo-H or site-directed mutagenesis to eliminate the glycosylation site at Asn-117 results in enhanced t-PA circulatory lifetime. In addition, t-PA may be simultaneously cleared by hepatocytes *via* a second mechanism which is apparently carbohydrate-independent and t-PA-specific.

A comparison of the rate of conversion of the single-chain into the double-chain form, the enzymatic activity of the different forms, and their susceptibility to plasma protease inhibitors led to the conclusion that these properties are affected by the glycan at Asn-184. Variable occupancy at Asn-184 is therefore one of the factors which controls the rate at which plasmin is generated, and this may be related to the differences in the affinity of type I and type II t-PA for a lysine site exposed on the degraded fibrin.

A study in which recombinant t-PA with modified glycans was produced in CHO cells grown in the presence of the processing inhibitor deoxymannojirimycin, has shown that the structure of the oligosaccharide at Asn-448 also affects the catalytic activity of t-PA.

Because of the high clearance rate in circulation of t-PA, quite high doses are needed and this in turn is associated with an increased risk of bleeding. In an effort to create molecules with increased plasma half-life and thrombolytic efficiency, a number of recombinant t-PA analogues have been made. One of these analogues, FK2P, lacking the EGF domain-like and kringle 1 domains, when expressed in CHO cells contains only two *N*-linked glycosylation sites (equivalent to Asn-184 and 448). However, only the site equivalent to Asn 448 is actually glycosylated (Aeed et al.,

1994) and the oligosaccharides were almost all of the complex sialylated bi-, tri- and tetraantennary type. Since the oligomannose structure is thus missing, it is anticipated that this molecule will have a longer circulatory life-time.

Pfeiffer et al. (1994) have constructed two t-PA variants carrying an additional *N*-glycosylation site in the EGF-like domain for expression in murine C127 cells. One of the newly-introduced glycosylation sites carried predominantly biantennary oligosaccharides while the second carried a range of di-, tri- and tetraantennary species. The results also revealed that substitution of oligosaccharides by α-Gal and intersecting GlcNAc was more pronounced at the newly-introduced sites and at Asn-117 than at Asn-184 and 448, while the original glycosylation sites carried the range of anticipated structures, including sulphated oligosaccharides in most instances. Preliminary results revealed that both variants are characterised by a prolonged initial half-life.
(See Spellman, 1990; Rasmussen, 1991; Lis and Sharon, 1993; Dwek, 1995).

2.2.5.3.2 Erythropoietin (EPO)
Erythropoietin (EPO) is a haemopoietic hormone responsible for regulating the growth and terminal maturation of erythroid progenitor cells. Plasma EPO levels increase in response to hypoxia and, as a result, regulate the production of mature, oxygen-carrying red blood cells. A reduction in EPO production, which results in anaemia, is a major complication arising from chronic renal failure. Recombinant EPO (rEPO), the first recombinant biomedicine produced in heterologous mammalian cells, results in the amelioration of anaemia and so has been used extensively in the treatment of chronic haemodialysis patients.

Each of the recombinant EPOs described to date has been shown to contain about 40% carbohydrate which probably covers most of the molecular surface. The *N*-linked oligosaccharides are distributed over three glycosylation sites (Asn-24, 38 and 83) while the *O*-linked oligosaccharides are present at Ser-126. In all urinary and recombinant rEPO variants studied to date, the *N*-linked structures consist of di-, tri- and tetraantennary oligosaccharides with a proximal fucose at each glycosylation site. However, a characteristic of these structures is the occurrence of tri- and especially tetraantennary oligosaccharides with additional lactosamine units (e.g. Fig. 2.2.4iib). A recent study of rEPO from CHO cells (Watson et al., 1994) has shown that all these structures are fully sialylated with NeuAc in α2-3 linkage to Gal. *O*-Linked tri- and tetrasaccharides with one and two sialic acids, respectively (see Fig. 2.2.5), were also identified in this study. In addition, a recent study by Nimtz and Conradt (1993) reports the characterisation of an unusual structure at Asn-24 when EPO was produced in a BHK cell line; a hexamannose oligosaccharide incorporating a phosphodiester bridged additional α-linked GlcNAc.

The Asn- oligosaccharides of EPO have a major effect on the *in vivo* biological activity of the molecule. Complete desialylation of EPO by neuraminidase treatment results in a total loss of *in vivo* biological activity as a result of clearance by the asialoglycoprotein receptor in the liver. A good positive correlation has been found between the *in vivo* activity of rEPO and the ratio of highly-branched oligosaccharides, suggesting that these structures have a role in decreasing a non-specific clearance mechanism, such as filtration by the kidney, or in enhancing homing to bone marrow, which is the target organ of EPO. Studies have also concluded that the trimannosyl core is required for *in vitro* activity by helping maintain the active onformation of the molecule.

The Asn-oligosaccharides also affect the physicochemical properties of EPO. Deglycosylated EPO shows a greater tendency to aggregate and to bind non-specifically to cells. Desialylated EPO or deglycosylated EPO is more sensitive to heat denaturation and proteolytic degradation and rEPO produced in the presence of tunicamycin is not secreted.

The observation that neuraminidase treatment of rEPO after prior removal of the Asn-oligosaccharides with PNGase-F treatment increased receptor affinity suggests that the *O*-linked oligosaccharide may modulate binding in some manner. (See Rasmussen, 1991; Takeuchi and Kobata, 1991; Watson et al., 1994).

2.2.6 Glycosylation Analysis

Glycosylation analysis is fundamentally much more complex than analysis of either proteins or nucleic acids. The complexity arises not only from differences in the type of covalent linkages and glycosylation sites (*N*-linked to Asn and/or *O*-linked to Ser/Thr) found in glycoproteins but also in the degree of occupancy of each of the glycosylation sites. The oligosaccharides at each *N*-site may be less or more processed to give oligomannose, hybrid or complex structures, have varying antennarity, and have a diversity of terminal sugars present, in a variety of linkages with different anomeric configurations, giving rise to a huge array of possible structures. Finally, substitution by groups such as sulphate may occur. In effect, the structure of a family of glycoforms is being analysed.

The downstream processing protocol may also influence the set of isolated glycoforms. The reproducibility of the glycoform distribution will be dependent upon the reproducibility of the purification procedure. Significant differences in solubility between glycoforms are also possible, as demonstrated by the reduced solubility of EPO in the absence of terminal sialylation. It should be noted that the possibility exists of adjusting the purification protocol to isolate a subset of glycoforms with enhanced therapeutic potential, for example, with regard to degree of sialylation. The goal should be to develop a process that produces the most desirable distribution of glycoforms. For example, it is generally desirable to have a high level of sialic acid to ensure maximum glycoprotein circulatory half-life and solubility in the body.

The approach to be taken by the carbohydrate chemist in analysing a glycoprotein varies depending on the sample under investigation. For example, the strategy taken in the initial structural analysis of a novel glycoprotein isolated in small amount from a biological sample will be different from that taken in the routine batch analysis of a glycoprotein produced in large amount in an industrial setting where the fundamental information about the glycosylation is already established. In other words, the analysis may require the full structural characterisation of each oligosaccharide at each glycosylation site or may simply necessitate an overall glycosylation profile, or "fingerprinting".

The approach will obviously be dependent on the facilities and instrumentation available, either on-site or through collaboration; very few laboratories have a com-

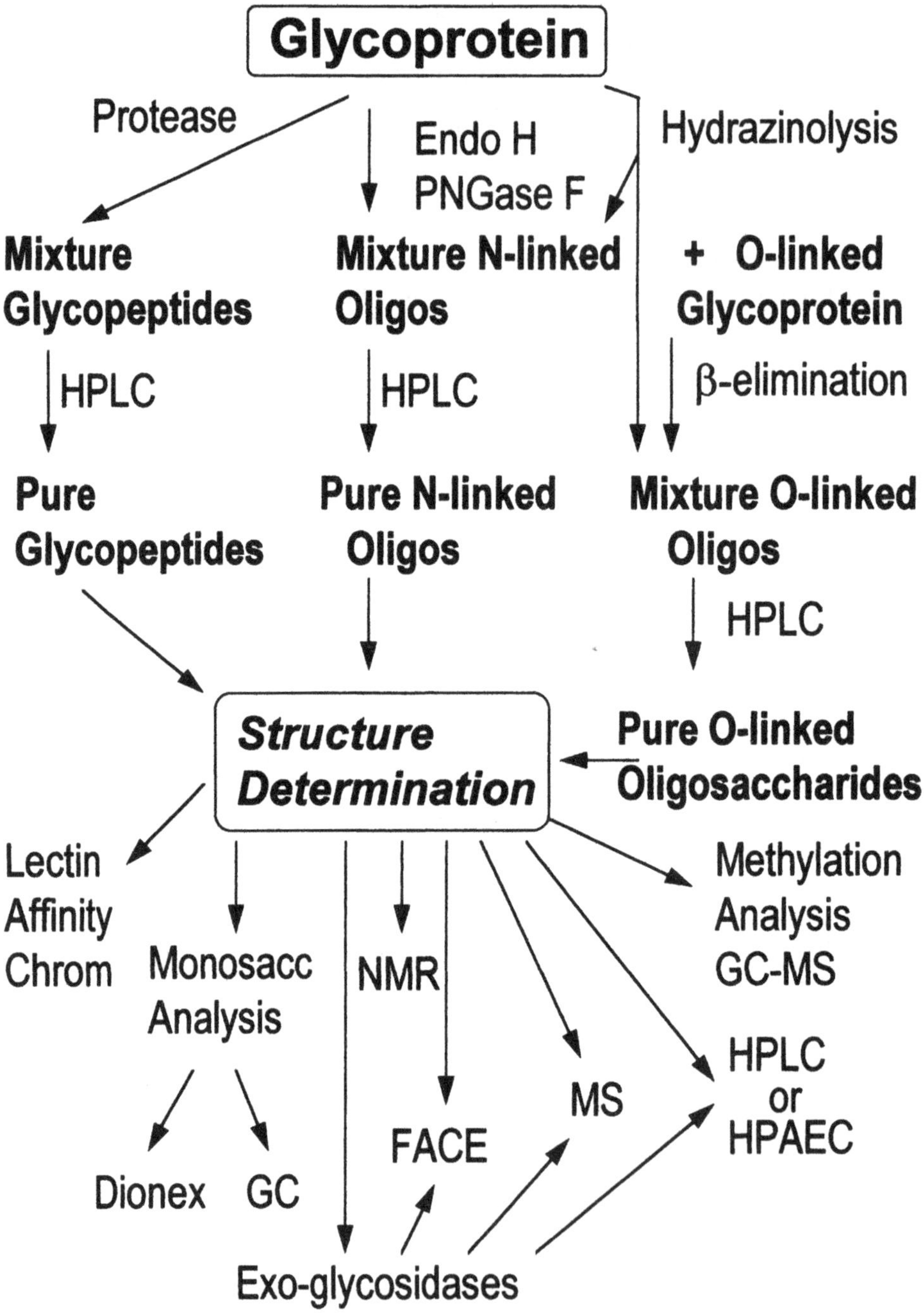

Scheme 2.2.3: Strategy for the isolation and structure determination of oligosaccharides and glycopeptides from glycoproteins.

plete range of state-of-the-art instrumentation available. However, by judicious use of different techniques and, if necessary, the use of instrumentation in specialised centres for complex analyses (such as high-field NMR or MS analyses), the complete glycoform analysis can be accomplished in an ever-decreasing length of time. A general strategy is outlined in Scheme 2.2.3.

In the past three years, a number of excellent monographs (Allen and Kisailus, 1992; Lenarz and Hart, 1994; Hounsell, 1993; Fukuda and Kobata, 1993; Verbert, 1995.) and review articles (Spellman, 1990; Cumming, 1991; Kobata, 1992; Dwek et al., 1993) have appeared, which detail the various techniques which may be used. Since an in-depth treatment of the many techniques is beyond the scope of this chapter, the various overall strategies which may be employed will be outlined and their relative merits discussed.

2.2.6.1 Is the Protein Glycosylated?

An initial step in the investigation of many proteins is to determine if the protein is in fact a glycoprotein. Glycosylation may conveniently be ascertained by staining a PAGE gel after electrophoresis, by staining a transblotted membrane after electroblotting or by removal of the oligosaccharides using enzymes and comparison of the molecular weight, before and after deglycosylation, using SDS-PAGE.

2.2.6.2 Does the Glycoprotein Contain *N*- and/or *O*-Linked Oligosaccharides?

Investigation of the amino acid sequence will indicate the possible glycosylation sites (see Tab. 2.2.2). However, the presence of the correct glycosylation sequon does not guarantee glycosylation. Monosaccharide analysis can give some indication of the type of glycosylation present, e.g. the absence of GalNAc suggests the absence of *O*-linked structures; the converse does not necessarily hold as GalNAc may appear as a terminal residue in *N*-linked structures. A high proportion of Man residues may indicate the presence of oligomannose structures. A molecular weight decrease, as shown by SDS-PAGE, when the glycoprotein is treated with an enzyme such as PNGase F (see section 2.2.6.3.2) indicates the presence of *N*-linked oligosaccharides. A further reduction in molecular weight on treatment with endo-GalNAcase (see section 2.2.6.3.2) indicates the presence of the disaccharide Galβ1-3GalNAc. The use of a lectin kit may also provide valuable information on the possible nature of the oligosaccharides present.

2.2.6.3 How Do I Release Oligosaccharides for Structural Analysis?

Before analysis of the primary chemical structure of glycoforms is undertaken, sufficient quantities of oligosaccharides are generally cleaved from the protein backbone prior to purification into homogeneous compounds using various chromatographic techniques. The following criteria should be satisfied in their release; (i) the oligosaccharides should be released in a nonselective manner, so that individual structures are present in the pool of released glycans in the same relative molar proportions in which they occurred in the glycoprotein and (ii) the oligosaccharides should be released in a structurally intact form. Alternatively, analysis of purified glycopeptides may give the required information.

2.2.6.3.1 Chemical Release of Oligosaccharides

Chemical release of intact *N*- and *O*-linked oligosaccharide chains is generally accomplished by hydrazinolysis and *β*-elimination, respectively, and, under strictly controlled conditions, each of these methods may release both *N*- and *O*-linked oligosaccharides. Chemical methods have the advantage of being less selective than enzymatic methods. However, they suffer from the disadvantage of potential for chemical degradation of the oligosaccharides. Therefore, it is imperative that the sample and reagents are initially purified and that the reaction conditions are strictly controlled.

Chemical release of *N*-linked oligosaccharides. During the reaction of hydrazine with glycoprotein or glycopeptides, the asparaginyl-oligosaccharide bond, as well as other ester and amide bonds, is cleaved, whereas the glycosidic linkages are stable. The terminal reducing sugar of the oligosaccharide is converted to its hydrazone derivative while the peptide is degraded during the reaction to yield amino acid hydrazides. In order to cleave the glycosylamine linkage and to recover the original oligosaccharides, the hydrazinolysis product is *N*-acetylated. Since hydrazinolysis may also result in the removal of non-carbohydrate substituents, including the NAc group, re-*N*-acetylation also rectifies the latter problem. To avoid the formation of artefacts, which may be present in large amounts and which mainly originate from the hydrazone derivative, the treatment of the reacetylated material with copper (II) acetate has been proposed to quantitatively convert the acetyl hydrazino-sugar into *N*-acetylglucosamine. Scheme 2.2.4 shows a synopsis of the reaction.

It should be noted that although sialic acid linkages themselves are usually stable to hydrazinolysis, substitutions at various positions in the sialic acid ring, and particularly *O*-acyl substitutions, are not expected to be retained.

Chemical release of *O*-linked GalNAc oligosaccharides. *O*-Glycosidic linkages to Ser/Thr residues of peptides or proteins are generally cleaved by the use of *β*-elimination under mild alkaline conditions (see Scheme 2.2.5). The alkaline conditions employed result in the loss of some non-carbohydrate substituents and degradation of the peptide. To prevent the degradation of alkali-labile linkages by a "peeling" reaction, the elimination is performed in the presence of excess reducing agent, $NaBH_4$, which leads to the production of stable oligosaccharide-alditols. Employment of NaB^3H_4 as reducing agent introduces a useful label.

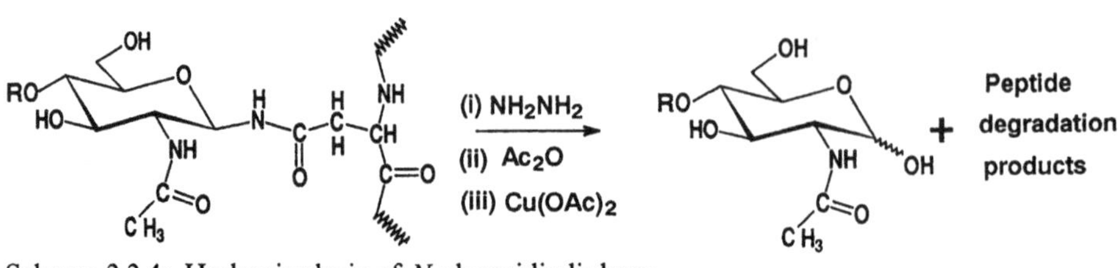

Scheme 2.2.4: Hydrazinolysis of *N*-glycosidic linkage.

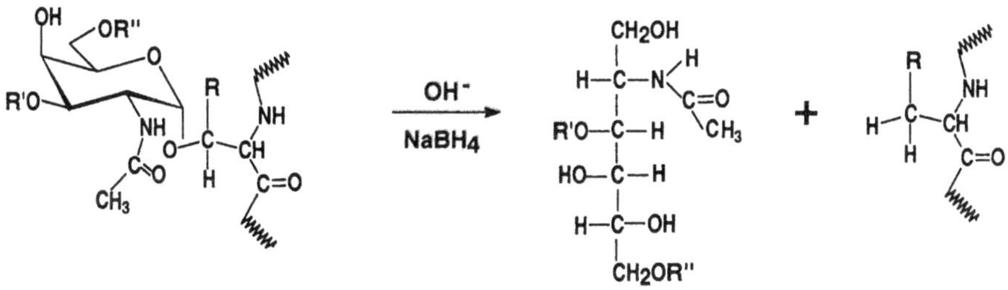

Scheme 2.2.5: Reductive β-elimination of the *O*-glycosidic linkage. R = H Ser; R = CH_3 Thr; R′,R″ = H or another sugar.

2.2.6.3.2 Enzymatic Release of Oligosaccharides

The use of enzymes for the release of oligosaccharides offers many advantages but also some disadvantages. Since enzymes usually exhibit strict substrate specificity, only well-defined structures can be released, so that some unusual or unknown structures may not be cleaved. However, this method offers ease of operation and good reproducibility where structures present can be predicted.

Enzymatic release of *N*-linked oligosaccharides. Two classes of enzymes, the endo-β-*N*-acetylglucoaminidases (or endo-glycosidases EC 3.2.1.96) and peptide-N^4-(*N*-acetyl-β-glucosaminyl)asparagine amidases (abbreviated as PNGases, *N*-glycopeptidases, glycoamidases or *N*-glycanases®, EC.3.5.1.52) are available for *N*-deglycosylation (see Tab.2.2.5).

Table 2.2.5. Examples of enzymatic cleavage of *N*-glycosidic linkages of glycans.

Enzyme	Specificity
Endo-β-N-acetyl-glucosaminidases	R-Manα1↘6 Manβ1-4GlcNAcβ1-▼4GlcNacβ1-Asn R-Manα1↗3
-Endo F_1 } Endo F	Oligomannose and some hybrid types
-Endo F_2 } Endo F	Oligomannose and biantennary complex types
-Endo F_3 } Endo F	Bi and tri antennary types
-Endo H	Oligomannose and hybrid types
-Endo D, C_1	Oligomannose type
Endoglycopeptidases	R-Manα1↘6 Manβ1-4GlcNAcβ1-4GlcNAcβ1▼-Asn R-Manα1↗3
-PNGase-F	All types, except Fucα1-3GlcNAc-Asn
-PNGase-A	All types

The endo-β-*N*-acetylglucosaminidases (e.g. endo F and endo H) cleave oligosaccharide chains at the di-*N*-acetylchitobiosyl linkage leaving a single *N*-acetylglucosamine linked to asparagine. The substrate specificities of the endo-β-*N*-acetylglucosidases can vary considerably as shown in Table 2.2.5.

The *N*-glycanase PNGase F from *Flavobacterium meningosepticum* cleaves the link between GlcNAc and asparagine and its broad specificity is due to its recognition of the polypeptide chain and the di-*N*-acetylchitobiosyl region (with or without Fuc linked α1-6) and not the peripheral oligosaccharides. Following the action of PNGases, the asparagine residues to which glycans were attached are converted into aspartic acid residues, thus creating a significant marker at the glycosylation site.

Oligomannose, paucimannose and complex oligosaccharides with di-, tri-, and tetraantennary arms are substrates for PNGase F, in addition to polysialyl complex oligosaccharides and those containing sulphate residues. PNGase F has been found to be active against substrates with fucose attached α1-6 to Asn linked GlcNAc or with a bisecting GlcNAc or Xyl residues but is inactive against substrates with fucose attached α1-3 to Asn linked GlcNAc (e.g. Fig. 2.2.7). However, PNGase F does not release oligosaccharides located on the amino- or carboxy-terminal amino acid and acts most efficiently against either heat- or SDS-denatured glycoproteins.

Enzymatic release of *O*-linked oligosaccharides. Endo-α-*N*-acetylgalactosaminidase (EC 3.2.1.97, abbreviated as endo-GalNAcase or *O*-glycanase®) catalyses the hydrolysis of the linkage between GalNAc and Ser/Thr in glycoproteins. This enzyme has been isolated from a number of sources but in most cases its value is limited since the substrate specificity is restricted to the disaccharide Galβ1-3GalNAc (Fig. 2.2.8).

Galβ1-3GalNAcα1 ↓ Ser/Thr

Fig. 2.2.8: Substrate specificity and hydrolysis site for endo-α-*N*-acetylgalactosaminidases.

2.2.6.3.3 Which to Use; Chemical or Enzymatic Methods?

The answer to this question depends on a number of factors; (a) whether the glycoprotein is novel or has been previously well characterised, (b) if both *N*- and *O*-linked oligosaccharides are present, (c) the amounts available, (d) the number of samples to be analysed, and (e) the instrumentation available.

In the situation of having many samples with both *N*- and *O*-linked oligosaccharides, an automated chemical-based system which will cleave all oligosaccharides offers distinct advantages since difficulties in reproducibility may be encountered using a manual hydrazinolyis method. Such an automated system, using hydrazinolysis, is currently commercially available from Oxford GlcyoSystems. Where *O*-linked oligosaccharides are to be isolated following the release of *N*-linked oligosaccharides (using enzymes) and where oligosaccharide larger than the disaccharide are present, the β-elimination method has been well established.

Where the *N*-linked oligosaccharide structures have previously been characterised and shown to act as substrates for PNGase F, then the enzymatic method offers the advantage of ease of operation and good reproducibility. Since *O*-GalNAcase only cleaves the disaccharide Galβ1-3GalNAc, its use is limited. However, this structure along with its mono- and disialylated counterparts occurs frequently in recombinant products. Initial treatment with neuraminidase will, therefore, expose the disaccharide which may then conveniently be released by enzyme.

2.2.6.4 Fractionation, Purification and Mapping of Oligosaccharides

In most instances, the oligosaccharide or glycopeptide under investigation must be isolated in a pure form before structure determination can be carried out. To date a large array of chromatographic methods has been developed for the separation and purification of heterogeneous oligosaccharide and glycopeptide pools and include;

- Ion Exchange Chromatography
- Gel permeation chromatography (GPC), e.g. Bio-Gel P-4
- High performance liquid chromatography (HPLC) of derivatised (e.g. pyridylamino), or non-derivatised oligosaccharides
- Dionex High Performance Anion Exchange Chromatography with Pulsed Amperometric Detection (HPAEC-PAD) of non-derivatised oligosaccharides
- Serial Lectin Affinity Chromatography
- Capillary Zone Electrophoresis
- Fluorophore-assisted carbohydrate electrophoresis (FACE)

The strategy to be used will depend on the degree of heterogeneity of the oligosaccharides present in the pool and how closely-related the structures are, whether there are sialylated as well as neutral oligosaccharides present and the amounts of material available. For complex mixtures, several purification methods may need to be used, whereas in other instances the use of Dionex HPAEC-PAD would be sufficient (Townsend and Hardy, 1991).

A common strategy (see Scheme 2.2.3) for analysing complex mixtures would be to separate the pool into sialylated and neutral fractions using ion-exchange chromatography. Then a gross separation could be carried out using either gel-permeation chromatography (GPC), lectin affinity chromatography or amino-bonded HPLC, followed by analysis of mixtures which were not fully resolved using Dionex HPAEC-PAD.

Serial lectin affinity chromatography has been widely used for the separation and purification of oligosaccharides and glycopeptides. In addition, the binding to a particular lectin gives many clues about the structure of the oligosaccharide. The technique is relatively cheap and yields are usually high.

The HPAEC-PAD technique has the advantages of (a) increased selectivity and sensitivity of detection (picomol level) of oligosaccharides as oxyanions by using NaOH/NaOAc as the eluant and by detecting using an electrochemical method, (b) increased resolution with pellicular-based anion exchange columns and (c) short analysis times, without the need for derivatisation. However, the technique suffers from the disadvantage of low loading and yield compared with HPLC and GPC and, because of the non-volatile eluant, salts must be removed prior to further analysis.

2.2.6.5 Structure Determination of Oligosaccharides

Recombinant glycoproteins intended for pharmaceutical use are generally available in relatively large quantities. Initial characterization of such a glycoprotein involves full primary structure analysis of each oligosaccharide at each glycosylation site; glycoform analysis. Thereafter, what is needed is a method of assessing the lot-to-lot consistency of production; so-called fingerprinting which may often be accomplished by oligosaccharide mapping (section 2.2.6).

Two of the most powerful techniques for glycoform analysis are ^{1}H-nuclear magnetic resonance (NMR) spectroscopy and mass spectrometry (MS). In some respects these techniques are complementary, with different sets of advantages and disadvantages.

Primary structure determination of a pure oligosaccharide involves ascertaining the following,

- Monosaccharide analysis
- Linkage and branching positions
- Anomeric configuration (α or β)
- Sequence of sugars
- Non-carbohydrate substituent (e.g. phosphate or sulphate)

2.2.6.5.1 Monosaccharide Analysis

Gas liquid chromatography: Derivatisation, for example as the trimethylsilyl (TMS) derivatives, is essential to render the molecules volatile and thermally stable. However, they can produce several peaks for reducing sugars as the result of anomerisation and formation of both furanose and pyranose rings, a situation avoided by the use of alditol acetates.

Dionex HPAEC-PAD: This method, which does not require derivatisation, is becoming increasingly widely used because of its superior resolution of mono- and oligosaccharides and also its high sensitivity. However, it should be noted that monosaccharide composition analysis alone may provide insufficient information to detect differences in structure between two samples.

2.2.6.5.2 Linkage Analysis

Several methods have been developed to determine the linkage positions between sugar residues in a linear or branched oligosaccharide. The most widely used is methylation analysis which involves

- permethylation in the presence of base to protect all free hydroxyl groups
- hydrolysis under acidic conditions to release the monosaccharides
- reduction to alditols to remove the anomericity
- acetylation of the newly-released hydroxyl groups.

Separation of the resultant partially methylated alditol acetates using gas liquid chromatography (GLC) is often accompanied by analysis of each peak using electron impact mass spectrometry. A unique series of fragment ions is produced from most monosaccharide residues.

2.2.6.5.3 Nuclear Magnetic Resonance Spectroscopy (NMR)

NMR spectroscopy provides a powerful, nondestructive means to characterize both *N*- and *O*-linked oligosaccharides and glycopeptides derived from glycoproteins. In addition to obtaining information on the primary chemical structure, NMR is becoming increasingly used in conformational studies. ^{1}H-NMR spectroscopy is much more widely used than ^{13}C since the latter requires at least 10 times the amount of sample. An instrument with a magnet size of 300 MHz may be sufficient for smaller, well-characterized structures while larger magnets of 500 or 600 MHz are required for more complex structures. The technique requires about 10–50 nmol material, depending on the size of the magnet.

NMR parameters (chemical shift, δ, relative intensity of peaks, multiplet structure and line width) are very sensitive to changes in both chemical bonding and spatial structure. For relatively simple or well-known structures, the diagnostic signals from

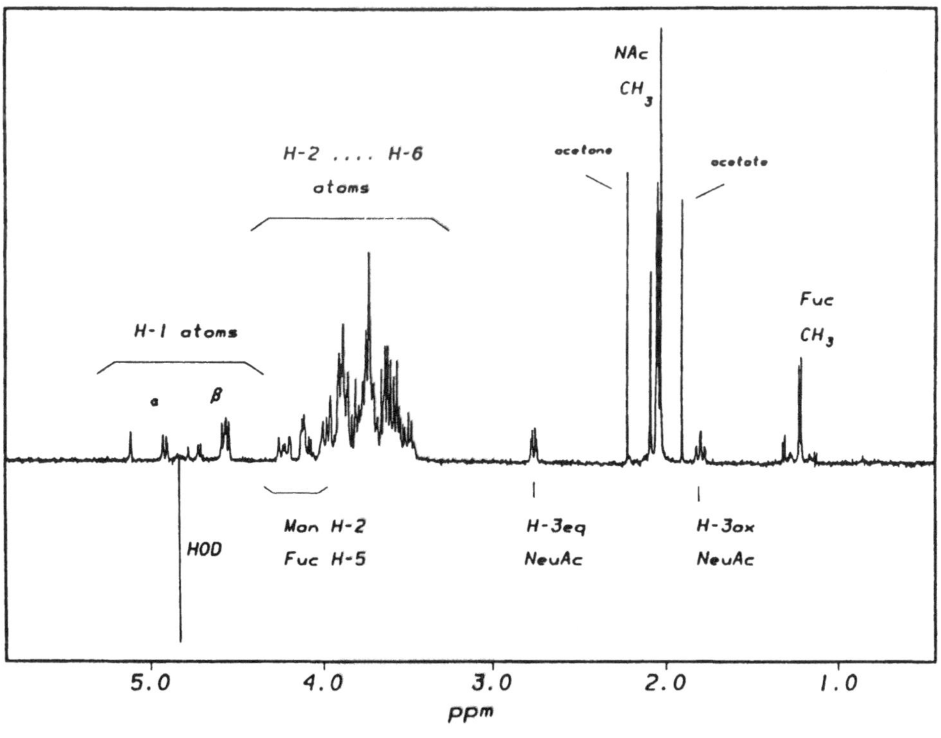

Fig. 2.2.9: A 500-MHz ^{1}H NMR spectrum of a diantennary *N*-type oligosaccharide alditol obtained from recombinant human erythropoietin expressed in CHO cells showing the structural reporter groups (from H. van Halbeek with permission; Copyright 1994 by Academic Press, In.).

"structural reporter groups" can be used as a fingerprint in the identification of the structure. The most important signals occur

- in the anomeric region from about 4.2 to 5.5 ppm
- at circa 1.2 ppm for the methyl group from deoxy hexoses
- in the region circa 2 ppm for methyl groups from acetyl substituents
- at circa 1.7, 2.7 and 4.2 ppm for the methylene resonances from sialic acids.

This information indicates the monosaccharides present and their ratios, along with the anomeric configuration and in many cases the linkage positions and sequence (Fig. 2.2.9, 2.2.10).
For more complex or novel structures it is often necessary to perform multiple-pulse techniques such as two-dimensional total correlation spectroscopy (TOCSY) and rotating-frame NOE spectroscopy (ROESY) provided much larger amounts of sample are available. These experiments yield a "subspectrum" for each sugar present and ultimately give information on the complete primary structure of the oligosaccharide. When ^{13}C and ^{1}H NMR spectroscopy are used together in correlation techniques, additional information may be obtained.

It should be noted that differences in chemical shift may result from differences in experimental conditions such as temperature and solvent thus causing difficulty in comparing data. Recently, a database computer program that uses ^{1}H-NMR

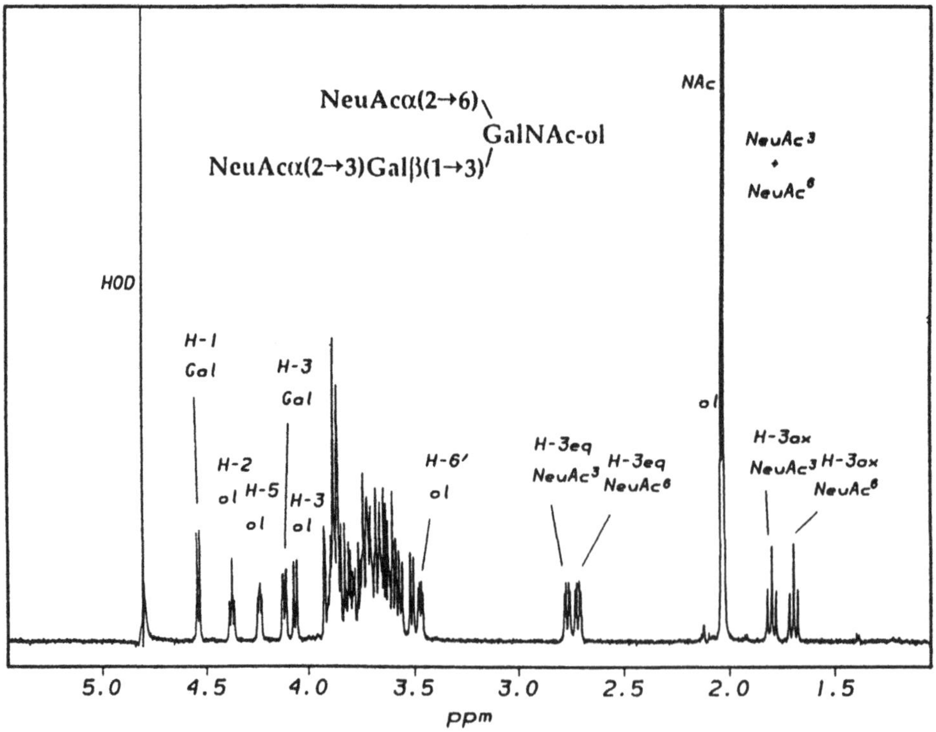

Fig. 2.2.10: A 600-MHz ^{1}H NMR spectrum of a disialylated *O*-type oligosaccharide alditol released by *β*-elimination from recombinant human erythropoietin expressed in CHO cells (from H. van Halbeek with permission).

chemical-shift values has been developed for computer-assisted identification of complex carbohydrates[2]. For novel compounds, the database can be used to recognize structural elements that occur in known carbohydrates.

The most important advantage of NMR over other techniques is its nondestructive nature. The sample can be recovered 100 % unimpaired and used for other analyses which confirm the NMR data, such as monosaccharide analysis, methylation analysis and mass spectrometry. The most important limitation of NMR spectroscopy is its sensitivity and even at 600 MHz, heterogeneity occurring in low abundance in the sample may escape attention.

2.2.6.5.4 Mass Spectrometry

In addition to its use in investigating monosaccharide derivatives in compositional analysis and methylation analysis, mass spectrometry is a powerful technique for the analysis of oligosaccharides, glycopeptides and glycoproteins. Several ionisation methods are available as outlined in Table 2.2.6.

Table 2.2.6. Mass spectrometric methods used for glycoform analysis.

Ionization Method	**Max MW**	**Mass Assignment Accuracy**	**Derivatization Needed**	**Amount**
FAB	10 kDa	± 0.01 % up to 5 kDa	Yes	1–1,000 pmol
ES	150 kDa	+ 0.01 % up to 50 kDa ± 0.03 % up to 150 kDa	No	0.3–50 pmol
MALDI	350 kDa	± 0.05 % up to 25 kDa ± 0.03 % up to 300 kDa	No	0.1–10 pmol

Fast atom bombardment mass spectrometry (FAB-MS) gives a considerable amount of sequence, branching and linkage information by virtue of the fact that the molecule is induced to fragment within the mass spectrometer ion source. Oligosaccharides generally give good FAB mass spectra, but derivatisation, usually in the form of permethylation or peracetylation, is necessary for the production of strong signals. Increased sensitivity can also be conferred by reductive amination of the reducing terminus and introduction of a hydrophobic chromophore. In order to obtain linkage information from FAB mass spectra, it is necessary to employ further derivatisation techniques. The most widely used is periodate oxidation which causes oxidative cleavage of cis-hydroxyl groups to give a di-aldehyde which is then reduced with sodium borodeuteride and permethylated to give a product whose mass differs from that of the parent sugar.

Electrospray Mass Spectrometry (ES-MS). In this technique a stream of liquid containing the sample of interest is injected directly into the ES source where the

[2] An NMR database on diskette is available from Dr. J. A. van Kuik, Department of Bio-Organic Chemistry, Bijvoet Center for Biomolecular Research, Utrecht University, P.O. Box 80.075, NL-3508 TB Utrecht.

sample molecules are stripped of solvent, leaving them as multiply charged species whose charges reflect the number of functional groups that can be protonated or deprotonated at the pH of the carrier liquid. In most ES-MS experiments the m/z values encountered are compatible with analysis by quadrupole analyzers, which are considerably less expensive than the magnetic sector analyzers required for FAB-MS. For glycopeptides, ES-MS, particularly when combined with capillary electrophoresis, is proving exceptionally valuable, and this technique coupled with MS/MS (which restores fragmentation) is seen as becoming a major method for glycoform analysis in the future.

Matrix-assisted laser desorption mass spectrometry (MALDI/MS) yields the molecular weight in only a few minutes for proteins, glycopeptides, or oligosaccharides. The sample is first mixed in solution with a large excess of a suitable UV-absorbing matrix and is particularly simple and quick to use. Unlike FAB-MS, no derivatisation is necessary. Mass separation usually employs a time-of-flight (TOF) mass spectrometer. Glycoform analysis of glycopeptides obtained from reverse-phase HPLC using MALDI-TOF/MS shows great promise.

2.2.6.5.5 Exoglycosidase Digestion

Sequential exoglycosidase digestion in conjunction with gel-permeation chromatography, is a very powerful method for characterisation in which the susceptibility of an oligosaccharide to purified exoglycosidases is inferred from changes in its elution position in gel-permeation chromatography. Because each exoglycosidase is specific for the monosaccharide and its anomeric configuration as well as, in some cases, the linkage position, a wealth of information may be derived. This concept has been extended by the Reagent Array Analysis Method (RAAM®) 1000 GlycoSequencer, marketed by Oxford GlycoSystems, in which samples are incubated with a number of sets of defined mixtures of exoglyosidases. The products are sized by GPC and the results from the various experiments analysed by software to yield the most likely structure. The products may also be analysed by FACE, MS or HPLC/HPAEC.

2.2.6.6 Oligosaccharide Mapping

To be suitable for fingerprinting, a technique must be reproducible, have sufficient resolution to distinguish between closely related structures, not be excessively labour-intensive, and if possible, not require expensive, specialised instrumentation. Mapping of released oligosaccharides in either one or two dimensions based on chromatographic separation using HPAEC-PAD (Hermentin et al., 1992), HPLC of 2-aminopyridine derivatives of asialooligosaccharides (Rice et al., 1992), as well as capillary electrophoresis (Hermentin et al., 1994) and fluorophore-assisted carbohydrate electrophoresis (FACE) (Jackson, 1994) show great promise as quick and facile methods for structural elucidation of *N*-linked oligosaccharides on the femtomolar level. Such oligosaccharide fingerprinting techniques could have many applications such as distinguishing different glycoproteins or cell lines, assessing consistency of production of a glycoprotein under varying cell-culture conditions, monitoring lot-to-lot consistency, or as a general characterisation tool.

2.2.6.7 Which Oligosaccharides Are Present at Each Glycosylation Site?

Glycosylation-site analysis is usually carried out on purified glycopeptides where occupancy of potentially glycosylated asparagine, serine and threonine residues can be determined during amino acid sequence analysis. During sequential Edman degradation, as usually performed, the labeled derivatives of glycoamino acids are not recovered to a significant degree unless the attached carbohydrate is only a monosaccharide. Glycosylated amino acid residues can therefore be revealed as "blanks" at the appropriate cycles.

Glycosylation site analysis has been performed for both *O*- and *N*-linked oligosaccharides, using either a chemical or enzymatic method to "tag" the amino acid moiety involved in the linkage. The strategy works best where only one potential glycosylation site exists on the (glyco)protein or (glyco)peptide.
(See Spellman, 1990.)

2.2.6.7.1 Chemical Method

(i) *O-linked:* A sensitive mass spectrometric method for the specific identification of *O*-glycosylation sites in glycopeptides has recently been reported (Rademaker et al., 1993). Reductive β-elimination liberates a peptide containing a specifically modified amino acid at the former site of carbohydrate attachment; glycosylated serine is converted to alanine while glycosylated threonine is converted to β-amino-butyric acid. Mass spectrometry is used to determine the sequence of the modified peptide, in which the position of the modified amino acid marks the site of *O*-glycosylation. It should be noted that reductive β-elimination also causes some peptide bond cleavage
(ii) *N-linked:* Solvolysis with hydrogen fluoride (HF) splits all *O*-glycosyl bonds to yield the corresponding monosaccharide fluorides which are converted to the free monosaccharides in the presence of water. Meanwhile, the *N*-glycosyl bonds and peptide bonds are stable against HF, yielding a glycopeptide with a single *N*-linked monosaccharide at the glycosylation site. This "tagged" Asn moiety may be identified as outlined below.

2.2.6.7.2 Enzymatic Method

Another approach to determining glycosylation site occupancy is based on the use of a combination of enzymes to cleave the oligosaccharide so as to identify the Asn at the *N*-glycosylated site as shown in Scheme 2.2.6.
Cleavage of an oligomannose structure with endo H leaves a GlcNAc residue attached to the glycosylated asparagine; subsequent treatment with PNGase F has no effect on this residue. However, initial treatment with PNGase F cleaves the oligomannose structure and concomitantly transforms the asparagine to an aspartic acid.

Cleavage of a complex oligosaccharide with endo H has no effect, while treatment with PNGase F cleaves the oligosaccharide and concomitantly transforms the asparagine to an aspartic acid. If the glycosylation site is not occupied, treatment with either enzyme will have no effect.

Subsequent analysis of the peptide or "tagged" peptide using Edman degradation or by mass spectrometry will reveal a GlcNAc-Asn moiety at sites that carry endo-H

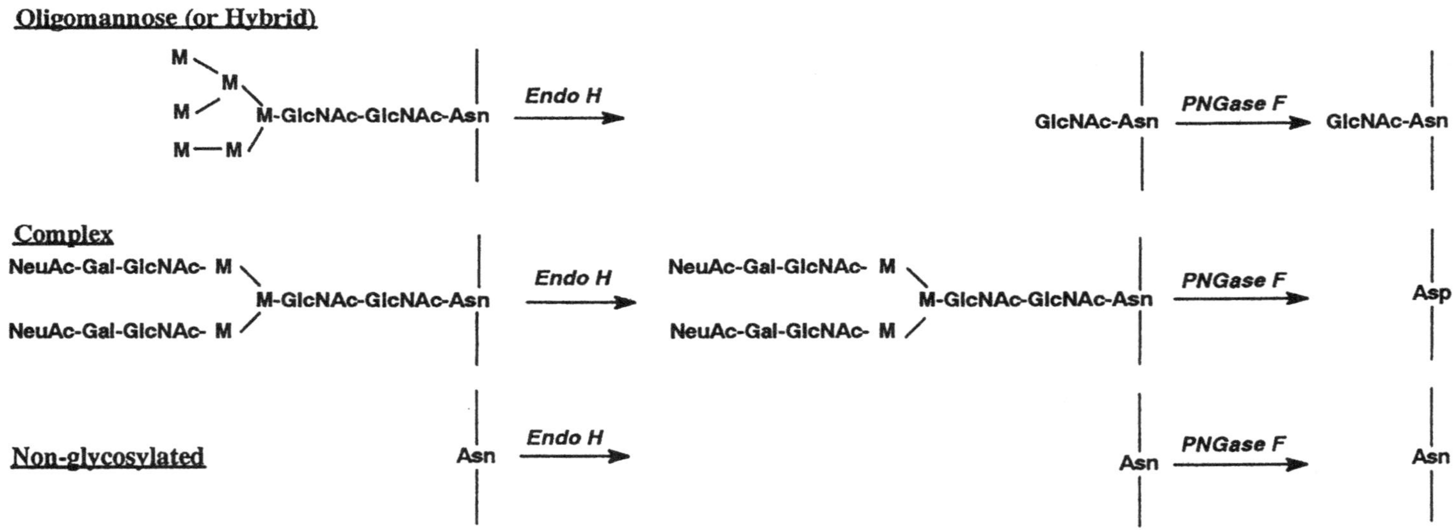

Scheme 2.2.6: The use of sequential digestions with endo H and PNGase F to characterise potential *N*-glycosylation sites (based on a scheme, with permission, from M. W. Spellman. Copyright 1990 by the American Society for Biochemistry and Molecular Biology).

sensitive oligosaccharides, while those that carry PNGase F-sensitive oligosaccharides will show aspartic acid and consequently the FAB-MS spectrum will show an increase by 1 amu over the theoretical (nonglycosylated) peptide.

The glycosylation site(s) of peptides bearing *O*-linked GlcNAc may also be investigated using FAB-MS of the propionylated derivatives prior to and following galactosylation.

2.2.7 Conclusion

Because many human therapeutics must be correctly glycosylated for full *in vivo* efficacy, it is imperative that cultured mammalian cells are used for their production. If a production process for a glycoprotein is to be considered developed and validated, it becomes relevant to establish not only that the glycosylation of the large-scale product is similar to that of the reference compound (usually produced on a laboratory scalc in small amounts), but also that a constant glycosylation pattern can be reproducibly obtained. Purely biological tests on different lots, such as *in vivo* activity in an animal model system cannot ensure constancy of therapeutic profile in humans; precise and sensitive structural comparisons with respect to polypeptide, glycosylation and other post-translational modifications are necessary.

With continued advances in instrumentation and structural methodologies, the once daunting and time-consuming task of oligosaccharide structure determination is now becoming increasingly simplified and a computer database of oligosaccharide structures is now available.[3] Although the era of automated total oligosaccharide sequencing has not yet arrived, its advent is eagerly anticipated, particularly by those involved in the routine analysis of glycoproteins of therapeutic importance.

[3] The Complex Carbohydrate Structure Database (CCSD) may be contacted at the Complex Carbohydrate Research Center, Univ. Georgia, Athens, GA30602, USA and CarbBank, the corresponding database-management program, may be contacted at 114 W. Magnolia St, Suite 305, Bellingham, WA 98225, USA for CD-ROM or by email at 76424.1122@compuserve.com.

Abbreviations

Carbohydrates

Gal	galactose
Fuc	fucose
GalNAc	*N*-acetylgalactosamine
Glc	glucose
GlcNAc	*N*-acetylglucosamine
Man	mannose
NeuAc	*N*-acetylneuraminic acid (sialic acid)
NeuGc	*N*-glycolylneuraminic acid
Xyl	xylose

Amino-acids

Asn	asparagine
Asp	aspartic acid
Cys	cysteine
Lys	lysine
Gly	glycine
Pro	proline
Ser	serine
Thr	threonine
Tyr	tyrosine

amu	atomic mass unit
BHK	baby hamster kidney
CHO	chinese hamster ovary
CMP	cytosine monophosphate
EGF	epidermal growth factor
endo-H	endo-β-*N*-acetylglucosaminidase
EPO	erythropoietin
E R	endoplasmic reticulum
ES	electrospray
FAB-MS	fast atom bombardment-mass spectrometry

Glossary

Anomers	Sugars containing different arrangements of atoms around C-1. For D pyranose sugars, the α anomer has the O atom lying in an axial orientation (below the plane of the ring) while the β anomer has the O atom in an equatorial orientation (above the plane of the ring) (see Fig. 2.2.1).
Antenna	Galβ1-4GlcNAcβ1- unit linked to Man of the pentasaccharide core of *N*-linked oligosaccaride (see Fig. 2.2.4).
Antigenic determinant	Carbohydrate or peptide moiety which reacts in a specific manner with antibody.
β-Elimination	Chemical method using sodium hydroxide for the cleavage of *O*-glycosidically linked oligosaccharides from peptide or protein.
Complex oligosaccharide	*N*-linked oligosaccharide containing an *N*-acetyllactosamine unit (Galβ1-4GlcNAc) on two or more antennae of the core pentasaccharide. This structure may be further substituted by other sugars (Fig. 2.2.4ii).
Epitope	Peptide sequence which is antigenic.
Erythropoietin (EPO):	A haemopoietic hormone responsible for regulating the growth and terminal maturation of erythroid progenitor cells. Plasma EPO levels increase in response to hypoxia and, as a result, regulate the production of mature, oxygen-carrying red blood cells. A reduction in EPO production, which results in anaemia, is a major complication arising from chronic renal failure.
Glycan	Oligosaccharide chain.
Glycoforms	A family of glycoproteins containing identical protein but different covalently-attached oligosaccharide structures.
Glycopeptide	Peptide with covalently attached oligosaccharide.
Glycoprotein	Protein with covalently attached oligosaccharide.
Glycosidic linkage	Linkage involving C-1 of a sugar.
Glycosidase	Enzyme involved in cleavage of a glycosidic linkage.
Glycosyltransferase	Enzyme involved in the transfer of a sugar from a nucleotide donor to an oligosaccharide acceptor.
Hybrid Oligosaccharide	*N*-linked oligosaccharide containing one or two mannose units on the 1-6 arm of the core along with at least one complex (*N*-acytyllactosamine) unit on the 1-3 arm of the core (Fig. 2.2.4iii).
Hydrazinolysis	Chemical method using hydrazine for the cleavage of *N*-linked oligosaccharides from protein or peptide.
Hypermannose	Oligosaccharide containing more than nine mannose units.
Kringle	Triple-looped peptide sequence containing three characteristically located disulphide bonds.

Lectin	Plant- or animal-derived protein which binds specifically to a carbohydrate moiety.
Mass spectrometry	Technique for the analysis of molecular structure based on the mass of the parent molecule or of its fragments.
Nuclear magnetic resonance spectrometry	Technique for the analysis of molecular structure based on the magnetic moments of nuclei such as ^{1}H or ^{13}C.
Neuraminidase	Enzyme used for the cleavage of neuraminic acid (sialic acid) from oligosaccharides.
N-glycan	Oligosaccharide linked *via N* to asn (Figs 2.2.2, 2.2.4).
Oligomannose oligosaccharide	*N*-linked oligosaccharide containing from two to six Man units attached to the core pentasaccharide (Fig. 2.2.4i).
Oligosaccharide	Two to about twenty monosaccharides covalently linked through glycosidic bonds.
O-glycan	Oligosaccharide linked *via O* to ser or thr (Figs 2.2.2, 2.2.5).
Paucimannose	Oligosaccharide containing only three or less mannose units linked to the two core GlcNAc residues.
Permethylation	Addition of methyl (CH_3) groups to all the unsubstituted OH groups of a sugar.
Methylation analysis	Chemical method involving permethylation used to deduce the linkage positions of monosaccharides in an oligosaccharide.
Protease	Enzyme used for cleavage of peptide bonds.
Sequon	Sequence of amino acids necessary for glycosylation.
Sialylated	containing sialic (neuraminic acid) in a terminal position.
Tissue plasminogen activator (t-PA)	A serine protease secreted by endothelial cells which is involved in the vascular thrombolysis system. It converts (by proteolytic cleavage) plasminogen into plasmin which in turn is able to degrade the fibrin network that holds blood clots intact.

FACE	fluorphore assisted carbohydrate electrophoresis
GalNAc-Tase	*N*-acetylgalactosaminyltransferase
GlcNAc-Tase	*N*-acetylglucosaminyltransferase
GDP	guanosine diphosphate
GPC	gel permeation chromatography
HF	hydrogen fluoride
HPAEC-PAD	high performance anion exchange chromatography with pulsed amperometric detection.
HPLC	high performance liquid chromatography
IgG	immunoglobulin G
IL-2	interleukin-2
MALD	matrix-assisted laser desorption
MS	mass spectrometry
NAc	*N*-acetyl
NMR	nuclear magnetic resonance
NOE	nuclear Overhauser enhancement

t-PA	tissue plasminogen activator
PAGE	polyacrylamide gel electrophoresis
PHGase F	peptide-N^4-(N-acetyl-ß-glucosaminyl)asparagine amidase
RAAM	reagent array analysis method
RER	rough endoplasmic reticulum
ROESY	rotating frame Overhauser enhancement spectroscopy
SDS	sodium dodecyl sulphate
Sialyl-Tase	sialyltransferase
TFMSA	trifluoromethane sulphonic acid
TMS	trimethyl silane
TOCSY	total correlation spectroscopy
TOF	time-of-flight
TSH	thyroid stimulating hormone (thyrotropin)
UDP	uridine diphosphate
UV	ultra-violet

References

Aeed, P.A. et al. (1994) Characterization of the oligosaccharide structures on a recombinant tissue-type plasminogen activator analog synthesized in Chinese hamster overy cells. Fibrinolysis *8*, 31–41.

Allen, H.J. and Kisailus, E.C. (eds.) (1992) Glycoconjugates. Marcel Dekker, New York, Basel, Hong Kong.

Andersen, D.C. and Goochee, C.F. (1994) The effect of cell-culture conditions on the oligosaccharide structures of secreted glycoproteins. Current Opinion in Biotechnology *5*, 546–549.

Conradt, H.S., Hofer, B. and Hauser, H. (1990) Expression of human glycoproteins in recombinant mammalian cells; toward genetic engineering of *N*- and *O*-glycoproteins. Trends in Glycoscience and Glycotechnology *2*, 168–180.

Cumming, D.A. (1991) Glycosylation of recombinant protein therapeutics: Control and functional implications, Glycobiology *1*, 115–130.

Cummings, R.D. (1992) Synthesis of asparagine-linked oligosaccharides: Pathways, genetics, and metabolic regulation. In: Glycoconjugates: Compostion, Structure and Function (Allen, H.L. and Kisailus, E.C., eds.), pp. 333–360. Marcel Dekker, New York.

Dwek, R.A. (1995) Glycobiology: Towards understanding the function of sugars. Biochem. Soc. Trans. *23*, 1–25.

Dwek, R.A. et al. (1993) Analysis of glycoprotein-associated oligosaccharides. Auu. Rey. Biochem. *62*, 65–100.

Fukuda, M. and Hindsgaul, O. (eds.) (1994) Molecular Glycobiology. IRL Press at Oxford University Press, Oxford, New York, Tokyo.

Fukuda, M. and Kobata, A. (eds.) (1993) Glycobiology: A Practical Approach. Oxford University Press, Oxford.

Goochee, C.F. and Monica, T. (1990) Environmental effects on protein glycoylation. Bio/Technology *8*, 421–427.

Goochee, C.F. et al. (1991) The oligosaccharides of glycoproteins: Bioprocess factors affecting oligosaccharide structure and their effect on glycoprotein properties. Bio/Technology *9*, 1347–1355.

Goochee, C.F. et al. (1992) The oligosaccharides of glycoproteins. Factors affecting their synthesis and their influence on glycoprotein properties. In: Frontiers in Bioprocessing II, (Todd, P., Sikdar, S.J. and Bier, M., eds), pp. 199–240. American Chemical Society, Washington, D.C.

Han, K.-K. and Martinage, A. (1991) Post-translational chemical modification(s) of proteins. Int. J. Biochem. *24*, 19–28.

Han, K.-K. and Martinage, A. (1992) Possible relationship between coding recognition amino acid sequence motif or residue(s) and post-translational chemical modification of proteins. Int. J. Biochem. *24*, 1349–1363.

Han, K.-K. and Martinage, A. (1993) Post-translational chemical modifications of proteins III. Current developments in analytical procedures of identification and quantitation of post-translational chemically modified amino acid(s) and its derivatives. Int. J. Biochem. *25*, 957–970.

Hermentin, P. et al. (1992) A strategy for the mapping of *N*-glycans by high-pH anion-exchange chromatography with pulsed amperometric detection. Anal. Biochem. *203*, 281–289.

Hermentin, P. et al. (1994) A strategy for the mapping of *N*-glycans by high-performance capillary electrophoresis. Anal. Biochem. *221*, 29–41.

Homans, S. (1993) Conformation and dynamics of oligosaccharides in solution. Glycobiology *3*, 551–555.

Hounsell, E.F. (ed.) (1993) Glycoprotein analysis in biomedicine. Humana Press, Totowa, New Jersey.

Jackson, P. (1994) High-resolution polyacrylamide gel electrophoresis of fluorophorelabeled reducing saccharides. In: Techniques in Glycobiology, (Lennarz, W.J. and Hart, G.W., eds.), pp. 250–265. Academic Press, San Diego.

Jenkins, N. and Curling, E.M.A. (1994) Glycosylation of recombinant proteins:
Problems and prospects. Enzyme Microb. Technol. *16*, 354–364.

Joziasse, D.H. (1992) Mammalian glycosyltransferases: Genomic organization and protein structure. Glycobiology *2*, 271–278.

Kobata, A. (1992) Structures and functions of the sugar chains of glycoproteins. Eur. J. Biochem. *209*, 483–501.

Kornfeld, R. and Kornfeld, S. (1985) Assembly of asparagine-linked oligosaccharides. Annu. Rev. Biochem. *54*, 631–664.

Lennarz, W.J. and Hart, G.W. (eds.) (1994) Guide to Techniques in Glycobiology. Academic Press, Inc, San Diego.

Lis, H. and Sharon, N. (1993) Proteins glycosylation: Structural and functional aspects. Eur. J. Biochem. *218*, 1–27.

März, L. et al. (1995) Protein glycosylation in insects. In: Glycoproteins (Montreuil, J., Schacter, H. and Vliegenthart, J.F.G., eds), Elsevier, Amsterdam in press.

Montreuil, J., Schacter, H. and Vliegenthart, J.F.G. (eds.) (1995) Glycoproteins. Elsevier, Amsterdam, in press.

Nimtz, M. and Conradt, H.S. (1993) Characterization of a phosphorylated oligosaccharide from erytropoietin expressed in recombinant BHK cells. Glycoconjugate J. *10*, 259.

Opdenakker, G. et al. (1993) Concepts and principles of glycobiology. FASEB J. *7*, 1330–1337.

Pfeiffer, G. et al. (1994) Glycosylation of two recombinant human uterine tissue plasminogen activator variants carrying an additional *N*-glycosylation site in the epidermal-growth-factor-like domain. Eur. J. Biochem. *219*, 331–348.

Rademacher, T.W., Parekh, R.B. and Dwek, R.A. (1988) Glycobiology. Ann. Rev. Biochem. *57*, 785–838.

Rademacher, G., Haverkamp, J. and Thomas-Oates, J. (1993) Determination of glycoslation sites in *O*-linked glycopeptides – A sensitive mass-spectrometric protocol. Org. Mass Spectrom *28*, 1536–1541.

Rasmussen, J.R. (1991) Glycosylation of recombinant proteins. In: Biology of carbohydrates. JAI Press, London. pp. 179–285.

Rice, K.G. et al. (1992) Quantitative mapping of the *N*-linked sialyloligosaccharides of recombinant erythropoietin; combination of direct high-performance anion-exchange chromatography and 2-aminopyridine derivatization. Anal. Biochem. *206*, 278–287.

Schacter, H. and Brockhausen, I. (1992) The biosynthesis of serine (threonine)-*N*-acetylgalactosamine-linked carbohydrate moieties. In: Glycoconjugates, Compostion, Structure and Function (Allen, H.L. and Kisailus, E.C., eds), pp. 263–332. Marcel Dekker, New York.

Spellman, M.W. (1990) Carbohydrate characterization of recombinant glycoproteins of pharmaceutical interest. Anal. Chem. *62*, 1714–1722.

Stanley, P. (1992) Glycosylation engineering. Glycobiology *2*, 99–107.

Takeuchi, M. and Kobata, A. (1991) Structures and functional roles of the sugar chains of human erythropoietins. Glycobiology *1*, 337–346.

Tanigawara, Y. et al. (1990) Pharmacokinetics in chimpanzees of recombinant human tissue-type plasminogen activator produced in mouse C127 and Chinese hamster ovary cells. Chem. Pharm. Bull. *38*, 517–522.

Townsend, R.R. and Hardy, M.R. (1991) Analysis of glycoprotein oligosaccharides using high-pH anion exchange chromatography. Glycobiology *1*, 139–147.

Varki, A. (1993) Biological roles of oligosaccharides; all of the theories are correct. Glycobiology *3*, 97–130.

Verbert, A. (ed.) (1995) Methods on Glycoconjugates. Harwood Academic Publishers, city? in press.

Watson, E., Bhide, A. and van Halbeek, H. (1994) Structure determination of the intact major sialylated oligosaccharid chains of recombinant human erythropoietin expressed in Chinese hamster ovary cells. Glycobiology *4*, 227–237.

3 Cell Cultivation Technology

3.1 Bioreactors Designed for Animal Cells

Michiyuki Tokashiki and Seiichi Yokoyama

3.1.1 Introduction

The progress in gene manipulation and cell fusion technology has made it possible to produce a large amount of physiologically active proteins by cell culture in vitro. In the technology of gene recombination, microorganisms such as Escherichia coli have been chosen first as host cells because of their easy cultivation. The production of the proteins by these microorganisms, however, has come to reveal that some products fail to manifest their biological activities satisfactorily, due to the deficiency of carbohydrates and difference in their tertiary or quaternary structure. Thus, the production of recombinant proteins has been briskly investigated using mammalian cells as a hosts. In the meantime, hybridoma cells have been successfully isolated by cell fusion and energetic studies have been made on the culture of these fused cells. In recent years, on the basis of these technical advances, pharmaceuticals produced by mammalian cells have been developed for therapeutic treatment as well as diagnostic applications, for example, monoclonal antibodies by hybridoma cells (Spier, 1991), interferon-α by Namalwa cells and interferon-β by normal human diploid fibroblasts, and tissue plasminogen activator and erythropoietin by CHO (Chinese hamster ovary) cells which have undergone gene manipulation, thus mammalian cell culture has demonstrated its high usefulness as a production process on an industrial scale (Spier, 1991; Mizrahi et al., 1990; Griffiths, 1990).

Here, large scale mammalian cell culture technology which can be employed in industry as a means for producing useful substances will be illustrated as the topics are focussed on the bioreactors.

The large scale culture processes and the bioreactors are required to have

i) high economical efficiency
ii) excellent operability, simplicity and high reliability
iii) easy production scale expansion and high scale-up potential
iv) reduced number of equipment units and their compactness, and
v) production of the products of more preferable qualities.

The economical efficiency or profitability of cell culture depends on the properties and values of the objective useful product. Further, as the culture process becomes more economical, the application range extends wider and the production scale must be expanded. The extent of scaling up depends on the required amount of the target useful substance and the ability of the cells to produce the target substances. In other words, the process for culturing a large amount of cells should be constructed comprehensively in accordance with the production purposes. The recent situation of the technology, the matters to be considered, how to select and design the culture system, and the future problems will be discussed mainly from the view point of technology and engineering.

3.1.2 Environments Needed for Large-Scale Culture

Mammalian cell culture has been investigated and developed with reference to the precedent microorganism culture. But, mammalian cell culture has the following intrinsic disadvantages compared with that of microorganisms:

- Lower cell proliferation rate.
- Lower productivity of target products.
- More complicated composition of the medium required.
- Lower resistance to toxic metabolites such as ammonia or lactic acid.
- Higher sensibility to outer stimuli and higher susceptibility to shear stress because mammalian cells are enveloped with a thin and soft cell membrane of lipid bilayer.

Thus, mammalian cell culture is more difficult than that of microorganisms.

Therefore, it is very important to give careful consideration to the following matters for mammalian cell culture as a means of producing a useful substance on an industrial scale so that the culture conditions are optimally prepared for their normal growth and completely full expression of the ability to produce the target substance:

i) Supply of nutrients such as saccharides, amino acids, inorganic ions and very small amounts of components such as hormones, growth factors and the like necessary for cell metabolism, namely selection of medium composition.
ii) Supply of dissolved oxygen in the medium and control of its concentration.
iii) Control of accumulated concentrations of toxic metabolites (ammonia, lactic acid or the like).
iv) Control of the shear stress loaded to the cells when the culture medium is mixed, stirred or fluidized, or an oxygen-containing gas (air or pure oxygen) is directly sparged.
v) Control of other factors such as temperature, pH, osmotic pressure or the like.

The large-scale cell culture technology has been developed so that these conditions may be optimized.

Serum-containing media have been used since the beginning, but serum is expensive and not always constant in quality from lot to lot. Further complications in purification can be expected because of the large number of unknown substances included, which is especially disadvantageous in large scale cell culture for commercial applications. Thus, study has been actively and widely done on chemically defined serum-free medium and excellent media have been proposed. Nowadays, these serum-free media are commercially available and mainly employed. But, preferable formulation for the serum-free media depends on the cell lines and varies according to the culture process and conditions, thus case-by-case optimization of formulation including supplements is one of the most important matters.

ii) The oxygen demand of mammalian cells is $(0.04 \text{ to } 0.6) \times 10^{-12}$ mol $O_2 \cdot \text{cell}^{-1} \cdot \text{h}^{-1}$, considerably less than that of microorganisms (Fleischaker and Sinskey, 1981; Glacken et al., 1983; Spier and Griffiths, 1984; de Bruyne, 1988) and there are few problems with small-scale culture. But, it becomes one of the most important factors in large volume culture in industry because the solubility of oxygen is low in the medium, 6 to $7 \text{ mg} \cdot \text{l}^{-1}$ at 37 °C under atmospheric pressure.
Dissolved oxygen is supplied by either of the following methods:

- Direct sparging into the culture medium (de Bruyne, 1988; Katinger and Scheirer, 1985; Griffiths, 1988; Prokop and Rosenberg, 1989; Spier and Whiteside, 1990
- Indirect supply without bubbles through thin-wall tubes or hollow fibers made of a polymer material (hydrophobic porous polytetrafluoroethylene and polypropylene, or oxygen-permeable silicone tubes) (Katinger and Scheirer 1985; Griffiths, 1988; Prokop and Rosenberg, 1989; Lehmann et al., 1988).
- Indirect supply through a liquid carrier which is harmless to the cells, mutually insoluble in the culture medium and high in oxygen solubility (Hamamoto et al., 1987; Tokashiki, 1991).

Aeration from the gas-liquid interface on the surface of the culture medium (overlay aeration) (Katinger and Scheirer, 1985; Griffiths, 1988; Prokop and Rosenberg, 1989; Spier and Whiteside, 1990) can be also cited, but the application of this process is limited in on industrial large scale because the interface area per unit culture volume becomes smaller as the volume increases. If the damage of cells and foaming are permissible, direct sparging is the most preferable process because the reactor can be simplified in its geometry, with low costs and easy scale-up. If not, some contrivance is needed to prevent bubbles from contacting the cells (Katinger and Scheirer, 1985; Spier and Whiteside, 1984; Griffiths et al., 1987; Beyerer et al., 1990) or a method other than direct sparging should be applied (see also Chapter 3.2).

The dissolved oxygen concentration varies its optimal range depending on the cell lines, cell growth phases or the like and the range should be investigated beforehand for the optimal maintenance.

iii) The concentration of toxic metabolite built up is controlled in connection with other conditions. Theoretically, selective separation of noxious metabolites is possible, but the process is not always advantageous since the culture system and operations become intricate. Generally, the perfusion culture to be described later is tried first.

iv) Physical stress, such as agitation or direct sparging, should be reduced as much as possible or avoided completely, and further addition of a polymer such as pluronic polyols, methylcellulose or carboxymethylcellulose is recommended so as to protect the cells from physical stress (Katinger and Scheirer, 1985; Spier and Whiteside, 1990; Cherry and Papoutsakis, 1990; Honda-Corrigan, 1990; Kitano, 1991; Croughan and Wang, 1991) (see Chapter 3.2 in detail).

v) As for other factors, the pH and osmotic pressure are controlled by adding a buffer solution such as HEPES or inorganic solutions to the medium in advance and, when needed, further appending, during the culture, carbon dioxide or alkali such as sodium hydroxide or sodium bicarbonate to control the pH to near 7.0.

3.1.3 Cell Culture Processes

The processes are largely classified into batch and continuous types.

3.1.3.1 Batch Culture

Batch culture is a process where the whole fresh medium is charged into the bioreactor, seed cells are inoculated, but no additive medium is supplied to the system and the culture medium is not harvested during the cultivation (Fig. 3.1.1 A). Dissolved oxygen is adjustable, and the cells proliferate with the passage of cultivation time, but the nutrients are consumed and toxic metabolites accumulate. Thus, the cells are inhibited from growing before long, and dead when a certain cell density is reached. This process is simple with regard to system and operation in comparison with others, but the cell density is low (usually, 1 to 2×10^6 cells $\cdot$ ml^{-1}) and the concentration of the target substance is also low.

Fed-batch culture is a process where the culture medium is charged into the bioreactor, seed cells are inoculated, then ingredients necessary for the culture such as nutrients are continuously or intermittently fed into the system during the cultivation. The nutrients and dissolved oxygen can be suitably adjusted so that they may not inhibit the growth of the cells, but culture medium is not taken out of the system and the accumulation of waste metabolites proceeds gradually. As a result, the cells are inhibited from growing and dead when the cell density reaches a certain level. Compared with the case of batch culture where cell growth is inhibited by exhaustion of the nutrients, this process enables cell growth for a prolonged period with higher cell density, thus the concentration of the target product can be increased more than in batch culture. But, growth inhibition is caused by toxic metabolities soon and large increase in cell density is difficult with limitations to a lower level.

3.1.3.2 Continuous Culture

Continuous culture includes chemostat culture and perfusion culture.

Chemostat culture is a process where fresh medium is continuously supplied to the bioreactor containing the cells, as the same amount of the culture medium containing cells is taken out, so that steady state is always maintained. The feed rate of the medium is determined by the proliferation rate of the cells. In other words, the feed is set so that the amount of the proliferated cells is equal to that leaving the system with the culture medium, thus the culture system is kept in the steady state. In this process, nutrients and dissolved oxygen can be supplied and accumulation of waste metabolites can be controlled without prohibition of cell growth, thus the culture can be maintained and continued for a prolonged period of time. However, the loss of proliferated cells, due to their removal from the system, makes it difficult to attain sufficient increase in cell density and concentration of the desired product. Thus, the advantages of the continuous culture process are reduced.

Perfusion culture is a process where the medium is continuously fed into the bioreactor containing the cells, while the same amount of cell-free culture supernatant

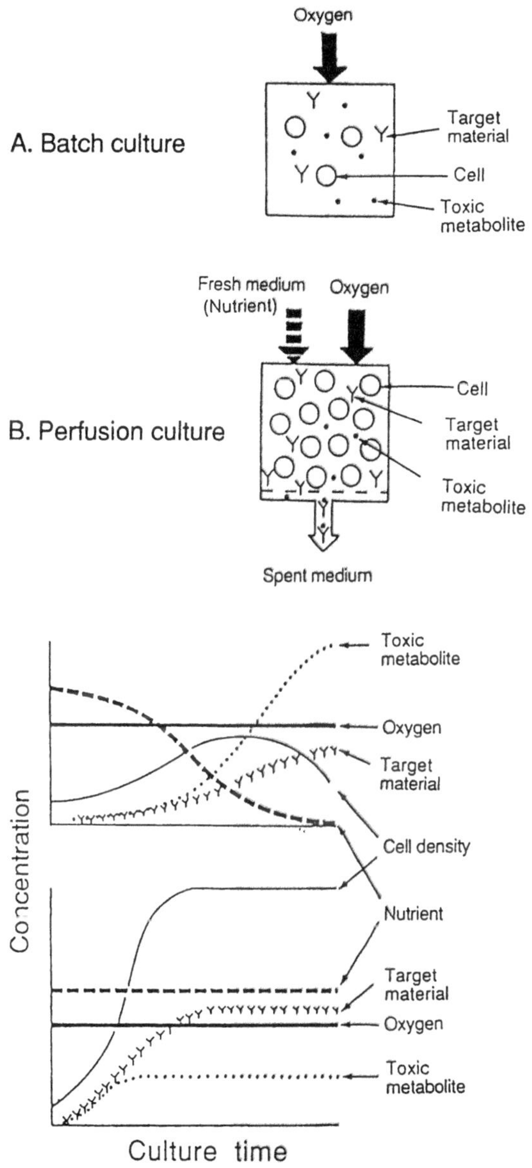

Fig. 3.1.1: Schematic batch culture (A) and perfusion culture (B).

is taken out of the system whereby the steady state is conserved (Fig. 3.1.1 B). The process can provide higher cell density (1×10^7 to 1×10^8 cells $\cdot$ ml^{-1}) and increased productivity of the target product for a long period of time in comparison with other processes if the culture environments can be maintained optimally, since the culture can be continued as the growing cells are allowed to stay in the system. But, the maintenance of high cell density requires optimal control of culture conditions and removal of the culture supernatant free from the cells, since they are kept appropriately viable in the system, and some contrivances are therefore required. Further, this process is more complicated in operation and system than the other processes and the technology has not yet been established. Nevertheless, it is one of the most promising processes for the industrial large-scale culture at present because of its high efficiency and economy.

3.1.4 Bioreactors

Mammalian cells are classified on the basis of their proliferation mode into anchorage-dependent cells and anchorage-independent cells, such as hemocytes or monoclonal antibody-producing hybridomas. The anchorage-dependent cells, to which most of cells belong, attach to solid surfaces and proliferate, while the anchorage-independent cells proliferate in suspension and need no attaching surface. The development of bioreactors has been made mainly on the basis of the techniques utilizing these properties, namely suspended-cell bioreactors where the cells themselves are allowed to suspend, and immobilized-cell bioreactors where the cells are allowed to attach to the solid surface or are entrapped in the gel support.

3.1.4.1 Suspended-Cell Bioreactors

3.1.4.1.1 Types of Reactors

The suspension culture process has been mainly applied to anchorage-independent cells such as monoclonal antibody-producing hybridomas. But, the possibility of suspension culture has been demonstrated even in intrinsically anchorage-dependent cells by sorting and breeding cells as optimal conditions are suitably selected (for example, CHO cells or BHK (Baby hamster kidney) cells, which have been widely used as a host for gene manipulation) and now the application of the process has been extended widely. It can be anticipated that the trials of this kind will increase more and more in future. Since this culture process is in no need of the support for immobilization, the operations and devices can be simplified to facilitate the scale-up with high economy. Additionally, homogenous mixing can also be readily attained so that the nutrients and dissolved oxygen may not localize to cause their partial shortage. There remain, however, the technical problems of avoidance of mechanical cell damage by shear stress and separation of cells from the culture medium (in the case of perfusion culture). The suspended-cell bioreactors are classified on the modes of cell suspension in the culture medium into the following two types:

i) Stirred tank bioreactor (Griffiths, 1988; Prokop and Rosenberg, 1989).
ii) Airlift (or bubble column) bioreactor (Griffiths, 1988; Birch et al., 1985; Lambert et al., 1987; Moo-Young and Chisti, 1988; Rhodes et al., 1991).

In the **stirred tank bioreactor**, the agitation unit or agitation blades have been contrived to improve the efficiency in mixing cells and medium and minimize the mechanical cell damages caused by, for example, the shear stress during agitation.

In the **airlift (or bubble column) bioreactor**, for example, an oxygen-containing gas is introduced through draft tubes inside the bioreactor into the bottom center of the culture vessel and allowed to bubble up, thereby density difference occurs in the culture medium between the outside and the inside of the draft tubes, and resultantly a loop flow is formed. Since the cells and the culture medium are mixed without any agitator, the influence of mechanical shear stress is further reduced. In addition, the system has the advantage of easy scale-up because of its simple configuration, but it is unsuitable in the case where bubbles cause extensive cell damage since bubbling is indispensable for this process.

3.1.4.1.2 Suspended-Cell Batch Culture

Batch culture has been mainly used as a large scale mammalian cell culture process to produce useful substances because the operations and the systems are simple with few technical problems to be solved. It was the earliest process developed and most technically established; stirred tank bioreactors of $8\,m^3$ and $10\,m^3$ volume (Pullen et al., 1985; Phillips et al., 1985; Nelson, 1988) and an airlift bioreactor of $2\,m^3$ (Rhodes and Birch, 1988) have been already operated in practice.
This process, however, has the following disadvantages:

- In mammalian cells, the proliferation rate is slow (doubling time: usually 15 to 30 hours) and it takes many hours to proliferate a large number of cells on an industrial scale, and
- the maximum density which can be attained by the cells is low (usually 1 to 2×10^6 cells $\cdot$ ml^{-1}) and the cell life is relatively short.

Thus, the productivity of the target substance is low with its concentration reduced in the culture medium and the bioreactor requires a large volume per unit production of the target substance. Further, this process also has the following disadvantage:

- The operations are not steady and may sometimes cause adverse influence on the qualitics of the target substance in comparison with the continuous culture, and increased load will be applied to the downstream processes including the purification.

3.1.4.1.3 Suspended-Cell Perfusion Culture

In order to overcome these problems, continuous culture should be performed with the cell density increased as the proliferated cells are kept actively viable as long as possible to minimize the influence of proliferation rate on the productivity. In the present situation, high-density perfusion culture can meet these requirements. However, this process is still under development and the technology has not yet sufficiently established in comparison with the batch culture process. The suspension perfusion culture has given good results in a variety of small scale trials. But the actual performance decreases as the capacity is increased. One of the reasons is the technical difficulty in cell separation from the culture medium. The separation capacity must be increased in proportion to the culture colume. For example, the culture tank of about m^3 order volume generally requires a separation capacity of about $m^3 \cdot d^{-1}$ order. Some conditions can be readily satisfied on a small scale, but difficulty increases as the culture volume expands. Further, the perfusion culture requires stabilized operations for a long period of time, at least two months, preferably several months. In the present situation, it cannot be said that the technology, which satisfies these requirements, has been established, namely the stabilized cultivation of an increased volume for a sufficiently prolonged period of time. Thus, the critical technical question is how to perform separation of the cells from the culture medium, while keeping the cells viable. Three processes have been developed for the separation, namely, filtration type, gravitational settling type and centrifugation type. Perfusion culture with these three types of separation will be described.

1) Filtration type
The development of perfusion culture was initiated with separation of cells from culture medium by filtration, which is still the predominant method chosen.

The filtration separation of cells from the culture medium causes clogging in course of time, until filtration becomes impossible. In order to avoid it, an attempt was made to rotate the filter so that clogging was relieved by centrifugal force. The development of this process was forestalled by Himmelfarb et al. (1969). They prepared a perfusion culture vessel provided with a built-in cylindrical filter, as shown in Figure 3.1.2A, to cultivate mouse leukemia cells and succeeded in high density

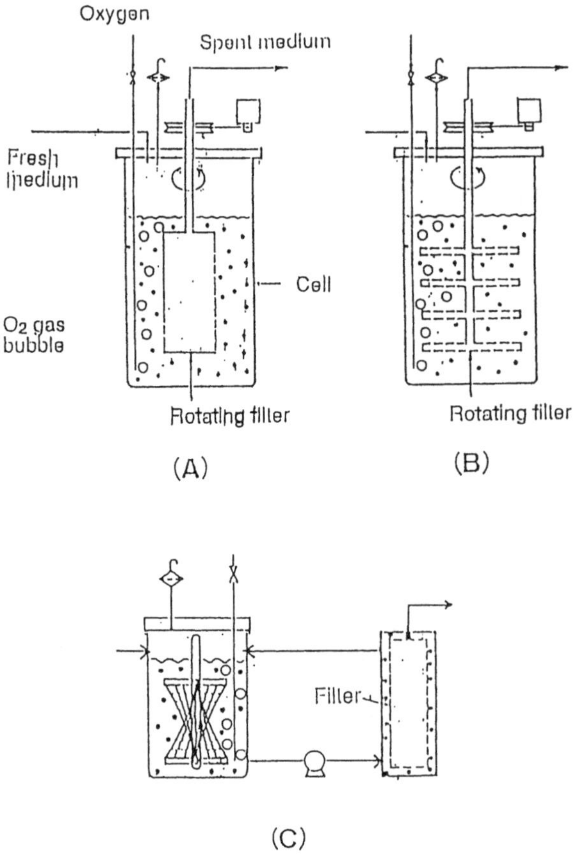

Fig. 3.1.2: Filtration type bioreactors

culture, with a maximum cell density of 10^8 cells · ml^{-1}. Tolbert et al. (1983; 1985) made a similar type of 40 liter culture vessel provided with a porecelain filter and cultured a variety of cells. In the scale-up phase of the process, the filtration area must be expanded in proportion to the culture volume, but the area expansion becomes more difficult as the capacity increases. Hamamoto et al. designed an apparatus in which disk filters are stacked upright and set to a single rotary shaft, as shown in Figure 3.1.2B (Hamamoto et al., 1985). Lehmann et al. (1988) equipped the bioreactor with a bundle of polypropylene porous hollow fibers which had been made hydrophilic as a filtration material and drove it as a stirrer, too. Simultaneously, the one operation in which fresh culture medium is fed from the inside of the hollow fibers into the reactor and the other operation, in which the culture solution is taken out of the bioreactor through fiber membrane in the reserve direction from the inside of the fibers while the filtration rate per unit filtration area is kept below a certain level, are alternately repeated to prevent the fibers from clogging. This process was expanded up to a 150 liter scale. A process in which the filter is separated from the culture tank, as shown in Figure 3.1.2C has also been proposed (Tolbert and Feder, 1983; Tolbert et al., 1985; 1988). Clogging is unavoidable in the filtration process, and the culture operation must be discontinued in the case that the built-in filter is blocked. On the contrary, when the filter is separated from the bioreactor, the filter can be regenerated or exchanged with advantage in commercial production. The process has been scaled up to 100 liters (Tolbert et al., 1988).

Thus, a variety of processes have been proposed for perfusion culture using a filter and their scale-up has reached 150 liters. It can be said that this process would be the main stream of research and development for conventional perfusion culture. The filtration process, however, has a critical problem when used as a culture process for commercial production, namely, filter clogging. Although the clogging situation varies based on the kinds of cells, the types of culture media and the culture conditions, the authors learned from their experiences that the clogging is inevitable as long as a filter is employed. The rotating filter and the Lehmann's method may relieve the clogging, but cannot remove it completely. Accordingly, the filtration type perfusion culture will present difficulty in maintaining stabilized operation for a prolonged period of time.

The advantage of perfusion culture over batch culture is high cell density and prolonged operation time, and the latter is more important in commercial production. In order to clarify the advantage, the stabilized culture should be continued for at least 2 months, preferably longer than 6 months, but it would be extremely difficult for these conditions to be satisfied with high reproducibility as long as a filter is used.

Consequently, the process with a filter cannot be always preferred in commercial production and the development of other, better systems is expected.

2) Gravitational settling type

In recent years, the perfusion culture has been studied where the cells are separated from the culture medium by gravitational settling. In gravitational settling, a part of the insoluble components originating from dead cells are taken out of the culture system together with culture supernatant without any problem of filter clogging,

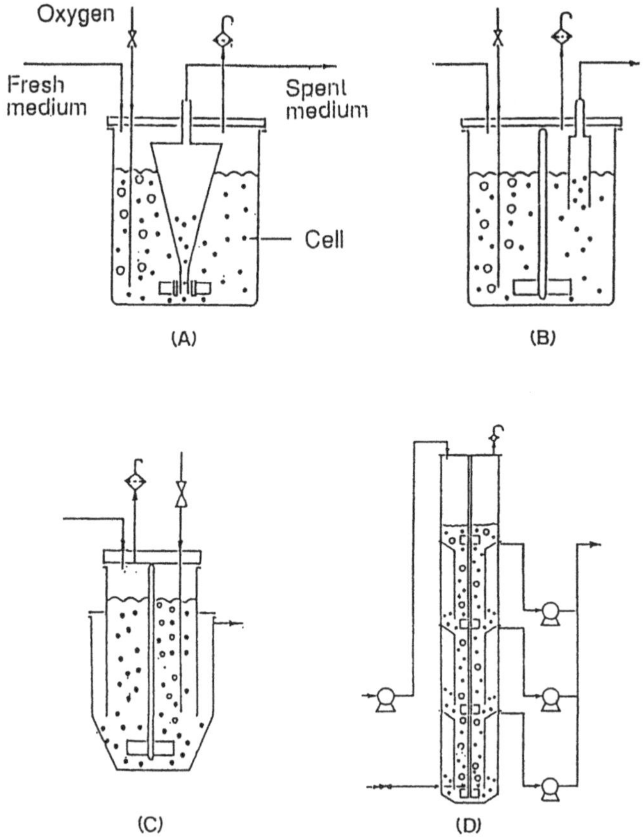

Fig. 3.1.3: Gravitational settling type bioreactors.

thus excessive accumulation of insoluble substances can be avoided and prolonged culture is readily achieved. The research and development situations will be outlined here.

Sato et al. proposed a perfusion culture process where the culture tank is provided with a cell-settling tube of a reversed conical shape inside as shown in Figure 3.1.3 A (Sato et al., 1983; 1985). The bottom end of the settling zone is opened toward the culture zone, while agitation blades are set to the outer periphery of the bottom end of the settling zone. Figure 3.1.3 B shows a bioreactor provided with a cylindrical settling zone inside and the bottom end of the settling zone is opened toward the culture zone. In any culture vessel, the cells are separated from the culture supernatant by gravity in the settling zone, while the supernatant is taken out of the culture system through the top of the settling zone, and fresh culture medium is supplied in the same amount as that taken out, thus the culture operations are continued. Sato et al. (1983; 1985) and Hosoi et al. (1991 a; 1991 b) used the bioreactor shown in Figure 3.1.3A to try perfusion culture of Namalwa cells and hybridoma cells and succeeded in high density culture. Kitano et al. and Shintani et al. used the bioreactor shown in Figure 3.1.3B to effect high-density culture of hybridomas producing a variety of human monoclonal antibodies (Kitano et al., 1986; 1991; Shintani et al., 1991). These studies achieved good results, but their culture scale was small, less than 1 liter. It can be readily understood from the system configuration depicted in the figures that it becomes more difficult to secure the settling area in proportion to the expansion of the culture volume.

Tokashiki et al. designed a perfusion culture system having the settling zone around the culture zone, as shown in Figure 3.1.3 C (Tokashiki, 1991; Tokashiki et

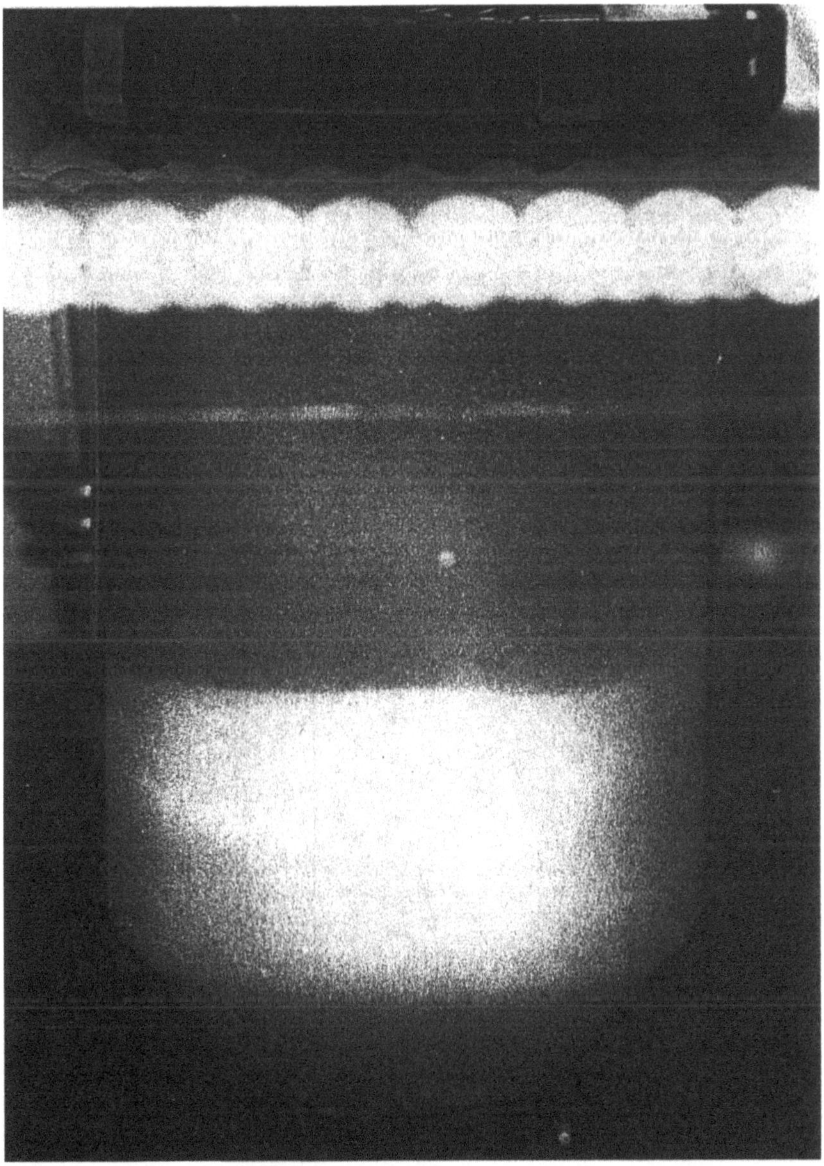

Fig. 3.1.4: Perfusion culture of mouse-human hybridoma cells in a gravitational settling type bioreactor (Net culture volume: 1.2 L)

al., 1988). The bioreactor comprises the inner cylinder and the outer one and the bottom end of the inner cylinder opens at the bottom of the bioreactor. The area surrounded with the inner and outer cyclinders is almost at rest, with largely reduced influence of agitation and the cells are separated from the culture medium by gravitational settling (this area is called the "settling zone"). Usually, the boundary between the zones containing and not containing cells can be grossly distinguished (see Fig. 3.1.4). The cells separated in the settling zone move downward, then are mixed with the culture medium by means of agitation blades at the bottom of the bioreactor, thus, the culture operations are continued. In this bioreactor, enlargement of the settling zone can be readily attained only by widening the diameter of the outer cylinder. Thus, it is easier to expand the settling area than in the conventional processes.

An example is given on the culture of hybridoma cells using the bioreactor. The cells were the mouse-human hybridoma X87 line which was prepared by fusing the mouse myeloma P3/X63-Ag8-U1 line with human B cells. These cells produce human IgG. First, the cells of the X87 line were cultured in a spinner flask batchwise and the result given in Figure 3.1.5 was obtained. The maximum cell density was 2.8×10^5 cell $\cdot$ ml^{-1} and the antibody concentration was 11 μg $\cdot$ ml^{-1}. The same cells were cultured by perfusion culture in a 120 ml volume using the same serum-free culture medium and the results are given in Figure 3.1.6. This culture maintained a viable cell density of 0.6 to 2.0×10^7 cells $\cdot$ ml^{-1} and continued for 230 days, and the cells stably produced the monoclonal antibody during the period, the concentration of

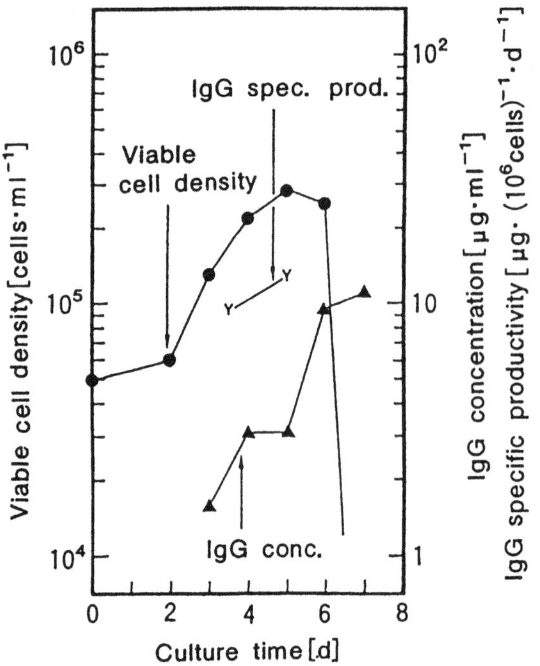

Fig. 3.1.5: Spinner culture of the mouse-human hybridoma X87 line (Medium: ITES + RDF). Reproduced with permission from Hamamoto, K. et al. (1989).

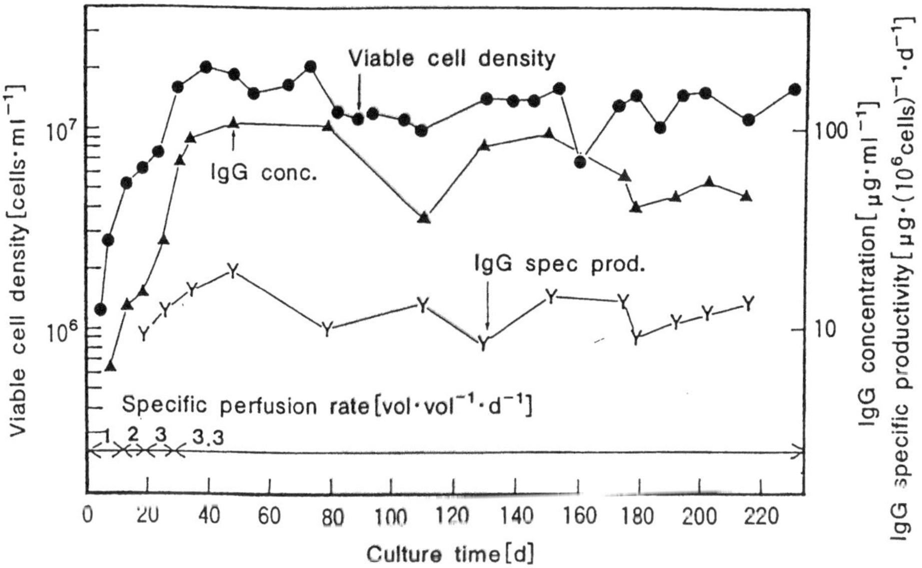

Fig. 3.1.6: Perfusion culture of the mouse-human hybridoma X87 line in a gravitational settling type bioreactor (Net culture volume: 120 ml, Medium: ITES + eRDF). Reproduced with permission from Tokashiki, M. (1991).

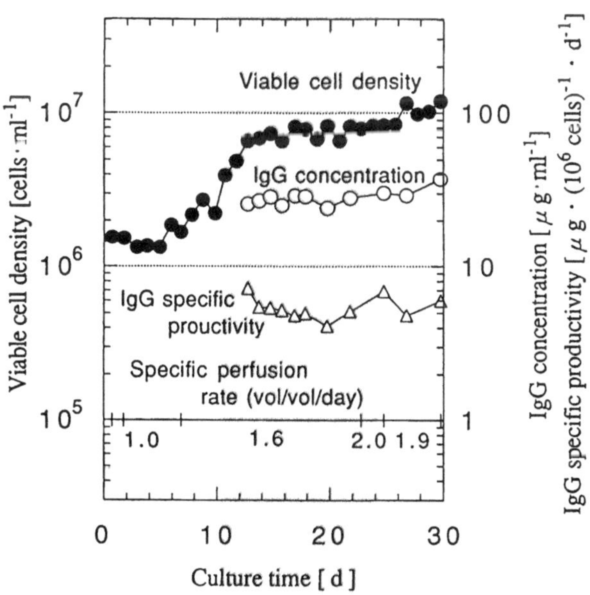

Fig. 3.1.7: Perfusion culture of mouse-human hybridoma X32 line in a gravitational settling type bioreactor (Net culture volume: 50 L, Medium: ITES + eRDF).

antibody in the culture medium was 35 to 100 $\mu g \cdot ml^{-1}$ and the specific antibody productivity was constantly about 10 $\mu g \cdot (10^6 \text{ cells})^{-1} \cdot d^{-1}$. In comparison of these results with the batchwise culture of Figure 3.1.5, the viable cell density increased 20 to 70 times, antibody concentration, 3 to 9 times and specific antibody productivity was almost equal. These data were obtained in small scale culture, but the culture of an expanded scale also was performed, and an example of a 50 liter-perfusion culture with the X32 line is given in Figure 3.1.7. Stabilized high-density culture of 0.8 to 1.2×10^7 cells $\cdot ml^{-1}$ of viable cell density was achieved with success in scale-up, in which the results were almost equivalent to those in small scale culture (published at the 5th Annual Meeting of the Japanese Association for Animal Cell Technology, Dec. 1992).

The perfusion culture combined with cell separation by gravitational settling enables very stable operations when the process can afford the separation capability. But mammalian cells settle at a rate of 2 to 10 $cm \cdot h^{-1}$ in the field of gravity, when they are separated individually, thus cell separability is very poor. When the bioreactor volume is expanded under the conditions that the specific perfusion rate is kept constant, the settling area must be provided in proportion to the culture volume, but it becomes more difficult, as the scale is expanded. Aiming at easier scale expansion, Tokashiki and Arai (1989; 1991) designed a perfusion culture bioreactor provided with a plurality of gravitational settling zones, as shown in Figure 3.1.3D. They cultured hybridoma cells with the reactor and showed that the same culture performance was attained as in the bioreactor provided with a single settling zone. This process would enable the scale expansion to some 200 liters even for cells such as hybridomas with small settling velocity, where every means available is tried, but further expansion seems to be very difficult.

The gravitational settling type perfusion culture has been described in the case of production of a monoclonal antibody by hybridoma cells, and the process can be widely applied to any anchorage-independent cells which can be cultured in a suspension state, and the authors have already cultured a variety of anchorage-independent mammalian cells subjected to gene manipulation (unpublished). Since some cells of this kind form aggregates, individual clumps have higher settling velocity than that of individually separated cells and the separation efficiency becomes higher. Further, this process can be applied to a culture with microcarriers to be described later.

3) Centrifugal settling type

In the filtration type and gravitational settling type perfusion cultures, stabilized culture operation is difficult if the volume is expanded over a certain level. It seems possible, though, if centrifugation can be applied. The perfusion culture where the cells are separated from culture medium by centrifugal settling will be outlined.

In this process, the cells must be separated from the culture medium under aseptic conditions and returned to the bioreactor as they are kept viable. Moreover, the operations should be stably continued for a long period of time. Such a centrifuge which can meet these requirements has not been commercially available and must be developed from the beginning. Hamamoto et al. (1989) have developed a centrifugal settling type perfusion culture process in which the cells and the culture medium are separated from each other intermittently and automatically. The flow

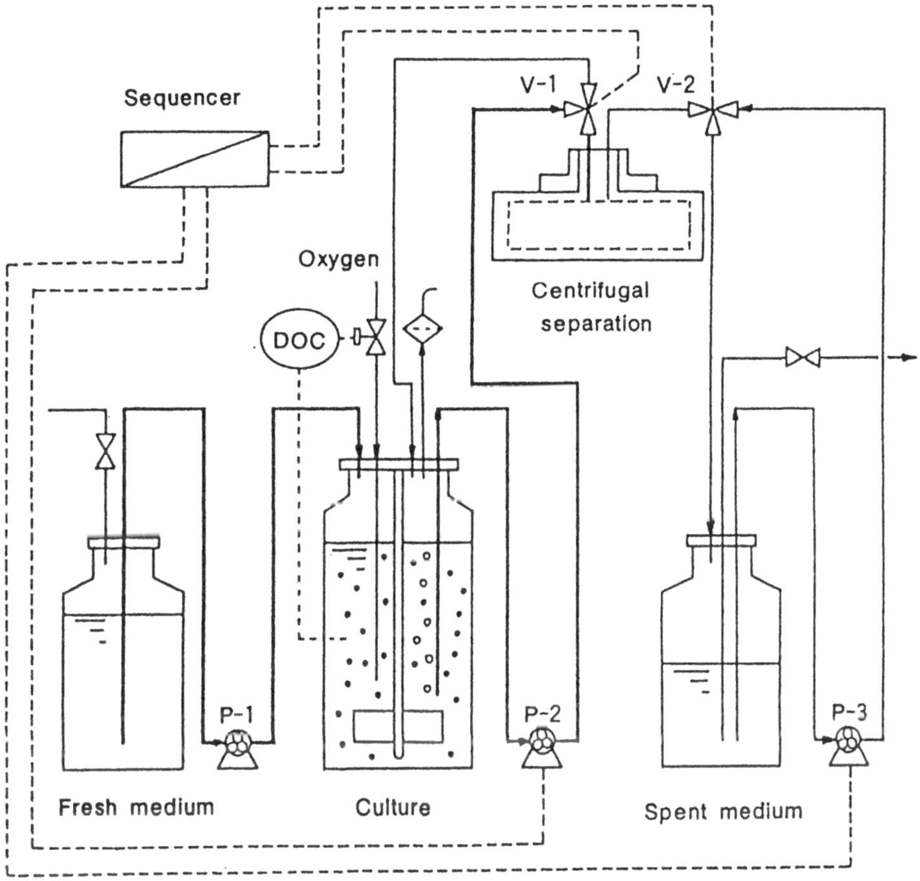

Fig. 3.1.8: Experimental apparatus for perfusion culture, in which cells are separated from medium intermittently by a centrifuge. Reproduced with permission from Hamamoto, K. et al. (1989).

sheet of the process and the operation procedures for cell separation are shown in Figure 3.1.8 and Figure 3.1.9, respectively.

In this process, the cycle including the following operations is repeated to separate cells from culture medium:

i) In Figure 3.1.9, the culture medium from the culture tank is withdrawn through line A into the rotor of the centrifuge. In the rotor, the cells and the culture medium are separated and the latter is withdrawn through line B out of the rotor. During the operation, the cells stay in the rotor.
ii) A part of the culture medium is fed through line B into the rotor to take out the cells remaining in the rotor and send back them to the bioreactor.

After these operations, some down-time is set, as needed, while the specific perfusion rate is set to a prescribed level.

i) feed the culture mixture to the rotor through line A, and discharge spent medium from the rotor through line B (cells are kept in the rotor)

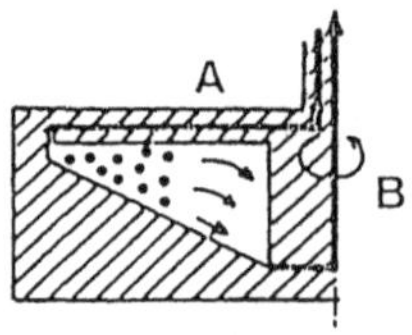

ii) recover cells from the rotor by feeding spent medium through line B

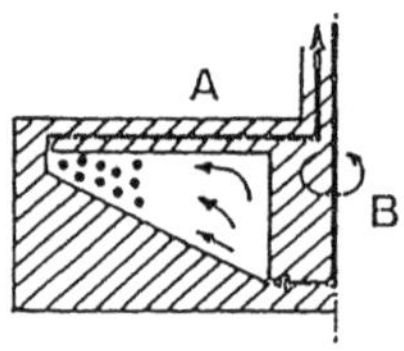

iii) suspension of operation

Fig. 3.1.9: Time cycle of intermittent centrifugal separation.

This culture system was applied to the culture of the mouse-human hybridoma X87 cells in serum-free medium and the results are given in Figure 3.1.10. In other words, under the condition of a specific perfusion rate of 2.0 vol · vol^{-1} · d^{-1}, viable density: 1.0 to 1.6×10^7 cells · ml^{-1}, antibody concentration: 40 to 80 μg · ml^{-1}, and antibody specific productivity: ca. 10 μg · (10^6 cells)$^{-1}$ · d^{-1} were achieved. They were substantially equivalent to the results in the gravitational settling type perfusion culture. The centrifuge was operated with 100 to 200 G in this process and no reduction in specific antibody productivity was noted in the continuous culture for 28 days.

Tokashiki et al. (1990) have developed a completely continuous separation process using a further contrived centrifuge to conduct the continuous culture of the mouse-human hybridoma for 37 days and obtain almost equivalent results to those in the intermittent centrifugation process described. Both of the described above processes have more than 200 l · d^{-1} separation capacity and further scale-up will be possible, if further improvement is applied.

Alfa-Laval Centritech AB (Sweden) has developed a centrifuge for cell separation in which a flexible plastic bag as shown in Figure 3.1.11 A is set to the rotor as shown in figure 3.1.11 B. The plastic bag is commercially available as a sterilized. The 50 liter-volume suspended perfusion culture of CHO cells achieved a high viable cell density of about 1.4×10^7 cells · ml^{-1} (Apelman, 1992).

Westfalia Separator AG (Germany) has recently developed a continuous flow centrifuge and applied it to the suspended perfusion culture of 6 liter-volume with

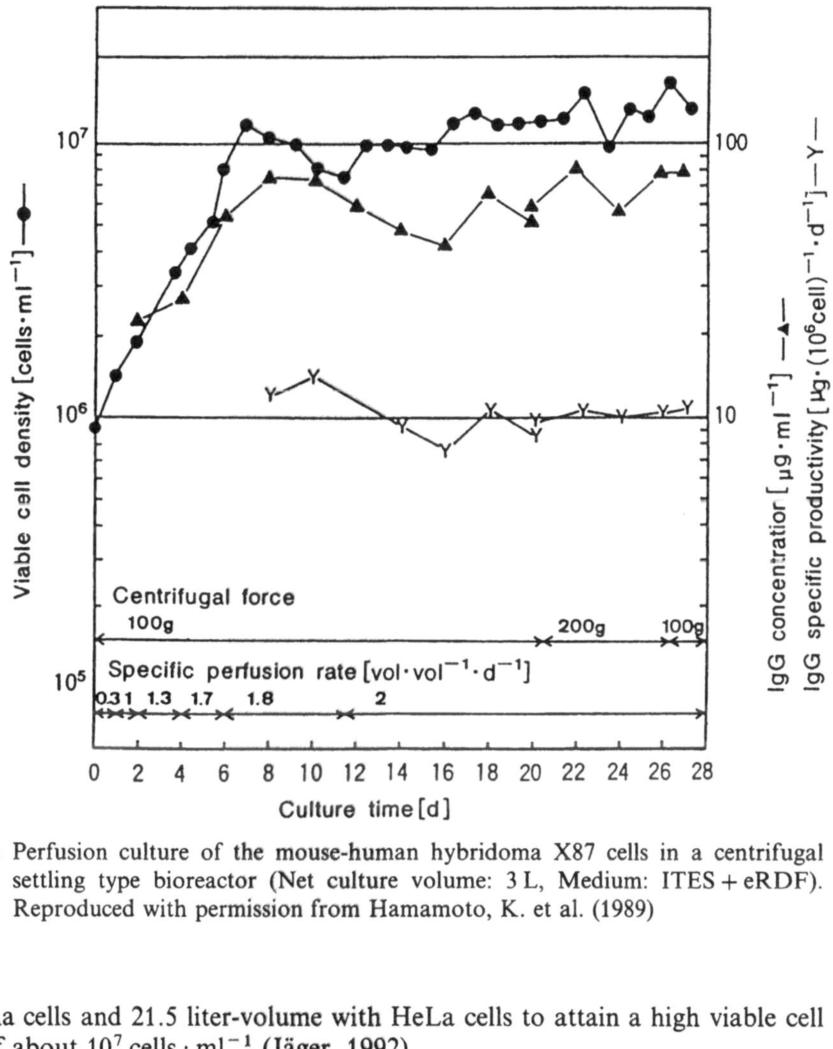

Fig. 3.1.10: Perfusion culture of the mouse-human hybridoma X87 cells in a centrifugal settling type bioreactor (Net culture volume: 3 L, Medium: ITES + eRDF). Reproduced with permission from Hamamoto, K. et al. (1989)

hybridoma cells and 21.5 liter-volume with HeLa cells to attain a high viable cell density of about 10^7 cells · ml^{-1} (Jäger, 1992).

It has not been long since perfusion culture using a centrifugal separator started and few reports have been published. But, this process has essentially high separation efficiency and operational troubles, such as filter clogging in the filtration type can be avoided if the geometry of the separator is suitably chosen, thus, it is one of the most promising processes for cell separation in future. One of the most important technical points in this process is how to take the cells out of the separator, and the authors have already solved this problem. But, when it is practically applied, the following matters should be considered. Namely, as long as the separation of the cells from the culture medium is effected by centrifugation, the exposure of cells to physical stress in the centrifugal field is unavoidable, and we must know how such stress influences cell proliferation and the production of the target substances. The culture of the mouse-human hybridoma X87 line is shown in Figure 3.1.10 as an example of the centrifugal settling type perfusion culture, and a preliminary

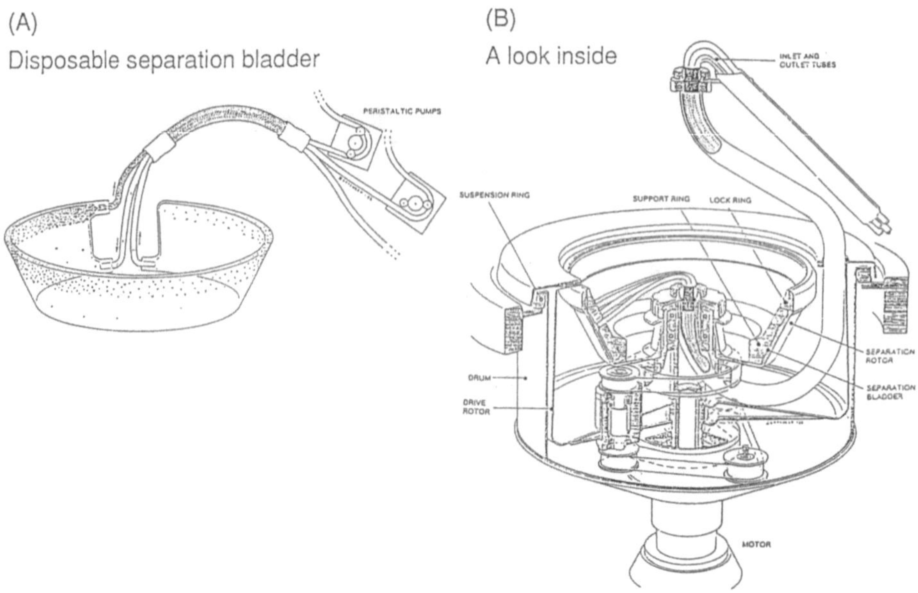

Fig. 3.1.11: ALFA-LAVAL CENTRITECH™ CELL separation system.

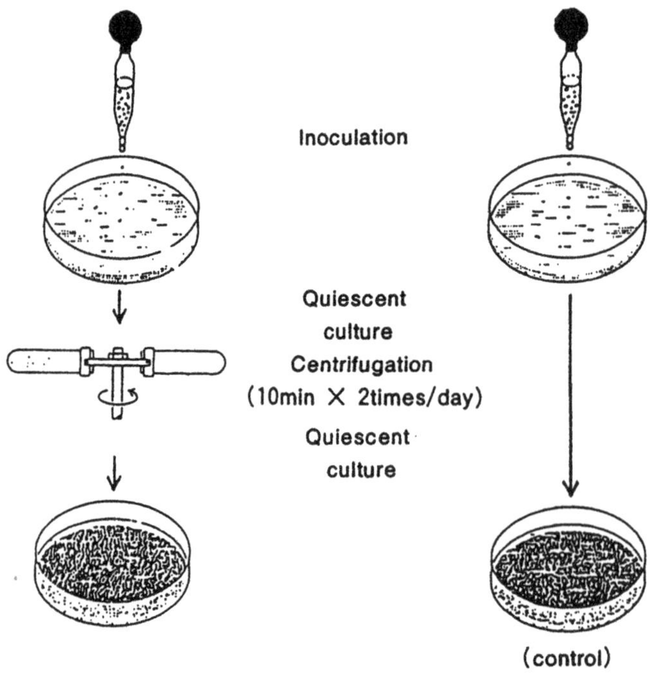

Fig. 3.1.12: Experimental method to study the effects of centrifugal force on cell culture.

experiment was performed in order to examine the effect of centrifugal force on the cell culture. The result are described below.

The experimental flow is given in Figure 3.1.12.

A dish was filled with the culture medium, the cells were inoculated and the static culture was started at 37 °C in a 5% CO_2 atmosphere in a CO_2 incubator. The dish was taken out of the incubator twice a day at certain times, placed in the rotor of centrifuge, and exposed to the centrifugal field for 10 minutes by driving the machine. After exposure, the dish was returned to the incubator to continue the cultivation. These conditions were decided as the practical operation of the perfusion culture was taken into consideration. In parallel, another experimental culture was performed without centrifugation as a control. The result of the culture of the mouse-human hybridoma X87 line is shown in Figure 3.1.13. Figure 3.1.13 A gives the cell proliferation when the centrifugal force was varied. Figure 3.1.13 B discloses the concentration of antibody in the culture medium and no effect of centrifugal force was noted up to 500 G.

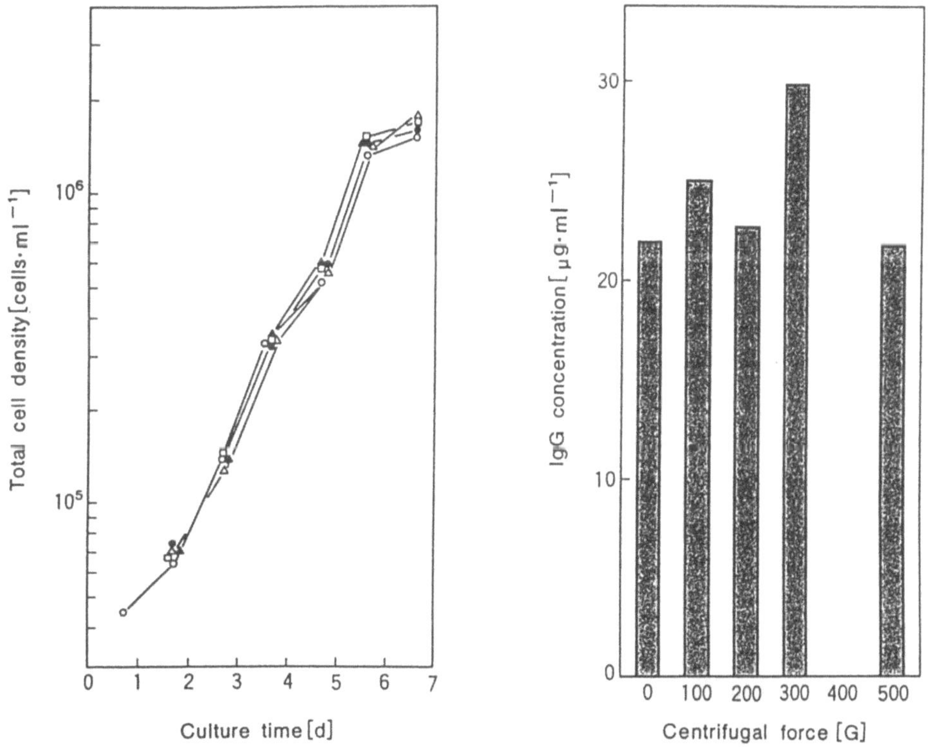

(A) Effect on the cell grouth
Symbols : ○ control, ● 100G
△ 200G, ▲ 300G
□ 500G

(B) Effect on IgG production
IgG was assayed at the 7th day after inoculation

Fig. 3.1.13: Effects of centrifugal force on the growth and IgG production of mouse-human hybridoma X87 cells (Quiescent culture exposed to centrifugal force 10 min × 2 times/d). Reproduced with permission from Tokashiki, M. et al. (1990).

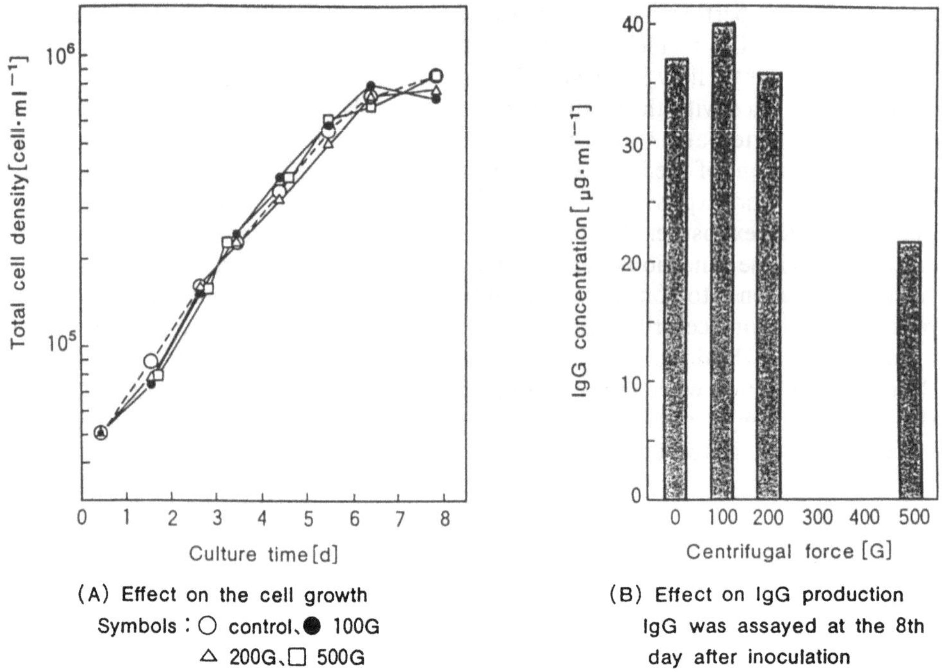

Fig. 3.1.14: Effects of centrifugal force on the growth and IgG production of mouse-human hybridoma P-20-3 cells (Quiescent culture exposed to centrifugal force 10 min × 2 times/d).

In another similar experiment with the human-hybridoma P-20-3 line, however, as shown in Figure 3.1.14, no effect was noted in the cell proliferation, but 50 % decrease was observed in the antibody production at 500 G, although no effect was observed at 200 G.

As these experiments show, the influence of centrifugal force on cell proliferation and substance production depends upon the kind of cells. Attention should be paid sufficiently to this point, but these adverse effects will be lowered if the configuration, the operation procedures and conditions of centrifugation are carefully selected.

The suspension perfusion culture has been described from the standpoint of the methods for separation of the cells from the culture medium, and the separation methods should satisfy the following parameters:

i) Easy scale expansion.
ii) Excellent operational stability.
iii) Reduced cell damage (lowered physical stress).

At present, the separation processes which have been applied to the perfusion culture are filtration, gravitational settling and centrifugal settling. When they are evaluated on the basis of these criteria i), ii) and iii), the conclusion is as follows (see Tab. 3.1.1).

The filtration type can readily expand its scale, but filter clogging is unavoidable and threatens the operational stability. Further, in order to lessen the clogging,

Table 3.1.1. Comparison of various perfusion culture systems

Cell separation	Filtration	Gravitational settling	Centrifugation
Process	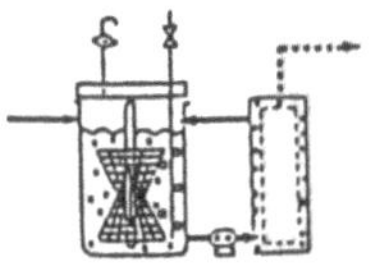	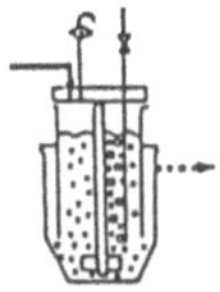	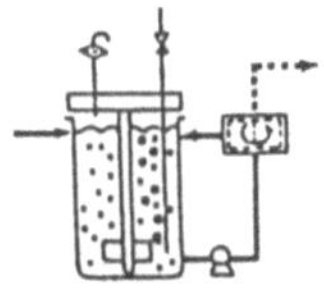
Advantage	Easy to scale up	· No clogging · Small shear stress to cells	Easy to scale up
Disadvantage	Clogging	Difficult to scale up	Difficult to develop

powerful agitation near the filtration surface or rotation of the filtration surface itself are needed, thus the cells cannot be spared from physical stress to same extent. One of the most serious problems is to ensure the stabilized operation, and gravitational settling is very favorable for operation stability with reduced physical stress. This process has, however, very small cell settling velocity in the gravitational field, thus resulting in poor separation efficiency, and limited scale expansion. The evaluation of this process will be settled by asserting the limit of scale-up potential.

The centrifugal settling is intrinsically high both in separation efficiency and in scale-up potential. The authors have already realized more than $200\ l \cdot d^{-1}$ and see the possibility of the expansion to more than $1\ m^3 \cdot d^{-1}$ and, moreover, a long-term culture of more than 2 months has been performed (unpublished). Appropriate type, selection and design of the centrifuge will also ensure stabilized operation. The problem of this process is that the cells must be exposed to the centrifugal force and static pressure in the centrifugal field, thus they cannot be spared a certain amount of physical stress.

As stated above, all of these methods, filtration, gravitational settling and centrifugal settling, have advantages and disadvantages, but on the basis of the reasons outlined above, it is preferred that the gravitational settling or the centrifugal settling is properly applied according to the purposes.

3.1.4.2 Immobilized-Cell Bioreactors

The development of cell-immobilizing carriers and devices used in the immobilized-cell bioreactor was mainly made using anchorage-dependent cells at the beginning, but later the application for anchorage-independent cells also has been investigated in order that the technical problems encountered in suspended-cell culture (for example, prevention of mechanical cell damage and separation of the cells from the culture medium in perfusion culture) may be avoided, resultantly the process will progress with advantage. The immobilized-cell culture can widely vary with the kinds of the carriers and the bioreactor types. Their characteristics and features will be described.

3.1.4.2.1 Cell-Immobilizing Carriers

1) Materials and Configurations

Cell-immobilizing carriers have been studied over a wide range, from the view points of the material and configuration that are suitable for cell attachment and entrapment, to proliferation.

As carriers, a variety of materials have been proposed and investigated, organic materials such as proteins, for example, collagen or gelatin, polysaccharides such as cellulose, cross-linked dextran, agarose, and synthetic polymers such as polyacrylamide, polystyrene, polyurethane or the like, further inorganic materials, for example, glass, ceramics and stainless steel and many other kinds of materials whose surfaces have been treated for improvement of the cell adhesion to the support.

Many configurations have been proposed with these materials and immobilization matrices of the following configurations have been investigated:
small-size carrier particles of 100 to several hundred micrometers suspendible in the culture medium such as microcarriers, cell-entrapping gel beads and microcapsules which are obtained by forming a polycationic semi-permeable membrane on the gel bead surface, glass beads of 1 to several millimeters to be used as a packed bed, multi-stage trays and roller bottles made of glass or plastics, plate or spiral modules with plastic films or flat sheet membranes, hollow fibers, polyurethane foam carriers of 1 to 10 mm sizes, cylindrical ceramic cartridge of honeycombed structure (Opticore), small ceramic pieces of 5 to 10 mm sizes, non-woven fabrics made of fine fibers of several tens of micrometers diameter, glass tubes, springs of stainless steel and so on. Moreover, in microcarriers and glass beads, porous matrices with improved volumetric efficiency have been developed by forming fine voids inside to enlarge cell attachment surface area per carrier volume.

2) Requirements for Cell-Immobilizing Carriers

What are the properties and conditions required for a carrier practically applicable to industrial uses? Here, the main requirements are cited:

- It is composed of such a material and configuration that the cells can easily attach to proliferate on the surface.
- It is innoxious to the cells.
- It has a density and size such that adequate fluidization can be readily attained only by mild mixing in order to avoid mechanical cell damage, for example, collision between the support pieces and excessive shear stress*).
- It has a large surface area per unit volume.
- It has a certain range of configurations so that the flow channels of the culture medium are kept uniform without distribution**).
- It can stably maintain its physicochemical properties for a long period of time.
- It has excellent mechanical strength and durability***).
- In a porous support, it has a geometry and a dimension which can provide an optimal microenvironment for cell growth since the pore size and pore distance

are adjusted so that the cells are readily entrapped and the nutrients, particularly dissolved oxygen, can diffuse without local shortage.

- It can attain high cell proliferation from the inoculation of cells to their growth saturation.
- In the steps from the seed cell inoculation to the final cultivation stage, the proliferated cells are readily releasable from the supports as a seed for the next stage, as they are kept viable and aseptic, ideally they can be used as they are, attached to the supports, without release from the supports.
- It is adaptable for the culture system which is readily expandable in its size with high scale-up potential.
- It can be autoclaved.
- The qualities are in a certain range without fluctuation between lots.
- It is commercially available with ease and is reasonable in price.

Further, it is more preferable as a cell-immobilizing carrier if it is reusable.

*) in the case of an immobilized-cell suspension bioreactor.
**) in the case of a fixed bed bioreactor.
***) particularly in the case of a fixed bed bioreactor.

3.1.4.2.2 Immobilized-Cell Culture

1) Main Cell-Immobilizing Carriers and Their Cultivation Characteristics

The scale expansion of this culture process means the surface area expansion of the cell-immobilizing carrier for attachment in anchorage-dependent cells. As an extension of dishes and flasks which are used in laboratories for stationary culture, a glass-made or plastic-made multistage tray or a roller bottle has been contrived, but both of them are poor in competitiveness because of their low volumetric efficiency and poor scale-up potential. As a result of various studies, the cell-immobolizing carriers which have the basically required properties to open the way of practical use in a large-scale culture are the microcarriers for suspended-carrier bioreactors which were first developed by van Wezel (1967), then proposed and investigated by many other scientists (Griffiths et al., 1987; Sinskey, et al., 1981; van Wazel, 1985; Nilsson and Mosbach, 1987; Butler, 1988; Jacobs et al., 1991; Junker et al., 1992) and the glass beads used in fixed bed bioreactors (Griffiths et al., 1987; Spier and Whiteside, 1976; Whiteside and Spier 1981; Brown et al., 1985; Looby and Griffiths, 1987; Brown et al., 1988; Griffiths and Looby, 1991), and the surface area per unit volume of the carrier has been expanded to the largest ever. Particularly, microcarriers have been widely accepted, and large volume cultures of 1 m^3 and 5 m^3 are described (Nelson, 1988; Montagnon et al., 1984). The cell density is, however, still low, usually some 2×10^6 cells $\cdot$ ml^{-1}, which is comparable with that in the cell-suspended batch culture. A report described that a perfusion culture with microcarriers attained a high cell density of 1×10^7 cells $\cdot$ ml^{-1} order (Tolbert et al., 1987; Butler, 1988), but it is not always general.

On the contrary, in anchorage-independent cells, there are immurement cultures in which the cells are immured in a certain area of the support such as hollow fibers, which were first developed by Knazek et al. (1972) and flat membranes (Klement et al., 1987), gel beads of agarose or alginate for inclusive immobilization of the cells (Griffiths, 1988; Nilsson and Mosbach, 1987; Nilsson et al., 1983;

Griffiths, 1985; Familletti and Fredericks, 1988), microcapsules having a semi-permeable membrane formed on the surface by chemical treatment of gel beads (Griffiths, 1988; Lim and Sun, 1980; Rupp, 1985; Duff, 1985; Posillico, 1986), and another one in which the cells are entrapped in the base material such as a ceramic matrix of a honeycomb shape (Opticore) (Lydersen et al., 1985a; 1985b; Pugh et al., 1987; Berg and Bödeker, 1988) or sponge-like polyurethane foam (Lazar et al., 1987; 1988; Matsushita et al., 1990).

The membrane-type immobilization carriers such as hollow fibers or flat sheet membranes mainly made of ultrafiltration membrane have a structure allowing the cells to stay on one side of the membrane while the culture medium is allowed to pass through the membrane. The molecular weight cut-off size of the membrane is chosen so that individual components such as nutrients or dissolved oxygen in the culture medium may move from the medium side to the cell side, and the waste metabolites, in reverse, from the cell side to the medium side, as the target product of high molecular weight secreted from the cells is allowed to remain in the cell side. For example, in a bioreactor using hollow fibers, a high cell density of more than 10^8 cells $\cdot$ ml^{-1} has been attained (Hopkinson, 1985; Tyo et al., 1988). Further, this process can protect the cells from mechanical damage, accumulate the target product from the cells in a high concentration, resulting in easy downstream processing, and dispense separation of the cells from the medium by allowing the cells to grow in the area divided with the membrane. Several reports describe perfusion culture with hollow fiber bioreactors lasting for 2 to 5 months (Tyo et al., 1988; Schönherr and van Gelder, 1988). At the beginning, this process was mainly applied to anchorage-independent cells, but has also become applicable to anchorage-dependent cells, by improving the qualities of the base materials (Tyo et al., 1988; Ku et al., 1981; Tharakan and Chau, 1986a). But, this bioreactor has a configuration liable to cause a pressure or concentration gradient, leading to local shortage of nutrients and dissolved oxygen, particularly lack of dissolved oxygen, and these matters are added to the complication of the structure to restrict the scale expansion severely. This problem has been improved by investigating the direction of the medium flow or switching the flow direction of the liquid moving through the membrane by raising the pressures on both sides of the membrane alternately (Tyo et al., 1988; Tharakan and Chau, 1986b), but the configuration intrinsically inhibits the scale expansion of the bioreactor, remaining on a scale of the liter order.

The cultivation with gel beads has attained a relatively high cell density of 10^7 cells $\cdot$ ml^{-1} of beads, but the density corresponds to 10^6 cells $\cdot$ ml^{-1} order, when calculated in terms of culture volume, thus cannot be concluded to be efficient for the laborious operation of inclusive immobilization. Thus, the process is not competitive in the present situation as a practical culture of a large volume.

The microcapsules which are covered with a semi-permeable membrane and immure the cells inside the membrane are used in the immobilized-cell suspension culture and the low-molecular-weight substances, for example, nutrients and dissolved oxygen, pass through the membrane into the microcapsules to grow the cells. In this process the cells are protected from mechanical damage by the membrane and the high-molecular-weight target product secreted from the cells is accumulated in the capsules, thus, the downstream processing becomes advantageously simple. Damon Biotech., Inc. (U.S.A.) reported that hybridoma cells were cultured in a

40 liter bioreactor to attain a high cell density of more than 10^8 cells · ml^{-1} (it decreases by half or less, when calculated in terms of culture volume) (Duff, 1985; Posillico, 1986). This process has difficulty, however, in continuous culture for a prolonged period of time, basically only within the range of the batch culture.

Honeycombed ceramic matrix (Opticore) is a cylindrical cartridge in which a number of fine channels of 1 to 2 mm square are arranged in close proximity and run in parallel in the same direction and two types are prepared for both anchorage-dependent and -independent cells. These cartridges have sizes in diameter (mm) × length (mm) ranging from 41 ϕ × 134 for research to 286 ϕ × 610 for industrial use and fixed bed bioreactors including these cartridges (Opticell) have been developed for the perfusion culture of high cell density. Until now, the system of up to a 240 liter culture volume has been described (Nicholson et al., 1991).

Polyurethane foam has a sponge-like porous structure with a large surface area per unit volume and its use has been examined as an inexpensive material for immobilization. This material can be applied not only to anchorage-independent cells (Lazar et al., 1987; 1988) but also to anchorage-dependent cells (Lazar et al., 1987; Matsushita et al., 1990). But, this support has been applied only to the fixed-bed type perfusion culture of a small volume of several hundred milliliters and future studies are required to determine whether it is a competitive material which can be used in the practical operation. Polyester foam (Murdin et al., 1987) and polyvinyl formal resin (Yamaji and Fukuda, 1991) have also been examined, and regarded to be in the same state of testing as the polyurethane.

As stated above, no cell-immobilizing carrier has been found which can realize easy unit volumetric scale-up, increased volumetric efficiency (the cell density per unit volume of the bioreactor) and prolonged-term cultivation (several months), simultaneously. Porous microcarriers and glass beads have been developed and are being studied with the expectation to meet these requirements (Griffiths, 1990). Both conventional microcarriers and glass beads are solid particles having a dense core and only the surface is utilized for cell attachment and cannot provide a large surface area. In other words, microcarriers are used in the immobilized-cell suspension culture, but the concentration of microcarriers cannot be set high in order to avoid or reduce mechanical damage of the cells by the mutual collision and shear stress during the mixing (usually several g · l^{-1} and 15 g · l^{-1} at most) (Karkare et al., 1985; Butler, 1988), resultantly the surface area of the microcarriers per the culture volume cannot be made large, thus it becomes difficult to realize high cell density and maintain it for a long period of time. On the contrary, glass beads have been used in fixed bed bioreactors and, if the diameter of the beads are selected so that suitable flow channels are formed in order to allow the cells to attach properly and proliferate and avoid exfoliation and damage of the cells by the flow of culture medium or avoid clogging of the flow channels by the cells, the diameter should be set larger than that of the microcarriers, on the order of mm, thus the surface area of the support per unit culture volume would be smaller than that of the microcarriers. In order to solve these problems, porous microcarriers and glass beads have been developed to further enlarge the cell attachment surface area per carrier volume. These carriers also have the effect that the cells proliferating in the pore cavities are protected from mechanical damage by shear stress or the like occurring in the outside flow of culture medium. Moreover, these can be applied not only to the anchorage-dependent cells but also to the anchorage-independent cells by entrapping them in the pores, thus receiving attention as a promising cell-immobilizing carrier for enhancing the competitiveness with application fields expanded. Porous microcarriers are typically cited, Microsphere (Verax Corp.: U.S.A.) (Karkare et al., 1985; Dean et al., 1987; Hayman et al., 1987; Runstadler and Cernek, 1988; Vournakis and Runstadler, 1991), CultiSpher (Percell Biolytica AB: Sweden → HyClone Labs Inc.: U.S.A.) (Nilsson et al., 1986; 1990; Amgren et al., 1991), and porous glass beads, for example, Siran (Schott Glasswerke; Germany) (Griffiths and Looby, 1991; Looby and Griffiths, 1988; 1989; Racher et al., 1990; Looby et al., 1990).

Microsphere (Verax Corp.) is used in a fluidized bed bioreactor. This matrix is made of collagen and has 500 μ m average diameter, 20 to 40 μ m pore diameter and 85 % porosity. Small metal particles (for example, stainless steel) are added to the basic material and the specific gravity is adjusted to about 1.7 so that the fluidized bed is formed at the upward superficial liquid velocity of 70 to 75 $cm \cdot min^{-1}$ (Runstadler and Cernek, 1988). Microsphere volume corresponds to about 25 % of the fluidized bed volume. In this case, the outer surface area is $3\ m^2 \cdot l^{-1}$ of fluidized bed volume and the estimated internal surface area is $30\ m^2 \cdot l^{-1}$ of fluidized bed volume, thus, the surface inside the pores predominates (Runstadler and Cernek, 1988). In the perfusion culture with the Microspheres, more than 10^8 $cells \cdot ml^{-1}$ of Microspheres (about 1/4 per fluidized bed volume) has been attained (Runstadler and Cernek, 1988; Ray et al., 1991; Runstadler et al., 1992), and the result of a long-term cultivation of more than 2 months is also described (Vournakis and Runstadler, 1991; Ray et al., 1991; Runstadler et al., 1992). The reactor unit has been scaled up to 24 liters (Verax System 2,000) at present.

CultiSpher is made of gelatin and applied to anchorage-dependent cells in a stirred tank bioreactor. A standard type CultiSpher-G is announced to have 1.04 specific gravity, 220 μ m average particle size (170–270 μ m), 10–20 μ m pore size and about $4\ m^2 \cdot g^{-1}$ surface area (Almgren et al., 1991; Shiragami et al., 1991). In addition, there are CultiSpher-S, and -GL, having the same parameters except 50 to 70 μ m porous size, larger than that of Cultispher-G.

Siran, consisting of porous sintered glass beads, have a variety of specifications, widely ranging in particle size from 0.4 to 5 mm. Up to now 2 kinds of Siran porous beads, of 5 mm and 0.6 mm average particle size, have been mainly investigated. They are applicable for both anchorage-dependent and anchorage-independent cells and the studies have been made on both packed bed bioreactors and fluidized bed bioreactors. In the packed bed bioreactors, particles of 5 mm (Griffiths and Looby, 1991; Looby and Griffiths, 1988; 1989; Racher et al., 1990; Looby et al., 1990) and 0.6 mm (Yoshida et al., 1991) average size have been investigated, and of 0.6 mm in the fluidized bed bioreactors (Looby and Griffiths, 1988; Looby et al., 1990). These Siran porous beads of 5 mm and 0.6 mm average particle size have pore sizes of 60–300 μ m and < 120 μ m, surface area of $74\ m^2 \cdot l^{-1}$ and $90\ m^2 \cdot l^{-1}$ and porosity of about 60 % (Looby and Griffiths, 1988; Looby et al., 1990), respectively. These results showed a little fluctuation according to the cells and the culture processes, but high cell density over 10^7 $cells \cdot ml^{-1}$ of the reactor volume was attained in most of them except in the combination of the anchorage-independent cells and the fluidized bed bioreactor (Griffiths and Looby, 1991; Looby and Griffiths, 1989;

Looby et al., 1990; Yoshida et al., 1991). In general, the packed bed type has given better results than the fluidized bed type and there seems to be room for improvement toward the optimal structure such as in pore diameter for application to the fluidized bed bioreactor. These results tell us that future progress can be expected, although they have already been examined on a laboratory scale.

Additionally, Informatrix (Biomat Corp.: U.S.A.) made of collagen-glycosaminoglycan copolymer (Griffiths, 1990), Asahi microcarrier (Asahi Chemical Industry Co., Ltd.: Japan) (Shiragami et al., 1991) and Cellsnow (Kirin Brewery Co. Ltd.: Japan), (Ogata et al., 1992), which are made of cellulose, have been developed as a novel type of porous microcarrier.

Typical cell-immobilizing carriers are summarized in Table 3.1.2.

Table 3.1.2. Typical cell-immobilizing carriers

Cell-immobilizing carriers	Approximate size (μm)	**Surface area per unit volume ($m^2 \cdot l^{-1}$)** Based on the net carrier volume	Based on the net reactor volume
	Diameter		
· Microcarriers	100–300	20–60	3–9
· Glass beads	2,000–8,000	0.75–3.0	0.45–1.8
· Porous microcarriers	170–600	120–150	4–40
· Porous glass beads	600–5,000	120–150	74–90
· Hollow fibers	200–300*	13–20	3–10
	Channel size		
· Honeycombed ceramic cartridge (Opticore)	1,000–2,200		0.8–2.8**

* Outer diameter of fiber
** Geometric surface area

2) Types of the reactors

As partially stated in 3.1.4.2.1, the immobilized-cell culture systems are roughly classified into the suspension type and the fixed bed type. The suspension type is classified into the following three reactors according to the modes of immobilized-cell suspension:

i) Stirred tank bioreactor.
ii) Fluidized bed bioreactor.
iii) Airlift (or bubble column) bioreactor.

The stirred tank bioreactor and the airlift (or the bubble column) bioreactor are basically the same as already stated. The carriers on which the cells are immobilized are mixed and fluidized in the culture medium instead of the cells themselves. The stirred tank bioreactor is used in the culture with microcarriers, porous microcarriers (for example, Cultispher), gel beads and microcapsules as a cell-immobilizing carrier and frequently used in general. In comparison with the separation of the cells from the culture medium in the system where the cells themselves are suspended, the separation from the cell-immobilized support is easier. For example, the settling velocity of Cultispher is about 200-fold higher than that of hybridoma cells.

The airlift (or bubble column) bioreactor has application examples of culture on a laboratory scale using gel beads, microcapsules (Bugarski et al., 1989) and porous polyvinyl formal resin (Yamaji and Fukuda, 1991) as a cell-immobilizing carrier, but it is not so frequently used as the cell-suspended batch culture described in 3.1.4.1.

The fluidized bed bioreactor is a system in which the culture medium is externally pumped to fluidize the carriers with its upward flow, and as stated above, the porous microcarriers (for example, Verax Microsphere) or porous glass beads (for example, Siran) are used as cell-immobilizing carriers.

In this bioreactor, a sufficient circulation flow must be maintained so that a proper amount of nutrients (particularly dissolved oxygen) are supplied to the cells, which have proliferated at a high density in the pores of the carrier, to arrange a suitable microenvironment surrounding the cells, whereby the porous carriers are fluidized. For example, let me try calculating the circulation rate on the basis of the uptake of dissolved oxygen. When cell density, oxygen uptake rate of the cells, and the concentration difference in dissolved oxygen between the inlet and outlet are assumed to be 6×10^7 cells $\cdot$ ml^{-1} of reactor volume, 1.5 to 15 pg $\cdot$ $cell^{-1} \cdot h^{-1}$, and e.g. 1 $\mu g \cdot ml^{-1}$, 2 $\mu g \cdot ml^{-1}$ or 3 $\mu g \cdot ml^{-1}$, respectively, the circulation rate per hour is 90 to 900-fold, 45 to 450-fold or 30 to 300-fold the culture volume. It depends on the permissible concentration difference of dissolved oxygen between the inlet and outlet, but it can be expected that a pump of considerably larger capacity is needed in comparison with other types of bioreactor. In this bioreactor, the carriers can be allowed to stay in the fluidized bed without overflow from the bioreactor, and perfusion can be done without another separation of the cells from the culture medium. Moreover, since dissolved oxygen can be fed into the circulation loop outside the bioreactor, the system has the advantages that the problems which would be caused otherwise can be easily avoided and room for designing the oxygen supply system can be increased.

Further, the following advantages and disadvantages can be cited for the suspension type immobilized-cell bioreactors:

- Uniform mixing and fluidization can be readily done in the bioreactors, accordingly the nutrients and dissolved oxygen can be readily transferred to the carrier surfaces.
- Sampling can be optionally done from the homogeneous culture system at any time during the cultivation and the parameters of the immobilized cells (cell density, viability or the like) can be directly monitored.
- The bioreactors have high scale-up potential in principle, but the ease of scale expansion depends considerably on the ease of seed cell recovery up to the final stage of the cell cultivation, further, in the fluidized bed bioreactors it depends on the possibility of designing of circulation pumps with a large enough capacity to meet the requirements for the mammalian cell culture system.
- It is clear that the damage of cell-immobilizing carriers may be caused by their collision and friction, making long-term cultivation difficult.

The fixed bed bioreactors include:

i) Packed bed bioreactor
ii) Other reactors
  · Ceramic matrix bioreactor (Opticell bioreactor).
  · Hollow fiber bioreactor.
  · Flat membrane bioreactor.
  · Static Maintenance reactor (Invitron Corp.: U.S.A.) (Tolbert et al., 1988)

The packed bed bioreactor has been investigated using glass beads, porous glass beads (e.g., Siran), and sponge-like materials (e.g., polyurethane foam) as cell immobilizing carriers. As already stated, glass beads are poor in competitiveness while porous glass beads show promising results. In this process, the culture medium is externally pumped as in the former fluidized bed bioreactor, but the medium flows through the channels in the packed bed of the static cell-immobilizing carrier in the reactor. There are two types of the bioreactors classified by the flow directions of the culture medium, the parallel flow to the axis of the bed (axial flow: upward or downward) (Looby and Griffiths, 1988; 1989; Looby et al., 1990) and the radial flow (Yoshida et al., 1991). In this type with porous glass beads, for example, fluidization of the carriers is not needed, but circulation flow must be maintained to provide a proper microenvironment surrounding the cells so that the nutrients (particularly dissolved oxygen) are adequately supplied to the cells which have grown at a high density in the pores of the carrier. The circulation volume cannot be set high from a hydrodynamic point of view, such as pressure drop in the packed bed. As a result, the flow in the packed bed can be regarded as a plug flow, and a concentration gradient occurs in the nutrients and dissolved oxygen along the flow in the bed. Thus the inlet and the outlet of the bed, namely the scale of the bioreactors, are limited. The radial flow can improve the conventional axial flow in this point, but the packed bed type is poorer in the scale-up potential than the fluidized bed type. But, as the packed bed type holds the cells in the carriers, perfusion culture can be readily performed without separation of the cells from the culture medium, and dissolved oxygen can be fed with the external circulation loop, it is commonly advantageous with the fluidized bed type, so that the problems which would be caused otherwise can be avoided and the room for design of the oxygen supply system will be increased.

Other advantages and disadvantages of the packed bed bioreactors are cited as follows:

- There is no damage of the support matrices caused by collision or friction.
- The system geometry or configuration is simple.
- Since fluidization is not needed, the specification of the supports matrices (material, size, or the like) can be selected from more extended ranges.
- Since the packed carrier density can be set higher than in other processes, the the volumetric efficiency is higher.
- The carriers can be sterilized in situ.
- Since it is substantially impossible to sample the immobilized cells during the cultivation, the parameters (cell density, viability or the like) of the cells cannot be monitored directly.

- Attention should be paid to how to inoculate the cells for uniform distribution over the whole packed bed.
- Attention should be paid to the clogging of flow channels in the packed bed.

The **Ceramic matrix bioreactor** (Opticell) possesses basically the characteristics of the fixed bed bioreactors as does the packed bed type, although it has a different configuration. Whether it is a more preferable bioreactor or not is decided depending on what positive properties the support matrix (Opticore) has. Although the cultivation is on the scale of liter order, high cell density of 2×10^7 cells $\cdot$ ml^{-1} has been reported in the perfusion culture with both anchorage-dependent and -independent cells (Nicholson et al., 1991).

The **Hollow fiber bioreactor** is commercially available from several manufacturers. Due to its structure, the scale expansion of the bioreactor unit is limited, but the reactors have already been used in small-scale production.

The **Flat membrane bioreactor** is essentially the same as the hollow fiber bioreactor, but flat membranes are used instead of hollow fibers and "Membroferm" is known as a typical system (Scheirer, 1988).

The merits and demerits of these membrane systems were already discussed in the former section.

The **Static Maintenance Reactor** is a cultivator which has been developed by Invitron Corp. (U.S.A.) and a plurality of porous tubes are placed in the culture vessel as a cell-immobilizing carrier and perfusion culture is carried out by using a couple of porous tubes in the vinicity to supply the fresh medium and withdraw the spent medium. The reactor is equipped with a gas-permeable membrane for

Table 3.1.3. Classification of the bioreactors for mammalian cells

Culture processes	Types of bioreactors	Cell-immobilizing carriers
1) Suspended-cell culture	Stirred tank Airlift (or bubble column)	
2) Immobilized-cell culture · Immobilized-cell suspension culture	Stirred tank	Microcarriers Porous microcarriers Gel beads Microcapsules
	Fluidized bed	Porous microcarriers Porous glass beads
	Airlift (or bubble column)	
· Fixed bed culture	Packed bed	Glass beads Porous glass beads Polyurethane foam
	Miscellaneous	Honeycombed ceramic cartridge Hollow fibers Flat membranes Porous tubes

supplying oxygen. In this reactor, a large amount of cells grown up in another culture process are inoculated at a high density (the order of about 10^8 cells · ml^{-1}), immobilized to the porous tubes and are maintained for a long period of time. A report describes the result of a perfusion culture over more than 3 months using 1 to 2 kilograms of seed cells in this reactor (Tolbert et al., 1988).

The main bioreactors have been illustrated in 3.1.4.1 and 3.1.4.2. Conclusively, the bioreactors are classified into the tank and column types for suspension culture, and the fixed bed types (see Tab. 3.1.3).

3.1.5 Selection and Design for Bioreactors

The products from large-scale cultivation of mammalian cells will be limited to substances of high value, such as pharmaceuticals, for the time being. Therefore, the topics here will be focused on the bioreactors for mammalian cells producing pharmaceuticals. In such cases, there are matters to be kept in mind for selection of proper bioreactors. For example, when utilized as a therapeutic medicine, a product expressed by mammalian cells is regulated as a biological. In other words, the qualities of a final product vary depending on the production conditions and consistency of the production process is always required. The product manufactured by changing production conditions must be proved to be identical to the conventional one in quality. Accordingly, when the efficiency and timing of product development are considered, the production process should be desirably established as early as possible, before preclinical and clinical tests start. In the meantime, manpower and time will be needed for establishing a process of high competitiveness depending on the purposes. The development of the process including the bioreactors should be made, as these matters are well-balanced.

The process for producing a therapeutic medicine with large scale mammalian cell culture basically comprises cultivation, purification, and pharmaceutical preparation. In the development of the production process, the value of the target substance as a medicine (efficacy, safety and needs) are estimated at first and the production scale is forecast, then the goals of the qualities and costs are set on that basis. The goals are divided into those for each process stated above to promote the development.

The basis concept for culture process development is, as stated in 3.1.1 Introduction,

i) high economical efficiency
ii) excellent operability, simplicity and high reliability
iii) easy production scale expansion and high scale-up potential
iv) reduced number of equipment units and their compactness, and
v) product of more preferable qualities.

At first, the preparation and selection of the optimal cell line for the target substance should be done prior to the development of a cultivation process. As criteria for cell line selection, are cited, for example, the following items:

- Kind of cell (established cell lines, recombinant cells, hybridoma cells or primary cells).
- The structure and biological activity of the useful substance to be expressed.
- The expression level and its duration.
- High suitability for supposed large scale cell culture.
- The safety of the cell line.

The information about the qualities (efficacy and safety) of the useful target product, which will become clear conclusively in the preclinical and clinical tests *in vivo*, is considerably limited in the first stages, but the prospect should be established as precisely as possible. As for the suitability for large scale cell culture, several promising cell lines are selected by preliminary stationary culture, then they are cultured on a small scale by the supposed process for industrial production thereby a preferable candidate or candidates are determined. Using the candidate, the culture process is developed. In this case, one of the critical points is to try reducing the load on the downstream processing as much as possible. These steps basically comprise:

i) The needed amount of production in the culture process is calculated from the estimated production amount of the final product and the yield of the target substance in the following purification process, and a rough culture scale is estimated from these values and the target substance productivity of the candidate cell line.

ii) The suitable culture process and bioreactor are proposed on the basis of the properties of the candidate cell line and the estimated production scale (batch or continuous, and suspended- or immobilized-cell culture): the optimal process largely depends on the properties of the cells.

iii) A small scale experiment on the proposed cell culture process is performed to determine proper operation conditions and obtain the following basic culture properties of the candidate cell line:
required medium composition, oxygen uptake rate, optimal range of dissolved oxygen concentration, dependence of temperature, pH and osmotic pressure, allowable concentration of toxic metabolites, specific perfusion rate (in the case of perfusion culture), carrier concentration per unit culture volume (in immobilized-cell culture), circulation volume per unit culture volume and superficial liquid velocity (in fluidized bed and fixed bed bioreactor), cell density and cell viability, cell growth rate, specific productivity of the target substance and stability of the productivity in subculture, concentration of the target substance and stability of the productivity in subculture, concentration of the target substance in the culture medium, metabolism rate of major ingredients, cell proliferation rate and yield per one stage of seed culture (particularly in immobilized-cell culture), seed cell recovery procedure (immobilized-cell culture), culture period and so on.

From these culture parameters, the culture medium cost and installation cost are estimated at the industrial production scale, then rough production costs are calculated. Further, the target product is purified and identified on its qualities.

When the qualities and the costs meet the target values at this stage, the cell line to be used is confirmed and the next stage of development of scale-up technology is started (otherwise, going back to the stage of preparation and selection of the cell line).

In the development of scale-up technology, suspended-cell perfusion culture is exemplified. The equipment up to a bench or pilot scale is used to make study on:

- The separation of the cells from the culture medium.
- Supply of dissolved oxygen into the culture medium.
- Mixing of cells and culture medium.

To develop the optimal process, one should obtain the prospect on scale-up technology and collect the relationship between the operation conditions and the culture parameters, whereby design criteria are specified. These procedures are not explained here because they were already stated in the preceding sections. When the product qualities and the production costs have cleared their goals at this stage, the design of the bioreactor at the production scale will be started on the basis of the data obtained. Here, one of the most important matters not to be forgotten from the standpoint of plant operation and design is how to take measures against contamination by microorganisms.

Whether the goals of the qualities and production costs have been attained or not will become obvious conclusively after the production process is established as stated above, therefore, the goals should have allowance to some extent.

3.1.6 Conclusion

In large scale mammalian cell culture, the optimal process and the suitable bioreactor depend on the target product and its value, the kind of cells, the production scale, and the difficulty in technical development, all of which will be properly selected until a versatile technology of high competitiveness is established. For more increased versatility, as a basic concept, the cells are conditioned to the culture process in suspended-cell culture, while the process (or immobilization matrix) is conditioned to the cells in immobilized-cell culture. The authors have studied mainly the suspended-cell culture process with intense interest in future development and trends for large scale mammalian cell culture technology. In any case, the authors expect that large scale mammalian cell culture technology will be highly developed and refined and enably scientific contributions on a world-wide scale.

References

Almgren, J., Nilsson, C., Petersson, A. C. and Nilsson, K. (1991) In: Production of Biologicals from Animal Cells in Culture (Spier, R. E., Griffiths, J. B. and Meignier, B., eds.), pp. 434–438, Butterworth-Heinemann, Oxford.

Apelman, S. (1992) In: Animal Cell Technology: Basic & Applied Aspects (Murakami, H., Shirahata, S. and Tachibana, H., eds.) pp. 149–154, Kluwer Academic Publishers, Dordrecht.

Berg, G.J. and Bödeker, B.G.D. (1988) In: Animal Cell Biotechnology, Vol. 3 (Spier, R.E. and Griffiths, J.B., eds.), pp. 321–335, Academic Press Ltd., London.

Beyeler, W., Thaler, T. and Clements, R. (1990) In: Trends in Animal Cell Culture Technology (Murakami, H., ed.), pp. 225–230, Kodansha Ltd., Tokyo.

Birch, J.R., Thompson, P.W., Lambert, K. and Boraston, R. (1985) In: Large-Scale Mammalian Cell Culture (Feder, J. and Tolbert, W.R., eds), pp. 1–18, Academic Press, Inc., Orlando.

Brown, P.C., Costello, M.A.C., Oakley, R. and Lewis, J.L. (1985) In: Large-Scale Mammalian Cell Culture (Feder, J. and Tolbert, W.R., eds.), pp. 59–78, Academic Press, Inc., Orlando.

Brown, P.C., Figueroa, C., Costello, M.A.C., Oakley, R. and Maciukas, S.M. (1988) In: Animal Cell Biotechnology, Vol. 3 (Spier, R.E. and Griffiths, J.B., eds.), pp. 251–262, Academic Press Ltd., London.

Bugarski, B., King, G.A., Jovanovic, G., Daugulis, A.J. and Goosen, M.F.A. (1989) Perfomance of an external loop air-lift bioreactor for the production of monoclonal antibodies by immobilized hybridoma cells. Appl. Microbiol. Biotechnol. *30*, 264–269.

Butler, M. (1988) In: Animal Cell Biotechnology, Vol. 3 (Spier, R.E. and Griffiths, J.B., eds.), pp. 283–303, Academic Press Ltd., London.

Cherry, R.S. and Papoutsakis, E.T. (1990) In: Animal Cell Biotechnology, Vol. 4 (Spier, R.E. and Griffiths, J.B., eds.), pp. 71–121, Academic Press Ltd., London.

Croughan, M.S. and Wang, D.I.C. (1991) In: Animal Cell Bioreactors (Ho, C.S. and Wang, D.I.C., eds.), pp. 213–249, Butterworth-Heinemann, Stoneham.

de Bruyne, N.A. (1988) In: Animal Cell Biotechnology, Vol. 3 (Spier, R.E. and Griffiths, J.B., eds.), pp. 141–176, Academic Press Ltd., London.

Dean, Jr. R.C., Karkare, S.B., Phillips, P.G., Ray, N.G. and Runstadler, Jr., P.W. (1987) In: Large Scale Cell Culture Technology (Lydersen, B.K., ed.), pp. 145–167, Hanser Publishers, Munich, Vienna, New York.

Duff, R.G. (1985) Microencapsulation technology: a novel method for monoclonal antibody production. Trends Biotechnol. *3*, 167–170.

Familletti, P.C. and Fredericks, J.E. (1988) Techniques for mammalian cell immobilization. Bio/Technology *6*, 41–44.

Fleischaker, R.J. and Sinskey, A.J. (1981) Oxygen demand and supply in cell culture. Eur. J. Appl. Microbiol. Biotechnol. *12*, 193–197.

Glacken, M.W., Fleischaker, R.J. and Sinskey, A.J. (1983) Mammalian cell culture: engineering principles and scale-up. Trends Biotechnol. *1*, 102–108.

Griffiths, J.B. (1985) In: Animal Cell Biotechnology, Vol. 1 (Spier, R.E. and Griffiths, J.B., eds.), pp. 49–83, Academic Press Inc. (London) Ltd.

Griffiths, J.B. (1988) In: Animal Cell Biotechnology, Vol. 3 (Spier, R. E. and Griffiths, J. B., eds.), pp.179–220, Academic Press Ltd., London.

Griffiths, B. (1990) In: Animal Cell Biotechnology, Vol. 4 (Spier, R. E. and Griffiths, J. B., eds.), pp. 149–166, Academic Press Ltd., London.

Griffiths, B. (1990) Animal Cells - the breakthrough to a dominant technology. Cytotechnology *3*, 109–116.

Griffiths, B. and Looby, D. (1991) In: Animal Cell Bioreactors (Ho, C. S. and Wang, D. I. C., eds.), pp. 165–190, Butterworth-Heinemann, Stoneham.

Griffiths, J.B., Cameron, D.R. and Looby, D. (1987) A comparison of unit process systems for anchorage dependent cells. Dev. Biol. Stand. *66*, 331–338.

Hamamoto, K., Ishimaru, K. and Tokashiki, M. (1989) Perfusion culture of hybridoma cells using a centrifuge to separate cells from culture mixture. J. Ferment. Bioeng. *67*, 190–194.

Hamamoto, K., Tokashiki, M. and Ichikawa, Y. (1985) Japanese patent Application, 60–224486.

Hamamoto, K., Tokashiki, M., Ichikawa, Y. and Murakami, H. (1987) High cell density culture of a hybridoma using perfluorocarbon to supply oxygen. Agric. Biol. Chem. *51*, 3415–3416.

Handa-Corrigan, A. (1990) In: Animal Cell Biotechnology, Vol. 4 (Spier, R.E. and Griffiths, J.B., eds.), pp.123–132, Academic Press Ltd., London.

Hayman, E.G., Ray, N.G. and Runstadler, Jr., P.W. (1987) In: Bioreactors and Biotransformations (Moody, G.W. and Baker, P.B., eds.), pp. 132–140, Elsevier, London, New York.

Himmelfarb, P., Thayer, P.S. and Martin, H.E. (1969) Spin filter culture: The propagation of mammalian cells in suspension. Science *164*, 555–557.

Hopkinson, J. (1985) Hollow fiber cell culture systems for economical cell-product manufacturing. Bio/Technology *3*, 225–230.

Hosoi, S., Miyaji, H., Satoh, M., Kurimoto, T., Mihara, A., Fujiyoshi, N., Itoh, S. and Sato, S. (1991a) Optimization of cell culture conditions for production of biologically active proteins. Cytotechnology *5*, S17–34.

Hosoi, S., Murosumi, K., Sasaki, K., Satoh, M., Miyaji, H., Hasegawa, M., Itoh, S., Tamaoki, T. and Sato, S. (199lb) Optimization of cell culture conditions for G-CSF (granulocytocolony stimulating factor) production by genetically engineered Namalwa KJM–1 cells. Cytotechnology *7*, 25–32.

Jacobs, M., Gummich, G., Keller, I. and Schwengers, D. (1991) In: Produdion of Biologicals from Animal Cells in Culture (Spier, R.E., Griffiths, J.B. and Meignier, B., eds.), pp. 439–442, Butterworth-Heinemann, Oxford.

Jäger, V. (1992) In: Animal Cell Technology: Basic & Applied Aspects (Murakami, H., Shirahata, S. and Tachibana, H., eds.), pp. 209–216, Kluwer Academic Publishers, Dordrecht.

Junker, B.H., Wu, F., Wang, S., Waterbury, J., Hunt, G., Hennessey, J., Aunins, J., Lewis, J., Silberklang, M. and Buckland, B.C. (1992) Evaluation of a microcarrier process for large-scale cultivation of attenuated hepatitis A. Cytotechnology *9*, 173–187.

Karkare, S.B., Phillips, P.G., Burke, D.H. and Dean, Jr., R.C. (1985) In: Large-Scale Mammalian Cell Culture (Feder, J. and Tolbert, W.R., eds.), pp. 127–149, Academic Press, Inc., Orlando.

Katinger, H. and Scheirer, W. (1985) In: Animal Cell Biotechnology, Vol. 1 (Spier, R.E. and Griffiths, J.B., eds.), pp.167–193, Academic Press Inc. (London) Ltd.

Kitano, K. (1991) In: Animal Cell Bioreactors (Ho, C.S. and Wang, D.I.C., eds.), pp. 73–106, Butterworth-Heinemann, Stoneham.

Kitano, K., Ichimori, Y., Sawada, H., Iwasa, S., Sasai, S. and Tsukamoto, K. (1991) Effective production of anti-tetanus toxoid and anti-HBsAg human monoclonal antibodies by serum-free culture of hybridomas. Cytotechnology *5*, S53–74.

Kitano, K., Shintani, Y., Ichimori, Y., Tsukamoto, K., Sasai, S. and Kida, M. (1986) Production of human monoclonal antibodies by heterohybridomas. Appl. Microbiol. Biotechnol. *24*, 282–286.

Klement, G., Scheirer, W. and Katinger, H.W.D. (1987) Construction of a large-scale membrane reactor system with different compartments for cells, medium and product. Dev. Biol. Stand. *66*, 221–226.

Knazek, R.A., Gullino, P.M., Kohler, P.O. and Dedrick, R.L. (1972) Cell culture on artificial capillaries: An approach to tissue growth in vitro. Science *178*, 65–67.

Ku, K., Kuo, M.J., Delente, J., Wildi, B.S. and Feder, J. (1981) Development of a hollow-fiber system for large-scale culture of mammalian cells. Biotechnol. Bioeng. *23*, 79–95.

Lambert, K.J., Boraston, R., Thompson, P.W. and Birch, J.R. (1987) Production of monoclonal antibodies using large scale cell culture. Dev. Ind. Microbiol. *27*, 101–106.

Lazar, A., Reuveny, S., Muzrahi, A., Avtalion, M., Whiteside, J.P. and Spier, R.E. (1987) In: Modern Approaches to Animal Cell Technology (Spier, R.E. and Griffiths, J.B., eds.), pp. 437–448, Butterworth, Guildford.

Lazar, A., Silberstein, L., Mizrahi, A. and Reuveny, S. (1988) An immobilized hybridoma culture perfusion system for production of monoclonal antibodies. Cytotechnology *1*, 331–337.

Lehmann, J., Vorlop, J. and Büntemeyer, H. (1988) In: Animal Cell Biotechnology, Vol. 3 (Spier, R.E. and Griffiths, J. B., eds.), pp. 221–237, Academic Press Ltd., London.

Lim, F. and Sun, A.M. (1980) Microencapsulated islets as bioartificial endocrine pancreas. Science *21*, 908–910.

Looby, D. and Griffiths, J.B. (1987) In: Modern Approaches to Animal Cell Technology (Spier, R.E. and Griffiths, J.B., eds.), pp. 342–352, Butterworth, Guildford.

Looby, D. and Griffiths, J.B. (1988) Fixed bed porous glass sphere (porosphere) bioreactors for animal cells. Cytotechnology *1*, 339–346.

Looby, D. and Griffiths, J.B. (1989) In: Advances in Animal Cell Biology and Technology for Bioprocesses (Spier, R.E., Griffiths, J.B., Stephenne, J. and Crooy, P.J., eds.), pp. 336–344, Butterworth, Guildford.

Looby, D., Racher, A.J., Griffiths, J.B. and Dowsett, A.B. (1990) In: Physiology of Immobilized Cells (de Bont, J.A.M., Visser, J., Mattiason, B. and Tramper, J., eds.), pp. 255–264, Elsevier, Amsterdam.

Lydersen, B.K., Pugh, G.G., Paris, M.S., Sharma, B.P. and Noll, L.A. (1985a) Ceramic matrix for large scale animal cell culture. Bio/Technology *3*, 63–67.

Lydersen, B.K., Putnam, J., Bognar, E., Patterson, M., Pugh, G.G. and Noll, L.A. (1985b) In: Large-Scale Mammalian Cell Culture (Feder, J. and Tolbert, W.R., eds.), pp. 39–58, Academic Press, Inc., Orlando.

Matsushita, T., Ketayama M., Kamihata, K. and Funatsu, K. (1990) Anchorage-dependent mammalian cell culture using polyurethane foam as a new substratum for cell attachment. Appl. Microbiol. Biotechnol *33*, 287–290.

Mizrahi, A., Lazar, A. and Reuveny, S. (1990) In: Animal Cell Biotechnology, Vol. 4 (Spier, R.E. and Griffiths, J.B., eds.), pp. 413–444, Academic Press Ltd., London.

Montagnon, B., Vincent-Falquet, J.C. and Fanget, B. (1984) Thousand litre scale microcarrier culture of Vero cells for killed polio virus vaccine; promising results. Dev. Biol. Stand. *55*, 37–42.

Moo-Young, M. and Chisti, Y. (1988) Considerations for designing bioreactors for shear-sensitive culture. Bio/Technology *6*, 1291–1296.

Murdin, A.D., Thorpe, J.S. and Spier, R.E. (1987) In: Modern Approaches to Animal Cell Technology (Spier, R.E. and Griffiths, J.B., eds.), pp. 420–436, Butterworth, Guildford.

Nelson, K.L. (1988) Industrial-scale mammalian cell culture, part 1: Bioreactor design considerations. Biopharm. Manufact. Feb. 42–46.

Nicholson, M.L., Hampson, B.S., Pugh, G.G. and Ho, C.S. (1991) In: Animal Cell Bioreactors (Ho, C.S. and Wang, D.I.C., eds.), pp. 269–303, Butterworth-Heinemann, Stoneham.

Nilsson, K. and Mosbach, K. (1987) Immobilized animal cells. Dev. Biol. Stand. *66*, 183–193.

Nilsson, K., Buzsaky, F. and Mosbach, K. (1986) Growth of anchorage-dependent cells on macroporous microcarriers. Bio/Technology *4*, 989–990.

Nilsson, K., Nilsson, L., Petterson, A.C. and Almgren, J. (1990) In: Trends in Animal Cell Culture Technology (Murakami, H., ed.), pp. 35–39, Kodansha Ltd., Tokyo.

Nilsson, K., Scheirer, W., Merten, O.W., Ostberg, L., Liehl, E., Katinger, H.W.D. and Mosbach, K. (1983) Entrapment of animal cells for production of monoclonal antibodies and other biomolecules. Nature *302*, 629–630.
Ogata, M., Arikawa, K., Matsukura, H., Yamazaki, Y. and Suzuki, A. (1992) In: Animal Cell Technology: Basic and Applied Aspects (Murakami, H., Shirahata, S. and Tachibana, H., eds.), pp. 201–207, Kluwer Academic Publishers, Dordrecht.
Phillips, A.W., Ball, G.D., Fantes, K.H., Finter, N.B. and Johnston, M.D. (1985) In: Large-Scale Mammalian Cell Culture (Feder, J. and Tolbert, W.R., eds.), pp. 87–95, Academic Press, Inc., Orlando.
Posillico, E.G. (1986) Microencapsulation technology for large-scale antibody production. Bio/Technology *4*, 114–117.
Prokop, A. and Rosenberg, M.Z. (1989) In: Advances in Biochemical Engineering and Biotechnology, Vol. 39 (Fechter, A., ed.), pp. 29–71, Springer-Verlag, Berlin Heidelberg.
Pugh, G.G., Berg, G.J. and Sear, C.H.J. (1987) In: Bioreactors and Biotransformations (Moody, G.W. and Baker, P.B., eds.), pp. 121–131, Elsevier, London.
Pullen, K.F., Jonson, M.D., Phillips, A.W., Ball, G.D. and Finter, N.B. (1985) Very large scale suspension cultures of mammalian cells. Dev. Biol. Stand. *60*, 175–177.
Racher, A.J., Looby, D. and Griffiths, J.B. (1990) Studies on monoclonal antibody production by a hybridoma cell line (C1E3) immobilized in a fixed bed, porosphere culture system. J. Biotech. *15*, 129–146.
Ray, N.G., Tung, A.S., Runstadler, P.W. and Vournakis, J.N. (1991) In: Production of Biologicals from Animal Cells in Culture (Spier, R.E., Griffiths, J.B. and Meignier, B., eds.), pp. 502–512, Butterworth-Heinemann, Oxford.
Rhodes, M. and Birch, J. (1988) Large-scale production of proteins from mammalian cells. Bio/Technology *6*, 518–523.
Rhodes, M., Gardiner, S. and Broad, D. (1991) In: Animal Cell Bioreactors (Ho, C.S. and Wang, D.I.C., eds), pp. 253–268, Butterworth-Heinemann, Stoneham.
Runstadler, Jr., P.W. and Cernek, S.R. (1988) In: Animal Cell Biotechnology, Vol. 3 (Spier, R.E. and Griffiths, J.B., eds.), pp. 305–320, Academic Press Ltd., London.
Runstadler, P.W., Ozturk, S.S. and Ray, N.G. (1992) In: Animal Cell Technology Basic & Applied Aspects (Murakami, H., Shirahata, S. and Tachibana, H., eds.), pp. 57–63, Kluwer Academic Publishers, Dordrecht.
Rupp, R.G. (1985) In: Large scale Mammalian Cell Culture (Feder, J. and Tolbert, W. R., eds.), pp.19–38, Academic Press, Inc., Orlando.
Sato, S., Kawamura, K. and Fujiyoshi, N. (1983) Animal cell cultivation for production of biological substances with a novel perfusion culture apparatus. J. Tissue Culture Methods *8*, 167–171.
Sato, S., Kawamura, K., Hanai, N. and Fujiyoshi, N. (1985) In: Growth and Differentiation of Cells in Defined Environment (Murakami, H., Yamane, I., Barnes, D.W., Mather, J.P., Hayashi, I. and Sato, G.H., eds.), pp.123–126, Kodansha Ltd., Tokyo & Springer-Verlag, Hidelberg.
Scheirer, W. (1988) In: Animal Cell Biotechnology, Vol. 3 (Spier, R.E. and Griffiths, J.B., eds.), pp. 263–281, Academic Press Ltd., London.
Schönherr, O.T. and van Gelder, P.T.J.A. (1988) In: Animal Cell Biotechnology, Vol. 3 (Spier, R.E. and Griffiths, J.B., eds.), pp. 337–355, Academic Press Ltd., London.
Shintani, Y., Kohno, Y., Sawada, H. and Kitano, K. (1991) Comparison of culture methods for human-human hybridomas secreting anti-HBsAg human monoclonal antibodies. Cytotechnology *6*, 197–208.
Shiragami, N., Ohira, Y. and Unno, H. (1991) In: Animal Cell Culture and Production of Biologicals (Sasaki, R. and Ikura, K., eds), pp. 121–126, Kluwer Academic Publishers, Dordrecht.

Sinskey, A.J., Fleischaker, R.J., Tyo, M.A., Giard, D.J. and Wang, D.I.C. (1981) Production of cell-derived products: virus and interferon. Ann. N.Y. Acad. Sci. 369, 47–59.

Spier, R.E. (1991) In: Animal Cell Culture and Production of Biologicals (Sasaki, R. and Ikura, K., eds.), pp. 41–46, Kluwer Academic Publishers, Dordrecht.

Spier, R.E. and Griffiths, B. (1984) An examination of the data and concepts germane to the oxygenation of cultured animal cells. Dev. Biol. Stand. *55*, 81–92.

Spier, R.E. and Whiteside, J.P. (1976) The production of foot and mouth disease virus from BHK21C13 cells grown on the surface of glass spheres. Biotechnol. Bioeng. *18*, 649–657.

Spier, R.E. and Whiteside, J.P. (1984) The description of a device which facilitates the oxygenation of microcarrier cultures. Dev. Biol. Stand *55*, 151–152.

Spier, R.E. and Whiteside, J.P. (1990) In: Animal Cell Biotechnology, Vol. 4 (Spier, R.E. and Griffiths, J.B., eds.), pp. 133–148, Academic Press Ltd., London.

Tharakan, J.P. and Chau, P.C. (1986a) A radial flow hollow fiber bioreactor for the large-scale culture of mammalian cells. Biotechnol. Bioeng. *28*, 329–342.

Tharakan, J.P. and Chau, P.C. (1986b) Operation and pressure distribution of immobilized cell hollow fiber bioreactors. Biotechnol. Bioeng. *28*, 1064–1071.

Tokashiki, M. (1991) In: Animal Cell Bioreactors (Ho, C.S. and Wang, D.I.C., eds.), pp. 327–356, Butterworth-Heinemann, Stoneham.

Tokashiki, M. and Arai, T. (1989) High density culture of mouse-human hybridoma cells using a perfusion culture apparatus with multi-settling zones to separate cells from the culture medium. Cytotechnology *2*, 5–8.

Tokashiki, M. and Arai, T. (1991) In: Production of Biologicals from Animal Cells in Culture (Spier, R.E., Griffiths, J.B. and Meignier, B., eds.), pp. 467–469, Butterworth-Heinemann, Oxford.

Tokashiki, M., Arai, T., Hamamoto, K. and Ishimaru, K. (1990) High density culture of hybridoma cells using a perfusion culture vessel with an external centrifuge. Cytotechnology *3*, 239–244.

Tokashiki, M., Hamamoto, K., Takazawa, Y. and Ichikawa, Y. (1988) High-density culture of mouse-human hybridoma cells using a new perfusion culture vessel. Kagakukogaku Ronbun-shu *14*, 337–341.

Tolbert, W.R. and Feder, J. (1983) In: Annual Reports on Fermentation Processes, Vol. 6 (Tsao, G.T., Flickinger, M.C. and Finn, R.K., eds.), pp. 35–74, Academic Press, Inc., New York.

Tolbert, W.R., Lewis, C., White, P.J. and Feder, J. (1985) In: Large-Scale Mammalian Cell Culture (Feder, J. and Tolbert, W.R., eds.), pp. 97–123, Academic Press, Inc., Orlando.

Tolbert, W.R., Srigley, W.R. and Prior, C.P. (1988) In: Animal Cell Biotechnology, Vol. 3 (Spier, R.E. and Griffiths, J.B., eds.), pp. 373–393, Academic Press Ltd., London.

Tyo, M.A., Bulbulian, B.J., Menken, B.Z. and Murphy, T.J. (1988) In: Animal Cell Biotechnology, Vol. 3 (Spier, R.E. and Griffiths, J.B., eds.), pp. 357–371, Academic Press Ltd., London.

van Wezel, A.L. (1967) Growth of cell-strains and primary cells on micro-carriers in homogeneous culture. Nature *216*, 64–65.

van Wezel, A.L. (1985) In: Animal Cell Biotechnology, Vol. 1 (Spier, R.E. and Griffiths, J.B., eds.), pp. 265–282, Academic Press Inc. (London) Ltd.

Vournakis, J.N. and Runstadler, Jr., P.W. (1991) In: Animal Cell Bioreactors (Ho, C.S. and Wang, D.I.C., eds.), pp. 305–326, Butterworth-Heinemann, Stoneham.

Whiteside, J.P. and Spier, R.E. (1981) The scale-up from 0.1 to 100 litres of a unit process system based on 3 mm diameter glass spheres for the production of four strains of FMDV from BHK monolayer cells. Biotechnol. Bioeng. *23*, 551–565.

Yamaji, H. and Fukuda, H. (1991) Long-term cultivation of anchorage-independent animal cells immobilized within reticulated biomass support particles in a circulating bed fermentor. Appl. Microbiol. Biotechnol. *34*, 730–734.

Yoshida, H., Mizutani, S. and Ikenaga, H. (1991) In: Animal Cell Culture and Production of Biologicals (Sasaki, R. and Ikura, K., eds.), pp. 329–334, Kluwer Academic Publishers, Dordrecht.

3.2 Hydrodynamic Properties in Bioreactors

Devamita Chattopadhyay, Miguel Garcia-Briones, Raghavan Venkat, and Jeffrey J. Chalmers

3.2.1 Introduction

The successful commercialization of the recent advances in the genetic engineering of animal cells is dependent on the ability to produce, in relatively large amounts, the product of interest. The term "relatively" is used here since the scale of cultivation can range from just a few to several thousands of liters depending on the product value and market demand. Obviously, for small scale production the use of well-characterized bioreactors (with respect to hydrodynamics) is possible; however, as the volume increases the optimization of the production process becomes more difficult in bioreactors because of the complex hydrodynamics in the vessels. For optimization, a better understanding of the hydrodynamics within the bioreactor as well as the effect that hydrodynamic forces have on the process itself is needed.

In regard to the large scale culture of cells that originate from tissue, two broad classifications can be made: cells that will only grow on surfaces, "anchorage-dependent", and cells that grow in suspension, "anchorage-independent". Because of this difference in cell characteristics and mode of growth, and the associated differences in large-scale cell culture, this chapter will be divided into several sections: the hydrodynamics associated with suspended cells, the hydrodynamics associated with anchorage-denpendent cells, the types of protective medium additives and their mechanisms of protection, and the non-lethal effects of hydrodynamic forces on cells.

3.2.2 The Hydrodynamics of Suspension Cell Culture

The first reports of suspended animal cell damage in agitated and/or sparged bioreactors can be traced back to the 1950s when the cultivation of mammalian cells at scales larger than a couple of liters was first attempted. The relationship of this damage to hydrodynamic forces was shown when increasing the agitation and, in particular, the addition of gas bubbles to satisfy the oxygen demand resulted in increased loss of cell viability (Bryant et al., 1960; Runyan and Geyer, 1963; Kilburn and Webb, 1968; Telling and Elswoth 1965). Along with these early observations of cell damage are reports that the addition of certain medium additives, with surface active characteristics, protect cells (Kilburn and Webb, 1968). While the early evidence seems to indicate that most of the cell damage was the result of gas sparging in the medium, a commonly held belief developed, and in some cases is still held today, that most of the damage is the result of mechanical mixing using agitators, hence the term "shear sensitivity".

New impetus for research to characterize the hydrodynamic environment in agitated and/or sparged bioreactors came with the development of the hybridoma technique and the associated 'shear sensitivity' of the hybridoma cell. In 1987, Handa et al. reported on pioneering work in the area of hydrodynamic damage to cells. They combined statistical and correlational work with photographic-microscopic techniques to determine the mechanisms of cell damage in bubble columns. This statistical work indicated that cell damage is cell line dependent and cell viability decreased with increasing bubble frequency and with decreasing bubble size. From their microscopic observations they suggested that the majority of cell damage takes place at the bubble disengagement region at the medium-air interface (the top of the culture). The above suggestions were based on observations of cells experiencing violent, turbulent, oscillations and surface deformations. They also observed cells entrained in the moving bubble surface interface and transported at high velocities through the draining bubble film. Summarize these visual observations, they suggested three mechanisms of damage: 1) damage due to shearing in draining liquid films in foams, 2) rapid oscillations caused by bursting bubbles, and 3) physical loss of the cells in the foam.

At the same time that hybridoma cell culture applications were expanding, the use of insect cell cultures at large scale to produce insect viruses and heterologous proteins was being investigated. Insect cells, however, were found to be extremely shear sensitive. On the basis of theoretical estimates of the level of laminar shear stress that damages insect cells, Tramper et al., (1987) and Jobes et al. (1991) suggested three regions that cells, in a bubble column bioreactor, could be damaging: the bubble injection and/or formation region, the bubble rising region, and the bubble disengagement region.

Tramper and coworkers further developed a model for cell damage which was based on the assumption that there exists a "hypothetical killing volume" associated with each bubble introduced into the system. This "hypothetical killing volume" was proposed to be independent of dimensions of the column and the number of bubbles, but dependent on the bubble diameter. A logical extension of this model is that since the "killing volume" is independent of the column dimensions, an improvement of cell growth would be experienced as the H/D ratio increases (column height/column diameter). This improvement in cell growth with H/D ratio has been verified experimentally for both insect and hybridoma cells. A logical implication of this work, as noted by Tramper et al., (1988), is that the detrimental effects of rising bubbles are not as important as the effects in either the bubble injection or bubble disengagement region. Using this model they calculated values for operation parameters (volumetric air flow rate, bubble diameter and bubble column dimensions) for which there is a positive growth rate and the surface area available for oxygen transfer is large enough to satisfy the oxygen uptake rate. At this point it is important to note that while this model predicts well the experimental observations obtained, it does not provide a mechanistic understanding of cell damage – it only indicates where to look.

The results reported by the research groups of Handa and Tramper were obtained in bubble columns without mechanical agitation. A significant question is whether mechanical agitation without sparging can also damage suspended cells. To address this question, Kunas and Papoutsakis (1990a; 1990b) proposed and tested two mechanisms by which cells are damaged in agitated bioreactors without sparging: 1) cell damage as a result of interactions with fluid eddies orginating from hydrodynamic turbulence, and 2) cell damage as a result of rupturing bubbles that are entrained within the medium due to vortexing created by mixing.

From a set of experiments where the agitation rates were varied from 0 to 220 rpm, Kunas and Papoutsakis (1990a; 1990b) observed cell damage only when bubbles were entrained (at 190–220 rpm) and disengaged at the liquid surface. This entrainment is the result of the instability of the center vortex when it reaches the impellers (This center vortex is usually provented when baffles are present in the vessel). These results suggest that cell damage in the area of air injection in sparged bioreactors as proposed by Tramper et al (1987) and Jobses et al. (1991) is not as important as the region of bubble breakup at the air-medium interface.

In a different experiment at high agitation speeds (600 rpm), in which the bioreactor was completely filled to prevent the formation of a vortex and the associated bubble entrainment, many (5000 bubbles/ml) small (30–50 micron) bubbles were observed. Since no head space was present and the bubbles were very small, these bubbles just continued to follow the fluid stream lines and never ruptured. Under these conditions no differences in the apparent growth rate was observed when compared to growth rates at lower agitation rates. They concluded that up to 600 rpm, and in the presence of a large number of fast moving bubbles, agitation is not detrimental to cell growth. Not until agitation rates of 800 rpm were reached did the apparent growth rate decrease. Since this degree of mixing is not necessary for a well-mixed condition, it can be concluded that suspended cell damage in sparged, agitated bioreactors is not the result of hydrodynamic forces resulting from inpeller agitation. The most likely cause of damage in suspension culture is cell interactions with rupturing bubbles at the air-medium interface.

Further evidence of the detrimental effect of sparging has been documented in the literature. Oh et al. (1989) studied the effect of agitation on three murine hybridoma cell lines at rpm's ranging from 100 to 450 and oxygen concentration greater than 20 % of saturation. They found no differences in cell growth, viability, antibody production, glucose consumption, lactate production, and metabolic activity in experiments using surface aerated bioreactors without bubble entrainment. In other experiments, however, sparging resulted in a reduction in total cell concentration and viability. This effect was even more pronounced at higher agitation speeds. Since the introduction of sparged air increases the oxygen concentration to near saturation, it was hypothesized that the above results could be related to an increased level of dissolved oxygen dO_2. To test this hypothesis, in a later publication, Oh et al. (1992) showed that experiments with headspace oxygenation where dO_2 was maintained at 5 % and 100 % of saturation resulted in similar growth and other metabolic parameters, thereby disproving this hypothesis. Further investigation in sparged bioreactors showed that cell damage increased as the volumetric air flow rate was increased. In addition, small bubbles were seen to be more detrimental while bubbles bigger than 5.0 mm (which were not further broken down by the impeller) did not cause cell damage.

Two other research groups also observed similar results with respect to sparging. Passini and Goochee (1989) used a mouse hybridoma cell line to study the effect of environmental stresses on the pattern of intracellular protein synthesis. No effects of agitation on growth rate, viability and monoclonal antibody production were observed for exponentially-growing cells in stirred cultures over a broad range of experimental conditions (50–400 rpm). However, sparging resulted in the disruption of cells in a few minutes. Similar results are presented by Murhammer and Goochee (1988) for SF−9 insect cells. Gardner et al. (1990) used a murine hybridoma cell line to study the effects of stirring and sparging. They found no difference in growth, lysis and antibody production for agitated, surface-aerated and stationary T flasks. Yet when gas sparging was used, a reduction in the apparent growth rate and an increase in LDH concentration (an indicator of cell lysis) was observed.

These results all indicate that the rupture of gas bubbles at a liquid interface is damaging to cells. The physical picture of a rupturing gas bubble was first introduced by MacIntyre (1968; 1972) and can be summarized as follows. When a bubble is injected, it rises to the interface driven by buoyancy forces. As the bubble approaches the interface, the still, air-liquid interface profile becomes distorted. Liquid from the top of the distorted profile starts to drain due to gravity forces and a thin liquid cap forms (film). Liquid from this film continues to drain not only by gravity but also by suction resulting from a negative curvature at the film boundaries. Theories on the thinning and rupture of liquid films have been extensively documented in the literature and it is known that the thickness at which a film will break as well as the thinning rate depends on the presence of surface active compounds (Hahn et al., 1985).

When the film breaks, it retracts collecting the liquid forming the film so that a toroidal rim develops. During this retraction process, this thickening toroidal rim can break into small droplets as a result of variations in surface tension, thickness irregularities, and air leaving the bubble cavity. It has also been demonstrated that there seems to be a minimum bubble size at which the formation of small doplets from the film are produced (Medrow, 1968). Thus, for sufficiently small bubbles the toroidal rim will not break into small droplets. The retraction of this toroidal rim is so fast that the sections of the film far from the exanding rim are unaware of the retraction process; consequently, very large levels of acceleration are experienced as the torus expands. This process is completed in tens of microseconds (Figure 3.2.1a).

Once the expanding toroidal rim reaches the edge of the bubble cavity, the driving force that results from surface tension and curvature accelerates the liquid close to the bubble interface down the cavity wall. The convergence of the liquid motion at the axis of symmetry results in two opposite jets. The upward jet eventually may break into several small droplets (Fig. 3.2.1b) while the bottom jet penetrates the liquid beneath the bubble cavity.

By capturing drops from these upward jets, a number of researchers have demonstrated that the concentration of compounds, particles, and bacteria in these drops can be significantly higher than that in the bulk (MacIntyre 1972; Quinn et al., 1975; Blanchard and Syzdek 1970; 1972). This concentration of compounds, particles and bacteria in the upward jet can best be explained by the work of Mac Intyre (1969; 1972) who demonstrated, through ink dye experiments, that the liquid in this upward jet originates in a thin layer surrounding the bubble cavity. He called this process a "boundary layer microtorne". Consequently, if a compound, particle or cell adsorbs to a bubble interface, there is a high probability that it will be ejected either upward or beneath a bubble.

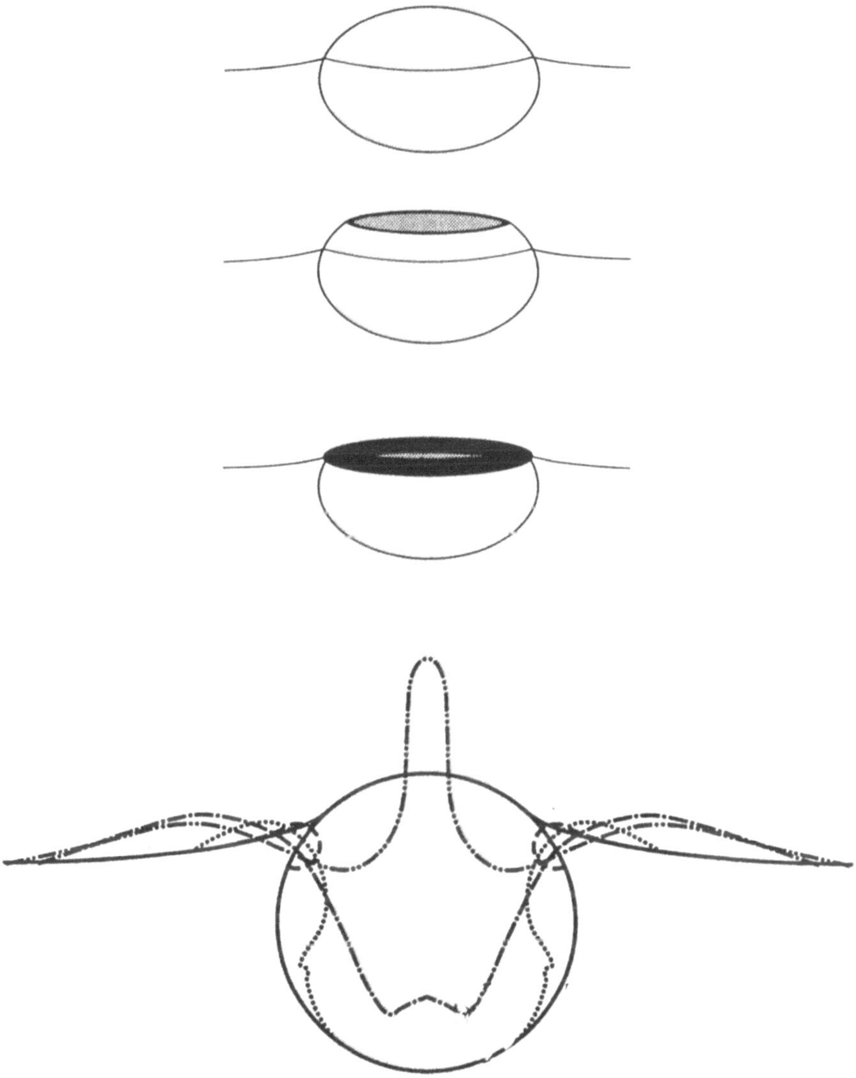

Fig. 3.2.1: Drawing of a bubble film rupturing (1 a) and a diagram of the process of a bubble rupture (1 b).

To determine the interaction of insect cells with bubbles, Bavarian et al. (1991) used a high-speed video and microscopic system to visualize these interactions in a thin bubble column. First, it was observed that individual as well as clumps of cells attach to rising bubbles. Second, cells attach to the medium-air interface with and without bubbles present. Third, similar to results previously discussed by Handa et al. (1987) and Handa-Corrigan et al. (1989) a large number of cells can become trapped in the foam layer. Figure 3.2.2 presents representative photographs of each of these observations. Fourth, when a bubble comes to rest at the air-medium in-

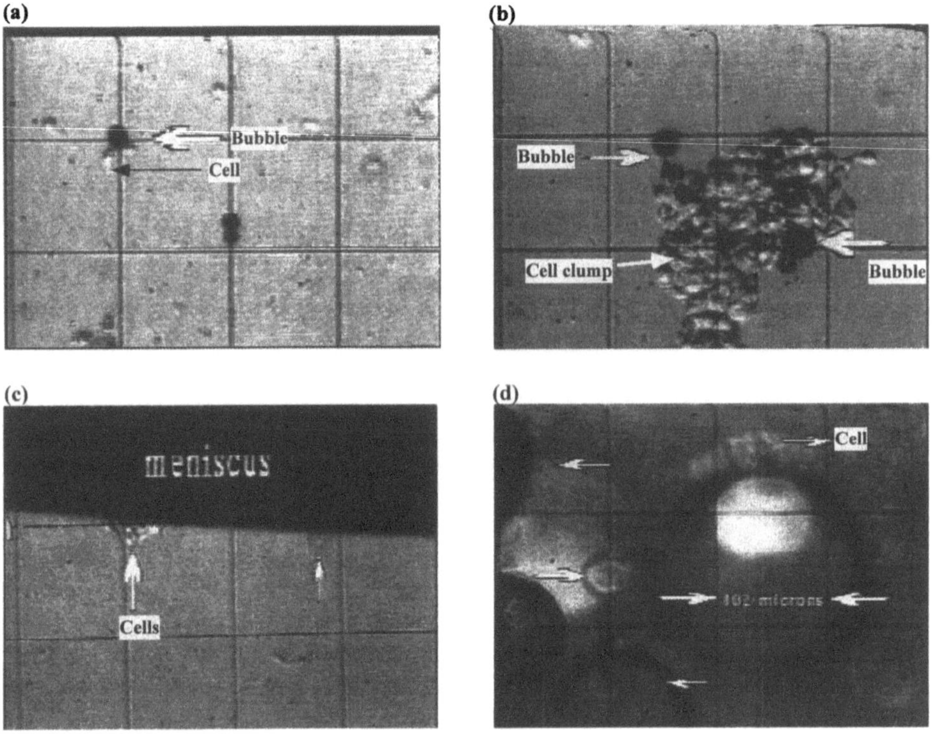

Fig. 3.2.2: Photographs of insect cells attached to individual bubblers, 2a; clumps of cells attached to bubbles, 2b; cells attached to the gas-liquid interface, 2c; and insect cells trapped in the foam layer, 2d.

terface, a large number of cells remain attached to the bubble film that separates the air within the bubble from the air above the interface (Fig. 3.2.3). While it is surprising that cells can attach to rising bubbles, what is of greatest interest in regard to cell damage is the attachment of cells to the bubble film and to the bubble cavity when bubbles are at rest at the air-medium interface.

Based on these observations, and the "boundary layer microtome" proposed by MacIntyre (1969; 1972), Bavarian et al. (1991) and Chalmers and Bavarian (1991) proposed two mechanisms of cell damage in sparged and agitated bioreactors with bubble entrainment: 1) Cells adsorbed on the bubble film are killed by the rapid acceleration of the bubble film after rupture, and 2) Cells attached or close to the bubble cavity are killed as a result of the high shear and possible normal stress associated with the bubble rupture and jet formation.

To determine if cells are present in the upward jet, Garcia-Briones and Chalmers (1992) collected a sample from the upward jet produced from a bubble rupturing in insect cell medium containing serum and cells. They observed in excess of one thousand cells in the approximately 1 microliter sample. Viability in this sample was found to be lower than 10 % as compared with 90 % for the bulk medium. Yet, when Pluronic F-68 was present in the medium, very few cells adsorbed to the

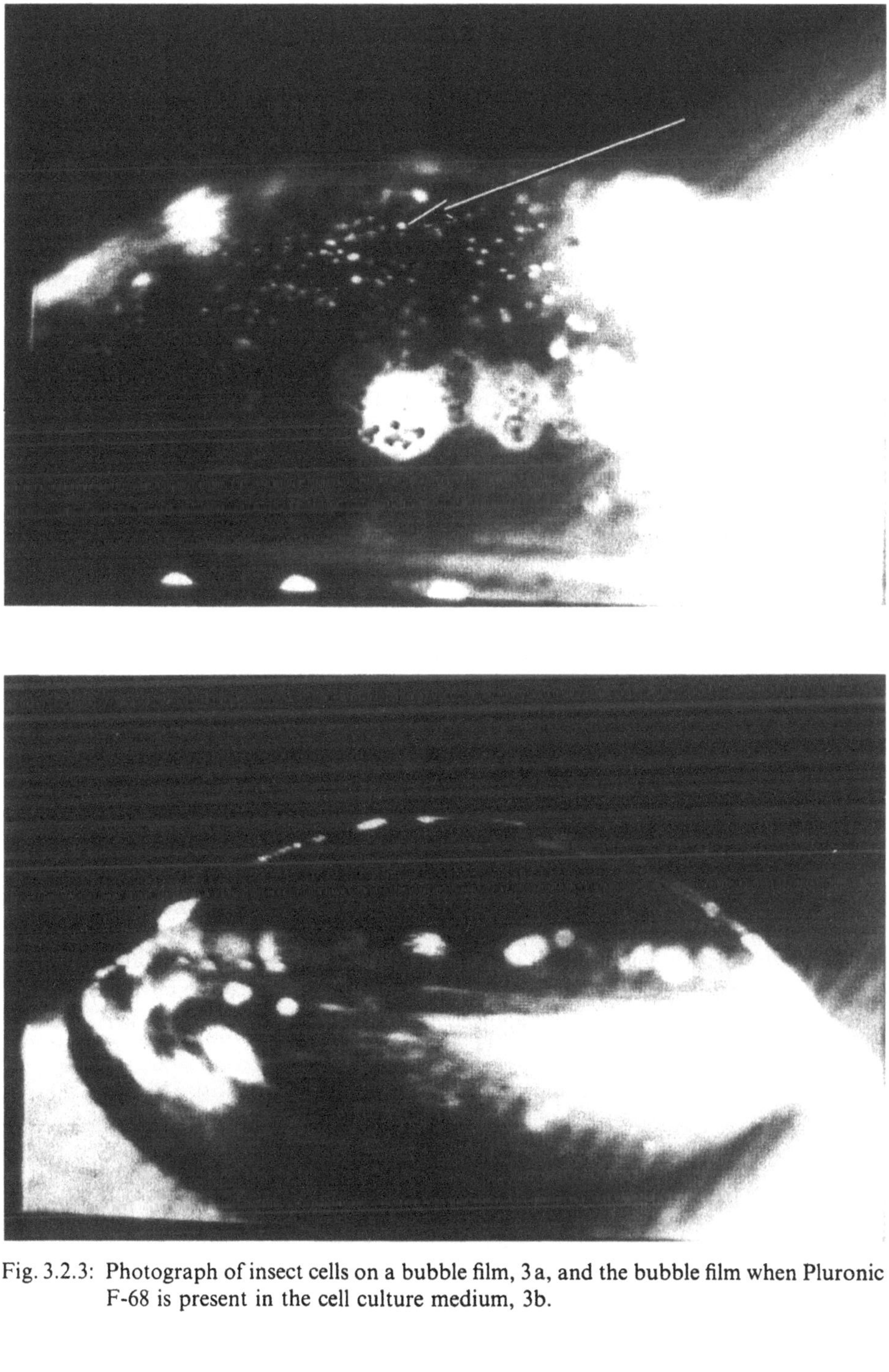

Fig. 3.2.3: Photograph of insect cells on a bubble film, 3a, and the bubble film when Pluronic F-68 is present in the cell culture medium, 3b.

bubble film and the collected sample from the upward jet contained very few cells. To quantify this observation, Trinh et al. (1994) developed a system in which a large number of single bubbles could be ruptured. In this system, four types of experiments were conducted: 1) the number of cells killed as a result of a single 3.5 mm bubble rupture, 2) the number and viability of cells in the upward jet that results when a bubble ruptures, 3) the number of cells on the bubble film, and 4) the fate of cells attached to the bubble film after rupture. The following observations were reported: 1) approximately 1050 cells are killed per single, 3.5 mm bubble rupture, 2) the concentration of cells in the upward jet was approximately 2 times higher than in the bulk (when Pluronic F-68 was not present), and 3) the average number of cells on the bubble film was approximately 292 and all of these cells were killed by the film rupture. It was also reported that Pluronic F-68 prevented this bubble rupture cell death (no statistically significant death was observed when Pluronic F-68 was present). It was suggested that this protection was the result of the Pluronic F-68 preventing cell adhesion to the bubble interface, thereby, preventing the cells from being transported into the upward and lower jet. These result lead Trinh et al., (1994) to suggest that the "hypothetical killing volume" introduced by Tramper et al., (1986; 1987; 1988) is in fact the medium and cells which make up the bubble film and a thin layer surrounding the bubble cavity.

The hydrodynamics of a bubble rupture is exceedingly complex. However, Boulton-Stone and Blake (1993) and Garcia-Briones et al., (1994) have reported on computer simulations of the rupturing of gas bubbles at an air-liquid interface. While these simulations cannot handle the film retraction steps, they accurately predict the remainder of the bubble rupturing process. The accuracy of these simulations were determined by comparing the position of the moving interfaces as a function of time to published photographic results and comparison of the calculated drop ejection velocities to experimentally determined velocities. These simulations predict that very high levels of shear stress are obtained during the bubble rupture process and that these levels of stress are a strong function of bubble size.

To compare the complex hydrodynamic forces in a bubble rupture to other types of systems in which cell disruption has been studied, both research groups calculated the energy dissipation rate when a bubble ruptures. Table 3.2.1 presents the rates of energy dissipation, as reported by Boulton-Stone and Blake and Garcia-Briones et al. as function of bubble size. Table 3.2.2 presents the rates of energy dissipation in which cell damage has been reported in well-defined flow systems. As can be

Table 3.2.1. Rates of Energy dissipation for the rupture of different size bubbles

Bubble Diameter (mm)	Total Elapsed time for bubble rupture (sec)	Maximum Energy dissipation rate ($J \cdot m^{-3} \cdot s^{-1}$) Garcia-Briones et al.	Bouton-Stone and Blake
0.77	5.5×10^{-4}	9.52×10^{7}	–
1.77	2.0×10^{-3}	1.66×10^{7}	4.0×10^{8}
6.32	1.0×10^{-2}	9.4×10^{4}	8.0×10^{3}

Table 3.2.2. Rate of energy dissipation in well-defined flow systems which kill cells

Cell Type	Instrument	Rate of cell damage (% min^{-1})	Rate of Energy Dissipation ($J \cdot m^{-3} \cdot s^{-1}$)	Reference
hybridoma	concentric cylinder	3.4	2.20×10^4	Schurch et al. (1988)
mammalian	capilary	16900	4.8×10^7	Augenstein et al. (1971)
insect	cone and plate	33.5	3.15×10^4	Goldblum et al. (1991)

observed, the rates of energy dissipation for the rupture of a small bubble (< 2.0 mm) are several orders of magnitude greater than what has been reported to kill cells.

In summary, all evidence to date indicates that the primary mechanism of suspended cell damage is the result of gas bubbles rupturing at the air-medium interface. It also appears that if the cells are not attached to the gas bubble interface through the action of protective additives, a majority of the cell damage is prevented. This indicates that the damaging forces are associated with a thin layer surrounding each bubble. Further discussion of these protective additives and their mechanism of protection is given in section 3.2.4.

3.2.3 The Hydrodynamics of Anchorage Dependent Cell Culture

3.2.3.1 History/Introduction

Since most animal cells grow in their native environment in the form of a tissue, it is not surprising that many animal cells need, for healthy growth, to be attached to a surface under *In-Vitro* conditions. Several theories exist to explain this requirement; however, a discussion of these concepts are beyond the scope of this work. There are also some cell lines which can grow either attached or in suspension, yet the preferred mode of growth is one attached to a surface. While these cells are not, strictly speaking, anchorage dependent, in the context of this section they will be assumed as ones that are.

Various methods have been developed for the growth of anchorage dependent cells (Murakami, 1989). The simplest, *In-Vitro* cell culture technique is the use of tissue culture flasks (T-flasks) coated with a protein which facilitates cell attachment to the surface. The most common material that these flasks are made of is hydrophobic polystyrene and one of the commonly used cell adhesion proteins is fibronectin. The next simplest method (or level of increased complexity) is the roller bottles in which cells attach to the inside surface of a cylindrical bottle which is slowly rotated on its major axis. However, for economical, large scale production of products from anchorage dependent cells, cultures which grow in a suspension type of environment (stirred tanks) are highly desirable since it provides a high cell-density to culture volume ratio.

To achieve the necessary surface area in a large, suspension type of vessel, it is necessary to attach cells to small, suspended particles. In general, two approaches have been used to attach cells to these small particles: 1) entrapment or attachment of cells within the particles or 2) attachment of cells onto the outer surface of the particles.

Regarding the first method, again two techniques have been used. The first technique consists of physically entrapping cells in a polysaccharide material such as agarose, while the second consists of cells attaching to a bead that has a sponge like texture (Nilsson, 1987). This "spongy" bead allows cells to grow both on and in the small passages within the particles. Verax corporation developed a continuous, fluidized bead system based on these beads and has marketed that system for large scale culture (Dean et al., 1987). Both of these methods provide physical protection of the cells since a majority of the cells are contained within the bead matrix (see also Chapter 3.1).

The second method, known as microcarrier cell culture (van Wezel, 1967), is commonly used and will be discussed for the remainder of this section. In this technique the mammalian cells are suspended in medium containing these microcarriers beads, the cells are allowed to attach to the beads, and then the cells grow and eventually cover the bead surface. The beads are nearly neutrally buoyant, are fabricated from polymer or glass material, and are available in a variety of compositions and sizes. A significant advantage of these beads is that standard bacterial fermentors can be used, or modified for use, as cell culture fermentors. However, since the cells are not protected, they experience significant, and in many cases lethal levels of hydrodynamic forces. Therefore, the remainder of this section will deal with these lethal forces. The non-lethal effects will be discussed in a later section.

3.2.3.2 Agitation Requirements

Mixing is essential for cell growth on microcarriers in a suspension bioreactor. This mixing is needed to distribute the nutrients throughout the culture and to keep the microcarriers suspended. The fundamental question is how to accomplish this mixing. Historically, impellers within a cylindrical vessel with a flat bottom, or a cylindrical vessel with a hemispherical bottom, have been used to grow bacterial and more recently animal and insect cells. This type of mixing apparatus is heavily used in the chemical industry (Nagata, 1966). In fact, the primary use of this type of vessel is not for biological production. Consequently, the characteristics that are essential for a well-mixed condition, on a molecular level, have been studied and estimates for agitation requirement have been made (Cherry and Papoutsakis, 1986). It has also been shown that the agitation needed to keep the microcarriers suspended are significantly lower than the requirements for a molecularly well-mixed environment (Cherry and Papoutsaki, 1986). Consequently, the challenge is to provide a molecularly well mixed system without damaging the cells on the microcarriers.

A significant amount of work has also been conducted on what is needed to obtain a homogeneous environment for microbial systems (Harrison et al., 1974; Hansford and Humphrey 1966; Cooney et al., 1981). The large number of large scale bacterial fermentations, and some of the unique fluid properties, such as high cell density and high viscosity, have driven this work. On the basis of this data it has been calculated that for a 1 liter culture vessel with a characteristic dimension of 0.1 m, the minimum liquid velocity needed is in the order of 0.05 m/s. This corresponds to a Reynolds number of 7410 which indicates a turbulent flow regime. Consequently, turbulent flow is desired for efficient operation.

3.2.3.3 Lethal Effects

While turbulent conditions are desired for a homogeneous culture, as will be reported below, sufficiently high levels of turbulence, and even laminar flow ($> 1.5-2.0\,N/m^2$), can be detrimental to cell growth. However, it has been reported that with at least some suspended animal cell cultures moderate turbulent conditions are not necessarily inhibitory (Kunas and Papoutsakis, 1990). All of these observations indicate the complexity of characterizing and understanding the effect of hydrodynamic forces on cells. However, the observation is reasonably clear that cells attached to the outside of microcarriers are significantly more sensitive to hydrodynamic forces than suspended cells. A review of the studies that have been conducted relating hydrodynamics forces and microcarriers as well as several of the proposed damage mechanisms will be presented below.

3.2.3.3.1 Correlational Studies

One of the first studies conducted to correlate the effect of agitation on cell growth was carried out by Sinskey et al. (1981) working with Chick embryo fibroblasts attached to Sephadex G beads (50–80 microns). In an attempt to characterize the damage observed, they developed a factor called the "Integrated Shear Factor" (ISF), which was calculated by dividing the impeller tip speed with the distance between the impeller tip and the vessel wall. They reported that the maximum cell density showed a steep drop when the ISF exceeded $60\,s^{-1}$.

Later Hu et al., (1985) attempted to correlate the effect of varying the bead concentration and agitation rates on growth rates of human fibroblasts, FS–4 cells. It was pointed out in this study that impeller tip speed was not a very good correlation factor; however, increasing the rpm beyond a corresponding ISF of $20-25\,s^{-1}$ resulted in a drop in growth rate.

Taking the correlational approach further, Croughan et al. (1987) and Cherry and Papoutsakis 1986; 1988) proposed a model which relates cell damage to the Kolmogoroff eddy length. Croughan et al. (1987) studied FS-4 cells on Cytodex-1 beads while Cherry and Papoutsakis (1986) studied BEK cells on Cytodex-3 beads.

3.2.3.3.2 Kolmogoroff Eddy Length Model

The Kolmogoroff eddy length scale (Kolmogoroff, 1941) is determined from a dimensional and order of magnitude analysis of experimentally measurable variables. This approach is based on the universal equilibrium or isotropic theory of turbulence which in turn is based on the statistical theory of turbulence (Hinze, 1975). According to this theory, the fundamental structure of the bulk turbulence is in the form of large fluid eddies which are very unstable and anisotropic in nature. These eddies generate smaller eddies and they in turn finer ones. The turbulent kinetic energy is passed on through this cascade of eddies until a characteristic size, called the Kolmogoroff-length scale, is reached. At this size the turbulent kinetic energy is released through viscous dissipation in the form of heat. A key assumption of this approach is that these small scale eddies are isotropic in nature and characterized as being independent of the bulk turbulence characteristics.

Kolmogoroff defined this energy dissipation eddy scale by the following set of equations:

$$\eta = \left(\frac{\upsilon^2}{\varepsilon}\right)^{1/4} \tag{1}$$

$$\nu = (\varepsilon \upsilon)^{1/4} \tag{2}$$

$$\varepsilon = \frac{P}{p_f V} \tag{3}$$

The power input can be calculated as follows:

$$P = N_p p_f n^3 d_i^5 \tag{4}$$

In the above equations P indicates the power consumed, V the dissipation volume, N_p the power number, v the Kolmogoroff velocity scale, η the Kolmogoroff length scale, ε the specific power dissipation, p_f the fluid density, u the kinematic viscosity, n the impeller rpm and d_i the impeller diameter.

Using the above correlations, Croughan et al. (1987, 1988) and Cherry and Papoutsakis (1986) observed that cell growth rate, for both FS–4 and BEK cells, was strongly dependent on this Kolmogoroff eddy length. When the eddy length was the size of the microcarrier bead or less, the growth rate decreased. To explain these observations, Cherry and Papoutsakis (1988) suggested that if this eddy length is greater than the bead diameter, the bead is carried by the fluid within the eddy and, correspondingly, the cells on the bead surface do not experience significant shear stress. However, once the eddy length is the size of the bead or less, significant shear stresses are exerted on the cells due to the viscous dissipation.

As can be observed in Equation 1, the Kolmogoroff eddy length is a function of kinematic viscosity. To determine the effect that this variable has on cell damage, Croughan et al. (1988), and Lakhotia and Papoutsakis (1992) varied the medium viscosity by adding Dextran and studied the corresponding change in cell damage. It was observed that both the cell death and cell removal from microcarriers were decreased for a given agitation condition as the medium viscosity was decreased.

Two other types of interactions, using the Kolmogoroff eddy length model, have been suggested to cause cell damage: bead-bead interactions and bead-wall interactions. The bead-bead collisions are postulated as being caused either due to interactions of independent small eddies propelling the particles towards each other or beads colliding with each other within a large eddy. A number of studies have been conducted in an attempt to correlate the above two mechanisms and a correlating factor, the turbulent collision severity term (TCS), has been developed. This correlation is based on the suggestions of Placek et al. (1985) who estimated the relative velocities of particles suspended in turbulent dispersions on the assumption that the turbulent scales are much larger than the particles. This has been incorporated into the eddy-length model by Cherry et al. (1988) due to the existence of a wide range of eddy sizes within an agitated reactor system.

3.2.3.3.3 Limitations of the Eddy-Length Scale Model

The statistical theory of turbulence which forms the basis of the Kolmogoroff model relies on the isotropy (not dependent on direction) of small scale turbulent eddies. This eliminates the effects of vessel and impeller geometry as long as the specific power dissipation is preserved. However, in a stirred tank just the opposite has been observed. Tatterson et al. (1985), using a high speed movie camera and a stereoscopic lens created three dimensional movies of the mixing in and around the impeller region. These movies clearly demonstrated the presence of vortices, high speed jets and recirculating flows. Some of these vortices form what are called "trailing vortex" which can extend significant distances from the impeller. On the basis of their observations they state that "the main energy transfer from the blade to the fluid is the vortex system which is supported by the high-speed flow contribution".

These large scale anisotropies arising out of vortices and recirculating flows have been studied by several researchers (Paleck and Tavlarides, 1985; Paleck et al., 1986; Gunkel and Weber, 1975). These studies indicate that the flow is much more complex than the simple approximation of isotropic turbulence used in the Kolmogoroff eddy length model. They also raise the question as to what volume, on which the Kolmogoroff eddy model is based, is the energy dissipated. Thus the empirical nature of the eddy length model makes a true estimate of the Kolmogoroff eddy, if it in fact exists, a very difficult task.

Oh et al. (1989) also discussed these issues and demonstrated that the Reynolds stresses resulting from the trailing vortices behind a Rushton Turbine are much higher than those calculated from a Kolmogoroff type of approach. They also pointed out that the work of Yianneskis et al. (1987), using Laser Doppler Anemometry, showed that the Reynolds shear stresses in regions immediately following the impeller are extremely position and direction/plane dependent further showing that the flow is very anisotropic.

All of these results indicate that while we are beginning to understand the flow within bioreactors, and how that flow interacts with cells, we are a long way from having a complete, fundamental understanding. This is especially apparent when researchers attempt to scale up a vessel beyond a few liters.

Most of the past research on agitation in stirred vessels, not just in animal cell culture, but in the area of mixing, has recommended the use of global parameters as scale-up criteria. Global parameters are popular because they are reasonably easy to determine and apply/work adequately in most circumstances. These global parameters do not account for "local" non-homogeneity of flow in mixing vessels. Non-homogeneity in flow is the result of structures produced by impeller and other internals within the vessels. Associated with these structures are high levels of hydrodynamic energy. These flow structures are a complex function of geometry and operating conditions of the vessels and cannot be described by global scale-up parameters. For proper scale-up of systems that are sensitive to energy levels, such as microcarriers cultures, these "local" structures need to be determined and the local environments conserved.

To begin to address this need for an understand and quantification of the "local" flow in various size bioreactors, Venkat et al. (1993, 1995, 1996) has used 3-D PTV (particle tracking velocimetry), a new methodology, to characterize the turbulence mixing conditions in several bioreactors. Applying this technique to the impeller region of several typical bioreactors, full 3-D velocity, local energy distributions and estimates of local, instantaneous shear rates were obtained. They also suggest that these "local" parameters can be used for scale-up criteria. Further work needs to be conducted to correlate vessel operating conditions and the energy distribution in the vessels. Future work in this area also needs to involve developing first, a fundamental, mechanistic understanding of the hydrodynamic flow around impellers and relating these flows to cell damage and death. With this as a basis, rational operating parameters can be developed for the operation of existing bioreactors.

3.2.3.4 Effects of Simultaneous Agitation and Sparging of Microcarrier Cultures

There have been very few studies conducted to determine the effects of sparging on large scale microcarrier cultures. Aunins et al. (1986) investigated the effect of sparging on FS-4 microcarrier cultures in spinner flask. This system was studied under low agitation conditions and the mechanism of cell death was not ascertained during the experiment. However, a clear increase in the death rate was attributed to sparging.

3.2.4 Protective Additives

3.2.4.1 Introduction

As was discussed previously, (section 3.2.2), a majority of the cell damage in sparged, suspension cell culture is associated with cell-bubble interactions (Kunas and Papoutsakis, 1990b; Handa et al., 1989; Tramper et al, 1986; Silva et al., 1987: Murhammer and Goochee, 1990a; Murhammer and Goochee, 1990b). It was also suggested by Chalmers and Bavarian (1991) that this damage is related to cells adhering to the medium-bubble interface. This suggestion was given further credibility when it was shown by Garcia-Briones and Chalmers (1992) that the presence of Pluronic F-68 in the medium prevents cell adhesion and this leads to a decrease in cell damage.

While Pluronic F-68 has recently received a large amount of interest as a protective additive, over the last thirty years a large number of medium additives have been reported to protect cells from damage associated with sparging: methylcellulose, serum, carboxymethylcellulose (CMC), tryptose phosphate (TPB), Primatone RL (a peptic digest of animal tissue), hydroxyethyl starch (HES), polyvinyl pyrrolidone, bovine serum albumin (BSA), and Polyethylene Glycol (PEG) to name a few. Table 3.2.3 attempts to summarize all of these additives by listing the type and concentration of protective agents added, cell type used, and method of cultivation.

3.2.4.2 Hypotheses for the Mechanism of Protection

Different mechanisms have been proposed for the protective action of the polymeric substances added to the growth medium. These mechanisms can be classified under three categories: 1) additives that change the interfacial phenomena governing cell adhesion (see discussion below), 2) additives that interact with the cell membrane, thus providing a protective interaction and/or layer, and 3) additives that have a nutritional effect, thus strengthening the cells.

A protection resulting from nutritional effect is not significant as it is highly unlikely that cells would be able to metabolize polymeric substances such as Pluronic F-68 and methylcellulose. In 1984, Mizrahi reported that 99% of several polymers added to the growth medium were present after the course of batch growth experiments. In 1991, Michaels et al. (1991) showed that the protective effect of fetal bovine serum (FBS), Pluronic F-68 and Polyethylene Glycol (PEG) tested with suspended animal cells (CRL 8018 Hybridoma cells) was almost immediate, thereby, ruling out a metabolic contribution.

Interactions between the added polymer and the cell membrane are more probable. Several authors have reported that several polymers such as: dextran, Pluronic F-68, and methylcellulose, adsorb onto cell membranes (Brooks and Seaman, 1973a; Brooks and Seaman, 1973b; Pfafferott et al., 1982, Pittz et al., 1977). In regard to providing a protective effect, Cudd et al. (1989) demonstrated that low pH hemolysis of erythrocytes can be inhibited by dextrans greater than 40,000. Electron microscopy was used to show that dextrans of molecular weight greater than 150,000 form a tight association with the membrane.

To determine if these additives strengthen cells, relative to one type of hydrodynamic force, Goldblum et al. (1990) subjected SF-9 and TN-368 insect cells to laminar shear stress. These two cell lines were grown in medium containing various types of protective additives. They observed that Pluronic F-68 strengthen cells, in a concentration dependent manner, by a factor ranging from 15 to 42 for a concentration of 0.2 to 0.5 percent, respectively. It was also observed that the greatest protection, or strengthening (by a factor of 76), was obtained with E4M Methocel.

Finally, Murhammer and Goochee (1990a) studied the structural features of Pluronic F-68 in an attempt to further understand it's protective effect. Pluronic F-68 is a member of a family of block copolymers, called Polyols, of hydrophobic/hydrophilic compounds produced by BASF Corporation (Parsippany, NJ). This family of Pluronics has a wide range of properties which are controlled by the Polyols structure and molecular weight. To determine the structural property responsible for cell protection, 19 different Polyols were studied. They found that the most hydrophobic compounds caused cell lysis indicating cell membrane interactions, while purely hydrophilic compounds caused no cell lysis and provided no protective effect. Only intermediate compounds, such as Pluronic F-68, with both hydrophilic and hydrophobic characteristics, provided cell protection.

Table 3.2.3.

Protective agent used	Cell type used	Optimum conc.*	Methods of cultivation	Reference
Pluronic F-68	L cells NCTC 929	0.1 %	Reciprocal agitation	Runyan and Geyer, 1963
	Mouse LS	0.02 %	Sparged	Kilburn and Webb, 1968
	Human Lypho-blasoid	0.1 %	Agitated, sparged	Mizrahi, 1984
	Hybridoma Ns1 myeloma	0.1 %	Bubble column	Handa et al., 1987
	SF-9	0.2 %	Sparged, agitated, airlift	Murhammer and Goochee, 1990
		0.1 %	Airlift	Maiorella et al., 1988
	Solanum dulcamara (plant cell)	0.1 – 1.0 %	Shaker flask	King et al., 1990
Methyl-cellulose	L cells NCTC 929	0.05 %	Reciprocal agitation	Runyan and Geyer, 1963
Carboxy-methyl-cellulose	Hamster and Kidney	2.4 %	Sparged	Tellings and Elsworth, 1965
	Human Lyphoblasoid	0.1 %	Agitated, sparged	Mizrahi, 1984
	Dunaliella	0.1 %	Roux bottle miniloop	Silva et al., 1987
TPB	Kidney cells	6.0 %	Agitated	Tellings and Elsworth, 1965
Hydroxy ethyl starch	Human Lyphoblasoid	0.2 %	Agitated, sparged	Mizrahi, 1984
BSA	Hybridoma		Bubble column	Heulscher and Onken, 1988
Methocel 65 HG	TN-368	0.3 %	Sparged, agitated	Hink and Strauss, 1979
Serum	Hybridoma	> 5.0 %	Agitated, sparged	Kunas and Papoutsakis, 1990

* See individual reference for units of the concentrations of particular compound.

These observations emphasize that not only do these additives interact with the cell membrane, but that interfacial phenomena are involved. This was indirectly indicated by the work of Garcia-Briones and Chaimers (1992) cited previously. When particular surface active compounds are present, cell adhesion to the gas-liquid interface as well as cell damage is prevented. It should be noted that most, if not all, of the additives listed in Table 3.2.3 exhibit surface active properties in growth medium.

3.2.4.3 The Thermodynamics of Cell Adhesion

Except for the observations of Chalmers and Bavarian (1991), there are no other reported causes of mammalian and insect cells attaching to gas-liquid interfaces. However, cell-cell adhesion has been observed for many systems. In regard to insect cells, Hink and Strauss (1979) observed a large amount of aggregation of TN-368 cells in spinner flask cultures. They also observed that the addition of 0.3% Dow 65 HG Methocel, a surface active additive, greatly reduced this adhesion. In 1988, Chalmers et al. (1988) reported that under well-defined shear stress, 0.5% 65 HG Methocel reduced cell-cell adhesion and 0.5% Pluronic F-68 completely prevented this adhesion. In both cases, the percentage of viable cells increased as the degree of aggregation decreased. Both of these reports provide indirect evidence that these additives interact and change the physico-chemical properties of the cell membrane.

The attachment of bacterial cells to interfaces, including gas-liquid interfaces, has been observed and studied. Several mechanisms have been suggested including electrostatic charge, and cell interactions with lipids on the air-medium interface (Kjellebergang and Stenstrom, 1980). However, the most probable mechanism for this adhesion is the result of hydrophobic-hydrophilic interactions (Absolom, 1988; Dahlback et al., 1981; Hermansson et al., 1982; Malmqvist et al., 1984; Absolom et al., 1983).

The observation that surface active additives provided protection from bubble ruptures lead Chattopadhyay et al. (1995b) to present a thermodynamic relationship to explain this protection. In Figure 3.2.4, the process of bringing a suspended cell, with a characteristic surface tension, γ_{cl} and a gas bubble, also with a characteristic surface tension, γ_{lv}, together such that a new interface between the cell and gas within the bubble, γ_{cv}, forms is presented. Mathematically, this process is represented by:

$$\Delta F^{adh} = \gamma_{final} - \gamma_{initial} = \gamma_{cv} - (\gamma_{lv} + \gamma_{cl}) \qquad (5)$$

where ΔF^{adh} is the change in free energy (thermodynamic feasibility) of the process creating the new interface (adhesion of the cell to the gas-medium interface). A positive value indicates that the process is unfavorable, while a negative value indicates a favorable process. To determine the validity of this relationship, Chattopadhyay et al. (1995b) determined the value of γ_{cl} and γ_{cv} from experimental data and the literature for 14 cell lines. The values of γ_{cv} were consistently high, ranging from 56 to 69 erg/cm^2, while the values for γ_{cl} was consistently low, 0.01 to 2.7 erg/cm^2. Since the surface tension of cell culture medium, γ_{lv}, also ranges from 56 to 69 erg/cm^2, by inspection of Equation 5, one can observe that only when

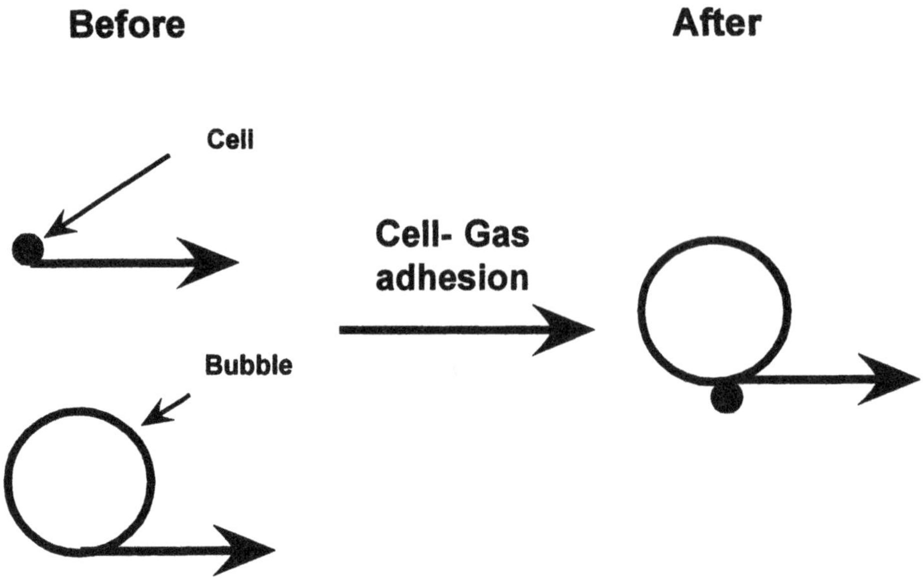

Fig. 3.2.4: The diagramm of the process of bringing a suspended cell in contact with a gas bubble.

γ_{lv} is less than γ_{cv} will the adhesion of cells to gas-liquid interfaces be thermodynamically unfavorable. This conclusions is consistent with the results of Chattopadhyay et al. (1995a) where it was reported that additives which rapidly and significantly lowered the medium interfacial also inhibited the adhesion of cells to gas-medium interfaces.

In unrelated, but similar studies, Michaels et al. (1995a; 1995b) investigated rheological and interfacial properties of a number of protective additives. Their studies focused on the kinetics of cell adhesion to gas-liquid interfaces and the affect these additives had on the interaction of cells and foam formed at the top of a bubble column/bioreactor.

Michaels et al. (1995a) determined the relationship between cell-bubble contact time, the number of cells which attached, and the presence of various media additives. It was observed that, like the studies of Chattopadhyay et al. (1995a, b), the additives which rapidly and significantly lowered the surface tension also had long induction times (low cell-bubble adhesion). However, unlike Chattopadhyay et al. (1995a, b), they observed that polyvinyl alcohol (PVA) also had a long induction time. With respect to cell entrainment in the foam layer above a bioreactor, Michaels et al. (1995a) determined a separation factor (the ratio of cell concentration in removed foam to that in the bubble column below the foam), as a function of media additive. Using this separation factor, various additives were rated with respect to their ability to prevent cells from being trapped in the foam layer. From least effective too must effective the additives are; polyvinyl pyrrolidone, polyethylene glycol (PEG), serum free medium with no additives, medium with 3% serum, Pluronic F-68, and Methocel A15LV.

Finally, Michaels et al. (1995 b) also investigated the interfacial properties of these additives. Like Chattopadhyay et al. (1995 a), the additives which lowered the dynamic surface tension the most (Methocel A15-LV, Pluronic F-68, and PVA) also reduced cell-to-bubble attachment the most.

3.2.5 Non-Lethal Effects of Hydrodynamic Forces on Cells

Previous discussions in this chapter have concentrated on cell lysis or loss of cell viability as a consequence of hydrodynamics in bioreactors; however, there has been wide speculation about the effect of non-lethal hydrodynamic forces that exist under a "normal" bioreactor operating conditions. Response of vascular endothelial cells to fluid hydrodynamic shear has been very well documented (Malek and Izumo, 1994); however studies on the effects of shear stress on other mammalian cell lines, especially the ones used for production of vaccines and protein therapeutics, recently have not very clear until recently.

Some of the previous studies (Al-Rubeai et al., 1993, Lakhotia et al., 1993) indicate that there are effects of sub-lytic shear, especially changes in DNA synthesis rate, while others (Passini and Goochee, 1989) indicate that no effects could be observed. A majority of the confusion/contradiction arises from the complex flow in the systems studied and the difficulty in characterizing and quantifying the effects. However, recent work (Ranjan et al., 1996, Mufti and Shuler, 1995, Mufti et al., 1995) indicates that there is a significant effect of sub-lytic hydrodynamic shear on the physiological responses of the cell lines studied. Ranjan et al. (1996) subjected four cell lines: two cell lines from endothelial tissue (primary human umbilical vein endothelial cells, HUVECs, and bovine aortic endothelial cells, BAECs), and two cell lines from non-endothelial tissue (HeLa cells that originated from a malignant tumor of the human cervix, and Chinese Hamster Ovary cells, CHO). All cell lines responded to moderate levels of laminar shear stress (25 dyne/cm^2) by expressing c-fos protein: a product of the c-fos gene, a protooncogene, which is a member of a family of transcriptional cofactors that mediate transcriptional stimulation. Mufti and Shuler (1995) observed the transient induction of a cytochrome P450 monoxygenase (CYP1A1) activity in human (HepG2) hepatoma cells attached to microcarriers in response to moderate levels of agitation in 50 ml spinner cultures. This response requires that the cell contain the cytosolic Ah receptor. CYP1A1 has been shown to be involved in the oxidation of arachidonic acid and has been implicated in intracellular responses including cell proliferation and c-fos expression. While at this stage it is not possible to know what are the implications of the above results to the expression of therapeutic proteins or vaccines, it is obvious that sub-lytic shear stress can have a significant effect on host cell physiology.

3.2.6 Conclusion

The question of hydrodynamic damage to cells in bioreactors has received a large amount of interest in the last several years and rapid progress in the understanding of mechanisms of damage has been obtained. It is not unreasonable to expect that in the next five to ten years the hydrodynamics within bioreactors will be well-defined and principles, based on fundamental understandings, will be available for both design and operation of bioreactors.

Acknowledgments

The authors would like thank Professor Robert Brodkey of the Department of Chemical Engineering at Ohio State and Professor Fred Hink of the Department of Entomology at Ohio State for their assistance and comments. The authors would also like to acknowledge the financial support of the National Science Foundation, Grant number BCS-9109151.

Nomenclature

V	Dissipated volume
P	Power consumed
v	Kolmogoroff velocity scale
N_p	Power number
p_f	Fluid density
n	Impeller speed (rpm)
u	Kinematic viscosity
η	Kolmogoroff length scale
d_i	Impeller diameter
ε	Specific power dissipation
V	Interfacial tension
dA	Change in area
ΔF^{adh}	Change in free energy of adhesion
γ_{cv}	Cell-vapor interfacial tension
γ_{cl}	Cell-liquid interfacial tension
γ_{lv}	Liquid-vapor interfacial tension
θ	Contact angle

References

Absolom, D. R. (1988) The role of bacterial hydrophobicity in infection: Bacterial adhesion and phagocytic ingestion. Can. Jou. Microbiol. *34*, 287–298.

Al-Rubeai, M., Emery, A. N., Chalder, S. and Goldman, M. H. (1993) A flow cytometric study of hydrodynamic and other stresses. J. Biotechnol. *31*, 161–177.

Aunins, J. G., Croughan, M. S., Wang, D. I. C. and Goldstein, J. M. (1986). Engineering developments in homogeneous culture of animal cells: Oxygenation of reactors and scaleup. Biotechnol. Bioeng. Symp. Ser. *17*, 699–723.

Aunins, J. G., Woodson, Jr., B. A., Hale, T. K., Wang, D. I. C. (1989). Effects of paddle impeller geometry on power input and mass transfer in small scale animal cell culture vessels. Biotechnol. Bioeng. *34*, 1127–1132.

Bavarian, F., Fan, L. S. and Chalmers, J. J. (1991). Microscopic visualization of insect cell-bubble interactions. I: Rising bubbles, Air-medium Interface, and the Foam layer Biotechnol. Progr. 7, 140–150.

Blanchard, D. C. and Syzdek, L. D. (1970). Mechanism for the water-to-air-transfer and concentration of bacteria. Science *169*, 626–628.

Blanchard, D. C. and Syzdek, L. D. (1972). Concentration of bacteria in jet drops from bursting bubbles. J. Geophys. Res. *77*, 5087–5099.

Boulton-Stone, J. M. and Blake, J. R. (1993) Gas-Bubbles bursting at a free surface. J. Fluid Mech. *154*, 437–466.

Brooks, D. E. and Seaman, G. V. F. (1973). The Effect of Neutral Polymers on the Electrokinetic Potential of Cells and Other Charged Particles: I. Models for the Zeta Potential increase. J. Colloid Interface Sci. 173a, *43*, 670–686.

Brooks D. E. and Seaman, G. V. F. (1973). The Effect of Neutral Polymers on the Electrokinetic Potential of Cells and Other Charged Particles: IV. Electrostatic Effects in Dextran-Mediated Cellular Interactions. J. Colloid Interface Sci. 173 d, *43*, 714–726.

Bryant, J. C., Schilling, E. L., Farle, W. R. and Evans, V. J. (1960). Massive fluidsuspension cultures of certain mammalian tissue cells. III. Exploratory supplementation experiments with chemically defined medium NCTC 109. J. Nat. Cancer Inst. *24*, 859–871.

Chalmers, J. J. and Bavarian, F. (1991). Microscopicvisualization of insect cell-bubble interactions. II: The bubble film and bubble rupture. Biotechnol. Prog. *7*, 151–158.

Chalmers, J. J., Bae, Y. K. and Brodkey, R. S. (1988). Microscopic Observations of Cells Subjected to Various Levels of Shear Stress, AIChE Annual Meeting, Washington D. C.

Chattopadhyay, D., Rathman & Chalmers, J. J. (1995). A Thermodynamic Explanation for Cell Adhesion to Gas-Liquid Interfaces, Biotechnol. Bioengin. *48*, 649–658.

Chattopadhyay, D., Rathman and Chalmers, J. J. (1995). The Protective Effect of Specific Medium Additives with Respect to Bubble Rupture, Biotechnol. Bioengin. *45* 473–480.

Cherry, R. S. and Papoutsakis, E. T. (1986). Hydrodynamic effects on cells in agitated tissue culture reactors. Bioproc. Eng. *1*, 29–41.

Cherry, R. S. and Papoutsakis, E. T. (1988), Physical mechanisms of cell damage in microcarrier bioreactors, Biotechnol. Bioeng. *32*, 1001–1014.

Cooney, C. L., Koplov, H. M. and Haggstrom, M. (1981). In: Continuous Culture of Cells (Calcott, P. H. Ed.) CRC Press, Boca Raton.

Croughan, M. S., Hamel, J. F. and Wang, D. I. C. (1988). Effects of microcarrier concentration in animal cell culture. Biotechnol. Bioeng. *32*, 975–982.

Croughan, M. S., Hamel, J. F. and Wang, D. I. C. (1987). Hydrodynamic effects in animal cells grown in microcarrier cultures. Biotechnol. Bioeng. *29*, 130–141.

Cudd, A., Arvinte, T., Schulz, B. and Nicolau, C. (1989). Dextran Protection of Erythrocytes from low-pH-induced hemolysis. FEBS Letters. *250*, 293–296.

Goldblum, S., Bae, Y.K., Hink, W.F. and Chalmers, J.J. (1990). Protective effect of methylcellulose and other polymers on insect cells subjected to laminar shear stress. Biotechnol. Progr. *6*, 383–390.

Dahlback, B., Hermansson, M., Kjelleberg, S. and Norkrans, B. (1981). The hydrophobicity of bacteria – An important factor in their initial adhesion at the air-water interface. Arch. Microbiol. *128*, 267–270.

Dean, Jr., R.C., Karkare, S.B., Phillips, P.G., Ray, N.G. and Runstadler, Jr., P.W. (1987). In: Large Scale Cell Culture Technology (Lydersen, B.K., ed.) pp. 163–192, Hanser Publishers, Munich.

Garcia-Briones, M. and Chalmers, J.J. (1992). Cell bubble interactions: Mechanisms of suspended cell damage. Ann. N.Y. Acad. Sci. *655*, 219–229.

Garcia-Briones, M.A., Brodkey, R.S. and Chalmers, J.J. (1994) Computer Simulations of the rupture of a gas bubble at a gas-liquid interface and its implications in animal cell damage. Chem. Eng. Sci. *49*, 2301–2320.

Gardner, A.R., Gainer, J.L. and Kirwan, D.J. (1990). Effects of stirring and sparging on cultured Hybridoma cells. Biotech. Bioeng. *35*, 940–947.

Goldblum, S., Bae, Y., Hink, W.F. and Chalmers, J.J. (1991) Protective effect of methylcellulose and other polymers on insect cells subjected to laminar shear stress. Biotechnol. Prog. *6*, 383–390.

Gunkel, A.A. and Weber, M.E. (1975). Flow phenomena in stirred tanks. AIChE Journal. *21*, 931–949.

Hahn, P., Chen, J. and Slattery, J.C. (1985). Effects of London-van der Waals forces on the thinning and rupture of a dimpled liquid film as small drop or bubble approaches a fluid-fluid interface. AIChE J. *31*, 2026–2038.

Handa, A., Emery, A.N. and Spier, R.E. (1987). On the evaluation of gas-liquid interfacial effects on hybridoma viability in bubble column bioreactors. Develop. Biol. Standard. *66*, 241–253.

Handa-Corrigan, A., Emery, A.N. and Spier, R.E. (1989). Effect of gas-liquid interfaces on the growth of suspended mammalian cells: mechanism of cell damage by bubbles. Enz. Microb. Technol. *11*, 230–235.

Hansford, G.S. and Humphrey, A.E. (1966). Effect of equipment scale and degree of mixing on continuous fermentation yield at low dilution rates. Biotechnol. Bioeng. *8*, 85–96.

Harrison, D.E.F. and Topiwala, H.H. (1974). Transient and oscillatory states of continuous culture. Adv. Biochem. Eng. *3*, 167–220.

Hermansson, M., Kjelleberg, S., Korhonen, T.K. and Stenstrom, T.A. (1982). Hydrophobic and electrostatic characterization of surface structures of bacteria and its relationship to adhesion to an air-water interface. Arch. Microbiol. *131*, 308–312.

Heulscher, M. and Onken, U. (1988). Influence of Bovine Serum Albumin on the growth of Hybridoma cells in air-lift loop reactors using serum free medium. Biotechnol. Lett. *10*, 689–694.

Hink, W.F. and Strauss, E.M. (1979). A serum-free medium for the culture of insect cells and production of recombinant proteins. J. Tissue Culture Methods. *5*, 1023–1025.

Hinze, J.O. (1975). Turbulence. McGraw-Hill, New York.

Hu, W.S., Meier, J. and Wang, D.I.C. (1985). A mechanistic analysis of the inoculum requirement for the cultivation of mammalian cells on microcarriers. Biotechnol. Bioeng. *27*, 585–595.

Jobses, I., Martens, D. & Tramper, J. (1991). Lethal events during gas sparging in animal cell culture. Biotech. Bioeng. *37*, 484–490.

Kilburn, D.G. & Webb, F.C. (1968). The cultivation of animal cells at controlled dissolved oxygen partial pressure. Biotechnol. Bioeng. *10*, 801–814.

King, A.T., Davey, M.R., Mulligan, B.J., Lowe, K.C. (1990). Effects of Pluronic P-68 on plant cells in suspension culture. Biotechnol. Letters *1*, 29–32.

Kjelleberg, S., Stenstrom, T.A. (1980). Lipid surface films: interaction of bacteria with free fatty acids and phospholipids at the air/water interface. J. General Microbiology *116*, 417–423.

Kolmogoroff, D.N. (1941). C.R. (Doklay) Acad. Sci. U.S.S.R., N.S. *30*, 301–305.

Kunas, K.T. and Papoutsakis E.T. (1990b). Damage mechanisms of suspended animal cells in agitated bioreactors with and without bubble entrainment. Biotechnol. Bioeng. *36*, 476–483.

Kunas, K.T. and Papoutsakis, E.T. (1990a). The protective effect of serum against hydrodynamic damage of hybridoma cells in agiatated and surface-aerated bioreactors. J. Biotech. *15*, 57–70.

Lakhotia, S., Bauer, K.D. and Papoutsakis, E.T. (1992) Damaging agaitiation intensities increase DNA synthesis rate and alter cell-cycle phase distributions of CHO cells. Biotechnol. Bioeng. *40*, 978–990.

Lakhotia, S., Papoutsakis, E.T. (1992). Agitation induced cell injury in microcarrier cultures. Protective effect of viscosity is agitation intensity dependent: Experiments and modeling. Biotechnol. Bioeng. *39*, 95–107.

MacIntyre, F. (1968). Bubbles: A boundary-layer „microtome" for micron-thick samples of a liquid surface. J. Phys. Chem. *72*, 589–592.

MacIntyre, F. (1972). Flow patterns in breaking bubbles. J. Geophys. Res. *77*, 5211–5228.

Maiorella, B., Inlow, D., Shauger, A., Harano, D. (1988). Large Scale Insect Cell-Culture for Recombinant Protein Production. Bio/technol. *6*, 1406–1410.

Malek, A.M. and Izumo, S. (1994) Molecular aspects of signal transduction of shear stress in the endothelial cell. J. Hypertension *12*, 989–999.

Malmquist, T., Thelestam, T.A., Mollby, R. (1984). Hydrophobicity of cultured mammalian cells and some effects of bacterial phospholopases C. Acta Path. Microbiol. Immunol. Scand. Sect. B *94*, 127–133.

Medrow, R.A. (1968). Floating bubble configurations and charges of jet drops produced by busting bubbles. Ph.D. Thesis, University of Illinois, Urbana, Illinois.

Michaels, J.D., Peterson, J.F., McIntire, L.V., Papoutsakis, E.T. (1991). Protection mechanisms of freely suspended animal cells (CRL 8018) from fluid-mechanical injury. Viscometric and bioreactor studies using serum, Pluronic F-68 and polyethylene glycol. Biotech. and Bioeng. *38*, 169–180.

Michaels, J.D., Nowak, J.E., Malik, A.K., Koczo, K., Wason, D.T. and Papoutsakis, E.T. (1995a) Analysis of Cell-to-Bubble Attachment in Sparged Bioreactors in the Presence of Cell-Protecting Additives. Biotechnol. and Bioeng. *47*, 407–419.

Michaels, J.D., Nowak, J.E., Malik, A.K., Koczo, K., Wasan, D.T. and Papoutsakis, E.T. (1995b) Interfacial Properties of Cell Culture Media with Cell-Protecting Additives. Biotechnol. and Bioeng. *47*, 420–430.

Mizrahi, A. (1984). Oxygen in human lymphoblastoid cell line cultures and effect of polymers in agitated and aerated cultures. Develop. Biol. Standard *55*, 93–102.

Mufti, N.A., Bleckwenn, N.A., Babish, J.G. and Shuler, M.L. (1995). Possible Involvement of the AH Receptor in the Induction of Cytochrome P-450IAI Under Conditions of Hydrodynamic Shear in Microcarrier-Attached Hepatome Cell Lines. Biochem. and Biophys. Res. Comm. *208*, 144–152.

Mufti, R.A. and Shuler, M.L. (1995). Induction of Cytochromoe P-450IAI Activity in Response to Sub-Lethal Stresses in Microcarrier-Attached HepG2 cells. Biotechnol. Progr. *11*, 659–663.

Murakami, H. (1989). Trends in Animal Cell Culture Technology. Proceedings of the Second Annual Meeting of the Japanese Association for Animal Cell Technology, VCH Publishers, Kodansha LTD, Tokyo.

Murhammer, D.W. and Goochee, C.F. (1988). Scale-up of insect cell culture: Protective effects of Pluronic F-68. Bio/Tech. *6*, 1411–1418.
Murhammer, D.W., Goochee, C.F. (1990a). Structural features of nonionic polyglycol polymer molecules responsible for the protective effect in sparged animal cell bioreactors. Biotechnol. Prog. *6*, 142–148.
Murhammer, D.W. and Goochee, C.F. (1990b). Sparged animal cell bioreactors: Mechanisms of cell damage and Pluronic F-68 Protection. Biotechnol. Progr. *6*, 391–397.
Nagata, S. (1996). Mixing – principles and applications. pp. 138–164. First edition, Halsted Press, New York.
Nilsson, K. (1987). In: Large Scale Cell Culture Technology (Lydersen, B.K., ed.) pp. 95–112, Hanser Publishers, Munich.
Oh, S.K.W., Nienow, A.W., Al-Rubeai,M. and Emery, A.N. (1992). Further studies of the culture of mouse hybridomas in an agitated bioreactor with and without continuous sparging. J. Biotech. *22*, 245–270.
Oh, S.K.W., Nienow, A.W., Al-Rubeai, M. and Emery, A.N. (1989). The effects of agitation intensity with and without continuous sparging on the growth and antibody production of hybridoma cells. J. Biotech. *12*, 45–62.
Passini, C.A. and Goochee, C.F. (1989). Response of a mouse hybridoma cell line to heat shock, agitation, and sparging. Biotechnol. Prog. *5*, 175–188.
Pfafferott, C., Wenby, R., Meiselman, H.J. (1982). Morphologic and Internal Viscosity Aspects of RBC Rheologic Behavior. Blood Cells. *8*, 65–78.
Pittz, E.P., Jones, R., Golberg, L., Coulston, F. (1977). Interactions of Polysaccharides with Plasma Membranes: I. Interactions of the Human Erythrocytes with Degraded Iota Carrageena and the Effect of Dextran and DEAE Dextran. Biorheology. *14*, 21–32.
Placek, J., Tavlarides, L.L., Smith, G.W., Fort, I. (1986). Turbulent flow in stirred tanks, Part II: A two scale model of turbulence. AlChE Journal. *32*, 1771–1786.
Placek, J., Tavlarides, L.L. (1985). Turbulent flow in stirred tanks, Part I: Turbulent flow in the turbine impeller region. AlChE Journal *31*, 1113–1120.
Quinn, J.A., Steinbrook, R.A. and Anderson, J.L. (1975). Breaking bubbles and the water-to-air transport of particulate matter. Chem. Eng. Sci. *30*, 1177–1184.
Ranjan, V., Waterbury, R., Xiao, Z. and Diamond, S.L. (1996) Fluid Shear Stress Induction of the Transcriptional Activator c-fos in Human and Bovine Endothelial Cells, HeLa, and Chinese Hamster Ovary Cells. Biotechnol. Bioeng. In Press.
Runyan, W.S. and Geyer, R.P. (1963). Growth of L cell suspensions on a warburg apparatus. Proc. Soc. Exp. Biol. Med. *112*, 1027–1030.
Silva, H.J., Cortinas, T., Ertola, R.J. (1987). Effect of hydrodynamic stress on Dunaliella growth. J. Chem. Tech. Biotechnol. *40*, 41–49.
Sinskey, A.J., Fleischaker, R.J., Tyo, M.A., Giard, D.J., Wang, D.I.C. (1981). Production of cell derived products: virus and interferon. Ann. N.Y. Acad. Sci. *369*, 47–59.
Tatterson, G.B., Yuan, H.S., Brodkey, R.S. (1980). Stereoscopic visualization of the flows for pitched blade turbines. Chem. Eng. Sci *35*, 1369–1375.
Telling, R.C. and Elswoth, R. (1965) Submerged culture of hamster kidney cells in a stainless steel vessel. Biotechnol. Bioeng. *7*, 417–434.
Tramper, J., Williams, J.B., Joustra, D. and Vlak, J.M. (1986). Shear sensitivity of insect cells in suspension. Enzyme Microb. Technol. *8*, 33–36.
Tramper, J., Smit, D. and Straatman, J. (1988). Bubble-column designed for growth of fragile insect cells. Biopr. Eng. *3*, 37–41.
Tramper, J., Joustra, D. and Vlak, J.M. (1987). In: Plant and animal cell cultures: Process possibilities (Webb, C., Ativuna, F., eds.), pp. 125–136, Chichester: Ellis Horwood.
Trinh, K., Garcia-Briones, M., Hink, F. and Chalmers, J.J. (1994) Quantification of Damage to Suspended Insect Cells as a Result of Bubble Rupture, Biotechnol. and Bioeng. *43*, 137–45.

van Wezel, A. L. (1967). Growth of Cell-Strains and Primary cells on Microcarriers in Homogeneous Culture. Nature *216*, 64–65.

Venkat, R. V. (1995) Study of hydrodynamics due to turbulent mixing in animal cell microcarrer bioreactors. Doctoral dissertation. Department of Chemical Engineering, The Ohio State University, Columbus, OH.

Venkat, R. V., Stock, R. L. and Chalmers, J. J. (1996) Study of hydrodynamics in microcarrier culture spinner vessels a particle tracking velocimetry approach. Biotechnol. and Bioengin. *49*, 456–466.

Venkat, R. V., Brodkey, R. S., Guezennec, Y. C., Chalmers, J. J. (1993) In: 3rd Interational Conference on Bioreactor and Bioprocess Fluid Dynamics (Nienow, A. W. ed.) pp. 483–501. BHR Group Conference Series, Publication No. 5, MEP, London.

Yianneskis, M., Popiolek, Z., Whitelaw, J. H. (1987). An experimental study and unsteady flow characteristics of stirred reactors. J. Fluid Mech. *175*, 537–555.

3.3 Kinetics and Simulation of Animal Cell Processes

Jean-Louis Goergen, Annie Marc and Jean-Marc Engasser

3.3.1 Introduction

The kinetics of growth and metabolism of mammalian cells in culture systems has received increasing attention. Major motivations of these studies were the design of improved culture media, the optimization of bioreactor operation to reach maximal productivity, as well as the deeper understanding of the basic phenomena controlling the metabolism of cells.

Experimental data have been collected for the culture of several cell lines in either batch or continuous culture systems, and under wide ranges of medium compositions. The most widely used approach is the so-called extra- cellular analysis. The rates of cell growth and death, ofnutrient uptake, of metabolite and protein excretion have been measured, and relationships established between the cellular activities and the composition of the culture medium. Based on these results, models capable of simulating the kinetics of mammalian cells have been developed.

More recently changes in cell morphology and intracellular content have also been investigated in an attempt to identify the main processes controlling cell growth and protein synthesis. Some of the results have been integrated into structured models aimed at a more quantitative description of cell physiology.

This review outlines the major experimental and quantitative approaches for the analysis of mammalian cell kinetics. It presents some of the main phenomena which have been found to control the behavior of cells in either batch or continuous culture systems and indicates rate expressions found applicable for many of the investigated cell lines. The potential of physiological or structured models is also briefly discussed.

3.3.2 Mammalian Cell Kinetics

The kinetics of several cell lines of industrial interest - hybridoma, CHO, BHK, Vero, human kidney cells - have been experimentally studied. Each cell line was found to have its own kinetic characteristics: its rates of growth, death and metabolism, its nutritional requirements, its sensitivities toward chemical and physical stress. Yet, comparing the different reported results, many common trends can be observed in terms of behaviors in batch and continuous culture systems, of rate limiting phenomena, and of quantitative relationships between the metabolic rates and the medium composition.

For instance the following phenomena have been reported to have a major influence on the metabolic activities of the various cell lines studied: i) the limiting levels of essential nutrients such as glucose, glutamine and other amino acids, oxygen, vitamins; ii) the accumulation of the inhibitory metabolites ammonia and lactate;

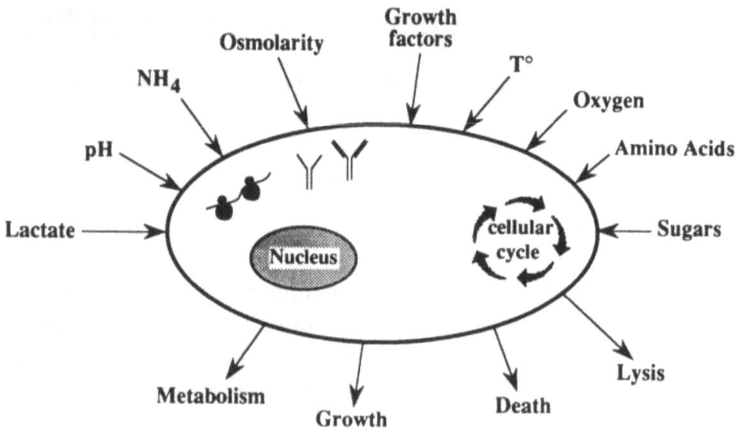

Fig. 3.3.1 Main factors of the cell environment influencing the kinetics of mammalian cell growth and metabolism.

iii) the physico-chemical characteristics of the culture medium, namely pH, temperature, osmotic pressure and ionic strength (Fig. 3.3.1). The rate of cell growth and metabolism is thus a complex function of the concentrations of the different medium components.

3.3.2.1 Kinetics of Cells in Batch Cultures

Most kinetic studies on animal cells have been performed in batch culture systems: in either simple culture flasks, or preferably in bioreactors under controlled pH and dissolved oxygen. They provided data on the time variation of the different components in the culture medium, on the evolutions in intracellular contents, on the possible changes in the composition of excreted proteins, on the different specific rates of cell metabolism during the culture.

3.3.2.1.1 Time Variation of the Concentration of Cells, Nutrients and Metabolites

Fig 3.3.2 shows a typical time variation of the concentrations of nutrients, metabolites, viable and dead cells for a batch culture of hybridoma. After an initial lag phase, cells grow rapidly and reach a maximal density of viable cells around $10^9 \, l^{-1}$. After about 100 hours of culture the level of viable cells declines, whereas the concentration of the dead cells rapidly increases.

The extent of cellular death is classically evaluated by the trypan blue method which counts the visible cells unable to exclude the blue dye (Cook and Mitchell, 1989). The number of dead cells rapidly increases when essential nutrients are completely consumed in the medium. Accumulation of too high concentrations of ammonia (above 2mM) and lactate (above 10 mM) may also be responsible for an increased rate of cellular death (see also Chapter 2.1).

The measured concentration of viable cells results from the two processes of formation of new cells and death of living cells. In order to evaluate the actual rate

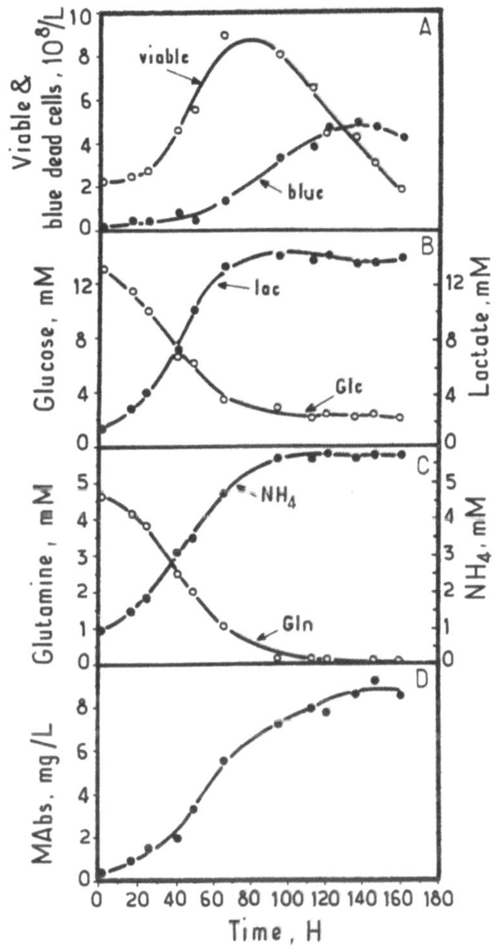

Fig. 3.3.2 Typical kinetics of a hybridoma culture in a batch bioreactor: evolution with time of the concentrations of viable (○) and blue stained (●) cells, lactate (●) and ammonia (●), glucose (○) and glutamine (○) and MAb (●), (from Goergen et al., 1992b).

of new cell production, one calculates the total number of cells (X_t) which have been produced during the culture. Since the phenomena of cell lysis is often negligible during batch cultures, the total number of produced cells can be calculated as the sum of the measured viable (X_v) and dead (X_d) cells. In many cases the total concentration of produced cells was found to slowly increase during the decline phase, which is indicative of a lasting cellular growth process.

3.3.2.1.2 Influence of Inoculum Age

For cells grown in batch cultures, the inoculum age can have an important effect on the subsequent culture kinetics. Indeed cultures containing an identical medium and seeded at the same initial cell density have been found to yield different growth

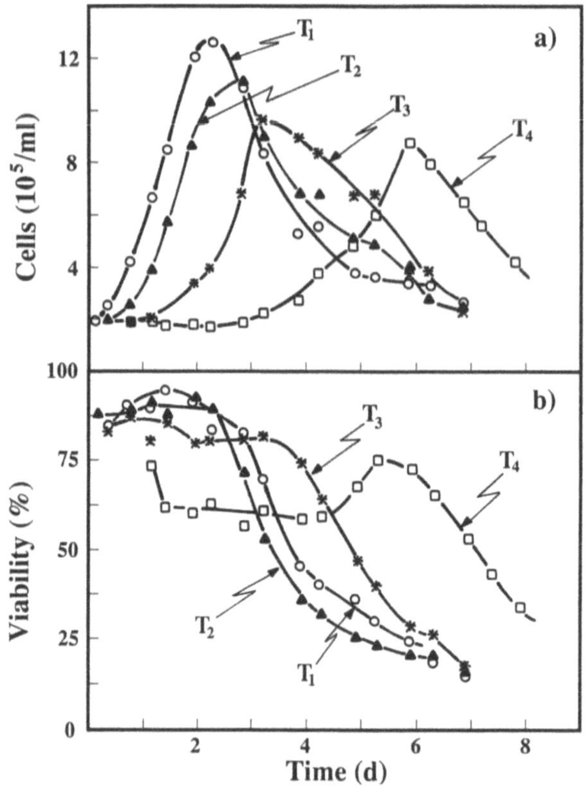

Fig. 3.3.3 Kinetics of hybridoma batch cultures seeded with cells of different ages of preculture: T1 = 43 h, T2 = 52 h, T3 = 62 h, T4 = 71 h (from Martial et al., 1991).

curves when starting with cells harvested at different times from the previous propagation culture (Fig. 3.3.3). Increasing the age of inoculum usually results in a longer lag phase, in a lower maximum specific growth rate and a reduced maximal cell density. The production of antibodies can also be delayed for an aged inoculum (Martial et al., 1991).

3.3.2.1.3 Variation of Cell Morphology and Intracellular Content

For a better basic understanding of the kinetic behavior of mammalian cells in bioreactors, more physiological studies have also been performed. They include investigations, during the cultures, of the cell morphology and intracellular content.

In batch cultures, the cell size usually increases during the rapid growth phase, then decreases during the decline phase. Time changes of cell sizes are generally accompanied by modifications in the cell membrane composition. It was observed that, at the end of the culture, the cell membrane undergoes significant structural changes resulting in the disappearance of microvilli and the appearance of blebs and deep indentations (Al-Rubeai, 1990). The bursting membrane tension and the compressibility modules of the cells were also found to strongly increase during the rapid growth phase, and later decrease in the death phase (Zhang et al., 1992).

Possible changes during cultures of the intracellular content in intermediary metabolites, proteins and nucleic acids, have been investigated either by cell lysis and subsequent analysis of the different metabolites, or by *in situ* flow cytometry or NMR analysis (Al-Rubeai and Emery, 1993). The pool size of ATP and UTP, as well as the adenylate energy charge, was thus found to change during different growth phases (Ryll and Wagner, 1992) (see Chapter 2.1.6.2). Elsewhere, some correlations have been established between the extracellular concentration of secreted protein and either its intracellular and surface levels monitored by flow cytometry (Sen et al., 1990) or its intracellular mRNA content (Leno et al., 1992). *In situ* monitoring of intracellular pH with the pH sensitive dye BCECF, using the flow cytometry method, has also been reported (Cherlet et al., 1993). In batch culture with an extracellular pH controlled at 7.0, the intracellular pH was found first to increase from 7.55 to 7.6 during rapid growth, then to decline to 7.35 during decline phase.

3.3.2.1.4 Determination of Cell Population Heterogeneity

Kinetic investigations classically yield the time variation of the total viable cell concentrations and the average specific rates of nutrient consumption and metabolite production. Actually, among the whole cell population, one can expect significant differences from one cell to the other, as they can be in different phases of the cell growth cycle and thus have different metabolic activities.

Flow cytometry represents a powerful technique to analyze heterogeneity in cell populations. For example staining cells with fluorophores, such as Propidium iodide or Acridine orange, that bind stoichiometrically to nucleic acids allows the visualization of the progress of cells through the cell cycle (Al-Rubeai and Emery, 1993). It was thus observed that during batch cultures, the proportion of the cells in the G_0, G_1, G_2/M phases significantly changes. As expected, the highest proportion of cells in G_2/M phase corresponds to the fastest growth phase, whereas during the stationary and decline phases a majority of cells are in the G_1 state. The measured cell-cycle distribution can be directly related to the generation time of cells. As the duration of the G_2/M and S phases are found relatively constant for a growing cell (about 5 and 6 hours respectively), the observed variation in generation time from 17 to 30 hours often results from a variable duration of the G_1 stage from 6 to 16 hours.

Flow cytometry analyses also demonstrate that protein synthesis and excretion may change with the physiological state of the cells and the cell cycle. For instance it was reported that, for a hybridoma cell line, IgG is mainly synthesized during the S phase, but secreted into the culture fluid as the cells go through the G_1 phase (Coco-Martin et al., 1992) (see also Chapter 2.1.6.2.2).

3.3.2.1.5 Time Variation of the Excreted Protein Composition

Recently, changes during batch cultures of the amino acid structure or the glycosylation pattern of the protein have been reported. These changes have been attributed either to modifications in the intracellular protein maturation processes due to some kinetic limitations, or to the extracellular action of proteases or glycosidases excreted into the medium (see Chapter 2.1.7.1).

The batch culture of CHO cells for the production of recombinant human interferon-γ provides an example of an observed time variation of the glycosylation

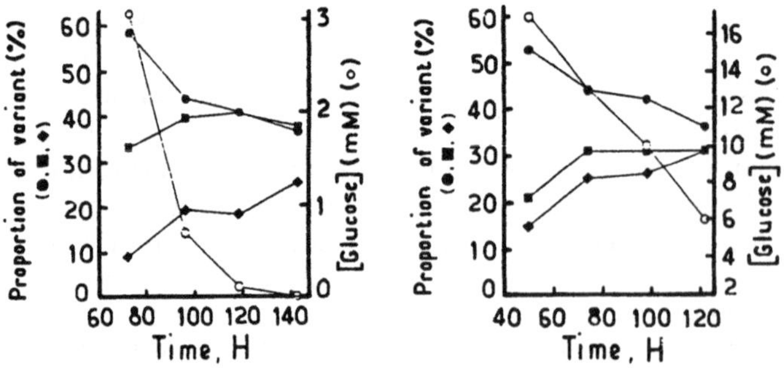

Fig. 3.3.4 Time variation of the glycosylation pattern of γ-interferon during a batch culture of a recombinant CHO cells at two initial glucose concentrations of 10 and 20 mM. Symbols : γ-interferon with ●, two-sites glycosylated; ■, one-site glycosylated; ◆, no sites filled (from Curling et al., 1990).

pattern of a produced recombinant protein. The proportion of the non-glycosylated form of interferon increases during the culture and is related to the initial level of glucose in the medium (Curling et al., 1990) (see Chapter 2.2).

3.3.2.1.6 Specific Rates of Cell Growth, Death and Metabolism

For a more quantitative description of cellular kinetics, the most appropriate approach is to calculate, during the culture, the specific rates of cell growth and metabolism. The specific rates are defined as following:

- the specific rate of cellular growth, μ, is the number of billions of new cells produced per billion viable cells present in the culture medium and per hour. In a batch culture it is calculated as:

$$\mu = \frac{dX_t}{X_v \cdot dt}$$

- the specific rate of cellular death, k_d, is the number of billions of dying cells per billion viable cells present in the culture medium and per hour. It is calculated by the relationship:

$$k_d = \frac{dX_d}{X_v \cdot dt}$$

- the specific rate of nutrient uptake, v, is defined as the number of mmole of nutrients consumed per billion viable cells and per hour. Similarly the specific rate of metabolites or protein production, π, is calculated as the number of mmole or mg of products excreted per billion viable cells and per hour.

The major advantage of determining the specific rates is to provide a quantitative evaluation of the intrinsic activity of a cell as a function of the medium composition.

Fig. 3.3.5 shows the typical variation of the specific rates of cellular growth, death and metabolism during a batch hybridoma culture. One first observes an initial lag phase during which the specific rates of growth, nutrient uptake and metabolite production progressively increase. During the following growth phase the specific

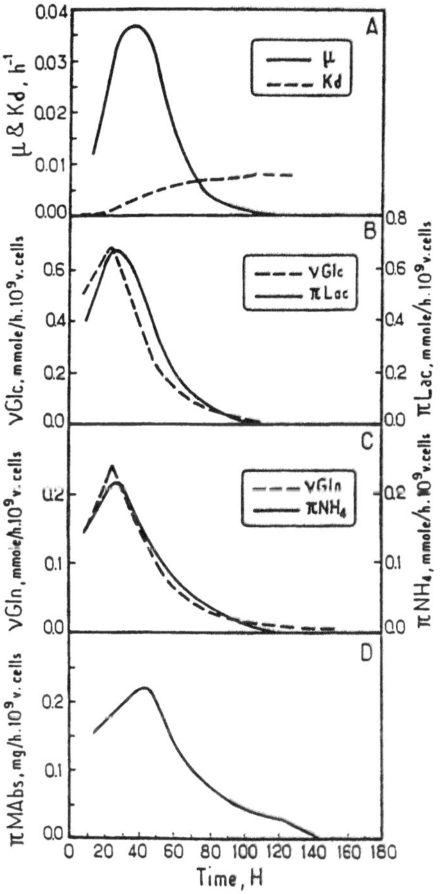

Fig. 3.3.5 Evolution with time of the specific rates of growth, μ, death, k_d, glucose and glutamine consumption, ν_{Glc} and ν_{Gln}, lactate and ammonia production, π_{Lac} and π_{NH4}, and antibody production, π_{MAb}, during a batch culture of hybridoma cells (from Goergen et al., 1992b).

rates gradually decrease because of either a depletion of essential nutrients or an accumulation of inhibitory metabolites. The specific death rate, on the other hand, is the lowest at the start of the culture, and then gradually increases. This can also be attributed to a limiting level of nutrients or to a toxic level of metabolites.

3.3.2.1.7 Relationships Between the Specific Rates of Growth and Metabolism
From the comparison of the time variations of the different specific rates, several similarities and differences can be noticed from one cell line to the other:

- the specific rates of nutrient uptake and metabolite production always increase with the specific growth rate of the cell;
- the production rate of lactate is directly related to the consumption rate of glucose. Yet the metabolic ratio of the produced lactate over the consumed glucose can

be very different depending on the cell and the culture conditions, with values generally between 0.7 and 1.5 mole lactate/mole glucose;
- similarly the production of ammonia is related to the consumption of glutamine with values of the metabolic ratio ammonia over glutamine between 0.5 and 1.0 mole/mole;
- the relationship between the specific rate of protein production and the specific rate of cellular growth follows different patterns. Depending on the cell line and on the culture conditions, the specific rate of protein production was found to increase or to decrease with the specific growth rate. In some cases, the protein production rate was found to be independent of the growth rate.

3.3.2.2 Cell Kinetics in Continuous Perfusion Cultures

In addition to traditional batch cultures, continuous bioreactors are increasingly used to investigate the kinetics of animal cells. They are either simple culture systems, with continuous feeding of nutrients and removal of cells and metabolites, or perfusion reactors where substrates are continuously fed and metabolites removed, but where cells are totally or partially retained inside the culture vessels. Several configurations of perfusion reactors are commercially available, with cell retention by membranes, centrifuges or settling devices, or with cells immobilized inside macroporous particles or membranous compartments.

3.3.2.2.1 Interests of Continuous Cultures

Continuous perfusion processes may offer several performance advantages with respect to batch cultures: higher cell densities and protein concentrations, increased reactor productivity, lower nutrient requirements. For example cell densities over $10^{11}\,l^{-1}$ and protein concentrations exceeding $1\ g\,l^{-1}$ have been achieved for continuous cultures maintained for several months.

Continuous cultures are also powerful tools for kinetic analysis since cells can be maintained in growth limiting conditions for extended periods of time. With these systems the effect of a broad range of environmental factors can be investigated by simply changing the dilution rate and/or the feed medium composition. A prolonged exposure of cells to a growth environment limited by either glucose, amino acids, lipids, vitamins, oxygen or inhibitory metabolites, can result in significant differences in cell physiology and kinetics.

3.3.2.2.2 Time Variation of the Concentration of Cells, Nutrients and Metabolites

A typical example of kinetics of a perfusion culture of hybridoma in a stirred bioreactor equipped with a microfiltration module for cell separation and recycling is shown in Fig. 3.3.6. After an initial batch propagation of cells, the culture medium is continuously fed to the reactor and removed through the microfilter (Pinton et al., 1991).

The perfusion culture kinetics show an initial rapid growth phase, a second period of reduced growth because of nutrient limitations, and finally a stationary phase during which cells can be maintained for weeks at a constant concentration. High cell densities between 1 and $3 \times 10^{10}\,l^{-1}$ are commonly reached with this perfusion

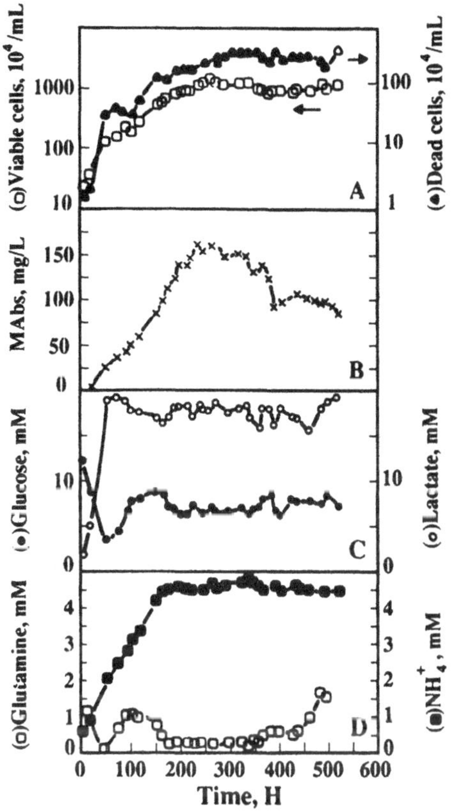

Fig. 3.3.6 Hybridoma culture kinetics in a continuous perfused bioreactor with recycling cells through a filtration module: evolution with time of the concentrations of viable (□) and blue stained (●) cells, of lactate (○) and ammonia (■), of glucose (●) and glutamine (□) and of MAb (X). (V = 1.2 L, D = 1 d^{-1}, Feed medium: DMEM/F12 (3/1), 5% FCS, 2% EAA, 1% NEAA, 25 mM glucose, 4 mM glutamine) (from Pinton et al., 1991).

technology. The concentration of monoclonal antibodies in the effluent reaches a maximal value of 150 mg/l, which represents a threefold increase with respect to the level obtained in batch cultures. The levels of glucose, glutamine, lactate and ammonia in the culture medium are relatively stable during the 500 hours of the perfusion culture.

Similar kinds of time variation have been reported for other suspension and adherent cells. In the case of adherent cells, after a first period of microcarrier colonization, formation of clumps or clusters have been observed (Wagner et al., 1990).

Prolonged continuous cultures can also result in important morphological or intracellular modifications. For example, during the stationary state of a perfusion culture the hybridoma volume was found to be lower than in the initial batch culture. The concentration of the intracellular amino acids and Krebs cycle intermediates was also reported to change with the level of glutamine in the medium (Schmid and Keller, 1972)

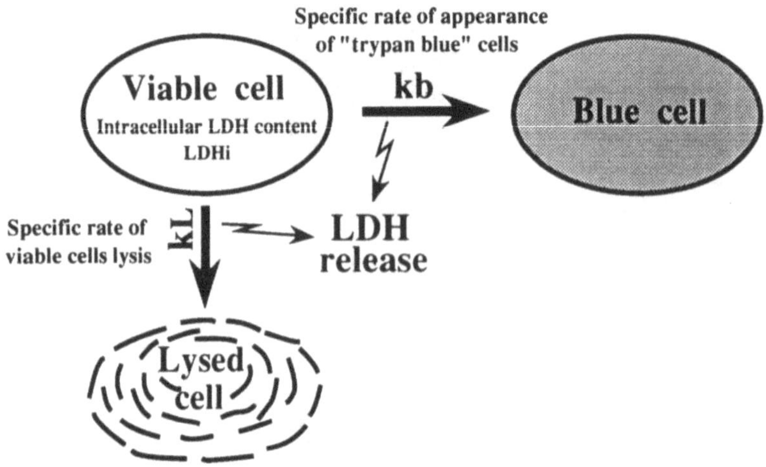

Fig. 3.3.7 Schematic representation of the processes of cell death and LDH release for hybridoma cells.

3.3.2.2.3 Evaluation of Actual Cell Death and Growth Rates

The stationary high concentration of cells which is generally obtained in continuous perfusion cultures can be the result of the two simultaneous processes of cell growth and death occurring at the same rate. A correct interpretation of cell kinetics in these systems thus necessitates an adequate evaluation of the actual rate of cell death.

The extent of cellular death is classically evaluated by the trypan blue method which, however, does not account for the cells which have been lysed. For several hybridoma cell lines cellular lysis was found to occur mainly on living cells, with a negligible phenomenon of lysis of cells counted as dead by trypan blue (Fig. 3.3.7). A different pattern of lysis of both living and dead cells has been observed with human kidney cells growing on microcarriers (Wagner et al., 1992).

When cell lysis becomes significant, a correct evaluation of total cellular death is obtained from the measurement of the release in the culture medium of the cytoplasmic enzyme lactate dehydrogenase (LDH) both by cells which have lysed and which are counted as blue by the trypan blue method (Racher et al., 1990).

From the measured time variation of the LDH concentration in the medium, an LDH balance can yield the cumulative number of dead cells, provided the possible influence of environment on intracellular LDH content and the denaturation of the enzyme in the medium are considered (Wagner et al., 1992; Goergen et al., 1993; Marc et al., 1991).

When this methodology was applied to the analysis of kinetic data for perfusion bioreactors, it was found that the cumulative number of lysed cells steadily increases during the whole culture and can greatly exceed the measured number of retained living cells (Fig. 3.3.8). Under these conditions, the total number of produced cells also increases during the apparent stationary phase, which means that cells continue to grow at a specific rate, which may be as high as $0.02\ h^{-1}$.

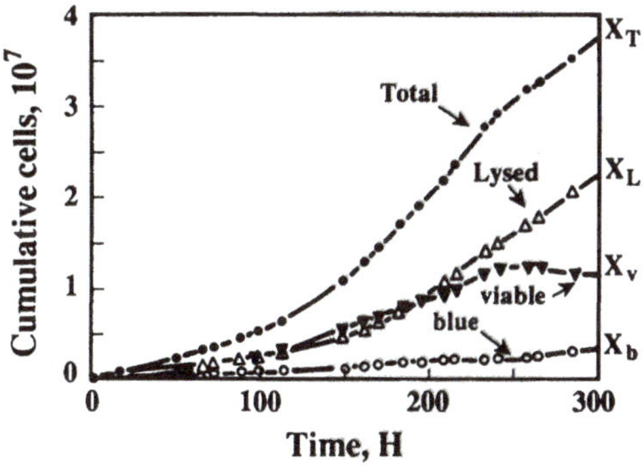

Fig. 3.3.8 Evolution with time of the amounts of viable (▼), blue staining(○), lysed (▲) and total (●) cells during a perfused culture of hybridoma cells (from kinetic data in Fig. 3.3.6).

3.3.2.2.4 Specific Rates of Cell Metabolism in Continuous Cultures

According to the few comparative studies of cell metabolism in batch and continuous cultures, there seem to be no major differences in specific rates of glucose, glutamine uptake, ammonia and lactate production at a given growth rate. Observed differences in measured metabolic rates can generally be attributed to differences in cell growth rates for the two culture systems. Nevertheless, after prolonged continuous operation significant reductions in protein production rates have been reported. They have been attributed either to the selection of non-producing clones or to genomic instabilities.

When cells are maintained in prolonged continuous cultures, important morphological modifications can be observed, together with the decrease of the specific

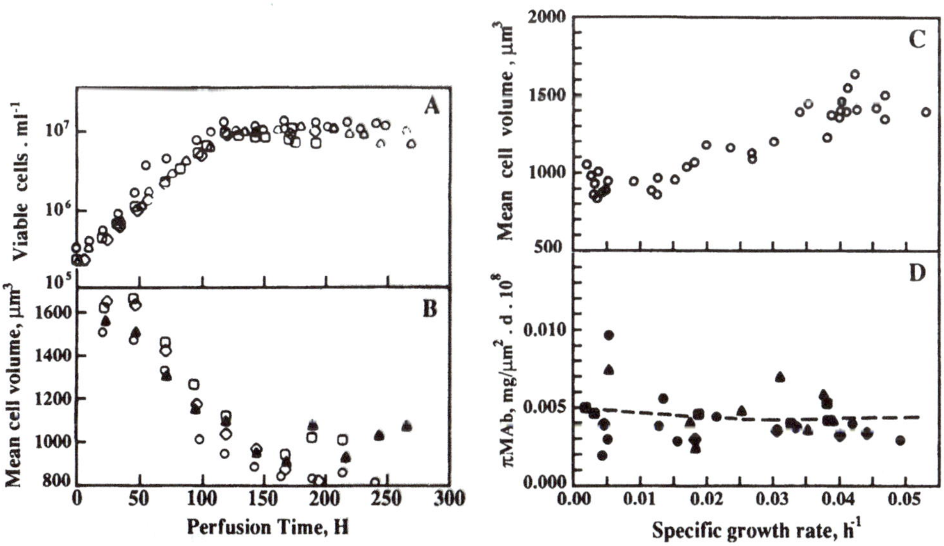

Fig. 3.3.9 Evolution with time of the concentration and of the mean cell volume of viable cells (A, B) and the specific growth rate (C, D), during a perfused culture of hybridoma cells (from Flickinger et al., 1990; Goebel et al., 1990).

growth rate. Under conditions of very slow growth, hybridoma microvilli density, observed by sequential SEM, appeared to be significantly reduced. The hybridoma volume has also been found to decrease with the specific growth rate. Interestingly, when calculating the specific rate of antibody production per unit of external surface area, a constant specific rate has been observed (Fig. 3.3.9) (Flickinger et al., 1990; Goebel et al., 1990).

3.3.2.3 Influence of Medium Environment on Cell Kinetics

The kinetic results collected on different mammalian cell lines grown in batch or continuous systems seem to indicate some general trends for the influence of the environment on cell growth and metabolism.

3.3.2.3.1 Influence of the Medium Composition

Most of the components of the culture medium may have an influence on cell kinetics. Generally, for each of them, there is an optimal concentration range where the cellular activity is at its highest. On the contrary, when the medium components are present either in insufficient quantities, and/or at a too high level, they have a rate reducing effect on the growth and metabolism of cells.

Glucose and Amino Acids

A too low concentration of the essential nutrients such as glucose, glutamine, and other amino acids, results in a growth rate reduction and an increase of the cell death rate. Critical concentrations which are cell dependent are classically around 0.1 mM for glucose and glutamine. In addition to glutamine, the fastest utilized amino acids, i.e. those which are more likely to limit cell growth during the culture, are generally the essential amino acids. However, considerable differences in amino acid utilization can be found depending on the cell line, the culture medium and the culture system (Büntemeyer et al., 1991).

Lactate and Ammonia

Accumulation of the two major metabolites, lactate and ammonia, often represents an additional growth limiting and death promoting factor. Critical inhibitory levels, which are cell dependent, are around 40 mM for lactate and 3 mM for ammonia (Ozturk et al., 1992). The production of metabolites can be reduced by the utilization of other energetic substrates such as fructose or galactose, and dipeptide or glutamate. The addition of potassium ions has also been found to reduce the ammonia level (Martinelle and Häggström, 1992).

Oxygen

Oxygen usually becomes a growth rate limiting factor at a level below 1 % of air saturation. Limitations in oxygen levels were found to increase the specific consumption of substrates such as amino acids or sugars and the production rates of metabolites (Ozturk and Palsson, 1990). Interestingly, the optimal oxygen level for protein production may be different from the optimal level for cellular growth.

Serum

Serum, especially fetal calf serum, has been traditionally added to the culture medium of mammalian cells as a source of several components: hormones, growth factors, lipids, protease inhibitors, etc. A too low level of serum decreases the cell growth rate but can also increase the consumption rate of other nutrients (Dalili and Ollis, 1989) and the production rate of metabolites (Ozturk and Palsson, 1991 a). Serum can also have a strong effect on the rate of protein production with an optimal level for protein synthesis different from the optimal level for cellular growth (Ramirez and Mutharasan, 1990).

3.3.2.3.2 Influence of the Physico-Chemical Parameters

Medium pH

Depending on the medium composition, the optimal pH value for cell growth is generally between 7.2 and 7.4 with extremum values until 6.8 and 7.9. Non-optimal pH can result in an increase of sugar and amino acid requirements and affects the lactate/glucose ratio (Schmid et al., 1990). For several hybridoma cell lines, a lower pH, around 6.8, has been found to stimulate the antibody production.

Medium Osmolarity

Osmolarity is one of the main physico-chemical parameters affecting the cell behavior. The optimal medium osmolarity for cell growth is generally between 260 and 330 mOsmole/kg (Ozturk and Palsson, 1991 b). A higher osmolarity often results in an inhibition of cell division, but can also induce a faster production of proteins (Reddy et al., 1992).

Shear stress

Mammalian cells are sensitive to shear stress caused by liquid circulation and bubble bursting. The shear stress sensitivity is strongly dependent on the cell line, on the medium composition and on the culture conditions. For instance hybridoma cells have been reported to be less sensitive during the rapid growth phase than during the later decline phase (Petersen et al., 1990) (see also Chapter 3.2).

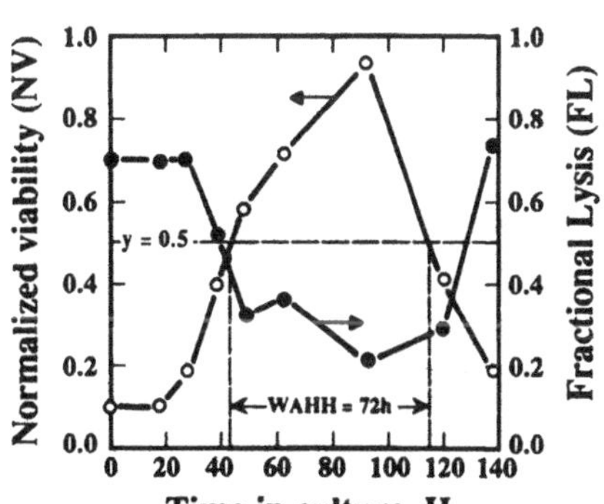

Fig. 3.3.10 Evolution of the normalized viability (○) and of the fractional lysis (●) during a batch culture of hybridoma cells (from Petersen et al., 1990).

3.3.3. Models for Kinetic Simulations

From the experimentally observed cell kinetics, quantitative relationships can be derived that relate the different rates of cellular growth and metabolism as a function of the medium composition. This set of rate expressions represents a kinetic model of the cellular process. When given to a computer, it can provide a simulation of the time variation of the concentrations of cells, nutrients and metabolites in the culture medium in different modes of culture.

The development of models of microbial and cellular transformations represents a central approach in modern biochemical engineering. As they can account for the simultaneous effect of several rate limiting factors, they are convenient tools for the quantitative evaluation of the different rate-limiting effects. Models also provide efficient tools of process simulations, which can facilitate the optimal design of culture media and selection of operating culture conditions. Moreover, models are more and more frequently used for the development of software sensors, which provide estimations of the medium composition with only limited on-line measurements.

3.3.3.1 Macroscopic Models for Cell Kinetics

The most widely used approach for the modeling of cell kinetics, which is also the simplest one, is the macroscopic or the extracellular approach. It describes the influence of the medium composition on the cell behavior, without any explicit consideration of intracellular events.

Fig. 3.3.11 illustrates the main phenomena which have been taken into account in the proposed models, namely the effect of nutrient limitations and metabolite accumulations on cell growth and death.

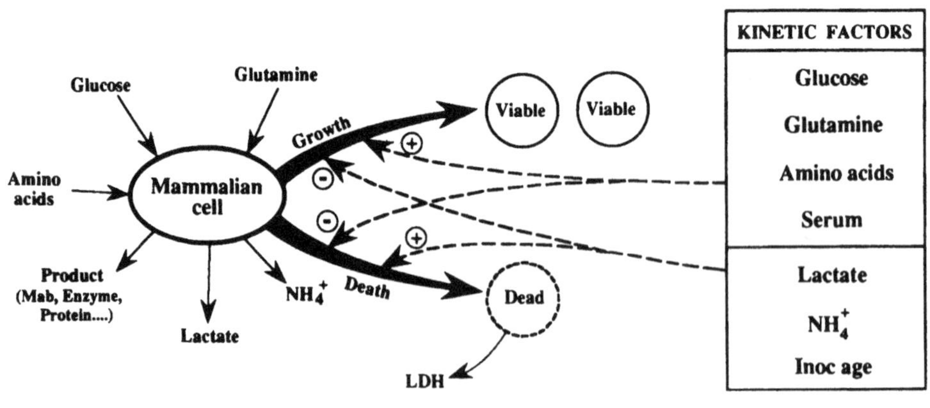

Fig. 3.3.11 Schematic representation of the main phenomena which have been taken into account in kinetic models of mammalian cells.

3.3.3.1.1. Kinetic Rate Expressions

An example of a kinetic model which takes into account the effect of the four main components of the culture medium - glucose, glutamine, lactate and ammonia - is presented in Fig. 3.3.12.

*** Specific rate of cell growth**

$$\mu = \mu_{max} \cdot \left(\frac{[Glc]}{K_{Glc} + [Glc]}\right) \cdot \left(\frac{[Gln]}{K_{Gln} + [Gln]}\right) \cdot \left(\frac{1}{1 + \frac{[Lac]}{K_{Lac}}}\right) \cdot \left(\frac{1}{1 + \frac{[NH4]}{K_{NH4}}}\right) \quad (1.1)$$

*** Specific rate of cell death**

$$k_d = A_d \cdot \left(\left(\frac{1}{1 + \frac{[Gln]}{C_1}}\right) + \left(\frac{1}{1 + \frac{[Glc]}{C_2}}\right) + \kappa_1 \cdot [NH4] + \kappa_2 \cdot [Lac]\right) \quad (1.2)$$

*** Specific rate of glucose consumption**

$$\nu_{Glc} = Y_{Glc/X} \cdot \mu + m_{Glc} \quad (1.3)$$

*** Specific rate of lactate production**

$$\pi_{Lac} = Y_{Lac/X} \cdot \mu + m_{Lac} \quad (1.4)$$

*** Specific rate of glutamine consumption**

$$\nu_{Gln} = Y_{Gln/X} \cdot \mu + m_{Gln} \quad (1.5)$$

*** Specific rate of ammonia production**

$$\pi_{NH4} = Y_{NH4/X} \cdot \mu + m_{NH4} \quad (1.6)$$

*** Specific rate of antibody production**

$$\pi_{MAbs} = Y_{MAbs/X} \cdot \mu + m_{MAbs} \quad (1.7)$$

*** Rate of glutamine degradation**

$$r_{Gln} = k_{deg} \cdot [Gln] \quad (1.8)$$

Fig. 3.3.12 Example of rate equations in a kinetic model for hybridoma cells.

Rate Expression for Cellular Growth

For a given medium, a cell line can be characterized by a maximum specific growth rate, mmax, inversely proportional to the doubling time of the cell. During culture the specific growth rate usually decreases because of either depletion of essential nutrients or accumulation of inhibitory metabolites. Equation 1.1 represents such a variation of specific growth rate as a function of the concentration of glucose, glutamine, lactate and ammonia in the medium. The introduced parameters K_{Glc}, K_{Gln}, K_{Lac}, K_{NH4} give the range of concentrations where either the nutrient becomes limiting or the metabolite becomes inhibitory. By modulating the values of these parameters the model can account for differences in cell sensitivities toward nutrient depletion and product inhibition.

Rate Expression for Cell Death

The specific rate of cellular death, kd, is also affected by the chemical composition of the medium and several physico-chemical parameters, such as pH, temperature, osmotic pressure. The rate expression for cell death in Equation 1.2. indicates the possible increase in the cell death rate due to limitations in glucose and glutamine or accumulations in lactate and ammonia. By proper adjustment of the values of the four parameters, C_1, C_2, κ_1 and κ_2, it is possible to account for differences in death kinetics from one cell line to the other.

Rate Expression for Nutrient Uptake and Metabolite Production

For most of the investigated cell lines, the specific rates of nutrient uptake, ν, and metabolite or protein production, π, have been found to increase linearly with the specific growth rate. Thus the specific rate of glucose uptake is expressed as a function of μ by Equation 1.3 which contains two parameters: the non-growth associated cell-specific glucose consumption rate, m_{Glc} (mmole glucose $10^{-9} h^{-1}$), and the glucose to biomass conversion yield, $Y_{Glc,X}$ (mmole glucose 10^{-9} cells). Similar expressions with two growth and non-growth terms have been found applicable to the specific rates of glutamine uptake, ammonia, lactate and protein production (Equations 1.4 to 1.7).

Rate Expression for Glutamine Decomposition

For the modeling of the time variation of the glutamine and ammonia concentration, the process of glutamine decomposition in the culture medium must also be considered. The spontaneous decomposition of glutamine into ammonia can be represented by a first order rate process with respect to the glutamine concentration as shown in Equation 1.8.

3.3.3.1.2 Simulation Capacities of the Model

The different rate expressions of the model have been found to correctly simulate batch or continuous cultures of several mammalian cell lines. An example of simulation of a batch culture of hybridoma at controlled pH and dissolved oxygen level is shown in Fig. 3.3.13.

Additional phenomena related to the composition of the culture medium have also been integrated into kinetic models: oxygen and serum limitations, effect of

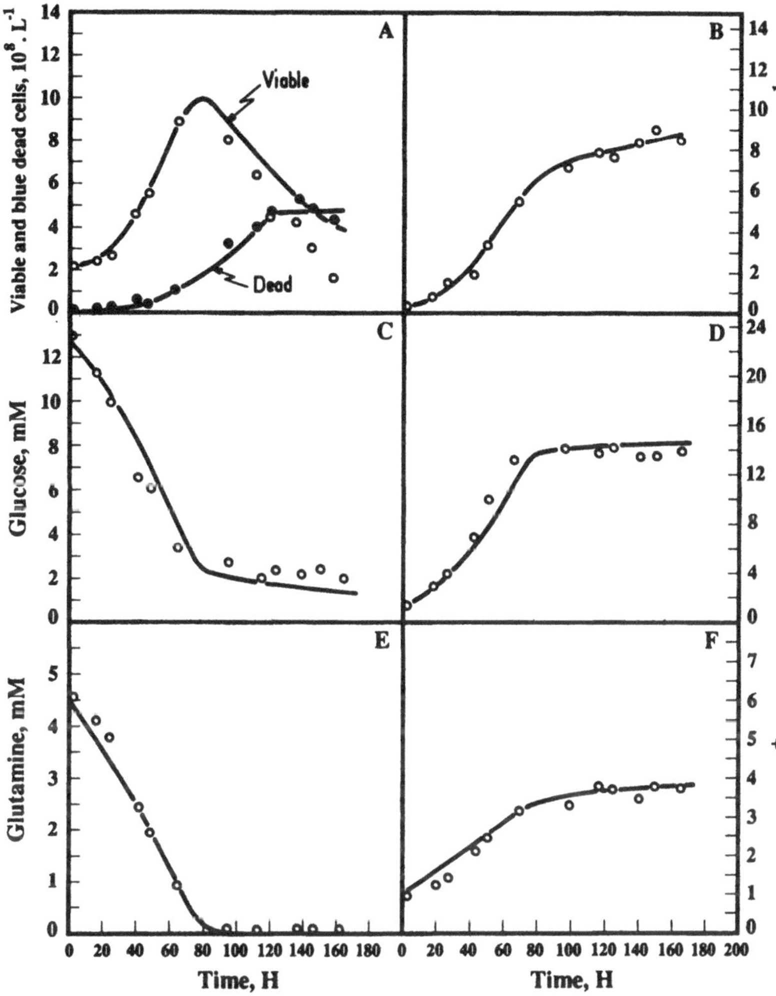

Fig. 3.3.13 Experimental (symbols) and modeling (lines) kinetic results of a batch culture of hybridoma cells at controlled pH and pO_2, using the model of Fig. 3.3.12 (from Goergen, 1992a).

pH and osmotic pressure, limitations by other amino acids. Fig. 3.3.14, for instance, illustrates the simulation of the growth promoting effect of the addition of a solution of essential amino acids to a continuous hybridoma culture.

Interestingly, the proposed model can be easily adapted from one cell line to the other, for instance hybridoma, CHO, BHK, human kidney cells, by keeping the same basic structure of the rate equations but by adjusting the values of the parameters.

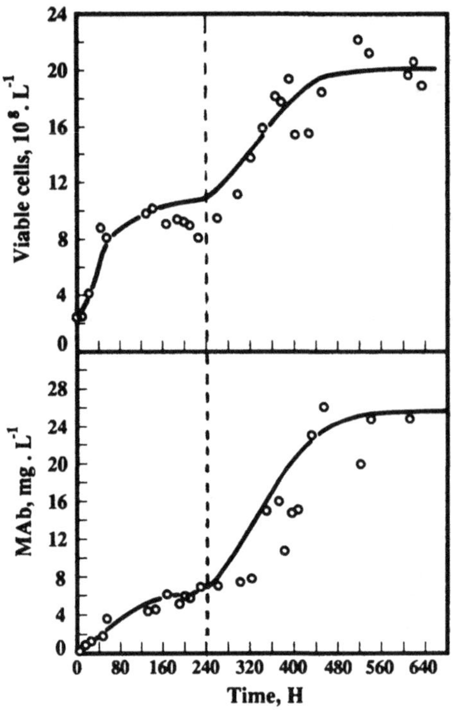

Fig. 3.3.14 Experimental (symbols) and modeling (lines) kinetic results of a continuous culture of hybridoma cells at a dilution rate of 0.02 h^{-1}. Amino acids have been supplemented 240 h. The model in Fig. 3.3.12 has been extended to account for amino acid limitations (from Goergen, 1992a).

3.3.3.1.3 Application of Models for Kinetic Analysis

As one of its major interests, a kinetic model such as the one presented in Fig. 3.3.12 can be used for the kinetic interpretation of experimental data. First it allows, identification of the rate limiting factors, second, quantitatively characterizing their relative importance. One can thus evaluate the kinetic effect of a depletion of glucose, glutamine and other amino acids or of an accumulation of ammonia and lactate on the measured decrease in specific growth rate or increase in specific death rate.

For instance for a hybridoma batch culture, it was found that under standard batch conditions with an excess of glucose, when ammonia does not exceed 5 mM, glutamine depletion was the main phenomenon responsible for the reduction of the growth rate and the increase in death rate. On the contrary, at higher initial ammonia concentrations, such as those which may be reached in high density fed-batch or

▷

Fig. 3.3.15 Simulation of the relative contributions of glutamine and ammonia on the reduction of the specific growth rate calculated by the model (■), and comparison with the experimental specific growth rate (□), during a batch culture of hybridoma cells.

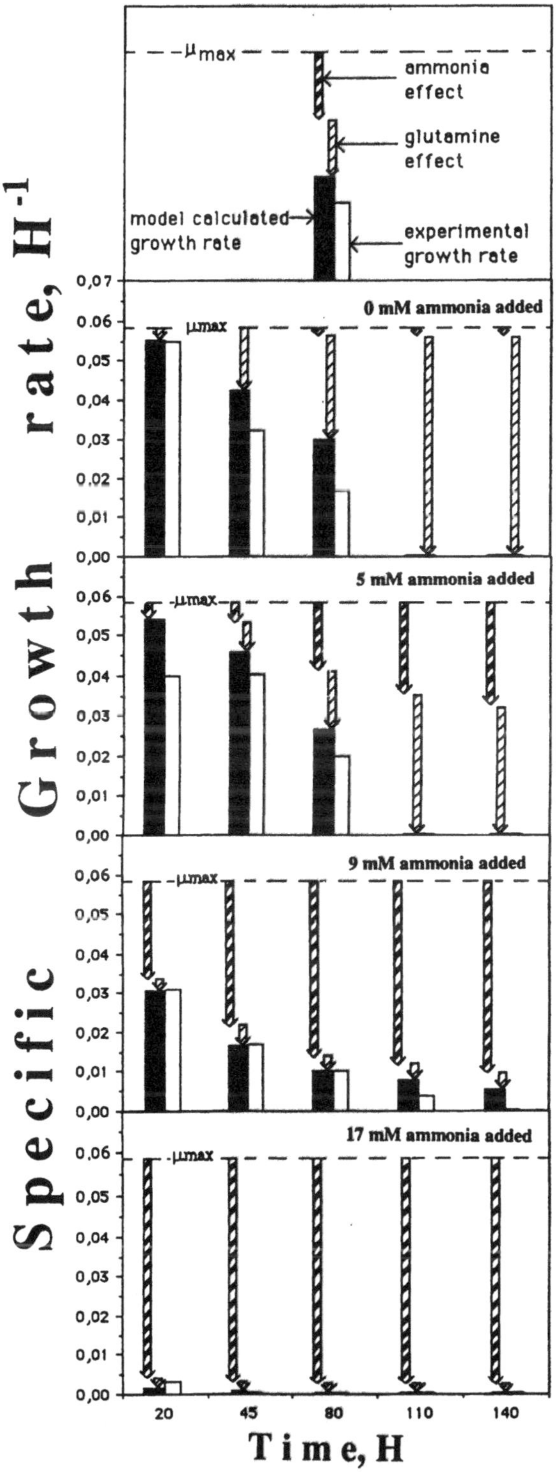
μmax
ammonia effect
glutamine effect
model calculated growth rate
experimental growth rate
0 mM ammonia added
5 mM ammonia added
9 mM ammonia added
17 mM ammonia added
μmax
0,07
0,06
0,05
0,04
0,03
0,02
0,01
0,00
20
45
80
110
140
Time, H
Specific Growth rate, H^-1

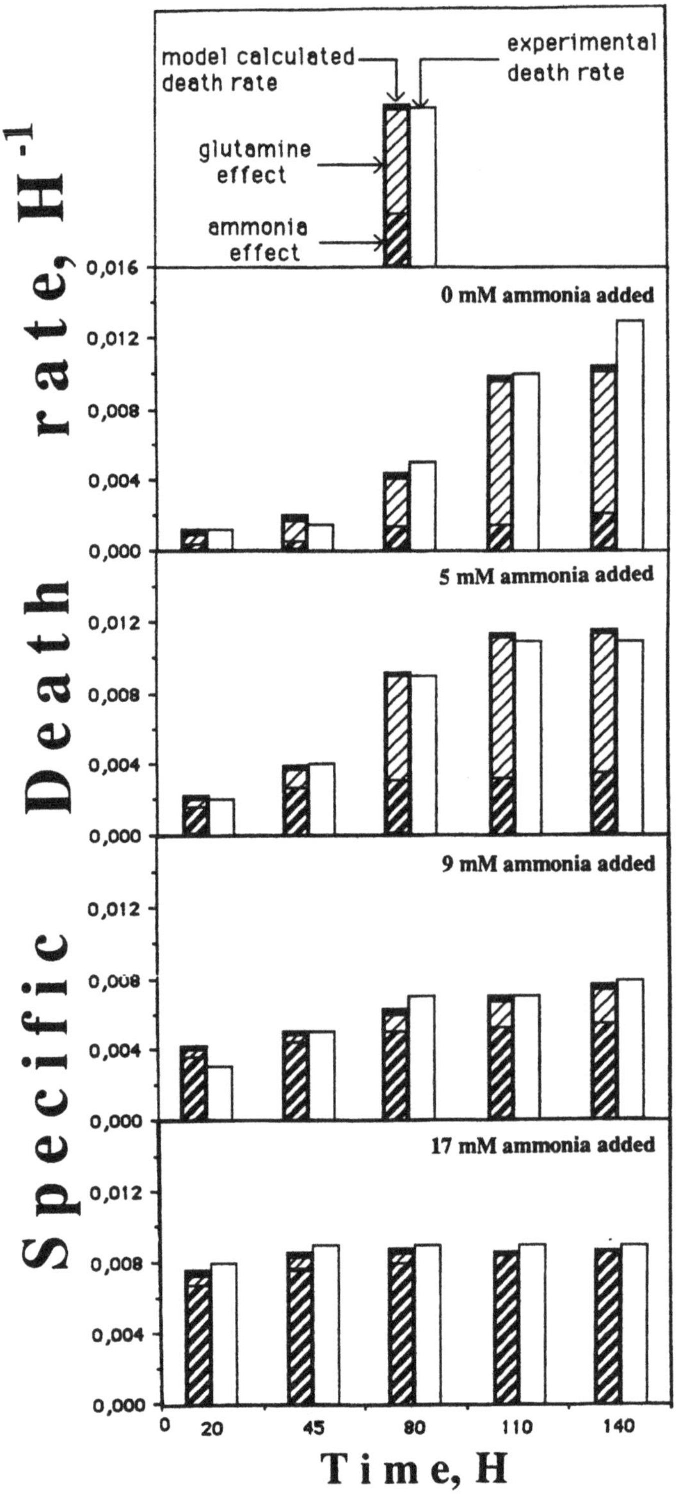

model calculated death rate
experimental death rate
glutamine effect
ammonia effect
0 mM ammonia added
5 mM ammonia added
9 mM ammonia added
17 mM ammonia added
Specific Death rate, H -1
0,016
0,012
0,008
0,004
0,000
0 20 45 80 110 140
Time, H

perfusion cultures, ammonia had an important, if not a dominant, influence both on the growth and death rates. Using the kinetic expressions of the models, the relative contributions of glutamine and ammonia can be evaluated and graphically visualized (Figs. 3.3.15 and 3.3.16).

3.3.3.2 Structured and Intracellular Models for Cell Kinetics

For a finer understanding of the intracellular events that control the growth and metabolism of cells, some intracellular or structured models have recently been developed. Several aspects of cell physiology have been quantified and integrated into the kinetic models of mammalian cells: the different phases of the cell cycle, the processes of protein synthesis and excretion, the binding of growth factors to cellular receptors.

Cell Cycle Models
A model has been developed that simulates the distribution of cells in the different phases of the cell cycle. Among its basic hypotheses, the death rate of the hybridoma is assumed proportional to the fraction of cells arrested in the G_1 phase. The model provides a correct simulation of the growing and arrested cell fractions in continuous cultures at different dilution rates (Linardos et al., 1992).

Model for Antibody Synthesis
The processes of antibody synthesis, maturation and trafficking have been described in a kinetic model of hybridoma growth and metabolism. The model is based on the intracellular balances of heavy and light chain coding mRNAs, the intracellular balances of heavy and light chains and the kinetics of heavy and light chains assembly (Bibila and Flickinger, 1991). In a more advanced model the interorganelle transport and secretion of antibodies between the endoplasmic reticulum, the Golgi and the extracellular medium have also been considered (Bibila and Flickinger, 1992a; 1992b; Sambanis et al., 1991).

Models of Growth Factor Effects
In a model aimed at the simulation of the effect of the growth factor EGF on the kinetics of fibroblastic cell proliferation, the binding of EGF to the EGF receptor and the intrinsic receptor signal transduction have been quantified (Starbuck and Lauffenburger, 1992). The derived model thus contains equations for binding, internalization, degradation, and recycling of EGF and EGFR, along with an expression relating the DNA synthesis rate to the EGF/EGFR complex levels (Fig. 3.3.17).

◁
Fig. 3.3.16 Simulation of the relative contributions of glutamine and ammonia on the increase of the specific death rate calculated by the model (■), and comparison with the experimental specific growth rate (□), during a batch culture of hybridoma cells.

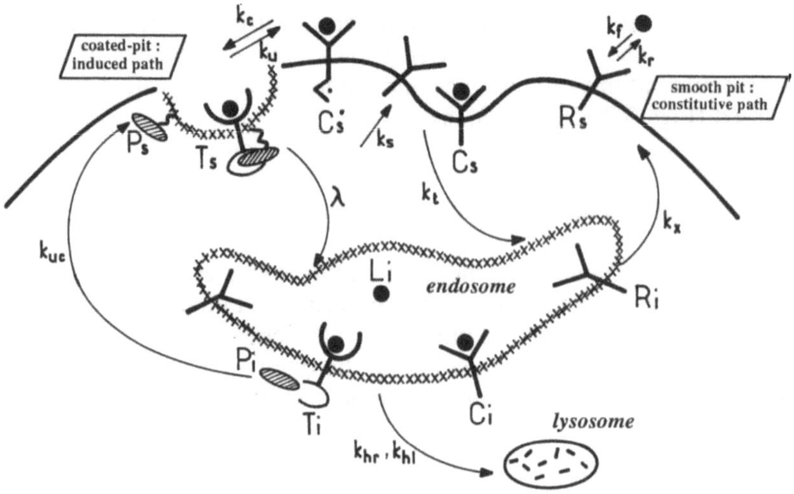

Fig. 3.3.17 Events involved in whole-cell kinetic model for binding and trafficking of the EGF receptor (from Starbuck and Lauffenburger, 1992).

3.3.4 Perspectives

In the future we will witness further progress in mammalian cell technologies, with the construction of more productive recombinant cells, the development of more intensive and better controlled processes and the production of new proteins. The design of improved process will require more quantitative data on the influence of the medium composition on cell growth and metabolism. Special attention should be given to the influence of amino acids, vitamins, lipids, and other effectors that stimulate the production of proteins. In the case of adherent cells, growing as mono or multilayers or forming aggregates, the influence of the solid matrix and of intracellular contacts should also be further investigated.

For the optimal design of high density perfusion or fed-batch processes lasting several weeks or months, more knowledge is required on the environmental factors controlling the rates of cell death by apoptosis or necrosis. Further physiological and kinetic studies should also address the questions of cell adaptation to prolonged exposure to different growth limiting media (nutrient limitations, inhibitory metabolite accumulation, low pH, high osmotic pressure, etc.). For the optimization of high density cultures, the possible influence of endocrine factors must also be established in more quantitative terms.

Concomitantly to this macroscopic approach, additional data are required on the control of the intracellular metabolism. Physiological or structured models, based on *in vivo* fluxes and concentration analysis, have to be constructed to describe the circulation of metabolites and energy inside the cell as a function of the composition of the external environment. The way external messengers or signal molecules control the growth and differentiation of cells should also be more quantitatively analyzed

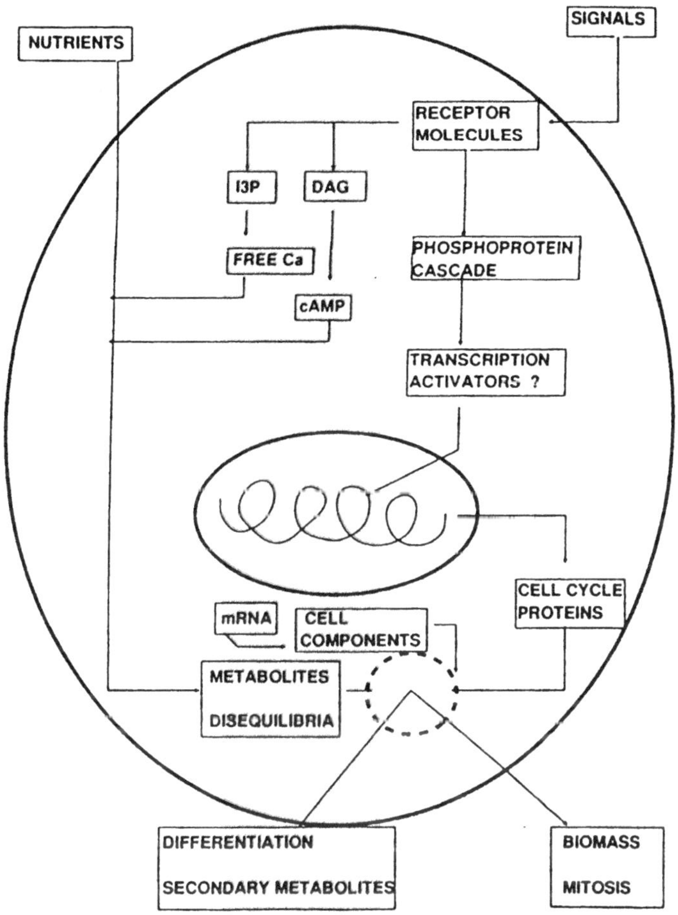

Fig. 3.3.18 Model depicting the interaction of nutritional and signal factors on the determination of the differentiation or division of a cell (from Spier, 1992).

(Fig. 3.3.18). More elaborate models taking into account the cell population distribution among different physiological states are also likely to bring key contributions to a better understanding of the control of cell growth and metabolite production (Spier, 1992).

A major challenge for cell engineers is the need to control the glycosylation profile of therapeutic glycoproteins. In the future, the development of kinetic models describing the processes of protein glycosylation as a function of the intracellular glycosyltransferase enzymes and of the external environment will assist engineers in the design of cell lines and culture conditions favoring the production of desired glycoforms or minimizing glycoprotein heterogeneity.

A growing interest of cell kinetic models is to offer a convenient framework for the organization of the available kinetic data. Models are likely to find increasing application for the analysis of experimental results, especially for the determination of rate limiting steps in cell growth and metabolism. Intracellular models are also expected to assist genetic engineers in a more rational and predictive approach toward gene transfer in mammalian cells.

For biochemical engineers an additional objective will be to create methodologies which exploit the powerful simulation capacities of the models for an improved design and operation of cell technologies. Hence, further progress is expected in the use of models for the design of culture media, for the optimization of operational parameters especially for high density fed-batch and continuous processes. One can also envision a larger use of models for the construction of software sensors allowing an on-line estimation of the medium composition and of the cellular activities from a limited number of on-line or off-line measurements. This will open the way to more advanced control procedures of cell cultures with optimal feeding strategies during the phases of cell growth and protein production.

References

Al-Rubeai, M. and Emery, A. N. (1993) Flow cytometry in animal cell culture. Biotechnology 11, 572–579.

Al-Rubeai, M., Mills, D. and Emery, A.N. (1990) Electron microscopy of hybridoma cells with special regard to monoclonal antibody production. Cytotechnology 4, 13–28.

Bibila, T.A. and Flickinger, M.C. (1991) A structured model for monoclonal antibody synthesis in exponentially growing and stationary phase hybridoma cells. Biotechnol. Bioeng. 37, 210–226.

Bibila, T.A. and Flickinger, M.C. (1992a) Use of a structured kinetic model of antibody synthesis and secretion for optimisation of antibody production systems: I. Steady state analysis. Biotechnol. Bioeng. 39, 251–261.

Bibila, T.A. and Flickinger, M.C. (1992b) Use of a structured kinetic model of antibody synthesis and secretion for optimisation of antibody production systems: II. Transient analysis. Biotechnol. Bioeng. 39, 262–272.

Büntemayer, H., Lütkemeyer, D. and Lehmann, J. (1991) Optimisation of serum-free fermentation process for antibody production. Cytotechnology 5, 57–67.

Cherlet, M., Petitpain, N., Dardenne, M., Franck, P., Engasser, J.M. and Marc, A. (1993) Intracellular pH evolution during batch cultures using flow cytometry. In "Animal Cell technology: Products for Today, Prospects for Tomorrow". Butterworth-Heinemann, London.

Coco-Martin, J.M., Oberink, J.N., Velden-de Groot, T.A.M. and Beuvery, E.C. (1992) The potential of flow cytometric analysis for the characterisation of hybridoma cells in suspension cultures. Cytotechnology. 8, 65–74.

Cook, J.A. and Mitchell, J.B. (1989) Viability measurements in mammalian cell systems. Anal. Biochem. 179, 1–7.

Curling, E.M.A., Hayter, P.M., Baines, A.J., Bull, A.T., Gull, K., Strange, P.
G. and Jenkins, N. (1990) Recombinant human interferon-γ: Differences in glycosylation and proteolytic processing lead to heterogeneity in batch culture. Biochem. J. 272, 333–337.

Dalili, M. and Ollis, D. (1989) Transient kinetics of hybridoma growth and monoclonal antibody production in serum-limited cultures. Biotechnol. Bioeng. 33, 984–990.

Flickinger, M.C., Goebel, N.K. and Bohn, M.A. (1990) Determination of specific monoclonal antibody secretion rate during very slow hybridoma growth. Bioprocess Eng. 5, 155–164.

Goebel, N.K., Kuehn, R. and Flickinger, M.C. (1990) Methods for determination of growth-rate-dependent changes in hybridoma volume, shape and surface structure during continuous recycle. Cytotechnology 4, 45–47.

Goergen, J.L. (1992a) Cultures continues d'hybridomes: Etudes cinétiques et modélisation de l'influence de paramètres physico-chimiques sur la croissance, le décès et le métabolisme des cellules. PhD Thesis, INPL, Nancy.

Goergen, J.L., Marc, A. and Engasser, J.M. (1992b) Comparison of specific rates of hybridoma growth and metabolism in batch and continuous cultures. Cytotechnology 10, 147–155.

Goergen, J.L., Marc, A. and Engasser, J.M. (1993) Determination of cell lysis and death kinetics in continuous hybridoma cultures from the measurement of lactate dehydrogenase release. Cytotechnology 11, 189–195.

Leno, M., Merten, O.W. and Hache, J. (1992) Kinetic studies of cellular metabolic activity, specific IgG production rate, IgG mRNAs stability and accumulation during hybridoma batch culture. Enzyme Microbiol. Technol. 4, 135–140.

Linardos, T.I., Kalogerakis, N. and Behie, L.A. (1992) Cell cycle model for growth rate and death rate in continuous suspension hybridoma cultures. Biotechnol. Bioeng. 40, 359–368.

Marc, A., Wagner, A., Martial, A., Goergen, J.L., Engasser, J.M., Geaugey, V. and Pinton, H. (1991) Potential and pitfalls of using LDH release for the evaluation of animal cell death kinetics. In: Production of Biologicals from Animal Cells in Culture (Spier, R.E., Griffiths, J.B. and Meignier, B., eds.), pp. 603–606. Butterworth, London.

Martial, A., Marc, A. and Engasser, J.M. (1991) Influence of inoculum age on hybridoma culture kinetics. Cytotechnology. 5, 165–171.

Martinelle, K. and Häggström (1992) Potassium ions affect the ammonia/ammonium toxicity in animal cell cultivations. In: Animal Cell Technology: Developments, Processes and Products (Spier, R.E., Griffiths, J.B. and MacDonald, C., eds.), pp. 163–165. Butterworth, London.

Ozturk, S.S. and Palsson, B.O. (1990) Effects of dissolved oxygen on hybridoma cell growth, metabolism, and antibody production kinetics in continuous culture. Biotechnol. Progr. 6, 437–446.

Ozturk, S.S. and Palsson, B.O. (1991a) Effect of medium osmolarity on hybridoma growth, metabolism and antibody production. Biotechnol. Bioeng. 37, 989–993.

Ozturk, S.S. and Palsson, B.O. (1991b) Growth, metabolic, and antibody production kinetics of hybridoma cell cultures: 2. Effects of serum concentration, dissolved oxygen concentration and medium pH in a batch reactor. Biotechnol. Progr. 7, 481–494.

Ozturk, S.S., Riley, M.R. and Palsson, B.O. (1992) Effects of ammonia and lactate on hybridoma growth, metabolism and antibody production. Biotechnol. Bioeng. 39, 418–431.

Petersen, J.F., McIntire, L.V. and Papoutsakis, E.T. (1990) Shear sensitivity of hybridoma cells in batch, fed-batch, and continuous cultures. Biotechnol. Progr. 6, 114–120.

Pinton, H., Rabaud, J.N., Engasser, J.M. and Marc, A. (1991) Cytoflow: A new perfusion bioreactor for research and production. Biotech Forum Europe 6, 344–347.

Racher, A.J., Looby, D. and Griffiths, J.B. (1990) Use of LDH release to assess changes in culture viability. Cytotechnology. 3, 301–307.

Ramirez, O.T. and Mutharasan, R. (1990) Cell cycle- and growth phase-dependent variations in size distribution, antibody productivity, and oxygen demand in hybridoma cultures. Biotechnol. Bioeng. 36, 839–848.

Reddy, S., Bauer, K.D. and Miller, W.M. (1992) Determination of antibody content in live versus dead hybridoma cells: Analysis of antibody production in osmotically stressed cultures. Biotechnol. Bioeng. 40, 947–964.

Ryll, T. and Wagner, R. (1992) Intracellular ribonucleotide pools as a tool for monitoring the physiological state of in vitro cultivated mammalian cells during production processes. Biotechnol. Bioeng. 40, 934–946.

Sambanis, A., Lodish, H.F. and Stephanopoulos, G. (1991) A model of secretory protein trafficking in recombinant AtT-20 cells. Biotechnol. Bioeng. 38, 280–295.

Schmid, G. and Keller, T. (1992) Monitoring hybridoma metabolism in continuous suspension culture at the intracellular level. Cytotechnology 9, 217–229.

Schmid, G., Blanch, H.W. and Wilke, C.R. (1990) Hybridoma growth, metabolism, and product formation in Hepes-buffered medium: II. Effect of pH. Biotechnol. Lett. 12, 633–638.

Sen, S., Hu, W.S. and Srienc, F. (1990) Flow cytometric study of hybridoma cell culture: correlation between cell surface fluorescence and IgG production rate. Enzyme Microbial. Technol. 12, 571.

Spier, R.E. (1992) Animal cell biotechnology in the 1990s: From models to morals. In: Animal cell Technology, Vol.5 (Spier, R.E. and Griffiths, J.B., eds.), pp. 1–46. Academic Press, London.

Starbuck, C. and Lauffenburger, D.A. (1992) Mathematical model for the effects of epidermal growth factor receptor trafficking dynamics on fibroblast proliferation responses. Biotechnol. Progr. 8, 132–143.

Wagner, A., Marc, A., Engasser, J. M. and Einsele, A. (1992) The use of Lactate dehydrogenase release kinetics for the evaluation of death and growth of mammalian cells in perfusion reactors. Biotechnol. Bioeng. 39, 320–326.
Wagner, A., Marc, A., Engasser, J. M., Villermaux, S. and Einsele, A. (1990) Continuous production of prourokinase in feed-harvest and perfusion cultures. Biotechnol. Bioeng. 36, 623–629.
Zhang, Z., Al-Rubeai, M. and Thomas, C. R. (1992) Mechanical properties of hybridoma cells in batch cultures. Biotechnol. Lett. 14, 11–16.

3.4 On-line and Off-line Process Analysis

Thomas Scheper, Ruth Freitag, Friedrich Srienc

3.4.1 Introduction

In biotechnology biological material is used for production purposes. A rather complex artificial environment has to be created to further this objective. Hardly any component of this mixture can be considered negligible for the well-being of the biological entity in question. The overwhelming importance of some substances, however, demands close surveillance. A detailed analysis of bioprocesses, especially of mammalian cell culture processes, is necessary for a better understanding. Based on this, process optimization, control, and most of all enhancement of product quality can be obtained. The number of papers about bioprocess monitoring has increased steadily during the last few years. However, most work has been performed on the monitoring of cultivation processes of yeast and bacteria. Most papers dealing with analysis of mammalian cell cultivation monitoring concentrate on off-line analysis or perspectives of on-line monitoring systems intended for future work (Meren et al., 1986; Twork and Yacynych, 1990; Kilburn, 1991; Griffiths, 1986; Cattaneo and Luong, 1993; Male et al., 1993; Konstantinov et al., 1992.)

In this paper we will try to focus on some examples of modern analysis techniques and their application on real cell cultivation processes. Bottlenecks and restrictions will be discussed as well as the interesting analytical potential of these techniques. Analysis can be carried out in line, *in situ*, on-line, off-line, or non-invasively (Callis et al., 1987). Except for the non-invasive methods and the few in-line sensors that combine resistancy to steam sterilization with long term stability and selectivity, all analyzers are located outside the bioreactor in a non-sterile area. While the attempts to increase the number of specific and sterilizable in-line sensors are applaudable, on-line analysis seems at the moment to be the most livable compromise between speed and reliability on one side and safety and flexibility on the other. For optimal analysis the analytical system must be supplied with a representative sample of the cultivation process. The sample should be ready for analysis, i. e. be diluted, concentrated, chemically modified, or otherwise conditioned. When on-line monitoring is to be performed, the sample must be provided continuously; the exact sampling procedure depends on the mode of analysis chosen and the character of the analyte. Thus appropriate sampling is the first step to reliable cultivation analysis.

Only if the interaction of the biological component with its environment is understood, and the fast, reliable, and accurate monitoring of vital parameters characterizing the state of the biological component and its surrounding is possible, then effective bioprocess design becomes possible. In this article we will therefore focus on three different topics: on the aseptic sampling procedures, on the on-line monitoring of cell number, cell characteristics, and cell environment, and on flow cytometric methods.

3.4.2 Aseptic Sampling Biotechnology

Off-Line Sampling Systems

Even though this article will focus on on-line sampling and detection procedures a brief detour on off-line sampling must be allowed here, since many culture parameters in mammalian cell culture, e. g. cell density and viability, are still ascertained off-line. Such samples are obtained manually e. g. by steam sterilizable bottom harvesting valves (for dissolved and suspended substances only) or by hodded samplers. Volatile components are determined in the exhaust or head-space. Glass traps (Ahlmann et al., 1986) and automatic aseptic samplers based on sterilizable 3-way valves were used in model (Imming et al., 1982), or actual (Seifert and Matteau, 1988) microbial fermentations. Mattiasson et al. (1983) used their valve-based sampling system together with an enzyme based biosensing analyzer to determine glucose. Romette (Romette, 1987) at last describes the determination of glucose, lactate, and glutamine in mammalian cell culture samples obtained via a steam sterilizable valve.

A computer-controlled automatic sampling device has been described (Lenz et al., 1985; Ghoul et al., 1986) which consists of a valve, which is automatically steam-sterilized between samplings, and an automatic filtration unit. Between measurements the strip of filter membrane is scrolled forward such that a fresh stretch of membrane is available for each sample. The apparatus was used to couple the bioprocess to an analyzer (here HPLC) and to evaluate the filtration characteristics of the fermenter content, i. e. the amount and morphology of the cells, from the pressure drop across the cake. The system has been used for biomass and glucose monitoring in various cultures and for the monitoring of the penicillin, precursor and by-product concentrations in batch cultures of *Penicillum chrysogenum*. The device is currently distributed under the name of BIO-FILTRATOR by Wazau (Berlin). A similar approach to sampling is used by Ghoul et al. (1986) to enable an automatic glucose analysis during a batch culture of *Candida utilis*.

Filtration Modules

The determination of substrates, products, and other chemical components in the broth, however, becomes biased in samples that include the culture organisms. They will continue to metabolize, they may well be able to react with substances added for analytical purposes, and they would with great likelihood attach themselves to various parts of the analyzer, forming a biologically highly active layer. The use of sampling systems that produce a cell-free aseptic sample stream are therefore preferred in biotechnology. Despite the fact that the molecular species mentioned above vary widely in molecular weight and physico-chemical character, the number of approaches to sampling for them is small. For the determination of low molecular weight substances ultrafiltration or dialysis modules are used, which exclude even cell debris and biomacromolecules from the sample. The upper limit in membrane pore size suitable for aseptic sampling in a bioprocess is determined by the size of the organism used and by the requirements for sterility. Only a membrane with pore size below 0.2 μm ensures complete separation of the suspended particles and the culture fluid. By the same argument only such a membrane will prevent the invasion of microorganisms from the outside, i. e. constitute a "sterile barrier".

Non-sterile sensors may simply be interfaced to the bioreactor content by such a membrane (Bradley et al., 1989).

The choice of using a filtration type sampling system does, however, entail the acceptance of a limited standing time, because of membrane fouling and pore clogging, and of the risk of membrane damage resulting perhaps in a contaminated culture or the escape of genetically modified organisms. Last but not least, the sampling module, including the membrane, has to be sterilizable, which reduces the number of suitable materials. The filter material should not lend itself easily to the adsorption of culture contents. The surface roughness is also important, it should be kept low in order to aid nonadhesion. Materials such as cellulose acetate, cellulose nitrate, nylon, acrylnitrile-copolymer, polyvinylidene fluoride, polytetrafluoroethylene, polysulfone, silicone and the most common polypropylene as well as asymmetric ceramic membranes are mostly used. Membranes made from cellulose are often less suitable, since many organisms are able to hydrolyze cellulose.

Generally the use of membranes in sampling modules requires good knowledge of the physico-chemical characteristics of the operation in question. If the factors influencing the process are not properly understood, correlation between the measured results and the development of the bioprocess are difficult to ascertain. Spohn and Voß (1989) give a review on the background of sampling in biotechnology. Bradley et al. (1991) include sampling devices in their general treatment of systems for bioprocess monitoring and control. If the data are to be used in relation to other important culture data, time lag and sample dispersion have to be considered. Here transmembrane pressure and pore geometry are most important (Christensen et al., 1991).

In-Line Filtration Modules

In-line modules have short answering times, the number of connections, valves, and pumps – all well known trouble spots in view of asepsis – is comparatively small. Their performance, however, depends heavily on their location within the bioreactor in respect to sources of turbulency, such as the stirrer or aerator, to keep from clogging. The modules described below are generally used in stirred tank or air lift bioreactors. If placed well, they show long-term stability and little membrane fouling. Unfortunately an internal module cannot easily be replaced during the culture. Early attempts to create in-line sampling modules included an ultrafiltration module designed by Kuhlmann (1983) for the on-line process analysis in a culture of *P. chrysogenum* and a disk-shaped device described by Lorenz (Lorenz et al., 1984). Similar *in situ* filters using flat ultrafiltration or microfiltration membranes were later used in various types of bioprocesses (Lorenz et al., 1987; Schügerl, 1988; Bayer et al., 1989). They could be steam-sterilized several times and were stable over extended periods of time (Niehoff et al., 1986). The disk-shaped modules can, however, only be inserted into the bioreactor from the inside, which renders them unsuitable for large scale industrial applications, they were never used in mammalian cell cultures.

In order to be able to insert the sampler from the outside various rod-shaped probes were designed that can be connected using a standard sensor port (e.g. Waarvik, 1985; Graf et al., 1991; Marck et al., 1990). The two modules first mentioned can be fitted with various tubular porous membranes both polymeric and ceramic,

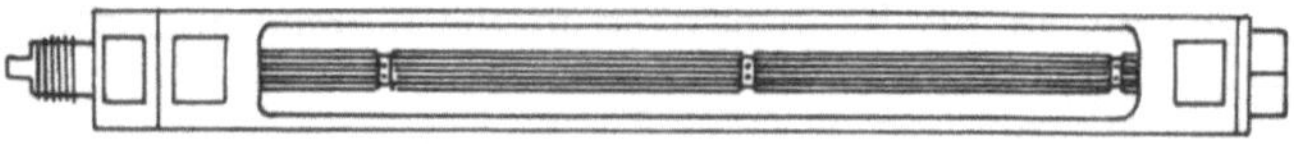

Fig. 3.4.1: In-line filtration module as distributed by ABC, Munich, and Eppendorf, Hamburg, both Germany

the last uses a tubular alumina membrane. While these modules are highly suitable for process monitoring in yeast and bacteria cultures, in-line modules do not come readily to the mind as a means for sampling in mammalian cell cultures with their sensitive organisms and the culture fluids rich in potential membrane adhesives. However, the module originally developed by the University of Hannover (Graf, 1989; Wentz, 1989) and currently distributed by ABC (Advanced Biotechnology Corporation, Munich; Germany), or Eppendorf-Netheler-Hinz (Hamburg, Germany, Fig. 3.4.1) has been successfully used not only for the monitoring of low molecular weight substances in cultures of *E. coli*, and fungi, but also in mammalian and plant cells (ABC Probennahmesysteme Produktinformation) Wentz (1989) and Hübner (1990) describe the use of a rod-shaped module fitted with a ceramic ultrafiltration tubing for sampling of small molecules in cultures of microorganisms and mammalian cells (BHK and hybridoma cells).

Any type of filter tubing developed for harvesting in perfusion cultures of mammalian cells (e.g. Fraune et al., 1988) would by the same reasoning be suitable to sample for low molecular weight components of these cultures.

External Modules

In principle, any type of sterilizable filter module available, e. g. hollow fiber modules, can be used as a sampling unit if its volume does not tax the response time of the analzyer too much. In order to prevent membrane fouling, tangential or cross-flow modules are used. The modules are placed into an external loop through which the bioreactor content is rapidly recirculated. The necessity for rapid recirculation, however, may in itself prohibit the use of the external sampling module in some cases, e.g. for cultures or biomolecules that are sensitive to shear stress. Moreover, the introduction of the external loop can change the process parameters of the bioreactor considerably. Oxygen depletion of the fluid, sometimes entailing changes of the cell metabolism, is also a common occurrence.

The reliability of an external system will be inferior to the in-line modules described before, since pumps tend to be among the least reliable parts of any continuously operated system. The danger of solid deposition with the possible consequence of tube erosion, clogging, and breaking cannot be neglected. The fact that external modules are easily replaced if need be relieves some of the misgivings mentioned above.

Disk-shaped modules

Several types of flat membranes in disk or cylindrical shaped modules have been described (e.g. (Graf, 1989; Riesmeier et al., 1987; Garn et al., 1989; Fenge et al., 1991). The modules require a high flow rate of the broth across the membrane to reduce membrane fouling. Graf (1989) used such a cross-flow module (Millipore Eschborn) equipped with a 0.2 μm membrane (Millipore GVWP 04700) to sample for glucose in continuous cultures of hybridoma cells, while Fenge et al. (1991)

used a small home-made tangential filtration unit fitted with a 0.2 μm hydrophilic membrane to aid the sampling for monoclonal antibodies from perfusion cultures of hybridoma cells. Since the cells were retained within the bioreactor by the perfusion membrane in Fenges case, rapid recirculation of the culture supernatant was no problem.

Tubular modules

An alternative to disk-shaped modules is the use of tubular ones. These modules have the additional advantage of behaving more stream-lined in a flow-system. The use of commercially available tubular microfiltration and ultrafiltration membranes fitted onto a rod-shaped cross-flow module (length 200 mm, i.d. 5.5 mm, dead volume 2.56 ml) has been described several times (e.g. Lorenz et al., 1987; Schmidt et al., 1986; Ross and Schügerl, 1988). The response time depends on the dead volume and the recirculation rate, while the permeate rate and the long term stability decreased with the viscosity and the solid content of the culture medium.

Mitzutami et al. (1987) describe the use of a tubular ceramic filter, Chotani and Costantinides (1982) the use of an Amicon filtration module with a cellulose membrane for sampling. A module consisting of a single hollow-fiber contained within a stainless steel holder has been described by Hoffman (1985), a module with seven

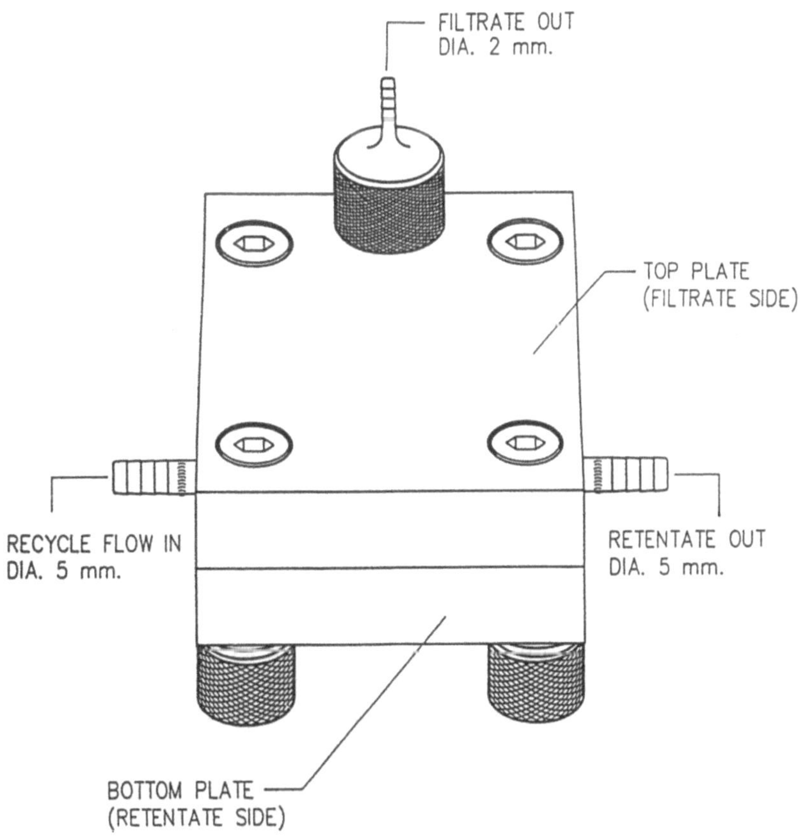

Fig. 3.4.2 External tangential flow module as distributed by Applicon

fibers by Bayer (1987) and Herold (1986). The successful use of a commercially available cross-flow module (Technomara, Fernwald) in monitoring pullulanases in continuous cultures of *Clostridium thermosulfurogenes* has been published by Freitag et al. (1991). External filtration modules for sampling have already been used in industrial type pilot plant fermenters. An example was introduced by Gram et al. (1990), who described a cross-flow module used in connection with an automated glucose analyzer in a pilot plant penicillin V production process.

Among the few commercially available samplers that have been used in mammalian cell culture is the A-sep module from Applicon (Fig. 3.4.2), a tangential flow module, where the bioreactor content is pumped through a meander shaped silicon tubing (i.d. 5 mm) at flow rates up to 50 ml · min^{-1} [44]. During a 19 d culture of HeLa cells, a continuous sample stream of 0.1 ml · min^{-1} was produced. Cell growth rate and size distribution depend on the filtration process parameters.

Modules with secondary flow

The modules described so far relied solely on a rapid tangential flow to reduce membrane fouling. As mentioned before, this remedy can only be used up to a certain limit. Better results, especially where sampling from fluids with high solid or protein contents or cultures of filamentous organisms is concerned, are obtained for modules where a secondary flow is induced, e.g. in the spiral module by Winzeler (1990), by vortices in the spiral flow channel as in the Biodruckfilter (Kroner et al, 1987), or by vortices due to an inner rotating cyclinder. The recirculation rate can be greatly reduced in these modules, while the protein transmission becomes independent of the transmembrane pressure.

Perhaps the commercially most successful attempt to design a membrane based sampling-module with secondary flow for biotechnology is the BIOPEM developed by the GBF (Gesellschaft für Biotechnologische Forschung, Braunschweig) and B. Braun Biotech International (Fig. 3.4.3). The module consists of a round filter-cell (200 ml) equipped with a magnetic stirrer, that can be lower or raised depending on the filtration process taking place. The stirrer causes a strong shear effect close to the filtration membrane, therefore membrane fouling is reduced. The membrane area is 45 cm^2, many conventional micro- and ultrafiltration membranes can be used in the system (Kroner et al., 1986). The cross-flow rate can be varied between 10 and 60 ml/h, the filtration rate between 0.1 and 1.5 ml · min^{-1}. Answering times as measured for a flow rate of 0.5 ml · min^{-1} were between 100 and 110 seconds. The module has been used as a basis for the monitoring of low molecular weight substances (Kroner et al., 1986), and also in measuring substances such as starch (Kroner et al., 1986), proteins (Recktenwald et al., 1985; Hustedt et al., 1985), enzymes (Kroner and Kula, 1984; Kroner (1982); Kroner et al., 1968; Stamm et al., 1989) and antibodies (Freitag et al., 1991).

A somewhat different approach to the successful separation of the culture organisms from the supernatant has come up recently (PERSEP by TECH-SEP GmbH Frankfurt a. M.). Here a small centrifuge is placed within the fermenter very much like the in-line sampling rods. Since the ceramic membrane is used at 3.000 to 5.000 RPM, membrane fouling is no problem. It remains to be seen whether this approach can also be used with sensitive organisms.

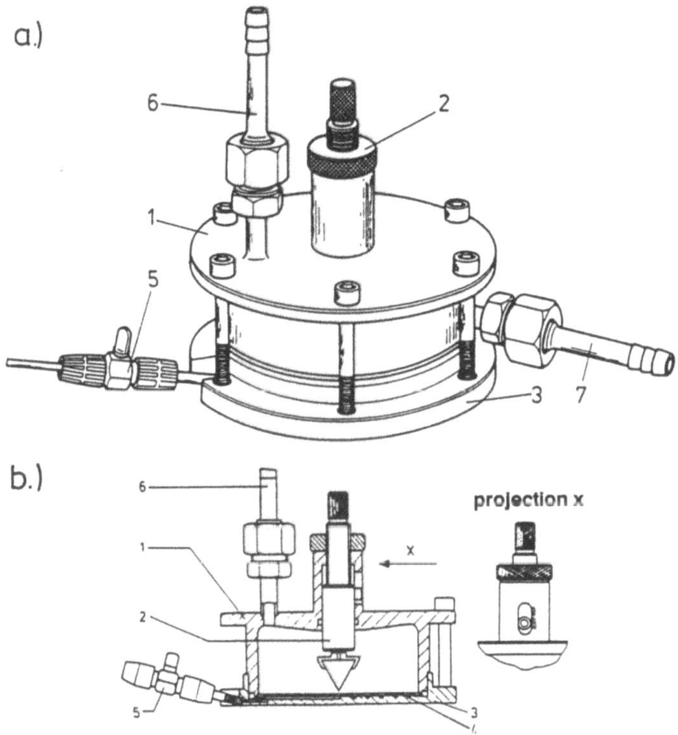

Fig. 3.4.3 BioPEM as distributed by B. Braun Biotech International, Melsungen Germany (a: whole module; b: schematic drawing with adjustable stirrer assembly; 1: top plate; 2: adjustable magnetic stirrer assembly; 3: bottom plate; 4: channels; 5: shut off valve; 6: outlet line; 7: inlet line)

Dialysis modules

For the monitoring of low molecular weight substances, the sampling modules can equally well be fitted with flat or tubular dialysis membranes. Here concentration difference on the two sides of the membrane are the reason for the molecules transgression. Two operation methods are feasible: static and dynamic. In the static mode, also called closed-loop, near equilibrium is reached for the concentrations on both side of the membrane between – discontinuous – sampling. The concentration measured is therefore directly related to the concentration within the bioreactor. In the dynamic or open-loop mode of operation, a suitable liquid is pumped by on the analyzer side of the dialysis membrane and drawn into the analyzer. Here mass-transfer properties of the analyte in question, as well as the volumetric flow rate of the sampling stream, have to be taken into consideration for the estimation of the analyte concentration within the bioreactor. Since the measuring range of many biosensors is rather small and sometimes way below the concentration range actually expected during the cultivation, the controlled sample dilution of the dynamic sampling method might even be beneficial.

Installation and operation of dialysis modules is similar to the procedures found for filtration modules. Zabriskie and Humphrey (1978) used a probe located in a baffle of their bioreactor to monitor glucose. The patent for a tubular dialysis module where the membrane is fitted over a groove milled into the tubular support is held by Glaxo Laboratories Ltd. (1974). Mandenius et al. (1984) describe a sterilizable cross-flow module where the dialysis principle is used to sample for low molecular weight molecules in model yeast "cultures". Gibson and Woodward (1988) describe the use of an in-line dialysis module for glucose and ethanol monitoring in a culture of *S. cerevisiae*. The dialysis membrane was stretched over the tip of the sampling rod. Isotonic sucrose solution was used to take up the sample in order to prevent excessive water transfer into the reactor.

Holst et al. (Hakanson et al., 1990; Holst et al., 1988) used an intriguing design for sampling. The probe itself consists of two coaxial concentric tubes, the inner of which is somewhat shorter than the outer one. Liquid containing a metabolic inhibitor to prevent changes in glucose concentration after sampling, is pumped

Table 3.4.1. Sampling systems

Type	Remarks	References
Off-line Sampling Modules	Glass trap	6
	3-way valve	
	model cultivation	7, 8, 9
	microbial culture	10, 11
	mammalian culture	12
	Valve/Filtration unit e.g. the BIO-FILTRATOR from Wazau, Berlin, Germany	13, 13
In-line Filtration Modules	Disk-Module	20, 21, 22, 23, 24
	Process Analysis *P. chrysogenum*	19
	Rod-Module (ABC, Munich, and Eppendorf, Hamburg, both Germany) Process analysis in cultures of *E. coli*, fungi, mammalian, plant cells	25, 26, 27, 28, 29, 30
External Modules (By-Pass)	Disk-Modules	28, 32, 33, 34
	Tubular Modules	37, 38, 39, 40, 41, 42, 43
	A-Sep Module (Applicon)	44
Modules with Secondary Flow	Spiral Module	45
	Biodruckfilter	46
	BIOPEM (B. Braun, Biotech International, Melsungen, Germany	47, 48, 49, 50, 51, 52, 53, 54
Dialysis Modules	In-line	55, 56, 59, 60
	Cross-Flow	57
	Process control (Yeast)	58

into the outer tube at a flow rate of 3 ml · min^{-1}. At the same time liquid is sucked out of the inner tube at twice that rate, such that some fermentation broth, the sample, is carried along. The sampler is interfaced to the non-sterile outside via a dialysis membrane (cuprophane flatfoil). Small molecules, such as glucose, pass the membrane and can be analyzed using conventional methods. Close control of the two flow rates is necessary to control the sample dilutions and to prevent the penetration of the inhibitor into the bioreactor. (Buttler et al., 1993; Van de Merbel et al., 1993)

While the problem of membrane fouling is less severe for dialysis modules, since no transmembrane pressure gradient is applied, a certain tendency of cells – especially mammalian cells – and biomacromolecules to adhere to the membrane remains. Since this layer is biologically active, frequent re-calibrations are still necessary.

All sampling devices introduced in this chapter have been compiled in Table 3.4.1.

3.4.3 Optical Sensors for Cell Characterization and Concentration Estimation

Two types of non-invasive optical sensors are available for on-line monitoring of bioprocesses: nephelometric sensors and fluorescence sensors. Nephelometric sensors monitor the turbidity of the cultivation medium. This turbidity results mainly from suspended cells, particles, and air bubbles in the liquid phase. The amount of turbidity is not only a function of the cell and particle concentration, but it also is extremely dependent on the physical size of the scattering centers (Vanous, 1978). No distinction between viable and non-viable biomass concentration is possible. Various designs can be used to measure the turbidity (Reardon and Scheper, 1991), but mainly two designs are used in biotechnology: the transmission sensor and the retroflective sensor.

Various types of transmission sensors are available (e.g., Bonnier Technology Group, Hach), some of them as a combined transmission and forward scatter sensor (Monitek, Wedgewood, Optek). Most of them work at wavelengths available from near infrared-emitting diodes as light source and with different light path lengths. Thus, high biomass concentrations (e.g., 5 OD's for the Wedgewood Model 610 probe) can be monitored. One disadvantage of transmission light sensors with a defined light path length is that sensors designed for high cell concentrations are often insensitive to small cell concentrations and vice versa. *In situ* versions of such sensors can sometimes only be interfaced into pipes, thus, they must be installed into a loop of the bioreactor (Monitek system). Others, designed as probes to be interfaced via standard electrode ports might cause problems with the hydrodynamic behavior of the medium in the flow-through cell. It cannot be guaranteed that the sample in the light path is representative in its composition in comparison to the homogeneously mixed cultivation medium.

Such disadvantages are overcome by retroreflective sensors. Here, the excitation light is guided into the bioreactor via a fiber optic (Aquasant, Mettler), and the backward scattered light is collected via a fiber optic and thus detected. The signal increases with increasing cell concentration up to a theoretical maximum value,

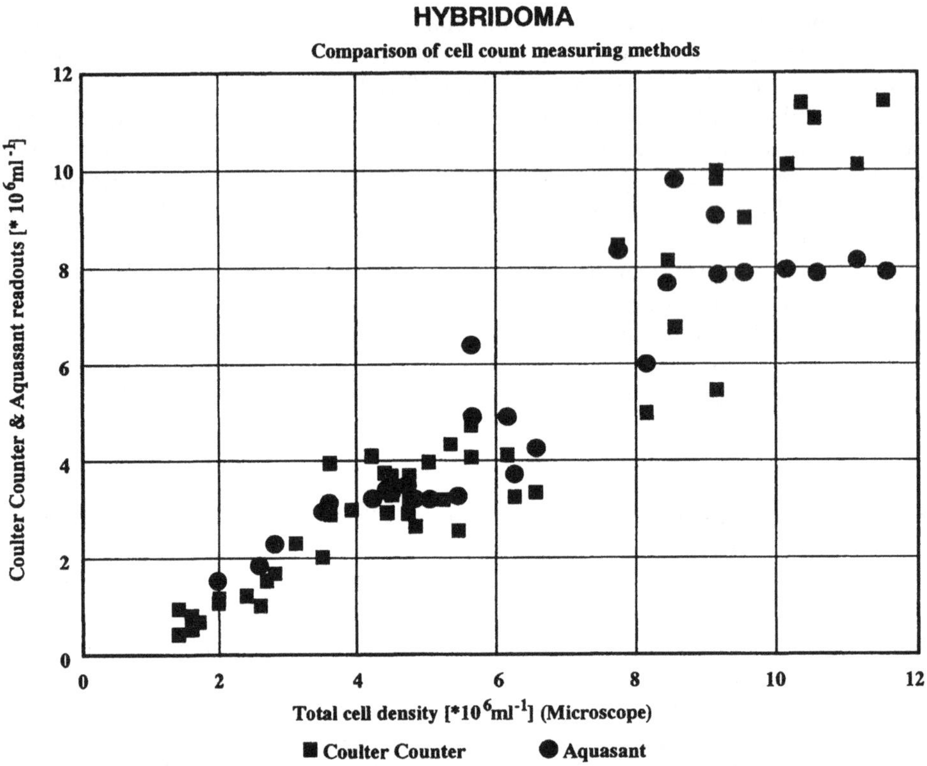

Fig. 3.4.4 Comparison of different cell counting methods during a cultivation experiment of hybridoma cells (microscopic counting, Coulter counter data and Aquasant data)

when all light guided into the reactor is scattered backwards again into the sensor. This kind of sensor is easy to install, and the cost is relatively low. It can be used especially for the control of perfusion cultivations in order to keep the whole process at a constant turbidity. This type of sensor is described in literature mainly for bacterial or yeast cultivations (Reardon and Scheper, 1991). However, it can be found in several industrial bioreactors used for mammalian cell cultivation (Konstantinov et al., 1992). Figure 3.4.4 shows the application of an Aquasant sensor for monitoring of a hybridoma cell cultivation. The Aquasant data are compared to microscopic and Coulter counter data. These data are nearly always slightly lower than the microscopic data but can be obtained much easier and only the Aquasant sensor can supply the data continuously.

The main advantages of nephelometric sensors are the small cost, the easy installation, the simple data interpretation, its nearly instantaneous response times, and its non-invasive monitoring principles. Disadvantages are the interference of other variables such as solid particles, bubbles with the sensor signal from cell turbidity, and the influence of the cell size on the turbidity (calibration must be performed for each type of cell line), and the impossibility to distinguish between viable and non-viable biomass.

A sensor monitoring metabolic active biomass concentration is the *in situ* fluorescence sensor. Here, the reduced dinucleotides (NADH and NADPH) in living cells can be irradiated with UV-light of about 340–360 nm to fluorescence at a wavelength of about 460 nm. By measuring this excitation light, the NAD(P)H pool inside living microorganisms can be monitored. *In situ* fluorescence sensors (BioChem Technology, Ingold) irradiate the cells in the bioreactor and monitor the NAD(P)H-dependent fluorescence light via the same optical system. The sensors can be interfaced via standard electrode ports. Various types of *in situ* fluoresence sensors are described in literature (Harrison and Chance, 1970; MacBridge et al., 1986; Beyeler et al., 1981; Scheper and Schügerl, 1986). The culture fluorescence intensity is affected by the viable biomass, the metabolic state of the cells, and also by non-biomass factors (e.g., bubbles, fluorescent medium components). The sensor's applications described in literature cover biomass concentration estimation, reactor characterization (Scheper and Schügerl, 1986; Beyeler et al., 1983; Einsele et al., 1978; Gschwend et al., 1983), medium composition Scheper et al., 1987), process monitoring, e.g., metabolic state of the cells (Reardon et al., 1986; Reardon et al., 1987; Srinivas and Mutharasan, 1987; Zabriskie and Humphrey, 1978), and process control (Meyer et al., 1984). Several applications of this type of sensor for animal cell cultivation monitoring are given in Table 3.4.2. Literature confirms that tremendous problems arise from low fluorescence intensities at relatively low cell concentrations. The interpretation of the signals is non-trivial and often in contrast to other cultivation data. However, Siano and Mutharasan (1991) were able to show impressively that metabolic studies such as aerobic-anaerobic transitions and substrate depletion experiments, can be carried out. All experiments were performed at concentrations with a viable cell count of $3-18 \cdot 10^6$ per ml. The NAD(P)H-dependent culture fluorescence gives a non-invasive, instantaneous insight into the metabolic state of the organisms under real cultivation conditions.

Table 3.4.2. Application of NAD(P)H-dependent culture fluorescence monitoring systems for mammalian cell cultivation processes

Organism	Phenomenon Studied	References
S. frugiperda	cell growth, biomass concentration	[70]
	virus infection,	[77]
human melanoma cells	metabolic studies, biomass concentration	[108]
mouse hybridoma cells	metabolic studies, biomass concentration	[109]
hybridoma cells	metabolic studies, biomass concentration	[110]
ATCC HB32	metabolic studies, biomass concentration	[76]

The advantage of this method is that only viable cells are monitored, while dead cells with no NAD(P)H do not contribute to the sensor's signal. This technique is based on simple and robust sensors, is non-invasive and has fast response times. However, several disadvantages becomes obvious from literature, such as interferences with other fluorescing compounds (background fluorescence) and the non-trivial signal interpretation.

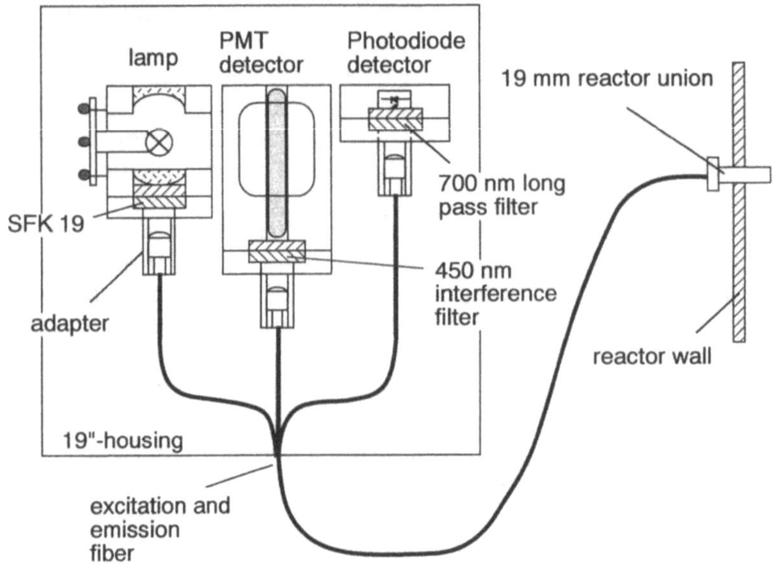

Fig. 3.4.5 Setup of a two-channel fluorescence/turbidity sensor. Modulat system for the simultaneous detection of the NAD(P)H-dependent culture fluorescence and the turbidity inside the bioreactor

The combination of turbidity sensors and fluorescence measurement should be a promising alternative to overcome several problems of both optical monitoring techniques. In Figure 3.4.5, the schematic setup of a two-channel optical system for the simultaneous detection of NAD(P)H-dependent fluorescence and the retroflective scattered light is shown (Anders et al., 1993). A quartz tungsten halogen lamp and simple condensing optics were used for the NAD(P)H excitation at about 360 nm and for the scattering measurements at higher wavelengths (above 700 nm). The light was focused on a bifurcated fiber bundle that guided the light into the bioreactor. Here, cells were irradiated with light of both wavelengths. A fraction of the NAD(P)H-dependent fluorescence and the scattered light is collected via the same fiber optic bundle and is guided backward to a photomultiplier tube. NAD(P)H fluorescence is monitored at 450 nm and the scattered light at wavelengths above 700 nm. The NAD(P)H-dependent culture fluorescence light is a measure of the viable cell mass and of the metabolic state of the cells, while the backward scattered light is a measure of the turbidity inside the bioreactor.

The two-channel system was used for the *in situ* monitoring of a batch cultivation of the insect cell line of *Spodoptera frugiperda*. The turbidity and fluorescence signals obtained during such a cultivation are given in Figure 3.4.6. The whole process is of interest for the production of baculoviruses which can be used as a bioinsecticide, since it infects the caterpillars of these butterflies. In the batch process, the *Spodoptera* cells were cultivated up to high cell densities and were then infected with the virus (after 120 h). In the second phase of the bioprocess, baculoviruses were produced by the infected cells which lysate in the third phase (about 200 h) and set the virus free at high concentrations. From the reciprocal plots of the turbidity and fluor-

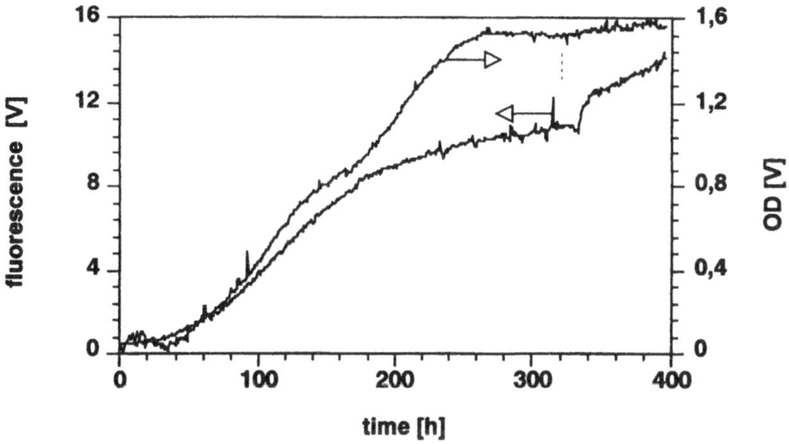

Fig. 3.4.6 Course of fluorescence and turbidity signals during a batch cultivation of *Spodoptera frugiperda*. Cells were infected after 120 hours with a baculovirus.

escence data shown in Figure 3.4.7, the different phases of the whole process become obvious by comparing the slopes of the graphs. During the pure exponential growth phase, a constant slope was achieved (65–120 h), which changed tremendously after infection during the virus expression phase (120–200 h). Another change in the slope is observed during the cell lysis phase (180–200 h). A direct insight into the state of the cultivation process becomes possible with this type of monitoring system and helps to determine the optimal harvesting point or to find control strategies for process optimization.

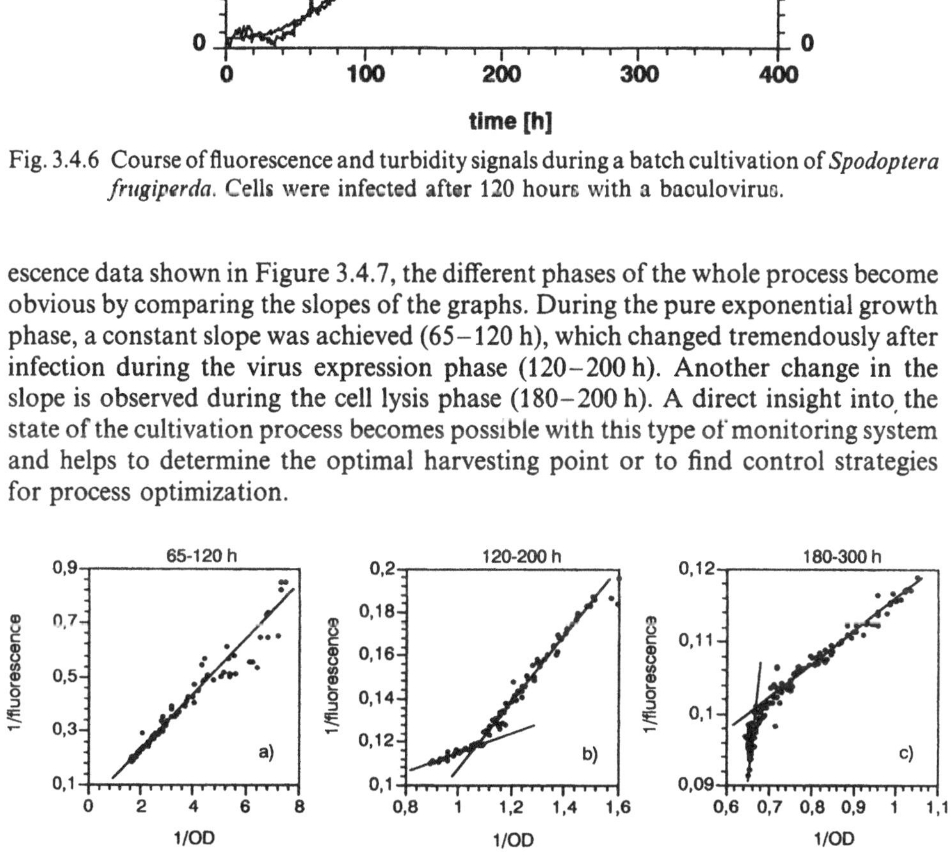

Fig. 3.4.7 Reciprocal fluorescence and turbidity data during a batch cultivation of *Spodoptera frugiperda* for the determination of growth phase, infection phase and cell lysis.

3.4.4 On-line Monitoring of the Cell Environment

Different sensors are used for on-line monitoring of the cell environment. Especially pH, pO_2, and redox probes are widely used for this purpose. The amount of literature about the application about such systems is vast (Kilburn, 1991; Griffiths, 1986;

Spier et al., 1991; Hwang and Sinskey, 1991). More complex analyses were mostly performed off-line up to now. However on-line systems for the monitoring of substrates, products, or interesting metabolites might be helpful for mammalian cell cultivation monitoring. However, it must be guaranteed that such systems cause no infections, that a validation is possible, and that the efforts for such high sophisticated on-line analyses really have an impetus on process optimization (Werner et al., 1992).

In this part we will focus on on-line biosensor systems for the analysis of low- and high-molecular weight components in the bioprocess medium. However, there are other techniques well known from bacteria and fungi cultivation processes with a huge potential for mammalian cell cultivation processes, such as on-line HPLC (Monseur and Motte, 1988; Mathers et al., 1986; Schügerl, 1991; Schügerl et al., 1991; Compton, 1990), on-line mass spectroscopy (Heinzle and Dunn, 1991; Kotiaho et al., 1991), or NMR measurement (Reardon and Scheper, 1991; Fernandez et al., 1988).

Biosensors are increasingly important for bioprocess analysis because they are intended to make complicated analysis procedures simple and easy to use. They combine the high specificity of biological components and the capabilities of modern transducer technology. The biological component can either be a catalytic one (e.g., enzymes, organelles) or one of an affinity reaction (e.g., receptors, antibodies). It reacts with the analyte to be measured, and during this reaction, variables in the microenvironment of the transducer are changed. Heat, mass changes, optical properties, proton, electron, gas, or ion concentrations are among these variables. The transducer records the changes during the biochemical reaction. The amount of the signal correlates to the analyte concentration. For detailed information we refer to the literature (Twork and Yacynych, 1990; Scheller and Schubert, 1991; Schmidt and Scheller, 1992; Turner et al., 1987).

Biosensors should be able to give a detailed insight into bioprocesses and would, thus, be extremely helpful for better process monitoring, understanding and optimization. Several publications deal with the applications of biosensor systems for bioprocess monitoring (Merten et al., 1986; Twork and Yacynych, 1990; Kilburn, 1991; Griffiths, 1986; Fernandez et al., 1988; Scheller and Schuber, 1991; Schmidt and Scheller, 1992; Turner et al., 1987; Scheper, 1992; Bradley et al., 1991; Mattiasson and Hakanson, 1992). The application of biosensors as *in situ* sensors is often problematic, since the non-sterilizable biological component must be applied to the sensor after sterilization procedure, and sample dilution, calibration, and conditioning must be performed inside the sensor. Thus, relatively complicated sensor arrangements are used. However, promising examples of such sensors exist in the literature for penicillin and glucose (Enfors et al., 1977; Bradley et al., 1989). Another aspect must be regarded: Will the sensor be approved by the FDA? Often, amperometric sensors use mediators, such as ferrocene derivatives not covalently bound to the sensor, and problems might arise if toxic mediators bleed out into the cultivation medium. Such aspects must be regarded during development of *in situ* sensors for cell cultivation monitoring.

Another way to overcome several problems in biosensor applications for bioprocess monitoring is to incorporate such sensors in flow injection analysis (FIA) systems. Figure 3.4.8 shows the principle of such an arrangement. A sampling device

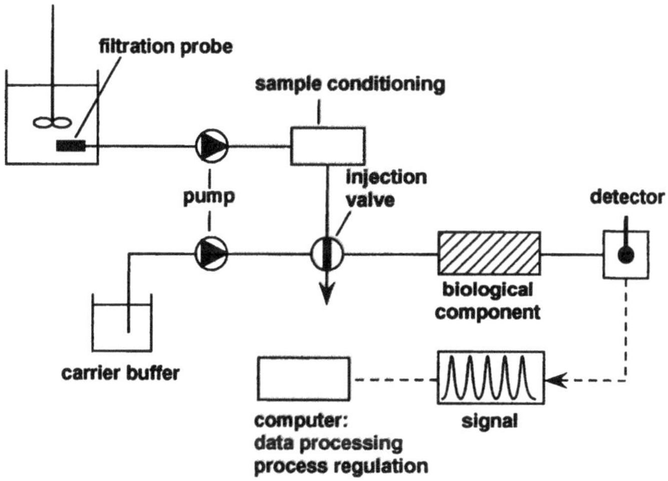

Fig. 3.4.8 Principles of flow-injection analysis systems with biosensors incorporated.

is installed at the fermenter and supplies the analysis sample continuously with a cell-free sample. This sample can be conditioned according to the analysis requirements (e.g., pH or buffer capacity adjustment, addition of cosubstrates and reactants, dilution). The conditioned sample is then injected via an injection valve into a continuous flowing buffer stream transporting the sample to the biosensor, where the analysis is performed under optimal conditions. Several biosensors or other analysis systems (e.g., HPLC) can be connected to the cell-free buffer stream.

This kind of FIA analysis does not allow real on-line monitoring. Discrete data are obtained, but normally with a high frequency (within a few minutes) resulting in a quasi on-line monitoring. One of the main advantages of these systems is that cultivation process and multi-component analysis are completely separated. One sampling system is necessary to provide the cell-free sample, and at the same time is the sterile barrier. Of course, it is also the bottleneck of the whole system, as shown above, because it must supply the analysis system with a representative sample from the cultivation process for the entire bioprocess and with a very small delay time. These criteria are frequently fulfilled in small-molecular weight components, but not in high-molecular weight components such as proteins. Different applications of such FIA-based biosensor monitoring systems for the monitoring of animal cell cultivations are given in Table 3.4.3. The most interesting analytes are glucose, lactate, and glutamine (Cattaneo and Luong, 1993; Male et al., 1993). Of course, not only biosensors can be used, but other sensors as well, such as ion selective electrodes (Graf and Schügerl, 1991). The major drawbacks of FIA systems for bioprocess monitoring are the demand of an appropriate sampling system, the often high investment costs, the more complicated data processing (results are obtained as peaks), and the need of a control for the entire FIA system. However, the advantages detailed above are tremendous and, thus, FIA-based analysis systems have an enormous potential for bioprocess monitoring, even in industrial processes.

Table 3.4.3 Application of biosensor systems for on-line process monitoring of mammalian cell cultivation processes

Analyte	Bioprocess	Sensor system	Duration of online application?	Reference
low molecular weight components:				
Glucose	BHK cells	FIA, optrode	220 h	[111]
		FIA, luminometer	220 h	[112]
		FIA, oxygen electrode	100 h	[98]
	Hybridoma cells	FIA, YSI analyzer	200 h	[97]
Lactate	BHK cells	FIA, optrode	220 h	[111]
		FIA, oxygen electrode	100 h	[98]
	Hybridoma cells	FIA, H_2O_2 electrode	270 h	[113]
Glutamine	Hybridoma cells	FIA, H_2O_2 electrode	270 h	[113]
high molecular weight components:				
Mouse IgG	Hybridoma cells	FIA, affinity chromatography	160 h	[100]
		streaming potential		[106]
		FIA, immunoassay	500 hours	[101, 102]
Antithrombin III	Hybridoma cells	FIA, immunoassay	35 hours	[101–105]
rt-PA	Hybridoma cells	FIA, immunoassay		[107]

Monitoring of high-molecular weight components

Especially in animal cell cultivation technology, the analysis of target proteins (product) is of interest. These proteins might be the product produced by the cultivated cells or metabolic intermediates indicating the physiological state of the organisms. The latter would be, for example, lactate dehydrogenase which is analyzed as a marker for cell damage (Wentz and Schügerl, 1992). The analysis of enzyme activities in FIA systems is described in literature in detail (Ruzicka and Hansen, 1988). Problems arise when proteins that should be determined precisely are not catalytically active or cannot be analyzed by a simple, specific enzymatic assay. In such a case, immunoanalysis is the best way to determine these proteins very specifically. Normally, immunoassays are used for the determination of proteins at very low concentrations. In animal cell cultivations, the target protein is mainly produced at a high level, or the result of each optimization should at least be the production of this protein at high concentrations. This fact entails the requirement that immunoanalysis systems which can be used for bioprocess monitoring should work in a protein concentration range of about 0.1 to 1000 mg/l. Up to now, mainly the conventional off-line ELISA is used for monitoring. Several samples are diluted and monitored in microtiter plates. Such system can hardly be used for on-line monitoring, because it is time consuming (dilution, several washing steps), error-inherent (dilution), and costs are too high when only one sample needs to be analyzed. Thus, different methods were developed during the last few years for automated immunoanalysis systems intended for bioprocess monitoring or even applied to different processes. These systems are described in detail in excellent reviews by Bradley et al. (1991) and Mattiasson and Hakanson (1992). Among the systems already employed for animal cell cultivations are affinity chromatography systems (Stöcklein et al., 1992) (analysis time: 10 min), turbidometric systems (1–3 min) (Fenge et al., 1991; Freitag et al., 1991; Freitag et al., 1991; Freitag et al., 1991), competitive, heterogeneous systems (Degelau et al., 1992) (analysis time: 15–30 minutes), and a streaming potential system described by Miyabayashi (Miyabayashi and Mattiasson, 1990).

The principles of the turbidometric system are shown in Figure 3.4.9. A sample (20–50 μl) from the bioreactor is injected into an FIA transport buffer stream simultaneous with a defined amount of antibody. Both are mixed in the FIA system and put through a mixing coil. Here, immunocomplexes are formed while the flow is stopped for a defined time. Afterwards, the turbidity is measured in a photometer. The turbidity is a measure for the amount of immunocomplexes and, thus, a measure of the antigen concentration of the sample. In order to reduce interference with the sample matrix, a blank value of the sample is analyzed in a second photometer channel. The system can be calibrated simply by injecting calibration solutions of the antigen. Assays for IgG, ATIII, rt-PA produced in mammalian cell cultivation processes are described (Fenge et al., 1991; Freitag et al., 1991; Freitag et al., 1991; Freitag et al., 1991; Degelau et al., 1992; Miyabayashi and Mattiasson, 1990; Middendorf et al., 1993). All systems work well in the range of 1–800 mg/l protein. During the development of such an assay, the ratio of antigen-to-antibody must be optimized for the expected concentration range. Mostly a 3–6-fold excess of the antibody is necessary for an optimal assay within an analysis time of 1–3 minutes per sample. The whole FIA system is designed in such way that it can be activated

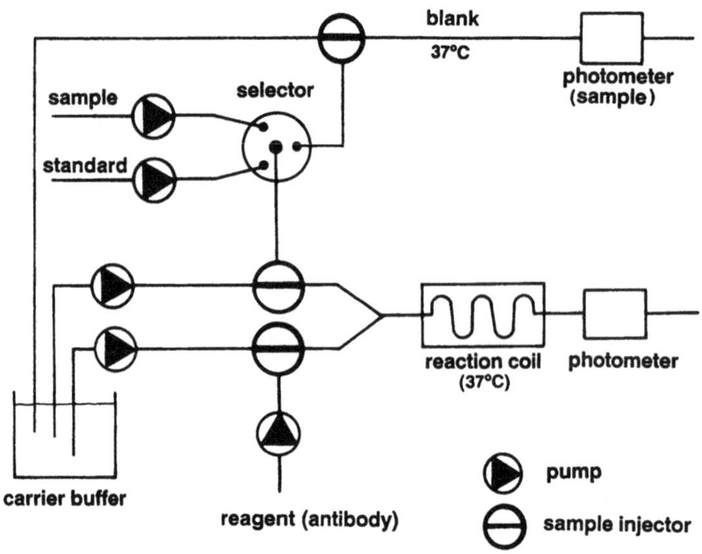

Fig. 3.4.9 Principles of an automated on-line immunoanalysis system based on the detection of the turbidity caused by the generation of immunocomplexes.

at different time intervals (e.g., every 5 hours) during which it analyzes three samples and runs a calibration procedure. Such a complete analysis cycle activated at the different time intervals requires about 15 minutes and gives data with a standard deviation below 1.5%. In Figure 3.4.10, the application of the turbidimetric system for the on-line analysis of a mouse IgG produced by a hybridoma cell line is presented (Fenge et al., 1991; Freitag et al., 1991; Schulze et al., 1994)). The fluctuations in the IgG concentration in the second part of the perfusion process are due to problems in the perfusion system and the sampling probe. Thus, the importance of a reliable

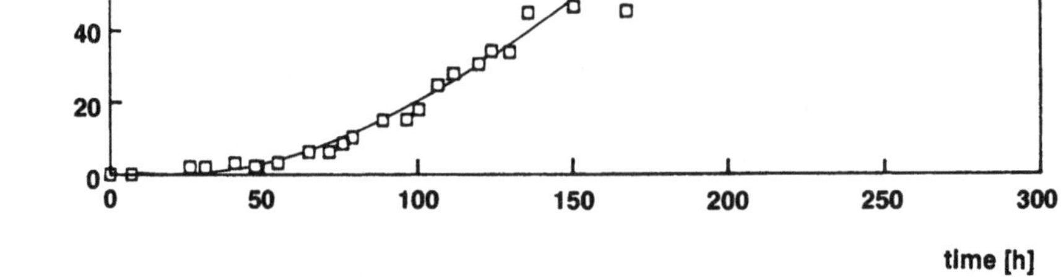

Fig. 3.4.10 On-line monitoring of mouse IgG production during the cultivation of a hybridoma cell line in a perfusion bioreactor system.

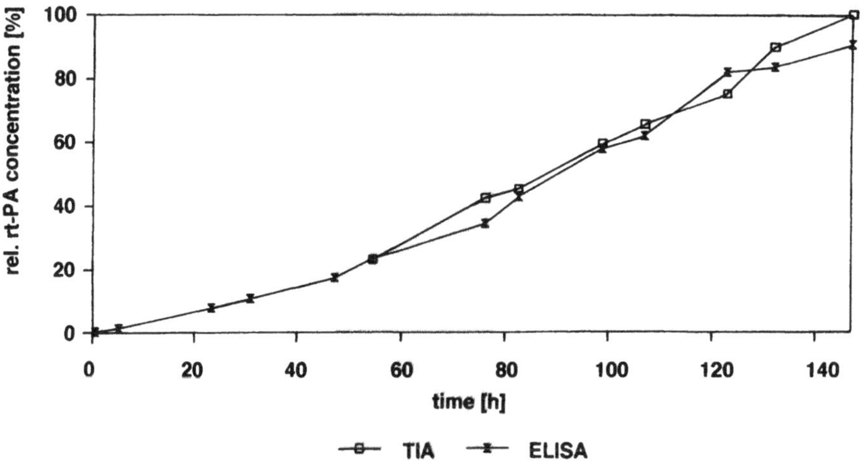

Fig. 3.4.11 Monitoring of rt-PA production in an industrial cultivation process of a production hybridoma cell line. Comparison of conventional ELISA data and the data of the turbidimetric system

sampling system for high-molecular weight components becomes obvious. In Figure 3.4.11, the comparison of turbidometric and conventional ELISA assays for a production process of rt-PA is shown. The turbidimetric data were obtained after the above-described assay cycles, while the ELISA assays were obtained with a delay of about 24 hours.

It becomes obvious from these results that on-line immunoanalysis is possible for direct cultivation monitoring. A major disadvantage of such turbidimetric systems is high consumption of antibody. It must be available at a low cost and in high amounts. However, the system runs as an automated stand-alone system with a high analysis rate. On the other hand, turbidimetric assays are relatively simple to perform and, thus, are interchangeable in this system. Another way to overcome the cost for antibodies is the immobilization of these immunocompounds, as described in the affinity and the heterogeneous, competitive assay systems.

3.4.5 Process Analysis with Flow Cytometry

Flow cytometry is an established measuring technique in basic biological and biomedical research. An important application of flow cytometry is in the area of cancer research, and an extensive body of literature exists, therefore, that describes many applications of this measuring approach in studies of basic cell biology, such as cellular immunology, and in investigations with a focus on cancer research.

Biotechnological applications have been rather rare. This is partially due to the significant cost associated with the purchase and maintenance of such an instrument. In addition, it requires considerable training in the operation of the instrument and some experience in the handling, analysis and interpretation of the large amount of data that can be obtained in a short period of time. Also, direct applications of flow cytometry that demonstrate an immediate benefit to bioprocesses are perhaps much less obvious than in the medical field where, for instance, the early detection of some aberrant cell forms might lead to increased chances of diagnosing and curing certain diseases. However, if one has the motivation and opportunity to elucidate the basic nature of problems associated with the dynamics of growth and product formation of cell cultures, then flow cytometry becomes an indispensible tool. The reward of its proper use is an insight into the mechanisms that govern the dynamics of cell culture processes.

Flow cytometry, along with the vast repertoire of cell preparation techniques, measuring procedures and data analysis techniques developed for mammalian cell systems by biomedical researchers, currently represents an under-utilized tool that could be exploited in mammalian cell culture technology. This is particularly true if it is realized that advances can be made only if one is able to understand phenomena on a mechanistic basis.

Cell culture deals with cell populations, and the complication one encounters is that cells within the culture are different from each other. This variability can be traced to different origins. First, cells might differ from each other because they represent different species as defined by inheritable, genetic material. This includes mutants or cell types altered with recombinant DNA. Second, a cell line with homogeneous and identical genetic information typically consists of cells that are different from each other. This variability is due to the inherent nature of cell growth, that is, the operation of the cell cycle which drives cells from one cell division to the other. Cells at different positions in the cell cycle typically have significantly different properties or physiological states. Finally, influences from the environment such as nonideal mixing conditions in a bioreactor might lead to different development and growth of certain cells, and this also contributes to heterogeneous cell populations in the culture (see also Chapter 2.1.6.2.2).

Heterogeneities in cell populations can be identified if cell cultures are studied at the single-cell level and if distributions of cellular properties are measured. All sensors discussed in the previous sections of this chapter are population average sensors, i.e. the sum of response obtained from each individual in the cell population is detected. Thus, the interpretation of the response on a mechanistic level is difficult, if not impossible, without additional information. From population average responses one can sometimes infer certain heterogeneities in a cell population if the effect of the heterogeneity is well established. In general, however, population average sensors preclude insight into the structure of a growing cell population. This makes it impossible to explain the mechanistic basis of cellular phenomena from population average data.

Principle of flow cytometry

Flow cytometry can only be applied to cell samples consisting of single cells. Cells grown in tissue must be dispersed. This is due to the principle of measurement (see Figure 3.4.12). A sample stream containing the individual cells is injected into a sheath stream, typically water, saline or a dilute buffer. The pressure difference between the two streams can be adjusted such that the laminar sample stream is reduced to a very small diameter and cells are spatially separated along the axis of this stream. A light source, generally a laser beam, is focused on this sample stream.

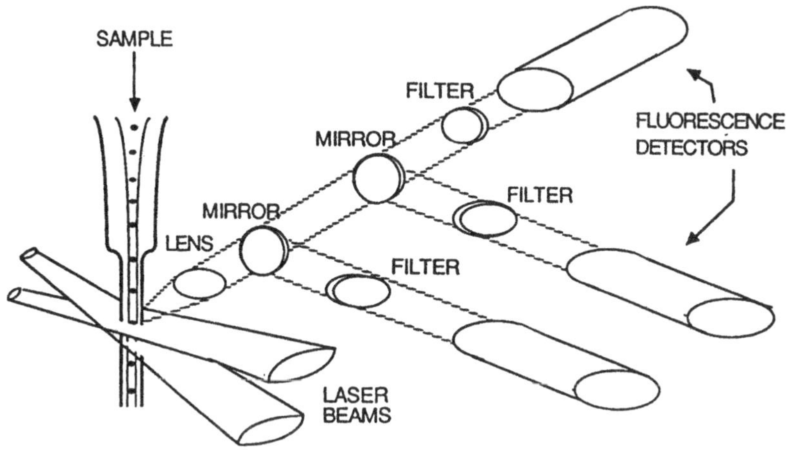

Fig. 3.4.12 Principle of dual laser, multiparameter flow cytometry. Cells are aligned in the flow cell by hydrodynamic forces and illuminated by two light sources. Fluorescence signals generated by individual cells are separated by combinations of dichroic mirrors and filters and detected in high sensitivity photomultiplier tubes.

Individual cells that flow through the light beam generate absorbance, scattering and fluorescence signals which are individually detected by photomultiplier tubes and registered in a pulse height analyzer. The cells can be analyzed at very high rates of up to several thousand cells per second. The result obtained from such analysis is a histogram that shows how the measured single-cell parameters are distributed within the cell population. If more than one parameter has been measured simultaneously for each cell, then multi-dimensional histograms, or frequency distribution functions, are obtained giving a detailed picture of the property space occupied by the cell population (Crissman et al., 1985) A more detailed description of the instrumentation can be found in several excellent books (Melamed et al., 1990; Shapiro, 1988; Watson, 1991) and review articles (Steinkamp, 1984; Degelau et al. 1992) and it will not be discussed further in this article.

Cell preparation

Flow cytometry is typically an off-line instrument. The analysis requires cell preparation procedures that can involve numerous staining steps and can last several hours. In principle, however, some of the simpler staining procedures could be automated and flow cytometry could be done on-line. Although such procedures are being discussed in several research laboratories, currently the main power of the instrument is as an off-line analyzer simultaneously serving multiple purposes.

The objective in cell preparation is to obtain a single-cell suspension with quantitative fluorescent staining of the cellular component of interest. This is straight forward if the cells are already growing as a single-cell suspension such as in the case of hybridoma cells. Cell tissues and cells growing in aggregates require disaggregation by mechanical or enzymatic means. In addition, a permeabilization step is usually required because most fluorescent dyes and large molecules such as antibodies do not cross the functional external membranes of viable cells. Cell permeabilization is typically achieved in fixation steps using either methanol, ethanol, acetone, or formaldehyde alone or in combination with detergents. During the cell staining step a particular component of the cells associates with fluorochromes in a specific manner and ideally in stoichiometric amounts. Antibodies or other proteins such as lectins covalently linked to fluorochromes are used for this purpose. Certain fluorescent dyes have by themselves a high affinity to certain cell components. Other fluorochromes might be generated *in situ* by the action of certain enzymes and enzyme activities can be measured in single cells in this way. Generally, the staining conditions are dependent on the cell material and on the cellular target to be quantified. Specific protocols can be found in several books (Melamed et al., 1990; Watson, 1991); Darynkiewicz and Crissman, 1990) or must be developed or adapted from original articles.

Which parameters are useful for process monitoring?
The outcome of a bioprocess is determined by the interactions of the cell population with its surroundings. The cellular environment is subject to external manipulations. Therefore, cellular activity can be directed if the interactions between the cells and their environment are well understood. The reactions of the cell population which govern interactions with the cellular environment are determined by the distribution of physiological states in a cell population. The physiological state is defined by the composition of a cell and by the spatial arrangement of cellular components within the cells (Fredrickson et al., 1967; Ramkrishna, 1979). In order to determine the physiological states of cells, it is necessary to obtain data on the cellular composition and how this composition is distributed within the growing cell population. In principle, determination of the amounts of all cellular components is of interest because they all contribute to the physiological state of the cell. In reality, however, the number of parameters must be reduced to the major, measurable components of a cell such as the total protein content, the DNA content, the RNA content, the amounts of individual proteins. These parameters, either individually or in combination, more or less describe the physiological state of a cell.

With flow cytometry one is able to measure for single cells (i) the intensity of scattered light at different angles from the incoming light beam, (ii) axial light loss (extinction) at different wavelengths, and (iii) the intensity of fluorescent light emitted by cells. While the intensity of scattered light at low angles can be related to cell size and indicates in many cases the cell volume, light scattered at a right angle serves as an indicator for the internal structure or granularity of a cell (Salzman et al., 1990). Light scattering measurements are especially attractive because no cell pretreatment is required and cells remain viable. Viable subpopulations can be physically separated with a cell sorter which provides a means for rapidly isolating certain desired cell types. However, in the light scattering measurement there is usually no clear correlation of the signal and a biological function of the cell. Fluorescence measurements, in contrast, enable a quantitative determination of specific cellular components related to well defined biological functions. Using appropriate staining procedures, subcellular components such as DNA, RNA, protein content, cytoplasmic enzyme content as well as the amount of surface components can be determined (Crissman et al., 1985). Some selected examples of flow cytometry applications are given in the next sections.

Cell count and cell viability

The increase in cell number over time is often the first parameter that is determined in a cell cultivation process. The light scattering properties of a cell is one of the most accessible parameters that can be measured with flow cytometry. The cells do not require any pretreatment and can be analyzed in growth medium if they grow in a single-cell suspension as do hybridomas. Two light scattering parameters, defined by the angle of the scattered light relative to the laser beam, are measured with a flow cytometer. The disadvantage of the parameters is that they cannot be associated with a clear biological function of the cell and their interpretation is therefore restricted. Also, the forward angle light scattering intensity (FALS) is affected by the shape and the refractivity of the cells. Therefore, a more easily interpreted volume distribution can be obtained with the Coulter counter. The parameter is still useful, however, as it allows identification of subpopulations which have distinctly different light scattering properties as shown in Figure 3.4.13. Here a two dimensional histogram with the two light scattering intensities as the single-cell parameters is shown for a hybridoma cell population. Clearly two subpopulations of cells can be recognized that have different light scattering properties. These can be identified as viable cells and dead cells. The cell population with the lower FALS intensity

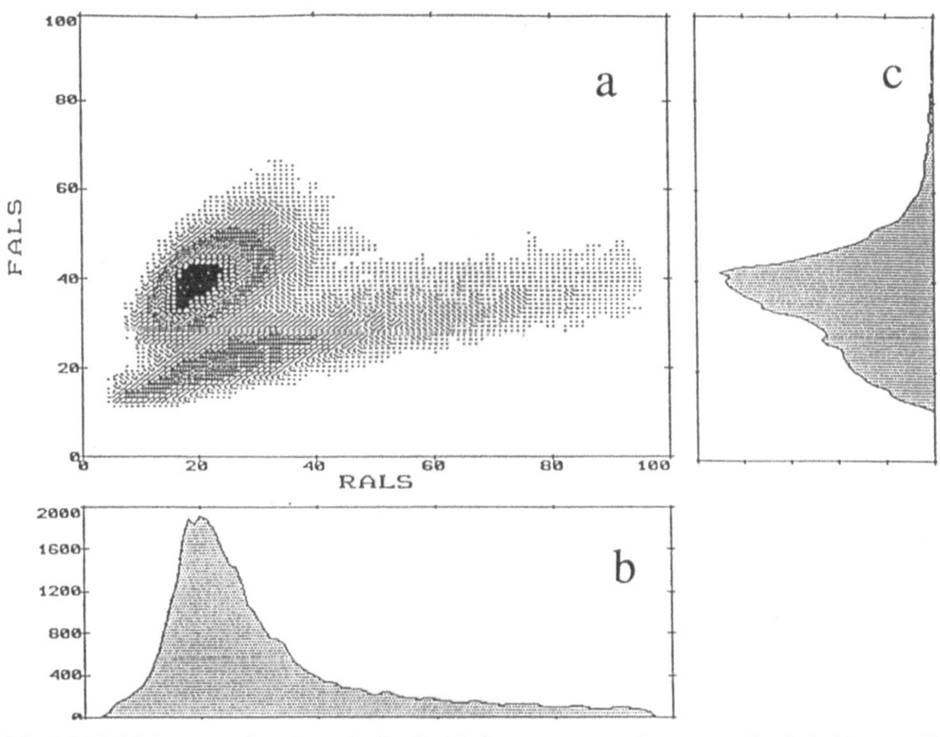

Fig. 3.4.13 Light scattering data obtained with flow cytometry from a murine hybridoma cell culture. (a) joint distribution of forward angle (FALS) and right angle (RALS) light scattering intensities; (b) one parameter FALS distribution; (c) one parameter RALS distribution. The shading in (a) varies according to the cell frequency in the population. The two subpopulations represent dead and live cells (see text).

and the right angle light scattering intensity (RALS) spread over a large range of relative values consists of dead cells. Dead cells typically have a disrupted cytoplasmic membrane such that the cell content is released into the medium. Such cells have a more granular appearence under the light microscope and generate a strong RALS signal when measured with the flow cytometer.

The example here also illustrates the advantage of dual-parameter analysis. The corresponding one-parameter histograms (Fig. 3.4.13b, c) do not suggest the existence of the two subpopulations that can be seen on the joint distribution of the two parameters (Fig. 3.4.13a). The identity of the dead cell population can be easily verified by addition of a fluorescent DNA stain such as propidium iodide to the sample (Sen et al., 1989). The dye is excluded from viable cells but can enter dead cells where it produces a bright fluorescence upon association with the nucleic acids. This is easily detected by simultaneous measurement of this third parameter. Those cells with a strong red fluorescence signal are identified as dead cells.

The light scattering parameters are also useful for quantifying fractions of viable and dead cells in a cell population during the time course of a cultivation. Because the measured sample volume is not known, the absolute cell concentration in the sample cannot be obtained from this measurement. The absolute cell count can be rapidly determined, however, by adding a defined amount of fluorescent tracer particles to the sample and by measuring the fluorescence together with the light scattering intensities (Sen et al., 1989). The detected number of fluorescent particles relative to the number of cells measured yields the concentration of cells in the sample. Further analysis of the data permits detailed analysis of the subpopulations occurring in the culture. Through interactive software, regions can be defined in a display such as Figure 3.4.13 and the individual subpopulations can be analyzed individually (Figure 3.4.14). One should note that this does not require previous

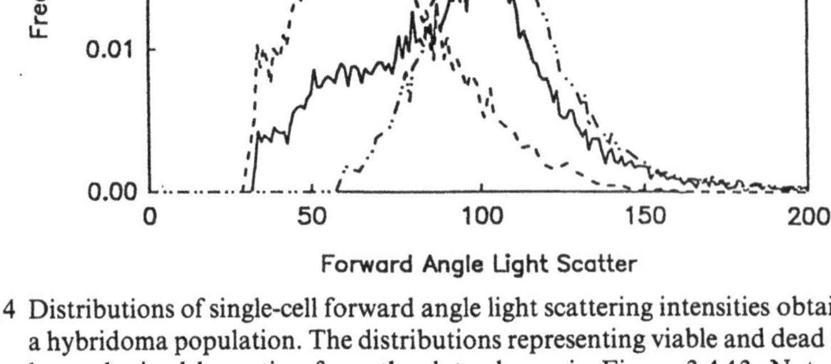

Fig. 3.4.14 Distributions of single-cell forward angle light scattering intensities obtained from a hybridoma population. The distributions representing viable and dead cells have been obtained by gating from the data shown in Figure 3.4.13. Note the large overlap of the two distributions. The two subpopulations can be identified only by two parameter analysis.

cell separation and that the analysis can be done immediately on the heterogeneous cell sample. Such information is not accessible by analytical instruments that yield population average values.

High viability is important for a cell cultivation process. Cell viability can be determined, with some effort and time, by microscopic cell count in combination with a dye exclusion staining method. Nevertheless, this simple example illustrates (i) that cell populations typically have distributed properties, (ii) that analytical sensors that operate on entire cell suspensions can yield only population average quantities, and (iii) that the knowledge of the property structure is necessary to understand the cellular activities in a cell cultivation process.

Genetically unstable cells and cell cycle analysis
It is essential that a productive cell culture maintain its characteristics throughout the cultivation period. Unfortunately this is not the case for many cell lines. Some cell's genetic makeup may change such that the cell line loses its high level of productivity (Frame and Hu, 1991; Ozturk and Palsson, 1990). Such changes can be

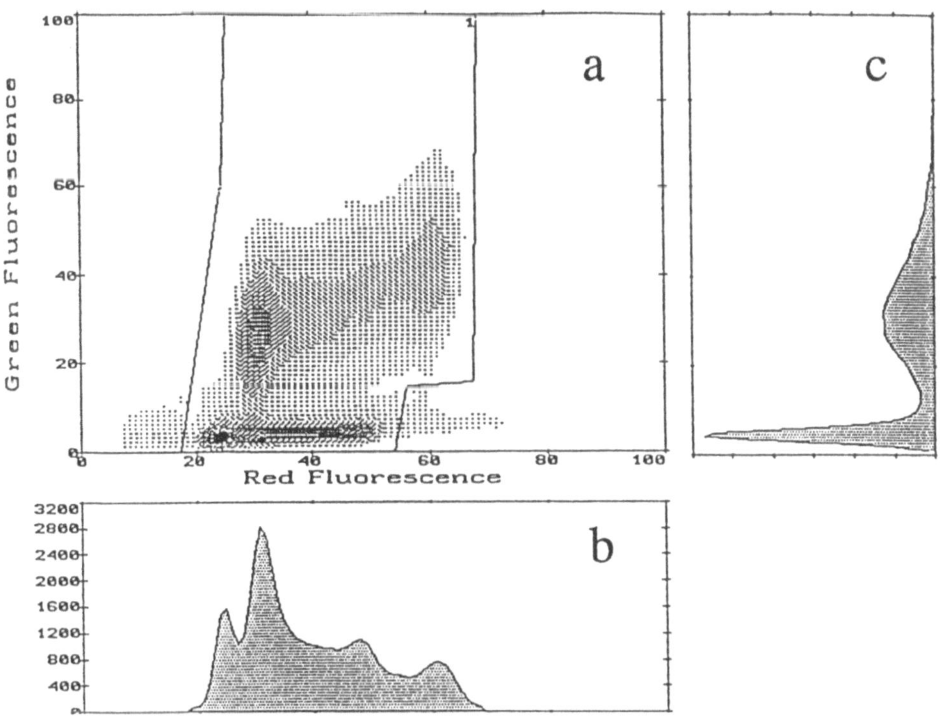

Fig. 3.4.15 Two parameter flow cytometry data obtained from a hybridoma cell population producing antibodies. The green fluorescence intensity is generated by FITC labeled goat antibodies against the antibody produced by the culture. The red fluorescence represents the cellular DNA which has been stained with propidium iodide. The joint distribution (a) is shown together with the corresponding marginal distributions (b) and (c). The data demonstrate heterogeneities in the population with respect to DNA content and intracellular IgG content.

detected with flow cytometry at a very early stage when the undesired, altered cells are still at a very low concentration in the population.

Multiple parameter flow cytometry can yield a significant amount of information on the integrity of such cell lines. Figure 3.4.15 illustrates this for the case of an antibody-producing murine hybridoma cell line. The data were obtained during the cultivation process. The cell sample was stained for cellular DNA content with propidium iodide and for the amount of intracellular IgG with an immunofluorescent staining protocol that requires a goat anti-mouse antibody conjugated to fluorescein isothiocyanate.

Several subpopulations of cells can be identified from this graph. First, the intracellular IgG staining discriminates the cell population into two fractions. The first subpopulation, with a high green fluorescence intensity represents the desired antibody-producing cells. The cells have a high intracellular pool of IgG because they are able to synthesize the antibody. The second cell population with a low green fluorescence intensity is a non-producing variant which evolved from the original cell line during the extended cultivation period of about 30 generations. Furthermore, the DNA distribution shows a very abnormal pattern. Instead of the two peaks expected for a normal eukaryotic cell population, four peaks are present. This alone indicates the existence of two different cell types of different ploidy. As can be seen from the joint distribution of green and red fluorescence the lower ploidy cell line represents a non-producing cell variant. Evidently the producing cell line generated a cell with a significantly lower DNA content. The loss of DNA also affected the expression of the antibody genes. The new cell variant had a growth advantage and eventually dominated the cell line. The productivity of the cell line decreased in proportion to the fraction of producing cells in the population. Moreover, the joint distribution permits immediate investigation of the non-producing and producing cell lines because their DNA distributions can be discriminated on the basis of the intracellular IgG content (Figure 3.4.16). Evaluation of the cell cycle

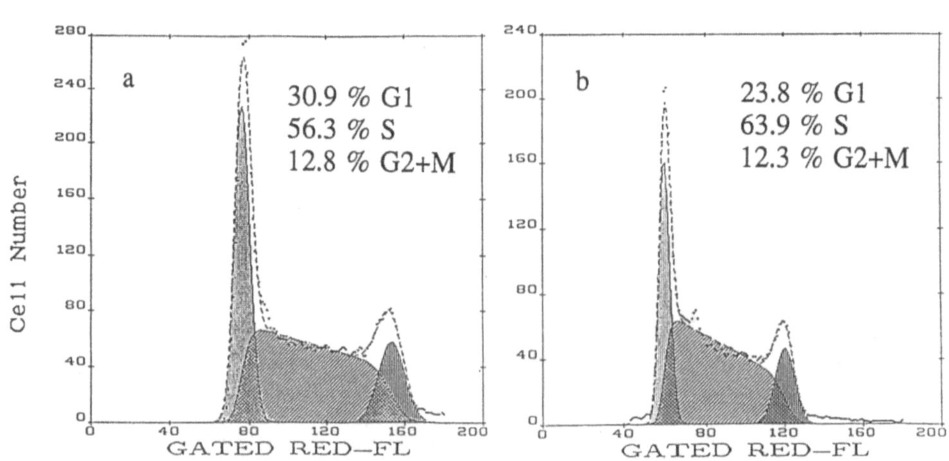

Fig. 3.4.16 Cell cycle analysis of the DNA histogram representing antibody producing (a) and non-producing (b) hybridoma cell populations. The histograms have been obtained by gating on the corresponding subpopulations of cells shown in Fig. 15a. The lower fraction of cells in the G1 cell cycle phase of the non-producer cells indicates a faster growth rate of this cell type.

fractions shows that in comparison to the producing cell line, the non-producing cell line has a significantly lower fraction of cells in the G1 cell cycle phase and a higher cell fraction in the S cell cycle phase which indicates a stronger proliferative activity of the non-producing cell population under the given cultivation conditions.

The measurements therefore have shown (i) the existence and quantification of a non-producing cell population, (ii) that a significant loss of the genome must have occurred as suggested by the lower DNA content of the non-producing cell population, and (iii) that the non-producing cell population had a significant growth advantage in the cultivation process which is indicated by the cell cycle fractions. This information is of course very valuable for the planning and operation of a production process with such cell lines.

Cell Surface fluorescence and protein secretion
Many proteins produced by mammalian cell culture are secreted into the medium. The basis for efficient production is a stable cell line with high specific productivity. Flow cytometry can be used in several ways to maintain high-productivity cell lines. It has been demonstrated that the antibody molecules secreted by and associated with the surface of hybridoma cells can be labeled immunofluorescently using antibodies directed against the hybridoma antibodies (Parks et al., 1979). This can be used in combination with cell sorting to isolate and clone specific cell lines producing specific isotypes of IgG (Dangl and Herzenberg, 1982). Moreover, it has been found that the measured surface-associated fluorescence is correlated with the specific antibody production rate (Sen and Srienc, 1990; McKinney et al., 1991). This observation might apply only to certain cell lines as other researchers were not able to find such a correlation (Meilhoc et al., 1989). However, it points to a potentially useful application of flow cytometry. The surface-associated fluorescence does not require cell fixation and thus can be obtained with live cells after short cell preparation procedure. The cell preparation takes less than 60 minutes. Using surface-associated cell fluorescence, higher-productivity variants can be rapidly isolated using cell sorting. Conventional cloning by limiting dilution has shown that a large clonal variation can exist in a hybridoma cell line (Kitano et al., 1988).

Furthermore, if surface-associated cell fluorescence is correlated with production rate then this provides a means to obtain information on the dynamic properties of the cell population from a single measurement. This measurement can be used to study effects of cultivation conditions on the kinetics of product synthesis.

Detection of intracellular pH, membrane potential, and calcium
Maintenance and regulation of intracellular pH is important for central biochemical pathways such as glycolysis, gluconeogenesis, and protein synthesis and secretion. The intracellular pH can be measured with flow cytometry using fluorescent dyes that are sensitive to hydrogen ion concentrations (Visser et al., 1979; Musgrove et al., 1986). It could represent a useful parameter for studying metabolic responses of cell populations. In particular, it has been shown that the intracellular pH changes depend on nutritional conditions of the cell population (Hernlem and Srienc, 1989). Similarly, membrane potential is an important physiological property of cells and some of the organelles. The electrical potential difference across the plasma membrane is directly involved in the transport of ions and nutrients across this membrane. The membrane potential dye Rhodamine 123 can be used for staining mitochondria

(Johnson et al., 1982) and has been used to characterize hybridoma cells in different growth stages (Al Rubeai et al., 1991). Also, intracellular calcium ion concentration can be measured with fluorescent probes and flow cytometry (Grynkiewicz et al., 1985; Rabinovitch and June, 1990). This would be important in studies of the effect of environmental conditions on cell growth as calcium has an important role in the transduction of signals across the cell membrane. The use of these dyes for monitoring the state of a cell population is somewhat restricted, however, as cells need to be loaded with the appropriate dye before the measurement can be made. The measured metabolic state of the cells may not represent the actual conditions in the bioreactor because the cells are most likely perturbed during the loading process.

Single-cell dynamics of accumulation of cellular components

The goal of mathematical models that describe growing cell populations is to connect the kinetics of cell populations with the biological processes in the cell and with the environmental conditions that surround the cells. Ultimately reliable mathematical models can be used to provide a quantitative basis for the design, optimization and control of bioprocesses. Many kinetic models of cell growth that have been develped for biotechnology have a limited utility as they are not based on a rigorous theory. In particular, many of these models neglect that a cell culture represents a segregated system and that cellular properties are distributed. An exact description of the dynamics of a cell population has to take into account the distribution of the physiological state of the cell population. Although the framework for a rigorous theory of cell growth was developed in the 1960's [120,193], little use has been made of this approach, presumably because of the lack of instrumentation that would provide the necessary experimental data.

Knowledge of three physiological functions are needed to exactly describe the evolution in time of a cell population. These are (i) the rate pattern during the cell cycle, (ii) the transition rates between temporally separated cell cycle stages, and (iii) the partitioning function which describes how cellular components are divided among the daughter cells that are generated at cell division. If the way in which these functions depend on the physiological state of a cell and the surrounding medium is known, the model is able to predict how a cell culture will develop in time. These three functions are unknown for most cell populations.

Recently it has been shown for growing hybridoma (Kromenakter and Srienc, 1991) and yeast (Srienc and Dien, 1992) cell populations that flow cytometry can provide the experimental data to determine these intrinsic physiological functions. In this approach it is essential that one realizes that cells move in time through the cell cycle. The compartmentalization of cellular functions in time establishes a temporal order for specific morphological and biochemical events that every cell must follow between birth and division. The relative temporal order of some of these events during the cell cycle can be determined for an asynchronous culture with flow cytometry. For instance cellular DNA content before and after the S phase can be quantified with flow cytometry and this provides a basis for classifying cells into three categories: cells that have not yet duplicated their DNA (G1 phase), cells which are in the process of duplicating their DNA (S phase) and cells which have completed DNA duplication (G2 and M phase). The beginning and ending of each of these three cell cycle compartments is defined by two transitions which represent the state of the cells at entry into and exit from that cell cycle phase. For instance, the G1 phase is bounded by cell birth and the initiation of DNA replication, and when cells are growing and dividing, they cannot enter or leave the G1 phase by any other biological means. Similarly the S phase is bounded by initiation of DNA synthesis and completion of DNA synthesis, and the G2 + M cell cycle phase is bounded by the completion of DNA synthesis and cell division. It has been shown in this work that distributions of cellular state properties in specific cell cycle phases and at transition points between cell cycle phases can be experimentally determined. These data in combination with population balance equations can be used to compute single-cell growth rates of the measured state properties.

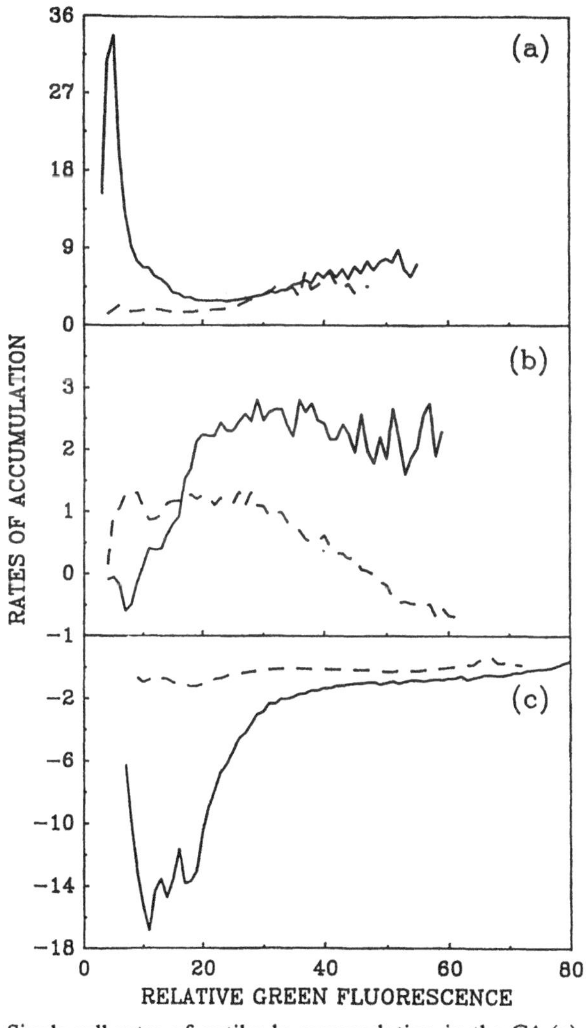

Fig. 3.4.17 Single-cell rates of antibody accumulation in the G1 (a), S (b), and G2 + M (c) cell cycle phases of a hybridoma population. The solid line shows the rate pattern for an antibody producing population while the dashed line represents non-producing cells (Kromenaker and Srienc, 1991)

Figure 3.4.17 shows how single-cell growth rates depend on the protein content of a cell for a hybridoma population growing under balanced growth conditions. The analysis suggests that the highest growth rates in terms of the accumulation of cellular protein content occur in the G1 cell cycle phase at the beginning of the cell cycle and that negative growth rates occur during the G2 + M cell-cycle phase. Negative growth rates indicate that the rates of secretion and degradation together exceed the rate of synthesis. The interpretation of this result is that during the cell cycle of hybridomas, antibodies are accumulated at the beginning of the cell cycle and large amounts of antibodies are secreted during later stages of the cell cycle. It is hypothesized that these periodic events are stimulated by momentary interruption of the secretion mechanism caused by vesiculation of the golgi apparatus during mitosis (Kromenaker and Srienc, 1991).

Synopsis
Flow cytometry is being increasingly used to answer numerous questions that occur in connection with the cultivation of mammalian cells. This includes monitoring of the cells during the cultivation process to obtain a better idea of the structure of the cell population (Al-Rubeai et al., 1991; Dalili and Ollis, 1990; Heath et al., 1990; Leno et al., 1991). The interaction of cells with their environment can significantly affect the productivity of a cell culture. For instance, the serum level in a growing culture of hybridoma cells can change the ratio between producing and non-producing cells (Chuck and Palsson, 1992). Flow cytometry is being used to isolate and characterize cell lines with a high copy number of a desired gene (Kaufman et al., 1978; Pends and Bailey, 1992), or to determine the harvesting point of a baculovirus-infected insect cell culture (Lazarte et al., 1992). It has been applied to study in great detail the effect of shear stress on DNA synthesis rates in populations of CHO cells exposed to different agitation intensities (Lakhotia et al., 1992). The questions that can be addressed using this instrumentation are numerous. All examples demonstrate that cell populations are inherently variable, and that an understanding of this variability is needed if we want to understand the behavior of cell cultures in a cultivation process.

3.4.6 Conclusions

The application of modern on-line and off-line analysis systems for a better cultivation process monitoring in the area of mammalian or animal cell cultivation is still in its beginning. This is not due to the lack of instrumentation but to the number of applications in research and industries in this area. While the systems and techniques such as biosensors, flow-cytometry, or on-line immunoanalysis are well described for the cultivation of bacteria, yeasts, and other fungi only few authors describe the use of such systems for cultivation monitoring of mammalian cells. Thus, little reliable data is available. However, even this small number of examples shows the high potential of these methods and the possibility of a better understanding and documentation of these processes. Based on this increasing knowledge, better process control, economy, and safety will be available.

References

Ahlmann, N., Niehoff, A., Rinas, U., Scheper, Th. and Schügerl, K. (1986) Continuous on line monitoring of intracellular enzyme activity, Anal. Chem. Acta 190, pp.: 221–226.

Al-Rubeai, M., Chalder, S., Bird, R., and Emery, A.N. (1991) Cell cycle, cell size and mitochondrial activity of hybridoma cells during batch cultivation. Cytotechnol., 7, 179–186.

Al-Rubeai, M., Emery, A.N., and Chalder, S. (1991) Flow cytometric study of cultured mammalian cells. J. Biotechnol., 19, 67–82.

Anders, K.D., Wehnert, G., Thordsen, O., Scheper, T., Rehr, B. and Sahm, H. (1993) Biotechnological applications of fiber optic sensing: multiple uses of a fiber optic fluorometer, Sens. Actuat., in press.

Appleqvist, R., Johansson, G., Holst, O. and Mattiasson, B. (1989) A chemically sterilizable barrier for the protection of fermentors during automatic sampling into a flow-injection system, Anal. Chim. Acta 216, pp.: 299–306.

Armiger, W.B., Lee, J.F., Montalvo, L.M. and Forro, J.R. (1985) Fed-batch control based upon the measurement of intracellular NADH, proceedings 190th ACS Meeting, Chicago, MBTD 40.

Bayer, Th. (1987) Reaktionstechnische Untersuchungen zur Produktion von Cephalosporin C im Mammutschlaufenreaktor, Ph.D. Thesis, University of Hannover.

Bayer, Th., Zhou, W., Holzhauer, K. and Schügerl, K. (1989) Investigations of cephalosphorin C production in an air lift towery loop reactor, Appl. Microbiol. Biotechnol. 30, 26–33.

Bell, G.I. and Anderson, E.C. (1967) Cell growth and division. I. A mathematical model with applications to cell volume distributions in mammalian suspension cutlures. Biophys. J., 7, 329–351.

Beyeler, W.; Einsele, A. and Fiechter, A. (1981) On-line measurements of culture fluorescence: method and application, Eur. J. Appl. Microbiol. Biotechnl., 13, 10–14.

Beyeler, W.; Gschwend, K. and Fiechter, A. (1983) In-situ Fluorometrie: Eine neue Methode zur Charakterisierung von Bioreaktoren, Chem. Ing. Tech., 55 (5) 869–871.

Bradley, J., Anderson, P.A., Dear, A.M., Ashby, R.E. and Turner, A.P.F. (1989) Glucose biosensors for the study and control of bakers compressed yeast production, In: Computer Applications in Fermentation Technology: Modelling and Control of Biotechnological Processes, N.M. Fish, R.I. Fox, N.F. Thornhill (Eds.) Elsevier Science Publishers Ltd., Amsterdam.

Bradley, J., Stöcklein, W. and Schmid, R.D. (1991) Biochemistry based analysis systems for bioprocess monitoring and control, Process Control and Quality, 1, pp.: 157–183.

Bradley, J.; Kidd, A.J.; Anderson, P.A.; Dear, A.M.; Ashby, R.E.; Turner and A.P.F. (1989) Rapid determination of the glucose content of molasses using a biosensor, Analyst, 114, 375–379.

Bradley, J.; Stöcklein, W.; Schmid, R.D. (1991) Biochemistry based analysis systems for bioprocess monitoring and control, Process Control Qual., 1, 157–173.

Buttler, T., Gorton, L. and Marko-Varga G. (1993) Characterization of a sampling Unit based on Tangential Flow Filtration for on-line bioprocess Monitoring. Elsevier, 27–38.

Callis, J.B., Illman, D.L. and Kowalski, B.R. (1987) Process Analytica, Anal. Chem. 59, pp.: 624 A-637 A.

Cattaneo, M.V. and Luong, J.H.T. (1993) Monitoring glutamine in animal cell cultures using a chemiluminescence fiber optic biosensor. Biotechnol. Bioeng., 41, 659–665.

Chotani, G.; Costantinides, A. (1982) On-line glucose analyzer for fermentation application, Biotechnol. Bioeng. 24, 2743–2745.

Christensen, L.H., Nielsen, J. and Villadsen, J. (1991) Delay and dispersion in an in situ membrane probe for bioreactors, Chem. Eng. Sci. 46/12, pp. 3304–3307.

Chuck, A.S. and Palsson, B.O. (1992) Population balance between producing and nonproducing hybridoma clones is very sensitive to serum level, state of inoculum, and medium composition. Biotechnol. Bioeng., 39, 354–360.

Compton, B.J. (1990) On-line monitoring of bioprocesses using HPLC, in: Sensors in bioprocess control (eds.: J.V. Twork, A.M. Yacynych), Bioprocess Technology, Marcel Dekker, Inc, New York, 39–46.

Crissman, H.A., Darzynkiewicz, Z., Tobey, R.A. and Steinkamp, J.A. (1985) Correlated measurements of DNA, RNA, and protein in individual cells by flow cytometry. Science 228, 1321–1324.

Dangl, J.L. and Herzenberg, L.A. (1982) Selection of hybridomas and hybridoma variants using the fluorescence activated cell sorter. J. Immunol. Meth., 52, 1–14.

Darzynkiewicz, Z. and Crissman, H.A. (1990) Flow cytometry. Methods in cell biology Vol. 33, Academic Press, San Diego.

Degelau, A., Freitag, R., Linz, F., Middendorf, C., Scheper, T., Bley, T., Müller, S., Stoll, P. and Reardon, K.F. (1992) Immuno- and flow cytometric analytical methods for biotechnological research and process monitoring, J. Biotechnol., 25, 115–144.

Dremel, B.A.A., Li, S.Y. and Schmid, R.D. (1992) On-line determination of glucose and lactate concentration in animal cell culture based on fibre optic detection of oxygen in flow-injection analysis, Biosen. Bioelectron., 7, 133–139.

Dubois, D., Masquelier, D., Menozzi, F.D., Mouyard, M.A. and Miller, A.O.A. (1991) On-line analysis of mammalian cels bioreactor's content, Bioteknowledge 2, 7–10.

Einsele, A., Ristroph, D.L. and Humphrey, A.E. (1978) Mixing times and glucose uptake measured with a fluorometer, Biotechnol. Bioeng., 20, 1487–1492.

Enfors, S.O., Molin, N.L., Mosbach, K.H. and Nilsson, H.J. (1977) Enzyme electrode, US-Patent, No. 4,024,042.

Fenge, C., Fraune, E., Freitag, R., Scheper, Th. and Schügerl, K. (1991) On line monitoring of monoklonal antibody formation in high density perfusion culture using FIA, Cytotechnol. 6, 55–63.

Fenge, C., Fraune, E., Freitag, R., Scheper, T. and Schügerl, K. (1991) On-line monitoring of monoclonal antibody formation in high density perfusion culture using FIA. Cytotechnol., 6, 55.

Fernandez, E.J., Mancuso, A. and Clark, D.S. (1988) NMR spectroscopy studies of hybridoma metabolisms in a simple membrane reactor, Biotechnol. Bioeng., 22, pp: 457–462.

Frame, K.K. and Hu, W.S. (1991) Comparison of growth kinetics of producing and non-producing hybridoma cells in batch culture. Enzyme Microb. Technol., 13, 690–696.

Fraune, E., Fenge, Ch. and Kuhlmann, W. (1988) High density hybridoma cell culture in stirred tank perfusion bioreactors, 8th International Biotechnology Symposium, Paris, Abstract book p.: 115.

Fredrickson, A.G., Ramkrishna, D. and Tsuchiya, H.N. (1967) Statistics and dynamics of procaryotic cell populatons. Math. Biosci., 1, 327–374.

Freitag, R. (1990) Entwicklung und Automatisierung von immunchemischen Nachweisverfahren zur On-Line-Detektion hochmolekularer Medienkomponentenin Fermentationsprozessen, thesis, University of Hannover.

Freitag, R., Scheper, T. and Schügerl, K. (1991) Development of a turbidimetric immunoassay for online monitoring proteins in cultivation processes, Enzyme Microb. Technol., 13, 969–975.

Freitag, R., Scheper, Th. and Schügerl, K. (1991) Development of a turbidimetric immunoassay for online monitoring of proteins in cultivation processes, Enz. Microb. Technol. 13, 969–975.

Freitag, R., Scheper, Th., Spreinat, A. and Antranikian, G. (1991) On-line monitoring of pullulanase production during continuous cultures of *Clostridium thermosulfurogenes*, Appl. Microbiol. Biotechnol. 35, 471–476.

Freitag, R., Fenge, C., Scheper, T., Schügerl, K., Spreinat, A., Antranikian, G. and Fraune, E. (1991) Immunological on-line detection of specific proteins during fermentation processes, Anal. Chim. Acta, 249, 113–122.

Freitag, R., Scheper, T., Spreinat, A. and Antranikian, G. (1991) On-line monitoring of pullulanase production during continuous culture of *Clostridium thermosulfurogenes*, Appl. Microbiol. Biotechnol, 35, pp.: 471–483.

Garn, M., Gisin, M., Thommen, Ch. and Cevey, P. (1989) A flow injection analysis system for fermentation monitoring and control, Biotechnol. Bioeng. 28, 423–428.

Ghoul, M., Ronat, E. and Engasser, J.-M. (1986) An automatic and sterilizable sampler for laboratory fermentors: Application to the on-line control of glucose concentrations, Biotechnol. Bioeng. 28, pp.: 119–121.

Gibson, T. D. and Woodward, J. R. (1988) Continuous, reliable on-line analysis of fermentation media by simple enzymatic/spectrophotometric assays, Anal. Chim. Acta 213, 61–68.

Glaxo Laboratories Ltd. U.S. Patent 3, 830, 106 (1974).

Graf, H. (1989) Untersuchungen zur Automatisation eines Perfusionssystems zur Kultivierung von tierischen Zellen, thesis, University of Hannover.

Graf, H., Wentz, D. and Schügerl, K. (1991) Performance of a tubular microporous membrane filter used for in situ sampling of mammalian cell culture medium, Biotechnol. Biotechniques, 5, 183–186.

Graf, H. and Schügerl, K. (1991) Some aspects of hybridoma cell cultivation, Appl. Microbiol. Biotechnol., 35, 165–175.

Gram, J., de Bang, M., Nikolajsen, K. and Holm, K. A. (1990) Automated sampling and glucose analysis for a pilot plant penicillin V. fermentation, Workshop "Aseptic sampling", 5th European Congress on Biotechnology (ECB5), Copenhagen.

Griffiths, B. (1986) Scaling-up of Animal cell cultures, in: Animal cell culture (Ed: R. I. Freshney) IRL Press, Oxford, pp.: 3–69.

Grynkiewicz, G., Poenie, M. and Tsien, R. Y. (1985) A new generation of Ca^{2+} indicators with greatly improved fluorescence properties. J. Biol. Chem., 260, 3440–3450.

Gschwend, K., Beyeler, W. and Fiechter, A. (1983) Detection of reactor nonhomogenities by measuring culture fluorescence, Biotechnol. Bioeng., 25, 2789–2793.

Hakanson, H., Holst, O. and Mattiasson, B. (1990) Application of a general sampling system for continuous monitoring in biological systems using biosensors, Symposium: "Sensors using biological reactions", 5th European Conference on Biotechnology (ECB5), Copenhagen.

Harrison, D. E. F. and Chance, B. (1970) Fluorometric technique for monitoring changes in the level of reduced nicotinamide nucleotides in continuous cultures of microorganisms, Appl. Microbiol., 19 (3) 446–450.

Heath, C., Dilwith, R. and Belfort, G. (1990) Methods for increasing monoclonal antibody production in suspension and entrapped cell cultures: biochemical and flow cytometric analysis as a function of medium serum content. J. Biotechnol., 15, 71–90.

Heinzle, E., Dunn, I. J. (1991) Methods and instruments in fermentation gas analysis, in: Biotechnology, Measuring, Modelling and Control, Vol. 4, 2nd. rev. edition, (ed.: K. Schügerl), VCH Weinheim, pp.: 27–73.

Hernlem, B. J. and Srienc, F. (1989) Intracellular pH in single *Saccharomyces cerevisiae* cells. Biotechnol. Techn., 3, 79–84.

Herold, Th. (1986) Cephalosporin Produktion im Rührkessel, Ph. D. Thesis, University of Hannover.

Hoffman, H. (1985) Intergration von Membranfiltrationseinheiten zur Optimierung der Reaktionsführung an biotechnologisch relevanten Prozessen, Ph. D. Thesis, University of Hannover.

Holst, O., Hakanson, H., Miyabayashi, A. and Mattiasson,B. (1988) Monitoring of glucose fermentation processes using a commercial glucose analyzer, Appl. Microbiol. Biotechnol. 28, 32–36.

Huang, Y.L., Li, S.Y., Dremel, B.A.A., Bilitewski, U. and Schmid, R.D. (1991) On-line determination of glucose concentration throughout animal cell cultures based on chemiluminescencent detection of hydrogen peroxide coupled with flow-injection analysis, J. Biotechnol., 18, 161–175

Hübner, U. (1990) Entwicklung neuer On-Line-Analysenmethoden zur Steuerung und Regelung von Proteasefermentation, Ph.D. Thesis, University of Hannover.

Hustedt, H., Kroner, K.H. and Kula, M.-R. (1985) On-Line-Bestimmung von Enzymen und Proteinen zur Prozeßkontrolle in der Enzymgewinnung, BTF-Biotech-Forum 2, 56–63.

Hwang, C. and Sinskey, A.J. (1991) The role of oxidation-reduction potential in monitoring growth of cultured mammalian cells, in: Production of biologicals from animal cells in culture (eds.: Spier, R.E., Griffiths, J.B., Meigner, B.), Butterworths-Heinemann, Oxford, 548–568.

Imming, G., Schaller, K. and Meiners, M. (1982) The use of an automated analyzer to control microbial processes, In: First IFAC (International Federation of Automatic Control), Workshop on Modelling and Control of Biotechnical Processes, Helsinki, (Holme, A., Ed.) Pergamon Press, pp.: 27.

Johnson, L.V., Walsh, M.L., Brockus, B.J. and Chen, L.B. (1982) Monitoring of relative mitochondrial membrane potential in living cells by fluorescence microscopy. J. Cell Biol., 88, 526–535.

Kaufman, R.J., Bertino, J.R. and Schimke, R.T. (1978) Quantitation of dihydrofolate reductase in individual parental and methotrexate-resistant murine cells: use of fluorescence activated cell sorter. J. Biol. Chem., 253, 5852–5860.

Kilburn, D.G. (1991) Monitoring and control of bioreactors, in: Mammalian cell biotechnology (ed: M. Butler), Oxford university Press, Oxford, pp.: 159–185.

Kitano, K., Iwamoto, K., Shintani, Y. and Akiyama, S. (1988) Effective production of a human monoclonal antibody against tetanus toxoid by selection of high productivity clones of a heterohybridoma. J. Immunol. Meth., 109, 9–16.

Konstantinov, K.B., Pambayun, R., Matanguihan, R., Yoshida, T., Perusich, C.M. and Hu, W.S. (1992) On-line monitoring of hybridoma cel growth using a laser turbidity sensor. Biotechnol. Bioeng., 40, 1337–1342.

Kotiaho, T., Lauritsen, F.R., Choudhury, T.K., Cooks, R.G. and Tsao, G.T. (1991) Membrane introduction mass spectroscopy, Anal. Chem., 63 (18), 875A–883A.

Kromenaker, S.J. and Srienc, F. (1991) Cell-cycle-dependent protein accumulation by producer and nonproducer murine hybridoma cell lines: a population analysis. Biotechnol. Bioeng., 38, 665–677.

Kroner, K.H. and Kula, M.-R. (1984) On-line measurement of extracellular enzymes during fermentations using membrane techniques, Anal. Chim. Acta 163, 3–15.

Kroner, K.-H., Nissinen, V. and Ziegler, H. (1987) Improved dynamic filtration of microbial suspension, Bio/Technology 5, 921–926.

Kroner, K.H. (1982) On line Bestimmung von Enzymen in der Bioprozeßtechnik unter Berücksichtigung von Probenahme und Fließinjektionstechnik. On line determination of enzymes in bioprocessing with the emphasis on flow injection analysis (FIA) and continuous sampling, Fresenius Z. Anal. Chem. 329, 718–725.

Kroner, K.H., Recktenwald, A. and Kula, M.-R. (1986) Membrantrennverfahren als Instrument zur Prozeßanalytik von Enzymen, In: Technische Membranen in der Biotechnologie. GBF-Monographien, Vol. 9 (Kula, M.-R., Schügerl, K., Wandrey, Ch. Eds.) VCH Verlag, Weinheim, 93–104.

Kroner, K. H., Stach, W. and Kuhlann, W. (1986) Kontinuierliches Probenentnahmeverfahren für die Bioprozeßanalytik, Chem. Tech. 15/9, 74–77.

Kuhlmann, W. (1983) Untersuchungen zur On-Line-Prozeßlenkung der satzweisen und kontinuierlichen Kultivierung von *Saccharomyces cerivisiae*, Ph. D. Thesis, University of Hannover.

Lakhotia, S., Bauer, K. D. and Papoutsakis, E. T. (1992) Damaging agitation intensities increase DNA synthesis rate and alter cell-cycle phase distributions of CHO cells. Biotechnol. Bioeng., 40, 978–990.

Lazarte, J. E., Tosi, P.-F. and Nicolau, C. (1992) Optimization of the production of full-length rCD4 in baculovirus-infected sf9 cells. Biotechnol. Bioeng., 40, 214–217.

Leist, C., Meyer, H. P. and Fiechter, A. (1986) Process control during the suspension culture of a human melanoma cell line in mechanically stirred loop bioreactors, J. Biotech., 4, 235–246.

Leno, M., Merten, O.-W., Vuillier, F. and Hache, J. (1991) IgG production in hybridoma batch culture: kinetics of IgG mRNA, cytoplasmic-, secreted- and membrane-bound antibody levels. J. Biotechnol., 20, 301–312.

Lenz, R., Boelcke, C., Peckmann, U. and Reuss, M. (1985) A new automatic sampling device for the determination of filtration characteristics and the coupling of an HLC to fermentors, In: First IFAC (International Federation of Automatic Control) Symposium on Modelling and Control of Biotechnological Processes, Noordwijkerhout, (Halme, A., Ed.), Pergamon Press, pp.: 55–60.

Lorenz, Th., Niehoff, J. Diekmann, J. and Schügerl, K. (1984) On-line analysis of Penicillin fermentation broths, Analytical Methods and Problems in Biotechnology, Symposium, 17.–19. April, Nordwijkerhout, The Netherlands, p.: 74.

Lorenz, Th., Schmidt, W., Schügerl, K. (1987) Sampling devices in fermentation technology: A review, Chem. Eng. J. 35, pp.: B15–B22.

MacBride, W. R., Magae, J. A., Armiger, W. B. and Zabriskie, D. W. (1986) Optical apparatus and method for measuring the characteristics of materials by their fluorescence, US-patent, 4577110.

MacMichael, G., Armiger, W. B., Lee, J. F. and Mutharasan, R. (1987) On-line measurement of hybridoma growth by culture fluorescence, Biotechnol. Techn., 4, 213–218.

Male, K. B., Luong, J. H. T., Tom, R. and Mercile, S. (1993) Novel FIA amperometric biosensor system for the determination of glutamine in cell culture system. Enzyme Microb. Technol. 15, 26–32.

Mandenius, C. F., Danielsson, B. and Mattiasson, B. (1984) Evaluation of a dialysis probe for continuous sampling in fermenters and complex media, Anal. Chim. Acta 163, 135–141.

Marc, I., Blanchard, F., Engasser, J.-M., Taha, S. (1990) New type of "in situ" sterile liquid autosampler for fermentors and cell culture, Workshop: "Aseptic sampling", 5th European Congress on Biotechnology (ECB 5), Copenhagen.

Mathers, J. J., Dinwoodie, R. C., Talarowich, M. and Mehnert, D. W. (1986) Computer-linked HPLC system for feedback control of fermentation substrates, Biotechnol. Lett., 8, 311–314.

Mattiasson, B., Mandenius, C. F., Axelsson, S. P., Danielsson, B. and Hagander, P. (1983) Computer control of fermentations with biosensors, Ann. NY. Acad. Sci. 413, pp. 193–196.

Mattiasson, B., Hakanson, H. (1992) Immunochemically based assays for process control, Adv. Biochem. Eng. Biotechnol., 46, 81–102.

McKinney, K. L., Dilwith, R. and Belfort, G. (1991) Manipulation of heterogeneous hybridoma cultures for overproduction of monoclonal antibodies. Biotechnol. Progr., 7, 445–454.

Meilhoc, E., Wittrup, K. D. and Bailey, J. E. (1989) Application of flow cytometric measurement of surface IgG in kinetic analysis of monoclonal antibody synthesis and secretion by murine hybridoma cells. J. Immunol. Meth., 121, 167–174.

Melamed, M.R., Lindmo, T. and Mendelsohn, M.L. (1990) Flow cytometry and sorting. Wiley-Liss, New York. Ed. 2nd.

Merten, O.W., Palfi, G.E. and Steiner, J. (1986) On-line determination of biochemical/physiological parameters in the fermentation of animal cells in a continuous or discontinuous mode, Adv. Biotechnol. Proc., 6, pp.: 111–178.

Meyer, H.-P., Beyeler, W., Fiechter, A. (1984) Experiences with the on-line measurement of culture fluorescence during cultivation of *Bacillus subtilis, Escherichia coli, Sporotrichum thermophile*, J. Biotechnol., 1, 341–349.

Middendorf, C., Schulze, B., Freitag, R., Scheper, T., Howaldt, M. and Hoffmann, H. (1992) On-line immunoanalysis in bioprocess control, J. Biotechnol. 31, 395–403.

Mitzutani, S., Iijima, S., Morikava, M., Shimizu, K., Matsubara, M., Ogawa, Y., Izumi, R., Matsumoto, K. and Kobayashi, T. (1987) On-line control of glucose concentration using an automatic glucose analyzer, J. Ferment. Technol. 65, 325–331.

Miyabayashi, A. and Mattiasson, B. (1990) A dual streaming potential device used as an affinity sensor for monitoring hybridoma cell cultivations, Anal. Biochem., 184, 165–171.

Monseur, X. and Motte, J.C. (1988) On-line monitoring during fermentation of whey based on ultrafiltration and liquid chromatography, Anal. Chim. Acta, 204, 127–134.

Musgrove, E., Rugg, C. and Hedley, D. (1986) Flow cytometric measurement of cytoplasmic pH: a critical evaluation of available fluorochromes. Cytometry, 7, 347–355.

Niehoff, J., Müller, J., Hiddessen, R. and Schügerl, K. (1986) The use of an automatic on-line system for monitoring penicillin cultivation in a bubble-column loop reactor, Anal. Chim. Acta, 190, 205–212.

Ozturk, S.S. and Palsson, B.O. (1990) Loss of antibody productivity during long-term cultivation of a hybridoma cell line in low serum and serum-free media. Hybridoma, 9, 167–175.

Parks, D.R., Bryan, V.M., Oi, V.T. and Herzenberg, L.A. (1979) Antigen-specific identification and cloning of hybridomas with a fluorescence-activated cell sorter. proc. Natl. Acad. Sci. US, 76, 1962–1966.

Pendse, G.J., Karkare, S. and Bailey, J.E. (1992) Effect of cloned gene dosage on cell growth and hepatitis B surface antigen synthesis and secretion in recombinant CHO cells. Biotechnol. Bioeng. 40, 119–129.

Rabinovitch, P.S. and June, C.H. (1990) Measurement of intracellular ionized calcium and membrane potential. In: Flow cytometry and sorting. Melamed, M.R., Lindmo, T. and Mendelsohn, M.L., eds. Wiley-Liss, New York, 2nd Ed., 651–668.

Ramkrishna, D (1979) Statistical Models of cell populations. In: Advances in biochemical engineering. Vol. 11, Ghose, T.K., Fiechter, A. and Blakebrough, N., Eds. Springer-Verlag, New York, 1–47.

Reardon, K.F., Scheper, T. and Bailey, J.E. (1986) *In situ* fluorescence monitoring of immobilized *Clostridium acetobutylicum*, Biotechnol. Lett., 8 (11), 817–822.

Reardon, K.F., Scheper, T. and Bailey, J.E. (1987) Metabolic pathway rates and culture fluorescence in batch fermentations of *Clostridinium acetobutylicum*, Biotechnology Progress, 3(3), 153–167.

Reardon, K.F. and Scheper, T.H. (1991) Determination of cell concentration and characterization of cells, in: Biotechnology, Measuring, Modelling and Control, Vol. 4, 2nd. rev. edition, (ed.: K. Schügerl), VCH Weinheim, 181–223.

Recktenwald, A., Kroner, K.H., Kula, M.R. (1985) Rapid on-line protein detection in biotechnological processes by flow injection analysis (FIA), Enzyme Microbiol. Technol. 7, 145–149.

Riesmeier, B., Kroner, K.H., Kula, M.-R. (1987) Studies on secondary layers formation and its characterization during cross-flow filtration of microbial cells, J. Membr. Sci. 34, 245–266.

Romette, J. L. (1987) Mammalian cell culture process control: sampling and sensing, In: Biosensors, GBF Monographie Vol 10, R. D. Schmid, Ed., Verlag Chemie, Weinheim, pp. 81–86.

Ross, A. and Schügerl, K. (1988) Tetracycline procution by Streptomyces aureofaciens: the time lag of production, Appl. Microbiol. Biotechnol. 29, 174–180.

Ruzicka, J. and Hansen, E.H. (eds) (1988) Flow injection analysis, 2nd edition, Wiley & sons, New York.

Salzman, G.C., Singham, S.B., Johnston, R.G. and Bohren, C.F. (1990) Light scattering and cytometry. In: Flow cytometry and sorting. Melamed, M.R., Lindmo, T. and Mendelsohn, M.L., eds. Wily-Liss, New York, Ed. 2nd 81–107.

Scheller, F. and Schubert, F. (1991) Biosensors, Elsevier Sci. Publ., Amsterdam, 1992

Scheper, T. (1992) Biosensors for process monitoring, Journal Industrial Microbiol., 9, 163–172.

Scheper, T., Lorenz, T., Schmidt, W. and Schügerl, K. (1987) On-line measurement of culture fluorescence for process control of biotechnological processes, Annals of the New York Acad. Sci., 506, 431–445.

Scheper, T. and Schügerl, K. (1986) Characterization of bioreactors by *in-situ* fluorometry, J. Biotechnol., 3, 221–229.

Schmidt, R.D. and Scheller, F. (eds) (1992) Biosensors: Fundamentals, Technologies and Applications, GBF Monographs, Vol. 16.

Schmidt, W.J., Azzopardi, G., Grabosch, M., Herold, Th., Hoffman, H., Kuhlmann, W., Lorenz, Th., Niehoff, J. and Schügerl, K. (1986) Probennahmesysteme mit Permeation und Pervaporation unter Fermentationsbedingungen, In: Kula, M.-R., Schügerl, K., Wandrey, Ch., Technische Membranen in der Biotechnologie, GBF-Monographien, Verlag Chemie, 105–115.

Schügerl, K. (1988) On-line analysis and control of production of antibiotics, Anal. Chim. Acta, 213, 1–9.

Schügerl, K. (1991) On-line analysis of broth, in: Biotechnology, Measuring, Modelling and Control, Vol. 4, 2nd. rev. edition (ed.: K. Schügerl), VCH Weinheim, 149–180.

Schügerl, K., Brandes, L., Dullau, T., Holzhauer-Rieger, K., Hotop, S., Hübner, U., Wu, X. and Zhou, W. (1991) Fermentation monitoring and control by on-line flow injection and liquid chromatography, Anal. Chim. Acta, 249, 87–100.

Seifert, G.K.E., Matteau, P.P. (1988) An automatic aseptic bioreactor sampling system, Biotechnol. Bioeng. 32, pp. 923–926.

Sen, S., Hu, W.-S. and Srienc, F. (1990) Flow cytometric study of hybridoma cell culture: correlation between cell surface fluorescence and IgG production rate. Enzyme Microb. Technol., 12, 571–576.

Sen, S., Srienc, F. and Hu, W.-S. (1989) Distinct volume distribution of viable and non-viable hybridoma cells: a flow cytometric study. Cytotechnol., 2, 85–94.

Shapiro, H.M. (1988) Practical flow cytometry. Alan R. Liss, Inc., New York. Ed. 2nd.

Siano, S.A. and Mutharasan, R. (1991) NADH fluorescence and oxygenuptake responses of hybridoma cultures to substrate pulse and step changes, Biotechnol. Bioeng., 37, 141–159.

Spier, R.E., Griffiths, J.B. and Meigner, B. (eds.) (1991) Production of biologicals from animal cells in culture, Butterworth-Heinemann, Oxford.

Spohn, U. and Voß, H. (1989) Probenahmesysteme in der Online Bioprozeßanalytik, BTF – Biotech-Forum 6, pp.: 274–288.

Srienc, F. and Dien, B.S. (1992) Kinetics of the cell cycle of Saccharomyces cerevisiae. Ann. N.Y. Acad. Sci., 665, 59–71.

Srinivas, S.P. and Mutharasan, R. (1987) Culture fluorescence characteristics and its metabolic significance in batch cultures of *Clostridium acetobutylicum*, Biotechnol. Lett., 9(2), 139–142.

Stamm, W.W., Pommerening, G., Wandrey, Ch. and Kula, M.-R. (1989) On-line measurement of extracellular proteins in the continuous cellulase production by flow injection analysis (FIA) Enz. Microb. Technol. 11, 96–105.

Steinkamp, J.A. (1984) Flow cytometry. Rev. Sci. Instrum., 55, 1375–1400.

Stöcklein, W., Jäger, V. and Schmidt (1991) Monitoring of mouse immunoglobulin G by flow injection analytical affinity chromatography, Anal. Chim. Acta., 245 (1) 1–6.

Trott-Kriegskorte, G., Renneberg, R., Pawlowa, M., Schubert, F., Hammer, J., Jäger, V., Wagner, R., Schmid, R.D. and Scheller, F. (1989) Enzyme sensors for process control of cell cultures, in: R.D. Schmid and F. Scheller (eds.), Biosensors – Applications in medicine, environmental protection and process control, GBF Monographs, Vol. 13, Verlag Chemie, Weinheim, 67.

Turner, A.P.F., Karuber, I. and Wilson, G.S. (eds) (1987) New approaches to electrochemical immunoassay, in: Biosensors – Fundamentals and Applications, Oxford Science Publication, Oxford.

Twork, J.V. and A.M. Yacynych (eds) (1990) Sensors in bioprocess control, Marcel Dekker, New York.

Van de Merbel, N.C., Lingeman, H., Brinkman, U.A.T., Kolhorn, A. and De rijke L.C. (1983) Automated monitoring of biotechnological processes using on-line ultrafiltration and column liquid chromatography. Analytical Biotechnology, 51–62.

Vanous, R.D. (1978) Understanding Nephelometric Instrumentation, American Laboratory, 6, 17–23.

Visser, J.W.M., Jongeling, A.A.M. and Tanke, H.J. (1979) Intracellular pH determination by flourescence measurements. J. Histochem. Cytochem., 27, 32–35.

Waarvik, T.L. (1985) Culture monitoring system, European Patent Application, Appl. No 85308834.2, Publ. No. 0184441, date filing: 4.12.1985, Applicant: Eli Lilly & Company, Indianapolis, IN, USA.

Watson, J.V. (1991) Introduction to flow cytometry. Cambridge University Press, Cambridge.

Wentz, D. (1989) On-Line-Prozeßüberwachung im Bioreaktor und Einfluß von Medienparametern auf das Wachstum tierischer Zellen, Ph.D. Thesis, University of Hannover.

Wentz, D. and Schügerl, K. (1992) Influence of lactate, ammonia, and osmotic stress on adherent and suspension BHK cells, Enzyme Microb. Technol., 14, 68–75.

Werner, R.G., Walz, F., Noé, W. and Konrad, A. (1992) safety and economic aspects of continuous mammalian cell culture, J. Biotechnol., 22, 51–68.

Winzeler, H.B. (1990) Membran-Filtration mit hoher Trennleistung und minimalen Energiebedarf Chimia 44, 288–291.

Zabriskie, D.W., Humphrey, A.E. (1978) Continuous dialysis for the on-line analysis of diffusable Components in fermentation broths, Biotechnol. Bioeng. 20, 1295–1301.

Zabriskie, D.W. and Humphrey, A.E. (1978) Continuous dialysis for the on-line analysis of diffusible components in fermentation broth, Biotechnol. Bioeng., 20, 1295–1301.

4 Down-Stream Processing

4.1 Principles of Product Extraction from Cell Culture and Purification for Pharmaceutical Proteins

Alois Jungbauer

4.1.1 Purity Criteria and Consequences for Downstream Processing

Biologics

"Mammalian cell culture protein" is not a defined term by the health authorities, which are responsible for licensing drugs. Instead of *mammalian cell culture proteins*, the term *biologics* is widely used but the meaning of biologics is not precise. In general *"biologics"* are substances derived from living organisms. Regarding medicinal products *biologics* is a well defined technical term. There is no definite distinction between animal cell proteins and other proteins. A strict discrimination between recombinant DNA products and proteins from other sources [guidelines for research involving recombinant DNA molecules (1986) published by the Center for Biological Evaluation and Research (CBER)] is not made. In fact, this makes no sense. A big difference between host cells for production of therapeutics is based on the susceptibility to human pathogenic viruses, mycoplasma etc. Therefore in practice, the applicants for a new drug have to present an extensive study which shows the clearance of adventitious agents when the protein is derived from mammalian cells, irrespective of whether is a recombinant DNA product or a conventional product secreted by a continuously growing cell line or a product from an animal (human) tissue or body fluid (see also chapter 4.2).

Additionally mammalian cells often show protein polymorphism. That means that a protcin with an identical amino acid backbone posseses varying side chain modifications, like glycosylation, phosphorylation etc. (see also chapter 2.2). Often the backbone itself shows minor chemical variations (like deamidation of Gln or Asn; oxidation of Met to Sulfo-met), resulting in significant changes of chemical and physical properties (isoelectric point, solubility etc.) which influence the conditions for protein purification. As already known for years, the side chain modi-

Table 4.1.1. Biological behavior of a monoclonal IgG in respect to carbohydrate moiety.

Biological function	Effect of carbohydrate depletion
Antibody dependent cellular cytotoxicity	absence
Complement activation	absence
Interaction with macrophage Fc-receptor	absence
Biological half-life of immune-complex	decreased
Binding to protein A	no effect
Affinity	no effect

fication may have significant consequences for the biological behavior of a protein. In Table 4.1.1 the effects of carbohydrates on the biological activity of a human antibody should demonstrate the importance of protein polymorphism. Therefore the consistency of a product is a critical quality parameter of a mammalian cell protein.
The consequence for downstream processing resulting from these requirements yields in a very complex in-process control, product control and validation of the process.

Definition of Biologics
In the USA biologics are defined in CFR 21 subchapter F as products with the following characteristics: "... "Biological product" means any virus, therapeutic serum, toxin, antitoxin, or analogous product applicable to the prevention, treatment or cure of diseases or injuries of man: ...". However the regulation affairs between drugs (CFR 21 subchapter C and D) are very similar to biologics. The guidelines of medicinal products derived from biotechnology of the Commission of the European Community (1988) are much more clear than the definition in CFR 21, but also do not consider future developments.

In some European countries biologics are defined as products derived from:

(1) human blood, other human body fluids or human tissue
(2) animal blood
(3) microorganisms or components of microorganisms
(4) animals or microorganisms for active or passive immunization
(5) monoclonal antibodies
(6) recombinant DNA products

The sense of the European definitions covers the sense of the definitions in CFR 21. In Germany, sera, monoclonal antibodies and vaccines are treated separately from biologics, but the general rules regarding production, purification, clinical testing etc. are identical to biologics.

For production, purification and quality control, preclinical testing etc. a series of recommendations and requirements must be obeyed. The office of biologics released *Points to Consider in the Production and Testing of New Drugs and Biologicals Produced by Recombinant DNA Technology and Points to Consider in the Manufacture and Testing of Mononclonal Antibody Products for Human Use.* Also the ad hoc Working Party on Biotechnology/Pharmacy, as a consulting committee of the Commission of the European Community released a series of recommendations concerning biologics. Purity, removal of adventitious agents, consistency, potency etc. are well defined and follow in general the same principles.

Briefly, absence of host proteins to the largest extent possible, absence of host DNA, absence of viral contamination (depends on the host cell) and absence of pyrogenes in the final product is required (Kozak et al., 1992). Consistency is also a very important issue, since it is known that cells may secrete proteins in different isoforms.

People responsible for purification have to meet all these requirements. Therefore a process has to be designed which is suited to purifying proteins and other biologicals up to a level which is much higher than generally obtained at the protein purifier's workbench.

4.1.2 General Strategy of Purification Processes for Biologics

There is a general purification strategy for biologics derived from microorganisms and animal cells (Leser and Asenjo, 1992). An overview is shown in Figure 4.1.1. The product can be secreted into the culture medium or deposited in the cells. Proteins can be either deposited in a paracristalline form, well known as inclusion bodies (or refractile bodies) or associated with membranes or DNA, or soluble in the cytoplasm. Today for mass production of proteins for therapeutic, diagnostic and research, in addition to recombinant DNA technology conventional sources like human and animal blood, body fluids and tissues are still used.
Secretion of proteins into the medium or deposition of proteins as inclusion bodies is the preferred strategy if possible. Association of proteins with membranes is prevented by using special constructs, which allow expression of a truncated protein depleted from the hydrophobic transmembrane region. Hydrophobic regions in proteins are often responsible for the association of proteins with membranes. However in some cases (e.g. vaccines) the association with membranes is desired to produce a synthetic vesicle, which may display improved immunogenicity.

The purification of membrane proteins is more an issue in research than in pilot production and industry. Two out of three general cases are discussed in the present publication:

1. product is secreted by the cell
2. product is deposited in solution
3. product is deposited as inclusion body

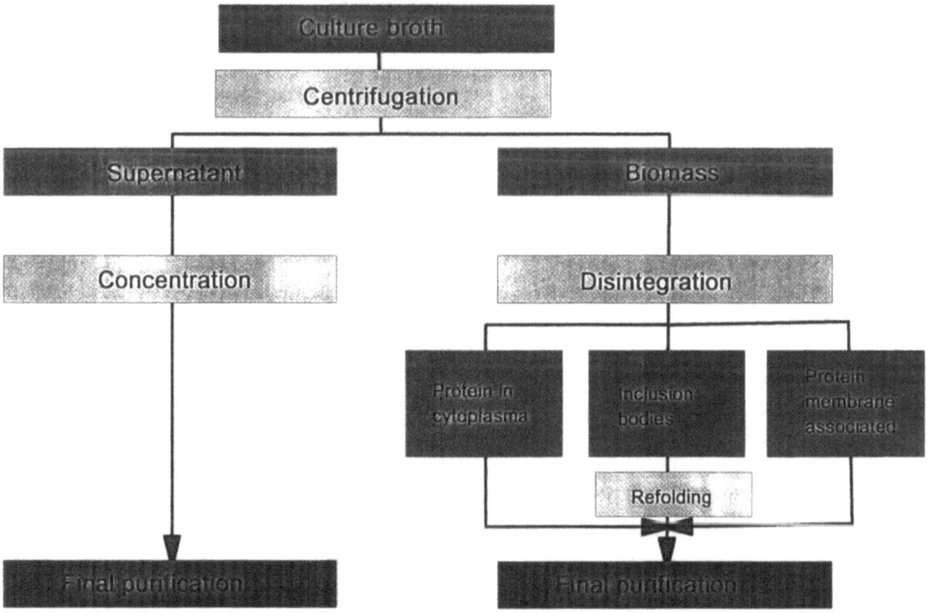

Fig. 4.1.1: General purification strategy for biologics derived from micro-organisms and animal cells. As final purification chromatography, electrophoretic methods, and other high resolution methods are considered. The concentration is achieved mainly by precipitation and ultrafiltration.

The third case is currently not used with mammalian cells, although inclusion bodies were first discovered in mammalian cells.

Secreted products

Product extraction can be accomplished by very simple means if the biologic is secreted into the culture medium. The broth is clarified either by microfiltration (Nagata et al., 1989; Patel et al., 1987; Warren et al., 1991; Sea and Bailey, 1986) or centrifugation (Datar and Rosen, 1987). Combinations of these processes are also used.

Animal cells in general secrete proteins and other metabolites into the culture medium. The very dilute culture broth can be easily harvested. Additives like antifoam oils or stabilizer e.g. pluronic, often decrease the flux in microfiltration and of course in ultrafiltration (Cautney MacGregor et al, 1988). To increase productivity the optimal time of harvesting the broth must be studied. Both growth dependent and growth independent product formation kinetics are observed. The production of monoclonal antibodies is very well studied regarding product formation. The product formation kinetics vary for different hybridomas (Al-Rubeai and Emery, 1990; Savinell et al., 1989; Flickinger et al., Ozturck and Palsson, 1990).

Animal cells and yeast are very attractive hosts for the production of heterologous proteins. Unlike prokaryotic systems, their eukaryotic subcellular organization enables them to carry out many of the post-translational folding, processing and modification events required to produce *authentic* and bioactive mammalian proteins (Buckholz and Gleeson, 1991). Viral contamination of the product is practically excluded and the productivity is high (Sleep et al., 1991; Fleer et al., 1991) as demonstrated by expression of albumin in *S. cerevisiae*.

Secreted biologics are usually not produced in *E. coli*. Low productivity and complex harvesting strategies do not allow a large scale application at the moment, although correctly folded proteins are obtained. Also secretion into the periplasm is not applicable for large scale due to the same reasons. An extensive comparison of process economics between animal cell and bacteria fermentation as well as product recovery was carried out by Datar et al. (Datar et al., 1993) examining t-plasminogen activator production in CHO-cells and *E. coli*-process. They conclude that the most cost effective part in the *E. coli*-process is the refolding. In animal cell culture it is the low expression level.

Induction versus constitutive expression

In principle two different methods of expression of heterologous proteins are applicable: Constitutive expression and induction of protein synthesis. Each expression system has its advantages and disadvantages (Zabriskie and Arcni, 1986). The method preferrably applied depends on the characteristics of the protein and the design of the whole process. In general both methods are used in industrial scale. Induction of heterologous protein synthesis is preferred if inclusion bodies are to be formed. The relevance of the expression of a protein and the influence on DP is demonstrated in Figure 4.1.1.

In mammalian cell culture constitutive expression systems are very popular in antibody production, interferon production (Sano et al., 1992) or production of interleukin-2 in BHK cells (Ryll et al., 1990) or CHO cells (Conradt et al., 1989).

A complete listing of all proteins constitutively expressed in mammalian cells would exceed the scope of this chapter.

In vaccine production the product formation is induced by addition of a virus inoculum, which very often is accompanied with cell lysis. Such systems like the vaccinia virus system are not suited for large scale production of proteins, but represent a valuable tool for screening of the optimal host cells (Pavirani et al., 1987) for a complex protein like F VIII.

The types of expression do not influence the strategies in downstream processing, because there are only a few continuous harvesting techniques applicable which fit into the scale of processes for protein therapeutics.

Products deposited intracellularly in solution

As already mentioned above the product-secretion to the periplasm in *E. coli* cells has the advantage that the protein is folded correctly but the process is not suited for technical scales. The lysis of the cell wall without damaging the cell itself is a very tricky process. The method is advantageous in bench scale for rapid production of correctly folded proteins in mg dimensions. Only a few examples of heterologous intracellular protein expression where a correctly folded protein can be extracted from the homogenized cells (Authors observation with Cu/Zn SOD and T7 polymerase expressed in *E. coli*) are observed.

Normally the intracellularly expressed proteins do not fold correctly and the protein must be transferred into the native conformation. Therefore large proteins and proteins with a biologically significant side chain modification should be produced in animal cells.

4.1.3 Removal of Cells, Cell Debris Initial Purification

Cells and cell debris are mainly removed by centrifugation and cross-flow microfiltration. Combination of both methods is also very popular. The big disadvantage of continuous centrifugation is the low throughput of bacterial suspension or homogenized cell suspensions. A combination of centrifugation and cross-flow microfiltration allows the improvement of the throughput, while small particles can remain in solution in the first step. In the subsequent step the fines are removed by microfiltration.

An important factor for efficient removal of animal cells from the culture supernatant is the viability of the cells and the supplements added to the medium. Very often viability is decreased when serum free media are used (Van der Pol et al., 1990). Instead of improved filtration performance a lower flux caused by the cell debris is observed. Similar problems in clarification are observed in centrifugation.

Screening for cell culture media for rapid downstream processing may circumvent this problem and contribute to process economics (Jäger, 1988).

4.1.3.1 Microfiltration

Microfiltration is a unit operation for separation of small particles. The separation limits are between 0.02 to 10 μm. According to Ripperger (1992) the membrane process can be divided into reverse osmosis, ultrafiltration and microfiltration (Fig. 4.1.2). The transmembrane pressure and the exclusion limits of the membranes are the criteria for the definition.

Microfiltration can be carried out in a dead-end mode and a cross-flow mode. In downstream processes (DP), the cross-flow filtration is carried out continuously or discontinuously.

The most important parameters determining the productivity of cross-flow microfiltration are

- transmembrane pressure
- velocity
- particle size and surface
- viscosity of the liquid
- additives like surfactants, changing the surface and surface tension

Many models are published to calculate the process. One important factor, the concentration polarization, is the limiting physical process. At higher particle concentration and with time a layer is formed on the membrane. This layer is responsible for the flux reduction. A comprehensive overview on this technique is given by Ripperger (1992) and Staude (1992). Often similar or identical module types are used in microfiltration and ultrafiltration.

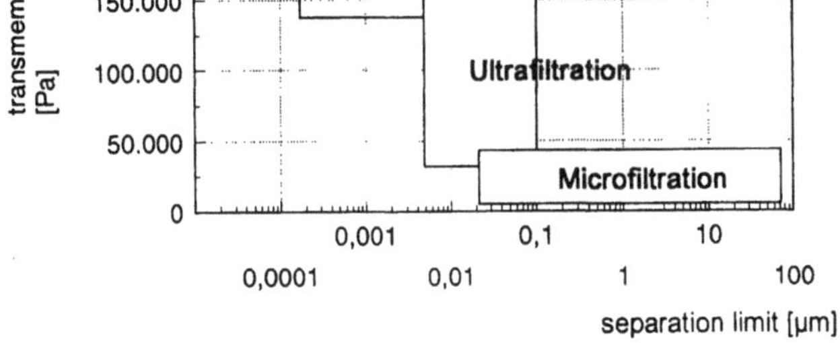

Fig. 4.1.2 Definition of membrane separation processes according to transmembrane pressure and exclusion limits.

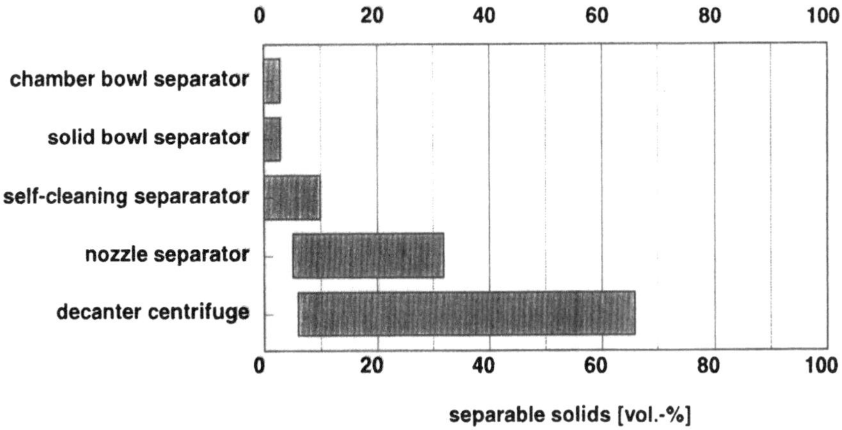

Fig. 4.1.3: Selection criteria for types of centrifuges.

4.1.3.2 Centrifugation

Cell harvesting is mainly carried out by centrifuges, either by disk bowl centrifuges, tube centrifuges or basket centrifuges. The sedimentation process can be described by Stoke's law

$$v_s = (d^2 \Delta\varrho\omega^2 r)/18\,\mu \qquad (1)$$

where v_s is the sedimentation velocity, d is the particle diameter, $\Delta\varrho$ is the density difference between medium and particle and μ is the dynamic viscosity. ω is the angular velocity and r is the radial distance of the particle from the axis of rotation. For removal of cell debris the angular velocity or the residence time must be increased. ω is limited by the stability of the construction material of the centrifuge. Tube centrifuges are better suited for removal of cell debris than disk centrifuges; higher g-forces are achievable.

For comparison of different centrifuges the static settling area is used. For scale-up the nozzle capacity must be taken into consideration (Brunner and Hemfort, 1988).

The selection criteria for the types of centrifuges are mainly the amount of separable solids present in the suspension (Fig. 4.1.3).

Since the amount of separable solids is very low in mammalian cell culture, decanters are never used. It is essential to operate the centrifuge at minimum g-forces. In practice animal cell culture centrifuges were developed in a manner which allow a gentle separation of the cells from the liquid to maintain the viability of the cells. In downstream processing very often viability is not a question. A clear supernatant is the aim. In the case when the centrifugation is performed in an insufficient way, nucleosomes (Franek et al., 1992) may remain in the supernatant and the nucleosomes interfere in the subsequent purification steps.

If lipids cause separation problems, the addition of silicon dioxide powder (e.g. Cab-O-Sil, Union Carbide, Danbury CN) may be helpful. A minimum time is required for complete reaction of the SiO_2-powder with lipids. Separation of lipid by very high speed centrifugation is only applicable in laboratory scale.

Table 4.1.2. Positively charged flocculants used for the separation of bacteria and animal cells, only materials used in the production of biologics are listed.

Trade name	compound	manufacturer	m_R	model cell	Ref.
–	Chitosan	various	4.0×10^5	bacteria	Hughes et al., 1990; Agerkvist et al., 1990
Zetag 88 N	Cationic Polyacrylamide Copolymer	Allied Colloids	1.0×10^7	bacteria	Agerkvist et al., 1990
Zetag 63	Cationic Polyacrylamide Copolymer	Allied Colloids	1.0×10^7	bacteria	Agerkvist et al., 1990
Zetag 92	Cationic Polyacrylamide Copolymer	Allied Colloids	1.0×10^7	bacteria	Agerkvist et al., 1990
–	Polygalacturonic Acid	various	–	animal cells	Kenny, 1986
	Polyethylenimin	various	$4.0–6.0 \times 10^5$	*E. coli*, bacteria	Agerkvist et al., 1990; Person and Lindman, 1994
Polymin SN	Polyethylenimin derivate	BASF	–	bacteria	Eriksson and Härdin, 1987

Addition of flocculents

To increase the lifetime of membranes and chromatography columns, for improving the performance of centrifugation and for removal of contaminants like DNA, endotoxins etc. flocculation is a very useful method. Since bacteria and cell debris have separation properties characteristic for the colloidal size range, the rate of recovery is largely improved if flocculation of the cell can be carried out (Bowden, 1985; Vorauer et al., 1993). With certain microorganisms and conditions flocculation may occur spontaneously as in activated sludge. However most of the microorganisms and homogenized cells used for production of biologics have lost the ability to flocculate. In such cases it is necessary to induce flocculation by adding flocculants. Flocculation of microorganisms relevant for production of biologics is accomplished by positively charged macromolecular flocculants (Tab. 4.1.2) or submicron particles, which are also positively charged (Tab. 4.1.3). A high charge density is important in reaction or competition with adsorbed material on the cell surface. Treeweek and Morgan (1977) studied flocculation with different polyethylenimines. The results can be explained in terms of charge neutralization and mosaic patch formation.

Table 4.1.3. Physical properties of the pellicular positively charged flocculent Biocryl Bioprocessing Aids, the data were obtained from the manufacturer Rohm and Haas Philadelphia PA.

Trade Name	Functionality	Charge Strength	Particle Size	Surface Area
Biocryl BPA-1000	Quaternary Amine	Strong Base	0.1 μm	55 m^2/g
Biocryl BPA 1050	Tertiary Amine	Weak Base	0.1 μm	55 m^2/g

According to C.S. Ho (1986) the relevant forces in flocculation can be divided into three main categories:

(1): Conservative forces or those arising out of the mutual proximity like the electrostatic and electrodynamic forces
(2): Hydrodynamic forces which determine such characteristics as particle diffusion and the rate of approach to equilibrium, and
(3): The forces due to externally applied fields such as the magnetic, hydrodynamic, shear and gravitational fields.

By addition of pellicular flocculants (Wha et al, 1987) or macromolecular charged compounds the sedimentation force (F_{sed}) increases due to the increasing diameter of the formed aggregate. The sedimental forces of particles can be expressed as

$$F_{sed} = (r_p - r) v_p \cdot g \qquad (2)$$

where $(r_p - r)$ is the density difference between particle and medium, v_p is the volume of the particle and g is the gravity.

According to the Stokes Law the sedimentation velocity

$$U = 2a^2 g(r_p - r)/9\mu \qquad (3)$$

depends on the viscosity (μ) of the medium and the square of the particle diameter formed during flocculation. Equations (2) and (3) show the most important parameters influencing flocculation.

In some cases (Vorauer et al., 1993; Agerkvist et al., 1990) dilution of the culture broth improves the overall performance of flocculation. In addition to clarification, charged flocculants also remove impurities such as pyrogenes (Vorauer et al., 1993; Wha et al., 1987) and DNA. In Figure 4.1.4 the effect of DNA removal from a tissue culture is demonstrated. Efficient removal of particles is the most important prerequisite for successful chromatography, or ultrafiltration, which could be the subsequent step. Lifetime of columns and membranes can be increased and since clogging by fines is avoided the flow rate is maintained at the highest possible level. The major drawback of flocculation in production of biologics is the quality control. Residual flocculant or abrasion of pellicular flocculants must be excluded in the final product. Several companies supply radiolabeled or fluorescent dye labeled flocculants as analytical tools for the analysis of residual material in the clarified supernatant. The shear strength of the pellicular flocculants is rather high.

The influence of shear forces in flocculation is well understood. Vigorous stirring should be avoided but nevertheless efficient mixing must be provided to exclude local inhomogeneity, which is responsible for loss of products.

In my experience, if clarification problems are observed, flocculations can circumvent these problems in most cases. As already pointed out, screening of media suited for downstream processing is a complementary strategy.

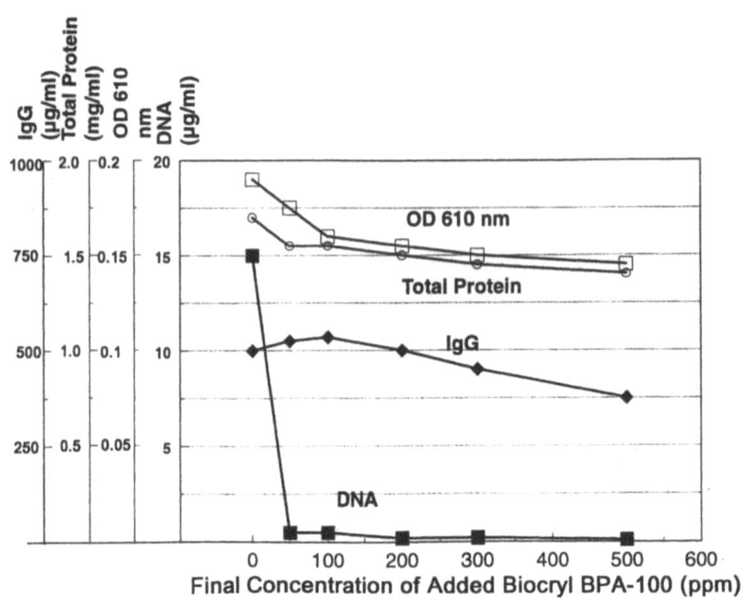

Fig. 4.1.4: Selective removal of DNA from hybridoma culture supernatant by flocculation with BPA 1000.

4.1.3.3 Ultrafiltration/Diafiltration

Ultrafiltration (UF) and Diafiltration (DF) are very popular methods for conditioning the clarified culture broth for chromatography. Especially in processing animal cell culture fluids the very dilute broth can be concentrated up to a factor of 20 and even more.

The ultrafiltration is a process in which a solution containing macromolecules is passed across a functional membrane, while macromolecules are retained by the membrane. The process is driven by the pressure drop across the membrane also called transmembrane pressure (ΔP_T) expressed as

$$(\Delta P_T) = [P_i + P_0)/2]\text{-}P_P \tag{4}$$

where P_i is the inlet pressure, is P_0 the outlet pressure and P_P is the permeate back-pressure of the membrane. The difference between the conventional dead-end filtration and the cross-flow filtration is the configuration of the system. For large scale operations only cross-flow filtration will be used. The membranes for micro as well as ultrafiltration are commonly utilized in a variety of filtration devices. There are three basic types of tangential flow filtration devices:

- plate and frame
- hollow fiber, and
- spiral wound.

The differences cannot be easily explained. The selection rules are often heavily influenced by economical considerations.

Technological parameters are:
- Flux
- Recirculation rate
- Chemical compatibility
- Pore size or molecular weight cutoffs
- Pressure and temperature limitations
- Cleanability

Once the feasibility has been established, the engineering design and scale-up come into play. Further details are described by E. Staude (1992)

The additives (antifoams or stabilizers) to the culture medium may strongly interfere with the UF process by reduction of the flux (Kloostermann et al., 1988). UF/DF in large scale is always carried out in cross-flow mode, discontinuously or continuously or in dynamic mode (Wronski and Rudniak, 1988) to reduce fouling of membranes.
A simple relationship between flux (J) and time (t) is the empirical relationship

$$J = J_0 t^n \tag{5}$$

where J_0 is the initial flux and n is an exponent less than zero (Jönsson and Trägårdth, 1990) also called fouling rate constant. This constant must be experimentally determined. The equation (5) shows that the flux is decreasing with time, because a layer is built up at the membrane surface.

Membranes in appropriate quality and characteristics are available to cover the separation problems in biotechnology. For further information look at Shiloach et al. (35?); or Staude (1992). A detailed description exceeds the scope of the chapter.

4.1.4 Precipitation

Protein fractionation on the basis of solubility in aqueous salt buffers or organic solvents is a well-accepted and documented method for prepurification of proteins (Green and Hughes, 1955; Englard and Seifter, 1990). The major drawback of precipitation is the aggregation of proteins during the process. As a result bioactivity is lost. Another disadvantage is the handling of wastes, especially the ammonium sulphate.

Ammonium sulphate and other salt precipitations follow the principle that at high concentrations the protein solubility decreases. This behavior is called salting out and is common to all non-electrolyte solutes and appears to be due to "dehydration" of the protein molecule. The simplified form of the Debye equation including the "salting out" term added by Hückel's is:

$$-\log\gamma = \log S - \log S_0 = \frac{0.5\,Z_1 Z_2 \gamma}{1 + A\mu} - K_s\mu \qquad (6)$$

in which γ is the activity coefficient, S is the solubility, S_0 is the solubility in the absence of the electrolyte, μ is the ionic strength, Z_1Z_2 is the valence type, Ks is the salting out constant and 0.5 is a theoretical constant. Ks depends on the type of the electrolyte, protein, pH etc.

Ammonium sulphate precipitation

Isolation of monoclonal antibodies from supernatant by $(NH_4)_2SO_4$-precipitation is widely used. A short protocol using saturated $(NH_4)_2SO_4$-solutions is given by Jonak (1984). The disadvantage of this procedure is the at least 1 : 1 dilution. Addition of solid ammonium sulphate is preferred. One has to take into consideration that $(NH_4)_2SO_4$ lowers the pH of the supernatant. Therefore a good buffered supernatant has to be used. Combination of ammonium sulphate precipitation with hydrophobic interaction chromatography or thiophilic adsorption is an elegant way to short cut a purification process. The two steps are connected together without any intermediate step. The purification sequence consists of following steps: clarification ⇒ $(NH_4)_2SO_4$-precipitation ⇒ washing of the precipitate ⇒ redissolution of IgG ⇒ clarification HIC or thiophilic adsorption.

IgG binds very high $(NH_4)_2SO_4$ concentrations (> 50 %) to HIC matrices, where IgG already precipitates. A repetitive injection of IgG dissolved in a low salt buffer followed by a solution containing high concentrations of $(NH_4)_2SO_4$ circumvents this problem. Direct HIC purification of monoclonal antibodies from a serum containing supernatant is not recommended. Albumin binds very strongly to hydrophobic matrices and reduces the capacity for IgG. Selective desorptions in the presence of albumin is also very complicated.

Caprylic acid

Caprylic acid (octanoic acid) precipitation is sometimes used for antibody purification from serum (Harlow and Lane, 1988) or hybridoma culture supernatant (Russ et al., 1983). Caprylic acid is a natural substance. In serum caprylic acid is in the range of 40 ppm (own results not published).

Organic solvents

Cohn's ethanol fractionation is still used for blood protein fractionation. The advantage of the method is the good recovery and the high efficiency in virus inactivation. The disadvantage of the method is the cooling below the freezing point of water to protect the proteins from inactivation. The whole precipitation procedure is very sensitive to changes in pH and ionic strength.

Polyethylenglycol

Polyethylenglycol-precipitation (PEG) is a very gentle purification procedure. The precipitation principles are related to ethanol fractionation. PEG is often considered as a polymerized organic solvent. Neoh et al. (1986) described PEG precipitation for monoclonal antibodies from mouse ascites fluid, the conditions could also be applied to culture supernatants from hybridoma culture produced by a dialysis reactor (Brooks et al., 1992). A short protocol for monoclonal antibodies (concentration between 2–4 mg/ml) looks like this: Polyethylenglycol 6000 (PEG) is dissolved to a concentration of 30% (w/v), PEG is added to a final concentration of 15% to the culture supernatant, gently mixed, incubated at least overnight at low temperature and centrifuged. The sediment is washed with 15% (w/v). An $(NH_4)_2SO_4$-precipitation may be added.

The removal of PEG from the precipitated protein causes problems in ultrafiltration and gelfiltration.

Isoelectric precipitation

Isoelectric precipitation and acid precipitation are also used in large scale. Isoelectric precipitation (also called euglobulin precipitation) uses the solubility properties of a protein near it's isoelectric point. When a concentrated protein solution in a low ionic strength buffer is titrated to it's isoelectric point the protein precipitates very slowly. In Table 4.1.4, a balance of a process for purification of murine monoclonal IgM antibodies is shown (Steindl et al., 1987). The success of precipitation can be followed by the cumulative IgM content in the centrifugate and the sediment. The procedure is very gentle but very sensitive to environmental conditions.

Table 4.1.4. Yields and purification factors during purification of monoclonal IgM; starting volume 150 l hybridoma culture supernatant. Kp, Kv are the purification factors related to protein or volume respectively.

Purification step	IgM (mg/ml)	IgM/Protein (%)	Kp	Kv	Yield (%)
Culture supernatant	0.039	1.42	1	1	100
Ultrafiltration	15.53	9.75	6.8	398	95
Precipitation	43.84	95.25	66.8	1124	66.5
Gel chromatography	4.67	99.0	69.5	115	40.3

4.1.5 Liquid-Liquid Extraction

Both extraction with organic solvents as well as the aqueous two-phase extraction are applicable for pretreatment of crude protein solutions.

Aqueous two-phase separation is also applicable for suspension, such as virus suspension (Walter and Johanson, 1986). Aqueous two-phase systems result from the incompatibility between two polymers (e.g. PEG and dextran) or between one polymer and an approppriate salt (e.g. potassium phosphate) which occurs above a certain reagent concentration. The high water content (approx. 65–90%) of such systems makes them much more suitable for gentle partitioning of proteins than water organic two-phase systems (Hustedt et al., 1985) and also have a potential for mammalian cell culture proteins (Sulk et al., 1992). Aqueous two-phase still has the disadvantage of wasting the phases, since the regeneration is very costly (Kroner et al., 1984).

Affinity partitioning

Affinity partitioning is a further improvement of partitioning (Johansson and Joelsson, 1985). A ligand is bound to the polymer to increase the partition coefficient. Purification of γ-interferon from cell culture is a key example for this technique (Menge et al., 1983).

The method of affinity partitioning is different from filtration or extraction with pellicular material, called affinity filtration (Pungor et al., 1987), which is related to the techniques using affinity-microspheres.

Reversed micelles

Partitioning with reversed micelles (Paradker and Dordick, 1991; Carlson and Magarakan, 1992) is also a useful method for prepurification of crude protein solutions. Reverse micelles are formed by a detergent and an organic solvent which are added to the protein solution. Aggregates of surfactant molecules are spontaneously generated in organic solvents as a result of molecular self-assembly. The aqueous phase is inside the micelle. By separation of the micelle the protein included in the aqueous phase can be extracted. As a result, substantial purification can be achieved. The selectivity can be tuned by the selection of cationic and anionic surfactants. There is an empirical relationship between isoelectric point, molecular weight, type of surfactant and the transfer of protein into the micelle.

Affinity-microspheres

Affinity-microspheres are particles in the micrometer and submicrometer range. They are covered with an affinity ligand and used for batchwise extraction of proteins. Harvesting of the loaded affinity microspheres is accomplished by microfiltration. Weiner et al. (1994a) described a novel type of affinity matrix which was prepared by binding Protein A to the crystal cell surface layer from the gram positive *Clostridium thermohydrosulfuricum* (Fig. 4.1.5). Monoclonal antibodies could be harvested by affinity cross-flow filtration from hybridoma culture supernatant (Weiner et al., 1994b)

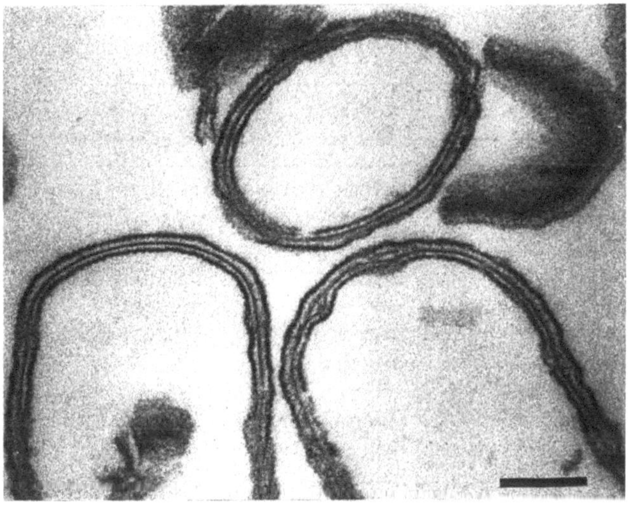

Fig. 4.1.5: Basic support for affinity microspheres: Electron micrograph from ultrathin-sectioned S-layer microparticles. The outer and inner layer represent two-dimensional protein crystals (S-layers), the middle layer is peptidoglycan (bar = 200 nm). (Courtesy of Margit Sara, Center for Ultrastructure Research, Vienna, Austria).

4.1.6 Conventional Chromatographic Methods

The high value of the products derived from mammalian cells justify expensive processes, which are required to reach the high standard in product purity. Chromatographic purification processes possess high resolution and high selectivity, and in many cases sufficient throughput can be achieved (Broschetti, 1988). The clearance factors for adventitious agents are up to 6 log steps.

In chromatography the distribution of a mixture of compounds between a mobile phase and a stationary phase is the driving force for separation. The separation process can be explained by a definite number of transitions of the solute between the mobile and the stationary phase, while the mobile phase is passing down the column. Due to the different distribution coefficients (K)

$$K = C_s/C_m \tag{7}$$

the solutes are retained to a different extent and a mixture of substances can be separated. C_s is the concentration of the solute in the stationary phase and C_m is the concentration in the mobile phase after equilibrium is reached. The mobile phase is modified by selected compounds called mobile phase modifiers (MPM, eluents or sometimes displacers). Depending on the MPM (eluent) concentration the solute is more or less retained in the column. Therefore in a lot of cases on/off mechanisms can be generated.

The separation usually takes place in a column made of glass, plastic or stainless steel. The solution is pumped over the column. Selected compounds are bound to the matrix by a defined action principle (Tab. 4.1.5). In prep. chromatography the separation resembles more an adsorption/desorption process (Jungbauer, 1993) than a chromatography.

Table 4.1.5. Types of chromatography distinguished according to the nature of retardation forces and the principles of separation (modified from Jungbauer et al., 1989).

Adsorption chromatography	Surface binding	Molecular structure	Low capacity, high clearance of adventitious agents
Ion-exchange chromatography	Ion binding	Surface charge	Highest capacity, high speed
Molecular sieve chromatography (Gel filtration)	Steric exclusion	Molecular size & shape	Low capacity, excellent for desalting and removal of aggregates
Affinity chromatography	Biospecific adsorption/ desorption	Molecular structure	Highest selectivity
Hydrophobic (interaction) chromatography	Hydrophobic complex-formation	Hydrophobicity & hydrophobic patches	Optimally suited for combination with other steps requiring high salt buffers
Covalent chromatography	Covalent binding	Functional groups	Sensitive to redox potential, limited use in preparative chromatography
(Metal) chelate chromatography	Coordination complex	Complex formation with transition metals	Requires accessible His on the protein surface
Reversed phase chromatography	Hydrophobic complex formation	Hydrophobicity	Limited use in preparative chromatography of proteins, frequently used in quality control

After washing out the unbound material the desired compounds are desorbed (eluted) by changing the composition of the mobile phase in a continuous mode (linear gradient) or discontinuous mode (step gradient). The quality of the packed bed is the major influence in large scale chromatography. Elution by step gradients are usually less sensitive than linear gradient elution.

Column liquid chromatography is the most powerful tool in purification of biologics. After initil purification final purification is achieved by a sequence of chromatographic steps composed from IEX, HIC, AC and finally by GF. The last steps are often called *polishing steps*.

The big advantages of column liquid chromatography are the simple scale-up, stable operation, the purity factors achieved, and the accumulated knowledge since the development of this type of chromatography, e.g. Sephadex G25 celebrated its 25th anniversary in 1984 (Janson, 1987) and the gel is still used in industrial chromatography.

4.1.6.1 Ion-Exchange Chromatography

Ion-exchange chromatography is a key chromatography step in most purification processes for mammalian cell culture proteins. The advantages of this methods can be summarized as follows:

– high capacity	up to 150 mg protein per ml packed resin
– rapid	exception conventional cellulose, agarose and dextran based beads without or with a low degree of cross-linking (like DEAE-Sephadex A25 from Pharmacia or DE52 from Whatman)
– high recovery	
– availability	over 200 different resins are commercially available
– simple operation	
– scale-up experience	scale-up to several m^3 is already carried out
– economic	gel can be discarded after several runs if the process requires it
– cleanability	many resins withstand NaOH-treatment over several days without significant change in the chromatographic performance, but only a few gels can be chemically sterilized by oxidizing agents.

Resins and resin-screening
Resins useful for mammalian cell culture protein purification do not differ from other applications for therapeutic protein purification. IEX-resins for protein purification are comprehensively described by Karlson et al. (1989). It is very difficult to keep track of the latest development of IEX-resins. The latest developments are:

- perfusion chromatography (Perspective Biosystems, Cambridge, MA)
- tentacle type gels (Merck, Darmstadt, FRG)
- highly diffusable gels, made from an organic skeleton, which is filled with polyacrylamide polymers, like Q-Hyper D and S-Hyper D (Sepracore Inc., Marlborough, MA)

The development of all listed types is directly for higher productivity and in the case of the tentacle type gels improvement of recovery.

It is not easy to select an IEX-resin for a particular separation problem. As already noticed, more than 200 different ion-exchangers for protein purification are available, the physical differences of the resins are:

- size, size distribution, content of fines and broken beads
- shape of the beads, deviations from the spherical shape
- density of the beads
- pressure drop versus flow rate
- porosity (transjecting pores, beads with an inner core)
- ligand density (amount of ionizable ligand per volume of packed gel)
- titration cure and shape of curve (sigmoid with one or two inflection points see Fig. 4.1.6)

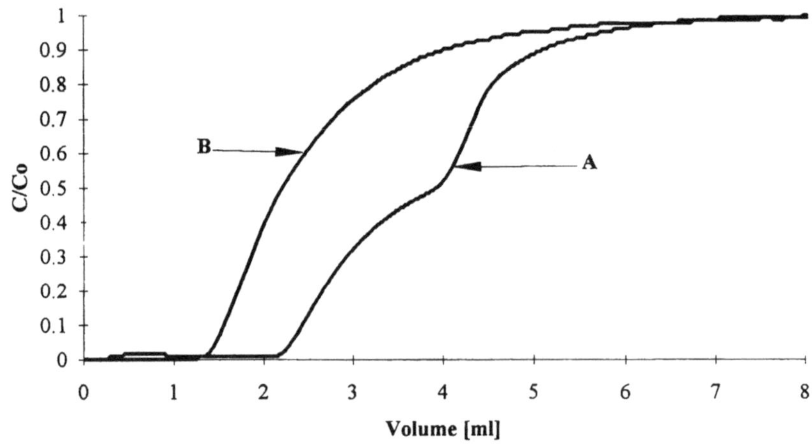

Fig. 4.1.6: Breakthrough curve of NaCl of a cation exchanger (Q-HyperD, Sepracore Inc. MA). C/Co is the dimensionless salt concentration of the column effluent. 1 ml gel was packed and operated at 1 ml/min. A biphasic breakthrough curve can be observed (**A**). A is the breakthrough curve without column to show the extra column effects (**B**).

- swellability in organic solutions and salt solutions (can be neglected when consisting of highly cross-linked dextran agarose etc. or beads with a core or skeleton consisting of an organic polymer)
- resistance to abrasive forces.

In addition to the physical properties, the chemical properties as well as some economic and regulatory issues also influence the selection of a particular resin. The main question is: Can the selected resin solve my purification problem in the desired manner? That means optimal resolution and high recovery under acceptable buffer conditions at a high flow rate. Additionally, the resin must be certified by a drug master file or equivalent certification. If such a certification does not exist, a lot of investigations have to be carried out (stability under working conditions, toxicological data, batch to batch variation etc.) to show that the resin is suited for manufacturing of a therapeutic agent.

Resin screening is a very critical step in the development of a protein therapeutic. A rational selection rule is only based on the simple rule of thumb which can be described as follows: Binding a protein to an anion-exchanger the pH of the loading buffer is more than 0.5 pH units above the isoelectric point, for cation-exchangers the loading buffer should be 1 to 1.5 pH units below the isoelectric point of the protein.

This rule can also be used for selection of appropriate buffers. Another approach is the determination of the titration curve of a protein solution (Bosisio et al., 1980). From this data a first selection of a resin with the corresponding buffer can be done. The selection of a resin is also influenced by the personnel experience, and the availability of a resin. It is still done by trial and error. Since the performance of IEX is determined by a variety of parameters, like buffer composition (pH, salt), protein load, velocity etc. no expert system or generalized fine structured protocol is available for optimization.

Optimization

After selection of a resin optimization of the running conditions is the next step. First of all it must be decided whether resolution or concentration is desired, maybe both. The biggest advantage of IEX is the high concentration factors which can be achieved. Up to factors 100 are reported (Östland et al., 1987) for purification of monoclonal antibodies from hybridoma culture supernatant. Since the animal cell culture supernatant is a very dilute protein solution this type of chromatography fits exactly to the purification strategy.

The elution conditions are also responsible for the performance of the IEX. Elution is accomplished either by instantaneous change of the salt concentration (stepwise elution or desorption) or by a continuous change of salt (gradient elution). Change of pH is not so popular for elution of a protein from an ion-exchanger, these conditions are more difficult to reproduce.

Applying stepwise elution one can obtain two elution types. Type I is elution of the protein together with the salt front resulting in a very sharp peak. In Type II elution the protein appears after the salt front. The peak is broader, but contaminants may by resolved from the product.

As already mentioned the particular buffer composition is also very critical. In Figure 4.1.7 the influence of the salt type on the recovery of a human monoclonal antibody from CM-Sepharose Fast Flow is shown (Jungbauer et al., 1988).

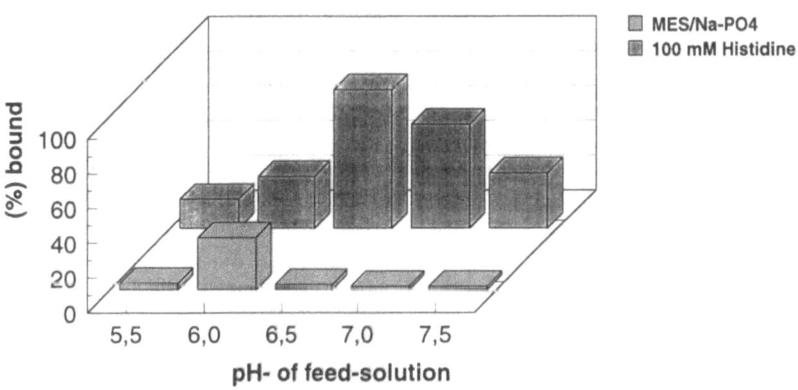

Fig. 4.1.7: Effects of salt type on the binding behaviour of a cation-exchanger (CM-Sepharose Fast Flow). An excess of human monoclonal antibody was applied on the column. The recovered antibody is expressed in % of total antibody and compared to different salts used for equilibration of the resin, ■ 100 mM MES or sodium-phosphate buffer, □ 100 mM Histidine buffer.

IEX is also very often combined with affinity chromatography. The affinity chromatography is considered as capture step, fishing out the protein from a dilute solution. The IEX step is used further for removal of DNA, and affinity ligand (Bloo et al., 1988) which is released into the product during the elution procedure (Table 4.1.6).

Graf et al. (1991) screened the chromatographic behavior of various mammalian cell culture media in order to facilitate the design of purification schemes. Reduction of serum proteins makes IEX a very attractive technique.

Table 4.1.6. DEAE-chromatography of a monoclonal antibody purified on immobilized protein A.

Fraction	IgG (mg)	Load (%)	Protein A (ng)	Protein A/IgG (ng/mg)
Load	8832	100	553104	62.6
Main Peak	4143	46.9	1678	0.0405
Tailing				
Fraction no. 1	1703	19.3	1243	0.730
no. 2	1118	12.7	3754	3.357
no. 3	668	7.5	9228	13.846
no. 4	281	3.2	19636	69.9900
no. 5	104	1.2	37055	356.6
no. 6	30	0.3	34915	1148.5
no. 7	6	0.07	55410	9381.0

4.1.6.2 Adsorption Chromatography

Adsorption chromatography based on silica derived material or hydroxyapatites or metaloxides was the original technique for purification of enzymes from biological material in the past. Hydroxyapatite and glass survived and they are once again becoming very popular, since the viral and DNA clearance seems to be the highest among all types of chromatography.

Hydroxyapatite is very often used for the purification of monoclonal antibodies. The mechanical instability of the particles is solved by copolymerization of hydroxyapatite with trisacryls like the HA-Ultrogel from Sepracore Inc. Paris, France. A result from this stabilization is a lower capacity compared to pure hydroxyapatite.

Hydroxyapatite also has the tendency to bind albumin and other serum proteins. Therefore the optimal performance of this purification method is utilized as a second step in a purification sequence. In the case of a human monoclonal antibody we completely failed to selectively desorb the antibody from the column. Either the antibody was eluted together with albumin or it stuck to the column (Jungbauer et al., 1989).

The mechanisms of adsorption chromatography are not completely understood, the capacity is rather low, compared to affinity chromatography or ion-exchange chromatography, that may be the reason why this technique is not used as frequently as other techniques.

4.1.6.3 HIC and Thiophilic Adsorption

The hydrophobic nature in hydrophobic interaction chromatography (HIC) was first demonstrated by Porath et al. (1973). The technique was also named salt promoted adsorption chromatography. The proteins are bound to the hydrophobic stationary phase in presence of a high salt concentration to promote the hydrophobic interaction between the protein and the stationary phase (Melander and Horvath, 1977). HIC is very well suited for combination with purification steps which are carried out in a high salt buffered solution like $(NH_4)_2SO_4$-precipitation or aqueous two-phase partitioning (Sulk et al., 1992). Sometimes protein denaturation is observed in HIC (Wu et al., 1986) by unfolding of the protein HIC is very sensitive to the ligand-density. On a highly substituted matrix the protein is eluted later than in a low substituted matrix of the identical ligand.

Thiophilic adsorption

Thiophilic adsorption is a further development of HIC. By introducing sulfur atoms into the hydrophobic ligand a specificity for IgG can be created. Antibodies can be specifically bound in presence of albumin and other serum protein contaminants. Sulk et al. (1992) describe monoclonal antibody purification from hybridoma culture supernatant by thiophilic adsorption.

The original ligand for thiophilic adsorption (Porath, 1986) looked like:

$$\text{Matrix} \leftarrow -SO_2-(CH_2)_2-S-(CH_2)_2-OH$$

Nopper et al. (1988) modified the ligand to improve the specificity to achieve a one step purification for monoclonal antibodies. They coupled the ligand to silica.

$$Si-(CH_2)_3-O-(CH_2)_3-O-(CH_2)_2-SO_2-(CH_2)_2-S-(CH_2)_2-OH$$

Thiophilic is the first step for creating generic ligands for protein purification.

4.1.6.4 Reversed Phase Chromatography

In the early eighties protein purification of large proteins (Mr > 50,000) by reversed phase chromatography (RPC) was first mentioned (Lewis et al., 1980). The separation principles of reversed phase chromatography are based on a very apolar stationary phase. The hydrophobic patches of a protein interact with this polar phase. The mobile contains water/buffer and an organic component like acetonitrile, methanol, propan-1-ol etc. (Coran, 1989) which are responsible for desorption from the mobile phase. Due to the composition of the mobile phase, proteins are dissolved optimally but very often denaturing conditions for proteins are generated. The protein may be unfolded following the proposed kinetic model (Sadana, 1992) assuming a first order deactivation kinetic for a zero incubation time.

$$P = P_0 e^{-kt_e} \tag{8}$$

The incubation time is the time from the injection of the sample to the time when the gradient is started. P_0 is the amount of protein injected into the column and t_e is the time from injection until elution of the native peak. Equation 8 demonstrates that the protein is rapidly inactivated (k-values are in the range of $10^4\ s^{-1}$).

RPC is a valuable method for in-process control or for separation of peptides from tryptic maps and is carried out in HPLC-mode.

RPC methods disappeared from the stage of large scale protein purification due to low recovery and the organic phases which require a lot of precautions in a production plant.

4.1.6.5 Gel Chromatography

Gel chromatography, also called size exclusion chromatography or gel filtration, is an essential method for desalting and separation of aggregates or degradation products. In general a gel chromatography step is preferred to be the final step in a protein purification sequence.

The simple relationship between the distribution coefficient and the elution volume can be expressed as:

$$\frac{v_e}{v_0} = 1 + K_{av}\varepsilon^1$$

where v_e is the evolution volume, v_0 is the void volume, K_{av} is the available distribution coefficient and ε is the fraction of the void volume of the total column volume (Laurent and Killander, 1964). Furthermore the K_{av} is related to the molecular mass (M_R) of a macromolecule by a logistic equation.

$$K_{av} = \frac{1}{1 + (M_R/c)^b} \tag{9}$$

where b and c are constants depending on the particular situation (Rodbard, 1976). Equations 8 and 9 show that the proteins are eluted in the order of their molecular mass or stokes radii, if interactions other than penetrating into the pores can be excluded.

The advantages of gel filtration can be summarized as:

- the product collected in the required buffer
- low molecular contaminants are removed

The disadvantages are:

- low sample volume (in general $< 10\%$ of total column volume)
- very sensitive to viscosity and salt content in the protein solution. With a high protein concentration or salt concentration of the sample "fingering" occurs causing reduced resolution.
- dilution of the final product by a factor 1.5 to 10

The gel chromatography can be carried out in two different modes, the desalting mode and the resolution mode.

In the desalting mode, the K_{av} value is close to zero and the protein does not enter the pores of the gel chromatography medium. The protein is eluted with the void volume and the contaminants which are substantially smaller than the product are retained. In that case the dilution of the product is less than 4-fold.

In the resolution mode the K_{av}-value is not zero, the protein is able to enter the pores and is retained. High molecular mass contaminants can be removed by this technique. They elute earlier than the product. In that case the product is diluted between a factor of 5–10.

A comprehensive description of the gels used for preparative protein purification was published by Hagel (1989).

4.1.6.6 Affinity Chromatography and Pseudoaffinity Chromatography

Affinity chromatography, first described by Cuatrecasas, Wilchek and Anfinsen (1968), became an indispensable method for mammalian cell derived protein purification. Affinity chromatography exploits the unique specificity inherent in a ligand biomolecule interaction. During the last decade affinity applications have increased readily in number and with them, the number of sorbents available for laboratory-scale purification. Support material and activation principles for covalent linkage of a ligand are comprehensively reviewed by Clonis (1987), Narayanan and Crane (1990) and Carlsson et al. (1989). Affinity chromatography has the highest resolution power among all chromatography modes.

By far the majority of the diagnostic and therapeutic products produced by mammalian cell culture belong to the group of immunoglobulins. Affinity purification with Protein A and Protein G as ligands turned out to be a standard method for antibody purification (Jungbauer and Wenisch, 1987). The varying binding specificity of Protein A to antibodies from different sera (Richman et al., 1982) and the low binding affinity of mouse IgG motivated people to introduce mutations to Protein A to increase the binding strength (Nilsson et al., 1987). In Figure 4.1.8 a typical example of an antibody purification by Protein A Sepharose Fast Flow is shown.

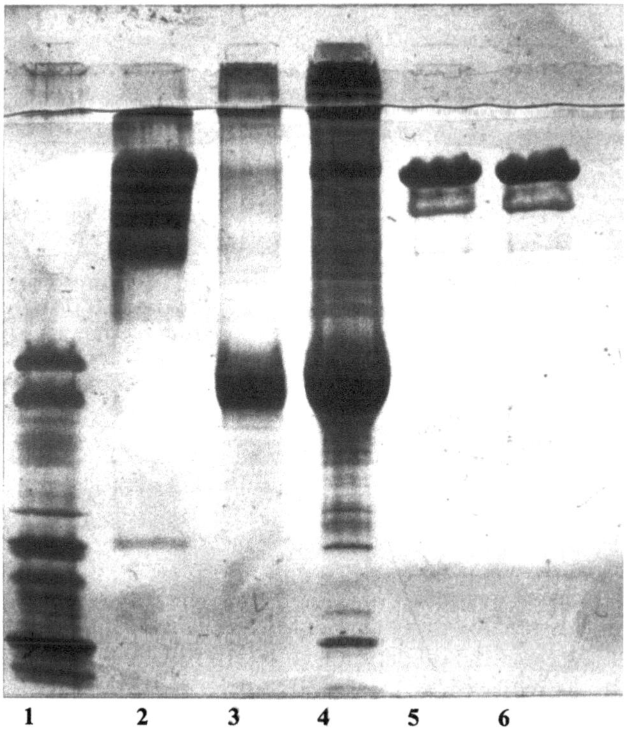

Fig. 4.1.8: SDS-PAGE with silver stain from purification of a human monoclonal antibody: lane 1, molecular mass markers (94 kD, 61 kD, 43 kD, 30 kD, 20.1 kD, 14.4 kD); lane 2, human IgG (Sigma), lane 3, culture supernatant, lane 4, Diaretentate, lane 5, eluate Protein A Sepharose Fast Flow, lane 6, eluate S-Sepharose Fast Flow.

Numerous companies offer material based on all known types of matrices suited for immobilization of proteins. Since productivity is very important one selection rule for the protein A affinity matrices should also be evaluated under the aspects of the dynamic capacity, ligand leakage, etc.

Heparin
Heparin, an acidic polysaccharide extracted from animal tissues, is a very useful ligand for isolation of DNA binding proteins because it closely mimics DNA or a protein shows unexpected affinity to the polysaccharide (Table 4.1.7) like extracellular superoxide dismutase (Tribell et al., 1987).

Table 4.1.7. Isolation of human recombinant extracellular Superoxide Dismutase (SOD) from CHO-cell culture supernatant (Optimeme (Gibco) supplement with 5% newborn calf serum).

Step	SOD activity Units	Total protein mg	Specific activity Units/protein	Recovery %
Culture Supernatant	1.260.000	16.390	77	100
Immunoaffinity chromatography	850.000	13.4	64.000	67
DEAE-Sephacel	913.000	12.7	72.000	73
Heparin-Sepharose				
Pool I*)	550.000	7.9	69.400	43.5
Pool II*)	170.000	3.22	55.800	13.5

*) Identical properties, pool II is only a side fraction of the main peak.

Another instructive example is the partial purification of topoisomerase II by Heparin-Sepharose chromatography from human leukemia cells. About 95% of nuclear contaminating proteins could be removed from Topo II applying a crude nuclear extract onto a Heparin-Sepharose (Boege et al., 1992).

Inhibitors
A very impressive example utilizing inhibitors as affinity ligand is the purification of urokinase using several types of inhibitors or substrate analogues:
p-amino-benzamidine (Holmberg et al., 1976)
β-naphthamidin (Äsdet et al., 1979)

Pseudoligand affinity chromatography:
Pseudoligand affinity chromatography also called dye ligand affinity chromatography was successfully used in the preparation of interferons and a variety of other mammalian enzymes as reviewed by Dean and Watson (1979).

Amino acids like Lys and His are also considered as pseudoaffinity ligands. Lysin-Sepharose became very popular in the late seventies for t-PA purification from melanoma cell culture supernatant and other sources (Radcliffe and Heinze, 1979). Monoclonal antibody can be purified on Histidine and Histamine-Sepharose (El-Kak and Vijayalakshimi, 1991).

Metal chelate affinity chromatography

Metal chelate affinity chromatography utilizes the accessible histidine residues on the surface of a protein (Porath et al., 1975). A heavy metal ion like Cu^{2+}, Zn^{2+}, Ni^{2+} or Fe^{3+} is complexed by an immobilized iminodiacetic acid or another complex-formation ligand. The metal ion selectively forms a coordinative bond with the histidine residue.

Immunoaffinity chromatography

One of the first examples for succesful application of monoclonal antibodies as ligands in immunoaffinity chromatography is the purification of interferon (Secher and Burke, 1980). A major drawback of the technique is sometimes the low cycle time caused by rigid denaturating elution conditions (Pickard et al., 1991; Yarmash et al., 1992) and the costs. Production of an antibody intended for use as affinity ligand has to be carried out under the same regulatory principle as a therapeutic protein.

Protein C purification is a very instructive example for the application of mild elution conditions. Protein C, which is an important anti-thrombolytic factor in the blood coagulation cascade, undergoes several post-translatorial modifications. γ-carboxylation on nine glutamic acid residues at the N-terminal region of the light chain is considered to be critical for full anti-clotting activity (Suigiura and Maruyama, 1992). It is known that when recombinant protein C is expressed in animal cells this particular modification is often lost. A monoclonal antibody must distinguishe the sufficiently γ-carboxylated protein from the rest (Table 4.1.8) by its specific affinity for Ca^{++} induced conformational change of Protein C. Sufficiently γ-carboxylated protein C expressed in CHO could be bound to the immunoaffinity column and could be desorbed by EDTA with high purity, and correct carboxylation and similar clotting activity to the wild type protein (Kamikira et al., 1992).

Table 4.1.8. Numbers of γ-carboxylated glutamic acids (Gla) and activities of enzymes bound and passed the immunoaffinity chromatography column.

Protein C	Gla*) residues per molecule)	Activity (% of plasma Protein C)	
		Amidolytic	Anti-clotting
Bound	8.1	90	67
Unbound	5.2	60	8

*) the number of Gla in plasma derived protein C is nine

An antibody with the same adsorption and desorption properties could be found for a Protein C mutant, where the Gla domain was replaced by that of factor X in principle this strategy also can be used for the design of generic purification processes.

4.1.6.7 Displacement Chromatography

Displacement chromatography may be an additional tool for polishing of proteins (Cramer, 1991). At the moment the method is more or less limited to analytical applications. Displacement chromatography takes advantage of the nonlinearity of the isotherms such that a larger feed can be separated on a column and a higher concentration can be achieved. A substance with a higher K-value than the feed component is introduced into the column. After loading according to their K-values, the different substances are displaced from the column. Rectangular shaped peaks are observed. In anion-exchange displacement chromatography carboxymethyldextrans (Petterson and Torres, 1983) chondroitin sulfate (Liao et al., 1987) and polyvinylsulfonic acids (Jen and Pinto, 1990) are used. In cation-exchange displacement DEAE-dextran, dextran sulfate displacers (Jamyaraman et al., 1984) and protamine (Gerstner and Cramer, 1992) are widely used.

4.1.6.8 Design of Generic Purification Methods

The design of generic purification methods is driven by the development costs and time, which are in general very high for proteins produced by mammalian cell culture. Especially validation of the process is very time consuming and is often felt as a boring activity. Following the concept of generic purification methods clearance data can be partially transferred to another product group. Protein A and protein G affinity chromatography using such group specific ligands are the first candidates for generic purification processes. There are not so many therapeutic proteins which show a characteristic group specificity like antibodies for protein A or protein G. The reversed strategy is the addition of a group specific tail to the protein, which allows the binding to a specific affinity tag. The protein is expressed as a fusion protein. Either small peptides or truncated proteins or whole proteins can be linked to the protein (Table 4.1.9). Depending on the intended use the affinity tail has to be removed by chemical or enzymatic cleavage or even can be used as a reporter in the field of diagnostics.

4.1.6.9 Productivity Concerns of Chromatography

Soft gels versus rigid gels

Most gels are composed of inert polysaccharides such as dextran or agarose. The polysaccharide is cross-linked to improve the mechanical stability of the gel beads. The flow rate can be increased by a factor of a hundred or even more. The chemical modification introduces hydrophobic or charged groups into the sorbent (Janson, 1967; Belew et al., 1978). In certain cases these additional functions may lead to substantial losses by irreversible adsorption. This is another explanation as to why soft gels are still in operation.

Modern gels allow an operation with a linear velocity up to 1500 cm/h (SP-Sepharose fast flow, data from supplier, Pharmacia Biotechnology) under an acceptable back pressure (> 2 bar, 25 cm bed height). Contradictory to these high

Table 4.1.9. Affinity tails used for purification of recombinant proteins and production of reporter systems for diagnostics.

Affinity tail	Affinity tag or type of chromatography	Limits	References
Peptides			
Arg-Gly-Arg-Gly-Gly-Arg	Anhydro-trypsin	Coelution of contaminants	Hirabayashi & Kasai (138)
Poly-Cys	Covalent chromatography	High Cys concentration downregulates the expression rate	Person et al. (139)
Poly-Phe	Hydrophobic interaction chromatography		Person et al. (139)
Poly-His	Ni-chelate chromatography	Decrease of pH may destroy protein	Hochuli (140)
Proteins			
Gluthation-S-transferase	Gluthatione		Davies et al. (141)
Protein A binding domain	IgG		Samuelsson E. et al. (142)
Glutamin-synthetase			
Alkaline phosphatase			
β-Galactosidase			Köhler et al. (143)
Streptavidin	Biotin		Sano & Cantor (144)
Protein A/luciferase	IgG	Immunoassays	Lindbladh et al. (145)
Protein A/glutathione-S-transferase	Glutathione, IgG	Immunoprecipitation	Lew et al. (146)

flow rates are the decrease of dynamic capacity and loss in resolution. Therefore many processes are carried out at much lower rates than the physical stability of the bed would allow. Materials with a high porosity and with transjecting pores of 6,000–8,000 Å pore size allow an operation up to several thousand cm per h linear velocity without a significant loss in resolution (Regnier, 1991).

The principles of preparative and industrial chromatography were comprehensively reviewed by L. Rydn and J.C. Janson (1989) and E. Boschetti (1994), where also an extensive list of packing materials can also be found. An overview of matrices for large-scale affinity chromatography was published by Clonis (1987).

HPLC modes are not often used in large scale protein purification. They are very popular in small scale and for protein analysis.

Productivity

Comparison of different chromatographic steps should be carried out on the basis of productivity (P). It is always important to maximize productivity in scale-up of separation processes.

Here, productivity is defined as

$$P = \frac{(\text{recovery}-\text{ratio})(\text{sample}-\text{concentration})(\text{sample}-\text{volume})}{(\text{column}-\text{volume})(\text{cycle}-\text{time})} = \frac{Q_R C_0 V_F}{V_t t_c} \tag{10}$$

at a specified purity ratio Q_p and a recovery ratio Q_r (Yamamoto et al., 1990). Vt is the column volume and tc is the separation (cycle time). The numerator of equation (11) represents simply the throughput of a recovered protein. Q_p increases with the linear velocity (u) as long as the specified purity ratio is not tangential. The prediction of P is still very difficult as it is a very complicated function of various parameters:

$$P = f(d_c, d_c, Z, V_F, c_0, u, Q_P, Q_R, \text{etc.}) \tag{11}$$

where d_p is the particle diameter and d_c is the column diameter, c_0 is the sample feed concentration, Z is the column length and V_F the feed volume. A short cut method for predicting the productivity of affinity chromatography is described by Yamamoto and Sano (1992).

4.1.6.10 Expanded Bed Operations

Expanded beds have been used in recovery of biologics for situations where conventional packed beds fail. An expanded, fluidized bed can tolerate suspended solids. In contrast, a packed bed would quickly clog. However sorption efficiencies of fluidized beds are generally lower than corresponding packed beds. The loss in efficiency is not only caused by an increase in the column void fraction, but also due to the fact that fluidized particles are free to move throughout the bed volume. The magnetic stabilization of fluidized beds containing magnetizable particles eliminates solid mixing (Nixon et al., 1991). During elution (desorption) the peaks undergo a bandbroadening and can be ranked in this order:

fluidized bed > magnetically stabilized FB > packed bed.

Therefore very often adsorption is carried out under fluidized conditions and elution is carried out in a packed bed mode. Affinity purification of proteins using expanded beds is described by Chase and Dräger (1992).

4.1.7 Continuous and Semicontinuous Chromatographic Processes

The conventional packed chromatography is a non-continuous process. There are several approaches to change the process to a continuous or semicontinuous mode:

- Counter current chromatography [CCC, (Arve and Liapis, 1988; Ito, 1992)] Moved bed chromatography

Simulated moved bed chromatography (Adachi, 1996)
Elutriation with angle rotor coil planet centrifuge (De la Poype, 1992)
- Annular chromatography (Charm and Matteo, 1974)

CCC is one solution to overcome the limitations of non-continuous packed bed processes. Already large scale applications in the conventional chemical industry are described. At the time there is no large scale application for purification of biologics. The quantities are too small.

Annular chromatography is a truly continuous chromatography but does not possess any industrial applications at the moment. Sample solution is applied to a slowly rotating annular column. As the sample moves down the rotating bed the components follow different helical paths. The various components leave the column at different points depending on their residence time in the column. Both sample application and collection of fractions are performed continuously.

4.1.8 Scale-Up Strategies for Purification of Biologics Using Adsorption/Desorption Techniques

For large scale operation for purification of biologics by reversible adsorption on a stationary phase, several strategies are applicable.

Repeated cycles and overloading
Overloading and repeated cycles are the simplest way to scale-up linear biochromatography. Gareil et al. (1980) first described maximum overloading without loss in resolution. Meanwhile a series of theoretical studies and computer programs are available to solve this problem. Overloading an analytical column or repeated cycles are in fact no real scale-up in the sense of chemical engineering. Often economical considerations as well as the required product quantity justify this mode of scale-up.

Increasing column diameter
Increasing the column diameter is the conventional way to scale-up preparative chromatography using adsorption/desorption mode, if the bead size or bead size distribution is not changed from laboratory scale to prep scale. If different particles are used in small and large scale, additional experimental work is necessary before the process size is increased. Due to the loss in resolution by larger particles or broader distribution of particle size the height must also be adjusted to the new diameter.

Increasing column with constant H/D
Using linear elution chromatography (e.g. size exclusion chromatography or frontal chromatography) an efficient way to scale-up is increasing the column size by constant height to diameter ratio.
In some cases scale-up by just increasing the column diameter is not successful. The insufficient result may be caused by irregularities in the bed, unequal distribution of the fluid over the gel surface or a propagation of disturbances in the gel (Fig. 4.1.9). HIC columns are more sensitive to those phenomenon than other packings (Vorauer et al., 1992).

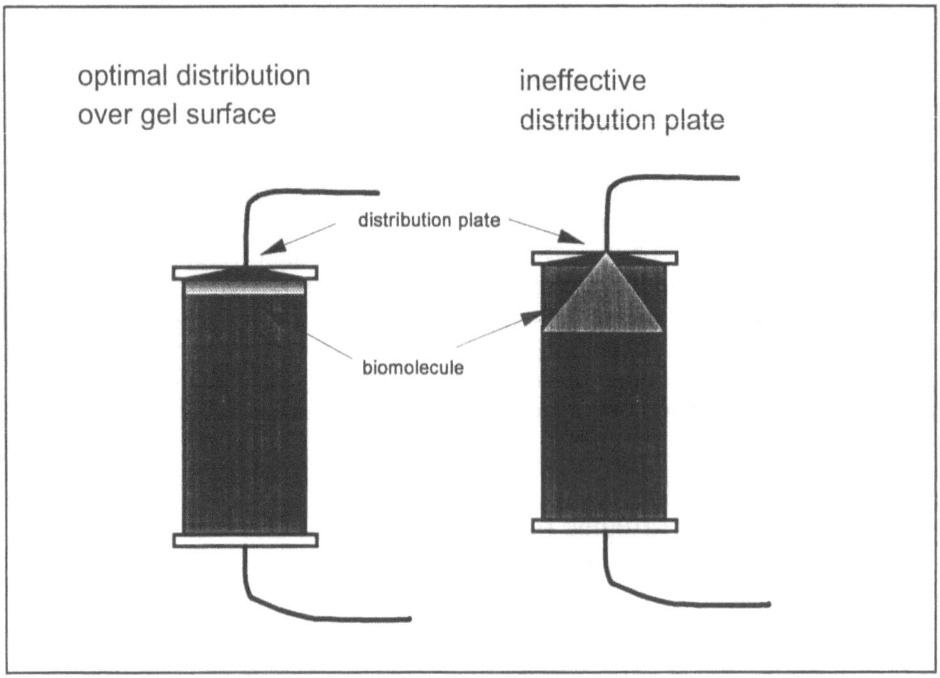

Fig. 4.1.9: Influence of the distribution plate on the distribution of a biomolecule in a chromatography column.

Radial chromatography
In radial chromatography the process solution is applied in radial direction onto the gel surface and the outlet is drained by a central tube (Jungbauer et al., 1988). The Separgen-design allows a fast operation with soft gels, because the surface is increased by the packing gel into an annular column.

Membrane adsorption
Increasing the kinetics of adsorption and circumvention of all problems with insufficient column packings led to the development of several types of "membrane chromatography".

Memsep® concept:
Membranes derivatized with different functional groups are tightly packed into a column and properly sealed (Langlotz and Kroner, 1992;Lütkemeyer et al., 1993). The performance of the chromatography is very similar to conventional chromatography with packed beds.

Sepracor concept:
Two potential rate-limiting factors in affinity chromatography are mass transfer and adsorption kinetics. In affinity chromatography with fast ligand protein interactions such as antigen-antibody and protein A-antibody interaction, mass transfer becomes the limiting process in packed bed processes. The diffusion time (t_D) must

be much smaller than the residence time (t_R). t_D is directly related to the diffusional distance (L) and the diffusion coefficient (D) of a protein.

$$t_D = L^2/D \quad (12)$$

The extremely short diffusional distance (< 1 μm) in membranes results in:

$$t_D \ll t_R$$

Sepracor Inc. (Marlborough, MA, USA) realized the concept by filtering a protein solution through a membrane (hollow fiber) with immobilized ligands (Brandt et al., 1988). The adsorption is very rapidly accomplished. A combination of washing steps allows the elution with moderate dilution of the product. The system is designed for handling large volumes of very dilute products (e.g. animal cell culture supernatants). The careful removal of fines from the process solution is a must. All membrane adsorption processes are very sensitive to clogging.

4.1.9 Electrophoretic Methods and Scale-Up

Electrophoresis is defined as migration of a charged particle or molecule in an electric field. The method can be carried out in tubes, vertical gels, horizontal beds or in free solution. Isoelectric focusing may be defined as electrophoresis in a pH gradient. The resolution power is obtained by the equilibrium which is reached during the electrophoretic process. The equilibrium is reached when the protein is migrated to the zone where the pI and the pH of the electric field are identical.

Preparative electrophoresis is carried out in flat beds with granulated gels or in columns using density gradients. Separation of gram-quantities of proteins is the scale of this type of operation.

Free flow electrophoresis is an old technique, but the scale-up and successful application on an industrial scale was not convincing until now. The "Bender Hobein" machine is the only equipment with notable resolution power and scale. Chamber electrophoresis became a promising strategy for scale-up (Polson, 1953; Kyhse-Andersen, 1989).

Multicompartment concept
Focusing is achieved in chambers separated by membranes covered with Ampholines. A detailed description of the apparatus and the mechanism is given by Righetti et al. (1990). So far this is a realistic candidate for scale-up. In fact one can imagine the purification of kg-quantities of proteins.

Rotophor
The main component of the Rotophor cell is a cylindrical focusing chamber divided into 20 discrete compartments. The compartments are separated by membranes with a pore size of 10 μm and offer resistance to fluid convection, but they do not hinder the flow of current or the transport of proteins. The Rotophor cell rotates during separation about its central axis at 1 rpm. There is a loading port and a collection port for each compartment (description obtained from the manfacturer, BioRad, CA, USA).

4.1.10 Removal of Adventitious Agents

Very high virus, DNA and pyrogen clearance factors can be achieved by column chromatography very high; up to a factor of 10^6. An example from antibody purification (Mariani and Tarditi, 1992) is shown in Table 4.1.10. Similar results are described by Matthews (1991). In general packed bed processes show the highest capacity for removal of adventitious agents.

Table 4.1.10. Virus Removal determined by spike-off experiments; data from M. Mariani and L. Tarditi, Bio/Technology, 10 (1992) 395.

Virus to be removed	Protein A clearance	Hydroxyapatite clearance	Total clearance
Polio Sabin 1	$10^{2.91}$	$10^{6.08}$	$10^{8.99}$ pfu*)
SV 40	$10^{1.63}$	$10^{4.17}$	$10^{5.80}$ pfu
Aujeszky herpes virus	$10^{1.98}$	$10^{4.52}$	$10^{6.50}$ pfu
Moloney murine leukaemia virus	$10^{1.76}$	$10^{8.84}$	$10^{8.84}$ ffu*)

*) pfu = plaque forming units
**) ffu = focus forming units

4.1.11 Conclusion and Future Prospects

The extraction of biologics has become a very important discipline in biotechnology and in the manufacturing of therapeutic peptides, proteins and oligonucleotides. Due to the strict regulations for therapeutic proteins produced by recombinant DNA technology the isolation and purification is often the most expensive part in the whole process. By combination of several unit operations the biologic is purified to homogeneity and high purity. In addition to the conventional technologies such as centrifugation, microfiltration and flocculation, chromatography such as affinity chromatography was optimized in the last decade and a list of specifically designed matrices are available for large scale operations. Electrophoretic methods may become very attractive in combination with chromatographic steps, because the separation principles complement each other. Design of ligands, modeling of processes to understand the direction for optimization, and scale-up issues such as continuous unit operations may be the dominating questions in process development for biologics in the future.

References

Ad hoc working party on biotechnology/pharmacy and the working party on safety of rned-icines (1988) Tibtech. 7 G13.

Adachi, S. (1996) Simulated moving-bed chromatography for continuous separation of two components and its application to bioreactors.

Agerkvist, I., Erikson, L. and Enfors, S. (1990) Selective flocculation with chitosan in Escherichia coli disintegrates: effects of pH and nuclease treatment. Enzyme Microb. Technol. 12, 584–590.

Al-Rubeai, M. and Emery, A.N. (1990) Mechanism and kinetics of monoclonal antibody synthesis and secretion in synchronous and asynchronous hybridoma culture. J. Biotechnol. 16, 67–86.

Arve, B.H. and Liapis, A.I. (1988) Biospecific adsorption in fixed and periodic contercurrent beds. Biotech. Bioeng. 32, 616–627.

Äsdet, B., Holmberg, L., Wagner, G., Richter, P. and Plough, J. (1979) Purification of Urokinase by beta-naphtamidine affinity column. Throbos. Haemostas 42, 924–928.

Belew, M., Porath, J., Fohlman, J. and Jansen, J.-C. (1978) Adsorption phenomena on Sephacryl S-200 superfine. J. Chromatogr. 147, 205–212.

Bloom, J.W., Wong, M.F. and Mitra, G. (1988) Detection and reduction of Protein A contamination in immobilized Protein A purified monoclonal antibody preparations. J. Immunol. Meth. 117, 83–89.

Boege, F., Gieseler, F., Müller, M., Biersack, H. and Meyer, P. (1992) Activation of topoisomerase II during partial purification by heparin-Sepharose chromatography. J. Chromatogr. 625, 67–71.

Boschetti, B. (1988) Liquid chromatography a compulsary step in the purification of proteins produced by animal cell culture. Chim. Oggi 6, 17–22.

Boschetti, E. (1994) Advanced sorbents for preparative protein separation purposes. J. Chromatogr 658, 207–236.

Bosisio, A.B., Loeherlein, C., Snyder, R.S. and Righetti, P.G. (1980) Titration curve or proteins by combined isoelectric focusing-electrophoresis in highly porous polyacrylamide matrices. J. Chrom. 189, 317–330.

Bowden, C. (1985) Recovery of microorganisms from fermented broth. Chem. Eng. 415, 53–54.

Brandt, S., Gaffe, R.A., Kessler, S.B., O'Conner, J.L. and Zale, S.E. (1988) Bio/Technol. 6, 779–785.

Brooks, D.A., Bradford, T.T. and Hopwood, J.J. (1992) An improved rnethod for the purification of IgG monoclonal antibodies from culture supernatant. J. Immunol. Meth. 155, 129–132.

Brunner, K.H. and Hemfort, H. (1988). In: Advances in Biotechnological Processes, Vol 8 (Mizrahi, A., ed.), pp. 1–50, Alan R.Liss, Inc., New York.

Buckholz, R.G. and Gleeson, M.A.G. (1991) Yeast systems for the commercial production of heterologous proteins. Bio/Technotogy 9, 1067–1072.

Carlson, A. and Nagarajan, R. (1992) Release and recovery of porcine pepsin and bovine chymosin from reverse micelles: a new technique based on isoprpyl alcohol addition. Biotechnol. Prog. 8, 85.

Carlsson, J., Janson, J.C. and Sparrman, M. (1989) In: Protein purification: principles, high resolution methods and applications (Janson, J.C. and Rydén, L., eds.), pp. 275–329, VCH, New York Weinheim Cambridge.

Charm, E.S. and Matteo, C.C. (1974) In: Methods of Enzymology Vol XXII (Colowick, S. P. and Kaplan, N.O., eds.), pp. 476–556, Academic Press New York.

Chase, H.A. and Draeger, N.M. (1992) Affinity purification of proteins using expanded beds. J. Chromatogr. 597, 129–143.

Clonis, Y.D. (1987) Large-scale affinity chromatography. Bio/technology 5, 1290–1293.
Conradt, H.S., Nimtz, M., Dittmar, K.E.J., Lindenmaier, W., Hoppe, J. and Hauser, H. (1989) Expression of human interleukin–2 in recombinant baby hamster kidney, Ltk- and Chinese hamster ovary cells. J. Biol Chem. 264, 17368–17373.
Coran, P.H. (1989) In: HPLC of macromolecules, a practical approach (Oliver, R.W.A., ed.), pp. 1527–156 IRL Press, Oxford New York Tokyo.
Courtney MacGregor, W., Weaver, J.F. and Tasney, S.F. (1988) Antifoam effects on ultrafiltration. Biotech., Bioeng 31, 385.
Cramer, S.M. (1991) Displacement chromatography. Nature 351, 251–252.
Cuatrecasas, P., Wilchek, M. and Anfinsen, C.B. (1968) Selective enzyme purification by affinity chromatography. Proc. Nat. Acad. Sci. USA 61, 636–643.
Datar, R. and Rosen, C.-G. (1987) Centrifugal separation in the recovery of intracellular protein from E. coli. Chem. Eng. 34, B49-B56.
Datar, R.V., Cartwrigth, T. and Rosen, C. (1993) Process economics of animal cell culture and bacterial fermentations. A case study analysis of tissue plasmoinogen activator. Bio/technoiogy 11, 349–357.
Davies, A.H., Jowett, B.M. and Jones, I.M. (1993) Recombinant baculovirus vectors expressing gluthatione-S-Transferase fusion proteins. Bio/technology 11, 933–936.
De la Poype, F. (1992) In: Proceedings of the 9th international symposium on preparative and industria1 chromatography, Nancy 1992 (Perrut, M., ed.), pp. 253–257, Institut National Polytechnique de Lorraine, Vandoeuvre-Les-Nancy.
Dean, P.D.G. and Watson, D.H. (1979) Protein purification using immobilized triazine dyes. J. Chromatogr. 165, 301–319.
El-Kak, A. and Vijayalakshimi, M.A. (1991) Study of the separation of mouse monoclonal antibodies by pseudoaffinity chromatography using matrix-linked histidine and histamine. J. Chromatogr. 570, 29–41.
Englard, S. and Seifter, S. (1990) In: Methods of enzyrnology, Vol. 182 (Deutscher, M.P., ed.), pp. 285–300, Academic Press Inc., San Diego.
Eriksson, L.B. and Härdin, A.-M. (1987) In: Flocculauon in biotechnology and separation systems (Attia, Y.A., ed.), Elsevier Amsterdam.
Fleer, P., Yeh, P., Amellal, N., Maury, I., Fournier, A., Bacchetta, F., Baduel, P., Jung, H., Becquart, J., Fukuhara, H. and Mayaux, J.F. (1991) Stabale multicopy vectors for high level secretion of recombinant human serum albumin by Kluyveromyces yeasts. Bio/technology 10, 969–975.
Flickinger, M.C, Goebl, N.K. and Bohn, M.A. (1990) Determination of specific monoclonal antibody secretion rate during very slow hybridoma growth. Bioprocess Eng. 5, 155–165.
Franek, F., Vomastek, T. and Dolinkova, J. (1992) Fragmented DNA and apoptotic bodies document the programmed way of cell death in hybridoma cultures. Cytotechnol 9, 117–123.
Gareil, P., Personnaz, L., Feraud, J.P. and Caude, M. (1980) Etude de lélution non-lineaire en chromatographie en phase liquide preparative. J. Chromatogr. 192, 53–74.
Gerstner, J.A. and Cramer, S.M. (1992) Cation-exchange displacement chromatography of proteins with protamine displacers: effect of induced salt gradients. Biotechnol. Prog 8, 540–545.
Graf, H., Rabaud, J.N. and Egly, J.M. (1991) Screening various mammalian cell culture media to establish a downstream purification scheme. J. Immunol. Meth. 139, 135–144.
Green, A.A. and Hughes, W.L. (1955) In: Methods in enzymology Vol I (Colowick, S. P. and Kaplan, N. O., eds.), pp.67–90, Academic Press Inc., New York.
Hagl, L. (1989) In: protein purification: principles high resolution methods and applications (Janson, J.C. and Rydn, L., eds.), pp. 63–103, VCH Publishers, Inc. New York Weinheim Cambridge.

Harlow, E. and Lane, D. (1988) Antibodies, a laboratory manual, pp. 300–301, Cold Spring Habour Laboratory Press, Cold Spring Habour.
Hirabayashi, J. and Kasai, K. (1992) Arginine-tail method, an affinity tag procedure utilizing anhydrotrypsin agarose. J. Chromato. 597, 181–187.
Ho, C.S. (1986) An undestanding of the forces in the adhesion of microorganisms to surfaces. Process Biochem. Oct., 148–152.
Hochuli, E. (1990) In: Proceedings of the 5th European Congress on Biotechnology (Christiansen, C., Munck, L. and Villadsen, J., eds.), pp. 1005–1009, Munksgaard, Copenhagen.
Holmberg, L., Bladh, B. and Äsdet, B. (1976) Purification of Urokinase by affinity chromatography. Biochem. Biophys. Acta 445, 215–222.
Hughes, J., Ramsden, D.K. and Symes, K.C. (1990) The flocculation of bacteria using cationic flocculants and chitosan. Biotech. Tech 4, 55–60.
Hustedt, H., Kroner, K.H., Menge, U. and Kula, M.-R. (1985) Protein recovery using two-phase systems. TIBS 6, 2–7.
Ito, Y. (1992) In: Journal of Chromatography Library (Heftmann, E., ed.), pp. A69-A 107, Elsevier, Amsterdam.
Jäger, V. (1988) Serum-free growth medium for the cultivation of a wide spectrum of mammalian cells in stirred bioreactors. Cytotechnology 1, 319–329.
Jamyaraman, G., Gadam, S. and Cramer, S.M. (1984) Ion-exchange displacement chromatography of proteins: dextran based polyelectrolytes as high affinity displacers. J. Chromatogr. in press.
Janson, J.C. (1987) On the history of the development of Sephadex. Chromatographia 23, 361.
Janson, J.-C. (1967) Adsorption phenomena on Sephadex. J. Chromatogr. 28, 12–20.
Janson, J.-C. and Rydn, L. (eds.) (1989) Protein Purification: Principles, High Resolution Methods, and Applications. VCH Publishers, Inc., New York.
Jen, S.C. and Pinto, N.G. (1990) Use of the sodium salt of poly-(vinylsulfonic acid) as a low-molecular-weight displacer for protein separations by ion-exchange displacement chromatography. J. Chromatogr. 29, 87–98.
Johansson, G. and Joelsson, M. (1985) Preparation of Cibacron Blue F3G-A (polyethylene glycol) in large scale for use in affinity partitioning. Biotech. & Bioeng. 27, 612–635.
Jonak, Z.L. (1984) In: Monoclonal antibodies and hybridomas: a new dimension in biological aspects (Kennet, R.H., McKearn, H. and Bechtol, K.B., eds.), pp. 405–406, Plenum Press, New York London.
Jönsson, A. S. and Trägårdh, G. (1990) Fundamental principles in ultrafiltration. Chem. Eng. Process 27, 67–81.
Jungbauer, A. (1993) Preparative chromatography of biomolecules. J. Chromatogr. 639, 3–16.
Jungbauer, A. and Wenisch, E. (1987) In: Monoclonal antibodies. Production and application (Mizrahi, A., ed.), pp. 161–192, Alan Liss Inc., New York.
Jungbauer, A., Tauer, C., Reiter, M., Purtscher, M., Wenisch, E., Steindl, F., Buchacher, A. and Katinger, H. (1989) Comparison of protein A, protein G and coplymerized hydroxyapatite for the purification of human monoclonal antibodies. J. Chromatogr. 476, 257–268.
Jungbauer, A., Tauer, C., Wenisch, E., Steindl, F., Purtscher, M., Reiter, M., Unterluggauer, F., Buchacher, A., Uhl, K. and Katinger, H. (1988) Pilot scale prodcution of a human monoclonal antibody against human immunodeficiency virus HIV-1.
Jungbauer, A., Unterluggauer, F., Uhl, K., Buchacher, A., Steindl, F., Pettauer, D. and Wenisch, E. (1988) Scale up of monoclonal antibody purification using radial streaming ion-exchange chromatography. Biotech. Bioeng 32, 326–333.
Kamikira, M., Iijima, S. and Kobayashi, T. (1992) Stabilities of antigen and antibody under elution conditions in immunoaffinity chromatography using monoclonal antibody. Bioseparation 3, 185–188.

Karlson, E., Rydn, L. and Brewer, J. (1989) In: protein purification: principles, high resolution methods and applications (Janson, J.C. and Rydn, L., eds.), pp. 107 146.
Kenny, C.K. (1986) U.S. patent 4,589,987.
Kloostermann, J., van Wassenaar, P.D., Slater, N.K.H. and Baksten, H. (1988) The effect of anti-foam agents on the ultrafiltration of a protease solution. Bioprocess Eng. 3, 181–185.
Köhler, K., Meide, A. and Enfors, S.O. (1990) In: Proceedings of the 5th European Congress on Biotechnology (Christiansen C., Munck, L. and Villadsen, J., eds.), pp. 1001–1004, Munksgaard, Copenhagen.
Kozak, R.W., Dufor, C.N. and Scribner C.L. (1992) Regulatory considerations when developing biological products. Cytotechnology 2, 203–210.
Kroner, K.H., Hustedt, H. and Kula, M.-R. (1984) Extractive enzyme recovery: economic considerations. Process Biochem. 10, 170–179.
Kyhse-Andersen, J (1989) Multi-chamber apparatus for preparative isoelectric focusing. Electrophoresis 10, 6–10.
Langlotz, P. and Kroner, K.H. (1992) J. Chromatogr. 591, 107–112.
Lauerent, T.C. and Killander, J. (1964) A theory of gelfiltration and its experimental verification J. Chromatogr 14, 317–330.
Leser, E.W. and Asenjo, J.A. (1992) Rational design of purification processes for recombinant proteins. J. Chromatogr. 584, 43–57.
Lew, A.M., Beck, D.J. and Thomas, L.M. (1991) Recombinant fusion proteins of protein A and protein G with gluthatione S-transterase as reporter molecules. J. Immunol, Meth. 136, 211–219.
Lewis, R.V., Fahon, A., Stein, S., Gibson, K.D. and Udenfried, S. (1980) Supports for reversed-phase high-performance liquid chromatography of large proteins. Anal. Biochem. 104, 153–159.
Liao, A., El Rassi, Z., LeMaster, D.M., Horvath, C. (1987) High performance displacement chromatography of proteins: separation of b-lactoglobulins A and B. Chromatographia 24, 881–885.
Lindbladh, C., Mosbach, K. and Bülow, L. (1991) Preparation of a genetically fused protein A/luciferase conjugate for use in bioluminescent immunoassays. J. Immunol. Meth. 137, 199–207.
Lütkemeyer, D., Bretschneider, M., Büntemeyer, H. and Lehmann, J. (1993) Membrane chromatography for rapid purification of recombinant antithrombin III and monoclonal antibodies from cell culture supernatant. J. Chromatogr. 639, 57–66.
Mariani, M. and Tarditi, L. (1992) Validating the preparation of clinical monoclonal antibodies. Bio/Technology 10, 394–395.
Matthews, B.R. (1991) In: Validation of chromatographic purification methods used for the manufacturing of therapeutic products (Hill, C.R., ed.), ESRA News and Views, Vol 3.
Melander, W. and Horvath, C. (1977) Salt effects on hydrophobic interactions in precipitation and chromatography of proteins: an interpretation of the lyothropic series. Arch. Biochem. Biophys. 183, 200–215.
Menge, U., Morr, M., Mayr, U. and Kula, M.R. (1983) Purification of human fibroblast hormone by extraction in aqueous two-phase systems. J. Appl. Biochem. 5, 75–90.
Nagata, N., Herouvis, K.J., Dziewulski, D.M. and Belfort, G. (1989) Cross-flow membrane microflitration of a bacterial fermentation broth. Biotech. Bioeng. 34, 447–466.
Narayanan, S.R. and Crane, L.J. (1990) Affinity chromatography supports. a look at performance requirements. Tibtech 8, 12–16.
Neoh, S.H., Gordon, C., Potter, A. and Zola, H. (1986) The purification of mouse monoclonal antibodies from ascites fluid. J. Immunol. Meth 91, 231–235.
Nilsson, B., Moks, T., Jansson, B., Abrahmsén, L., Elmblad, A., Holmgren, E., Henrichson, C. Jonrs, T.A. and Uhlén, M. (1987) A synthetic IgG- binding domain based on the staphylococcal protein A. Protein Engineering 1, 107–113.

Nixon, L., Koval, C.A., Xu, L., Noble, R.D. and Slaff, G.S. (1991) The effects of magnetic stabilization on the structure and performance of liquid fluidized beds. Bioseparation 2, 217–230.

Noper, B., Kohen, F. and Wilchek, M. (1988) A thiophilic adsobent for one-step high-performance liquid chromatography purification of monoclonal antibodies.

Östlund, C., Borwell, P. and Malm, B. (1987) Process-scale purification from cell culture supernatants: monoclonal antibodies. Develop. Biol. Standard. 66, 367–375.

Ozturk, S. and Palsson, B.O. (1990) Effect of initial cell density on hybridoma growth, metabolism, and monoclonal antibody production. J. Biotech 16, 259–278.

Paradker, V.M. and Dordick, J.S. (1991) Purification of glycoproteins by selective transport using concanavalin-mediated reverse micellar extraction. Biotechnol. Prog. 7, 330–334.

Patel, P.N., Mehaia, M.A. and Cheryan M. (1987) Cross-flow membrane filtration of yeast suspensions. J. Biotech 5, 1–16.

Pavirani, A., Meulien, P., Harrer, H., Schamber, F., Dott, K., Villeval, D., Cordier, Y., Wiesel, M., Mazurier, C., Van der Pol, H., Piquet, Y., Cazenave, J. and Lecoq, P. (1987) Choosing a host cell for active recombinant factor VIII production using vaccinia virus. Bio/technology 5, 389–392.

Person, I. and Lindman, B. (1994) Flocculation of cell debris for improved separation by centrifugation. In: Flocculation in biotechnology and separation system. (Attia, Y. A., ed.), pp. 457–466, Elsevier, Amsterdam.

Persson, M., Bergstrand, M.G., Bülow, L. and Mosbach, K. (1988) Enzyme purification by genetically attached polycysteine and polyphenylalanine affinity tails. Anal. Biochem. 172, 330–337.

Petterson, B.A. and Torres, A.R. (1983) Ion-exchange displacement chromatography of proteins, using narrow range carboxymethyldxtrans and a new index of affinity. Anal. Biochem. 130, 271–282.

Pickard Antonsen, K., Colton, C.K. and Yarmush, M.L. (1991) Elution conditions and degratation mechanisms in long-term immunosorbent use. Biotechnol. Prog. 7, 159–172.

Polson, A. (1953) Multi-membrane electrodecantation and its application to isolation and purification of proteins and viruses. Biochem. Biophys. Acta 11, 315–325.

Porath Carlsson, J., Olsson, I. and Belfrage, G. (1975) Metal chelate affinity chromatography a new approach to protein fractionation. Nature 258, 598–599.

Porath, J. (1986) Salt-promoted adsorption: recent developments. J. Chromtogr. 376, 331–341.

Porath, J., Sundberg, L., Fornstedt, N. and Olsson, I. (1973) Salting-out in amphiphilic gels as a new approach to hydrphobic adasoption. Nature 245, 465–468.

Pungor, E., Afeyan, N.B., Gordon, N.F. and Cooney, C.L. (1987) Continous affinity-recycle extraction: a novel protein separation technique. Bio/technology 5, 604–608.

Radcliffe, R. and Heinze, T. (1978) Isolationof plasmonogen activator rom human plasma by chromatography on lysine-Sepharose. Arch. Biochem. Biophys. 189, 185–194.

Regnier, F.E. (1991) Perfusion chromatography. Nature 350, 634–635.

Richman, D.D., Cleveland, P.H., Oxman, M.N. and Johnson, K.M. (1982) The binding behaviour of staphylococcal protein A by thesera of different animal species. J. Immunol. 128, 2300–2305.

Righetti, P.G., Wenisch, L., Jungbauer, A., Katinger, H. and Faupel, M. (1990) Preparative purification of a monoclonal antibody isoforms in a multicompartment electrolycer with immobiline membranes. J. Chromatogr. 500, 681–696.

Ripperger, S. (1992) In: Mikrofiltration mit Membranen, VCH, Weinheim New York Basel Cambridge.

Rodbard, D. (1976) In: Methods of protein separation (Catsimpoolas, N., ed.) pp. 145–179, Plenum press, New Nork.

Russ, C., Callegaro, I., Lanza, B. and Ferrone, S. (1983) Purification of IgG monoclonal antibody by caprylic acid precipitation. J. Immunol. Meth. 65, 269–275.
Ryll, T., Lucki-Lange, M., Jäger, V. and Wagner, R. (1990) Production of recombinant human interleukin-2 with BHK cells in hollow fibre and a stirred tank reactor with protein-free medium. J. Biotechnol. 14, 377–392.
Sadana, A. (1992) Inactivation of proteins and other biological macromolecules during chromatographic methods of bioseparation. Bioseparation 3, 145–165.
Samuelson, E., Wadenstein, H., Hartmanis, M., Moks, T. and Uhlén, M. (1991) Facilitated in vitro folding of human recombinant insulin-like growth factor I usinf solubilizing fusion partner. Bio/technology 9, 363–366.
Sano, E., Sawanda, R., Shimizu, H. and Kobayashi, S. (1992) In: Animal cell technology: basic and applied aspects (Murakami, H., ed.), pp. 453–462, Kluwer academic Publishers, Amsterdam.
Sano, T. and Cantor, C.S. (1991) A-streptavidin-protein A chimera that allows one-step production of a variety of specific antibody conjugates. Bio/technology 9, 1378–1381.
Savinell, J.L., Lee, G.M. and Palsson B.O. (1989) On the orders of magnitude of epigenetic dynamics and monoclonal antibody production. Bioprocess Eng. 4, 231–234.
Secher, D. and Burke, D.C. (1980) A monoclonal antibody for large-scale purification of human leucocyte interferon. Nature 285, 446–451.
Seo, J.-H. and Bailey, J.E. (1986) Continous cultivation of recombinant Escherichia coli: exisitence of an optimum dilution rate for maximum plasmid and gene product concentration. Biotech. Bioeng 28, 1590–1594.
Shiloach, J., Martin, N. and Moes, H. (1989) In: Downstream processes: equipment and techniques (Mizrahi, A., ed.), pp 97–125, Alan R. Liss Inc.
Sleep, D., Belfield, G.P., Balance, D.J., Steven, J., Jones, S., Evans, L.R., Moir, P.D. and Goody, A.R. (1991) Saccheromyces cerevisiae strains that overexpress, heterologous proteins. Bio/technology 9, 183–187.
Staude, E. (1992) Membranen und Membranprozesse, pp. 147–160, VCH, Weinheim New York Basel Cambridge.
Staude, E. (1992) Membranen und Membranprozesse, pp. 191–202, VCH, Weinheim New York Basel Cambridge.
Steindl, F., Jungbauer, A., Wenisch, E., Himmler, G. and Katinger, H. (1987) Isoelectric precipitation and gel chromatography for purification of monoclonal IgM. Enz. Microb. Technol. 9, 361–364.
Sugiura, T. and Maruyama, H.B. (1992) In: Animal Cell technology: Basic and Applied Aspects (Murakami, H., ed.), pp. 287–293, Kluwer Academic Publishers, Amsterdam.
Sulk, B., Birkenmeier, G. and Kopperschläger, G. (1992) Application of phase partitioning and thiophlic adsorption chromatography to the purification of monoclonal antibodies from cell culture fluid. J. Immunol. Meth. 149, 165–171.
Tibell, L., Hjalmarsson, K., Edlund, T., Skogman, G., Engström, Å. and Marklund, S.L. (1987) Expression of human extracellular superoxide dismutase in Chinese hamster ovary cells and charactenzation of the product. Proc. Natl. Acad. Sci. USA 84, 6634–6638.
Treeweek, G.P. and Morgan, J.J. (1977) Polymer flocculation of bacteria: the mechanism of E. coli aggregation by poly-ethyleneimine. Colloid Interface Sci. 60, 258–273.
Van der Pol, L., Zijlstra, G., Thalen, M. and Tramper, J. (1990) Effect of serum concentration on production of monoclonal antibodies and on shear sensitivity of a hybridoma. Bioproces Eng. 5, 241–245.
Vorauer, K., Skias, M., Trkola, A., Schönhofer, W., Unterluggauer, F. and Jungbauer, A. (1993) Floculation: an alternative process to-ion-exchange chromatography? (a scale-up study using recombinant human superoxide dismutase as model protein). Biosparation 3, 217–226.

Vorauer, K., Skias, M., Trkola, A., Schulz, P. and Jungbauer, A. (1992) Scale-up of recombinant protein purification by hydrophobic interaction chromatography. J. Chromatogr. 625, 33.

Walter, H. and Johanson, G. (1986) Partitioning aqueous two-phase systems: an overview. Anal. Biochem. 155, 215–242.

Warren, R.K., Macdonald, D.G. and Hill, G.A. (1991) Cross-flow microfiltration of Saccheromyces cerevisiae. Process Biochem. 26, 337–342.

Weiner, C., Sära, M. and Sleytr, U.B. (1994a) Novel protein A affinity matrix prepared from two-dimensional protein crystals. Biotech. Bioeng. in press.

Weiner, C., Sära, M. and Sleytr, U.B. (1994b) Affinity cross-flow filtration: purification of antibodies with a novel protein A affinity matrix prepared from two-dimensional protein crystals. Biotech Bioeng. in press.

Wha, C., Kim, S.-K., Rha, C. and Robertson, E. (1987) In: Flocculation in biotechnology and separation systems (Attia, Y.A., ed.), pp 429–439, Elsevier, Amsterdam Oxford New York Tokyo.

Wronski, S. and Rudniak, L. (1988) Dynamic filtraion in biotechnology. Bioprocess Eng. 4, 99–104.

Wu, L., Figueroa, A. and Karger, B.L. (1986) Protein conformational effects in hydrophobic interaction chromatography: retention charcterization and the role of mobile phase additives and stationary phase hydrophobicity. J. Chromtogr. 371, 3–27.

Yamamoto, S. and Sano, Y. (1992) J. Chromatogr. 597, 173–179.

Yamamoto, S., Nomura, M. and Sano, Y. (1990) Prediction the performance of gel-filtration chromatography of proteins. J. Chromatogr. 512, 77–87.

Yarmush, M.L., Pickard Antonsen, K., Sundanam, S. and Yarmush, D.M. (1992) Immunoadsorption: strategies for antigen elution and production of reusable adsorbents. Biotechnol. Prog. 8, 168–178.

Zabriskie, W. and Arcni, E.J. (1986) Factors influencing productivity of fermentations employing recombinant microorganisms. Enzym. Microb. Techn. 8, 706–716.

4.2 Validation of Downstream Processes

Joachim Walter and Hermann Allgaier

4.2.1 Introduction

The purification part of a biotech manufacturing process – downstream processing – came during the last years more and more to the attention of scientists and engineers dealing with the development, validation and manufacturing of biopharmaceuticals. The direct dependency on upstream processing determines the size and the course of the purification procedure. Within a given fermentation technology, e.g. stirred tank reactor, various cells can be grown within the same equipment. Recent developments with new expression systems allow for fermentation titers up to $2\,g\,l^{-1}$ compared to $0.1-0.2\,g\,l^{-1}$ with previous mammalian host systems. Processing volumes, column sizes and membrane areas in downstream processing are directly proportional to the absolute amount of protein at the end of the fermentation. The example above would require a 10–20-fold larger downstream processing area in a given facility than installed before. In addition, every protein represents a single molecule with unique features leading to dedicated and specific purification sequences. Therefore, the art of downstream processing represents a major milestone in the development, validation and manufacture of a protein used as a biopharmaceutical.

4.2.2 Master Validation Plan

The validation of pharmaceutical production processes is a general requirement of the Current Good Manufacturing Practice (cGMP) regulations for finished pharmaceuticals, 21 CFR 210 and 211 (U.S. Government Printing Office, 1991) and the Guideline on General Principles of Process Validation issued by the FDA (Division of Manufacturing and Product Quality, 1987). In this document, validation is defined as "establishing documented evidence which provides a high degree of assurance that a specific process will consistently produce a product meeting its pre-determined specifications and quality characteristics". It is the responsibility of the manufacturer to transfer these regulations and guidelines into appropriate validation concepts and schemes leading to a master validation plan describing all validation tasks and responsibilities in detail.

In general, a separation between the plant qualification and the process validation could be of advantage, in particular for those manufacturing plants where more than one product is produced (Fig. 4.2.1). Whereas the plant qualification can be carried out independently, for the most part from the features of the different products to be manufactured and can be amended for a new product, the process validation in detail considers all features of a new product.

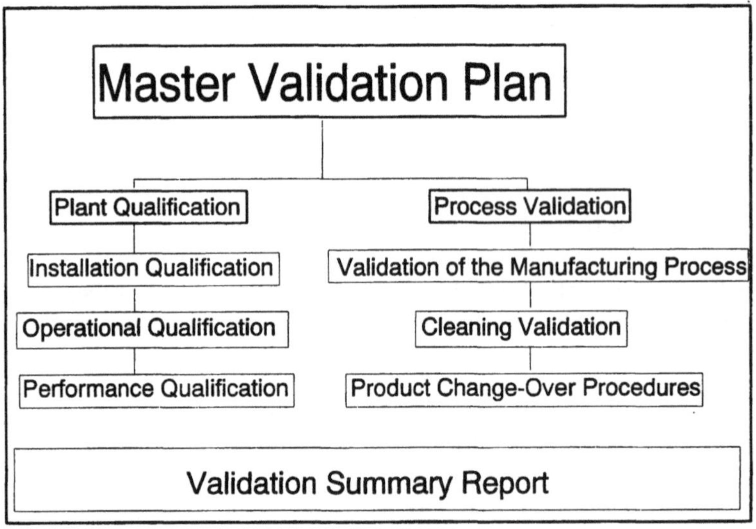

Fig. 4.2.1: Master Validation Plan with Major Elements of Plant Qualification and Process Validation

4.2.3 Manufacturing Plant Qualification

The plant qualification consists of the three major steps: installation qualification (IQ), operational qualification (OQ) and performance qualification (PQ, Fig. 4.2.1). The installation qualification is defined as "establishing confidence that process equipment and ancillary systems are capable of consistently operating within established limits and tolerances" (Division of Manufacturing and Product Quality, 1987). Major elements of the installation qualification are the verification of Piping and Instrumentation Diagrams (P&ID's) and isometric drawings with the "as built" reality and the documentation of any kind of discrepancies in punch-lists, the definition of spare part lists and subsequent procurement of spare parts, the definition of preventive maintenance measures and the review of data sheets and operational manuals.

The operational qualification defines the calibration of all instruments and their calibration intervals, defines alarms and their functional check, requires and documents first passivation and manual cleaning operations and requires the qualification of automated Cleaning-In-Place (CIP) cycles and Steaming-In-Place (SIP).

Whereas the SOP's for installation qualification and operational qualification for different types of equipment can be identical, the performance qualification requires different protocols for different types of equipment such as fermenters or chromatographic systems and focuses on the function of the whole unit consisting in the case of chromatographic systems of buffer vessels, transfer lines, product hold vessels and the chromatographic column with all measuring instruments. Therefore, a blank buffer run imitating the production reality as close as possible is at the heart of the performance qualification program.

4.2.4 Validation of the Manufacturing Process

4.2.4.1 Raw Materials Used for Downstream Processing

Raw materials used for downstream processing can be inserted for the most part into one of the subsequent classes: solvents, buffer salts and excipients, chromatographic matrices and micro- and ultrafiltration membranes. According to cGMP's, all raw materials used for manufacturing must be held in quarantine, investigated and released by an independant quality control unit. Excipients must comply to the methods and specifications laid down in the Pharmacopeia. Vendor audits should be carried out before the manufacturing process is finally fixed between phase II and phase III clinical trials. This procedure ensures the required quality of all employed raw materials and gives in addition valuable indications as to whether the selected raw material supplier will be able to deliver during the whole life cycle of a biopharmaceutical, about twenty years.

Water is the most commonly used solvent during downstream processing of smaller and larger biopharmaceutical proteins. Purified Water (PW) and Water for Injection (WFI) are the most important water qualities described in the Pharmacopeia used in different phases of the manufacturing process (United States Pharmacopeia Convention, 1990). Whereas Water for Injection in Europe can only be generated by distillation, other techniques like reverse osmosis can be applied in the United States (Council of Europe, 1990). Water for Injection in general has to be stored and distributed at 80 °C to ensure that the stringent specification of less than 10 Colony Forming Units/100 ml can be met (Table 4.2.1). Other solvents like ethanol and propanol are mainly used for hydrophobic interaction chromatography or, if smaller proteins are to be purified, other organic solvents may be used for High-Performance-Liquid-Chromatography or Liquid-Liquid-Extraction procedures.

Table 4.2.1. Chemical and Microbiological Requirements for Purified Water and Water for Injection.

	Chemical Requirements	Microbiological Requirement
Purified Water	according to USP	total microbial count < 10000 CFU 100 ml^{-1} absence of E. coli
Water for Injection	according to USP	total microbial count < 10 CFU 100 ml^{-1} absence of E. coli pyrogens < 0.25 EU ml^{-1}

Chromatographic matrices for downstream processing in general are the most critical raw materials used in manufacturing. Within a highly specialized, relatively small market, many of the offered matrices are unique due to different backbone and ligand chemistries. The exchange from one matrix to another even with the same ligand chemistry requires careful and thorough validation studies. Even though all suppliers deliver certificates of analysis for the provided resins careful in-house testing is necessary due to cGMP's. The lack of specifications laid down in the Pharmacopeia requires the set up of in-house assays and specifications. Identity testing must take the specific features of the resin into account. Flow rates and back-pressures of matrices are investigated in scale-down systems thereby considering wall effects of small systems. Functional tests, e.g. the separation of different sized proteins for gel permeation chromatographic matrices or the separation of differently charged proteins on ion exchangers, are carried out by the resin manufacturer but should be repeated in-house with the particular separation problem. Investigations with regard to microbial and endotoxin contamination round off the most important assays for chromatographic resins.

During the last years we observed, either during raw material testing or during development, the cross-contamination of a neutral gel chromatographic matrix with an ion exchanger, the inability of a resin manufacturer to supply consistent batches and the instability of a Protein G matrix during storage under conditions recommended by the manufacturer (Bill et al., 1995).

Ultrafiltration membranes used for concentration and buffer exchange/formulation purposes should be as thoroughly investigated as chromatographic resins. Not only the features of the membrane itself, like cutoffs, separation characteristics and fluxes, are of importance but also the integrity of the manufactured cassettes for membrane holding devices should be checked very carefully. However, only rough damage of the cassette system can be indicated by integrity testing of ultrafiltration systems. Sterile filters used for decades in the critical aseptic processing area are generally made under a high level of GMP compliance. The certificates of analysis are detailed and include tests for oxidizables, integrity testing by forward flow or bubble-point testing, challenge with Pseudomonas diminuta and freedom from pyrogens. Therefore, a vendor audit can allow for the transition of most dates of the certificate of analysis (Divisional Manufacturing and Product Quality, 1987).

4.2.4.2 Removal of Contaminants

In addition to a defined and correct amino acid sequence, which is determined by the genome of the producer cell (FDA, Office of Biologics Research and Review, 1987), the biosynthesis of small polypeptides as well as complex glycosylated human proteins demands reliable procedures and production conditions under which post-translational modifications of the product are reproducible (Walter and Werner, 1993). In addition, well-determined process conditions for the cultivation of microorganisms and mammalian cells reflect an essential backbone for reproducible downstream processing: not only the quality of the protein product in question is of significant importance but also the contaminants in the product fluid and their amount and composition may influence the quality of the final product, especially if certain contaminations – independent of whether they are of a chemical or biological nature –vary in a non-reproducible manner. Under those terms a validation of the downstream process would be difficult.

4.2.4.2.1 Features of Protein Product and Contaminations from Different Organisms

The protein product and types and nature of product contaminations are determined by the producer organism and the cultivation conditions (FDA, 1983; 1985; National Institutes of Health, 1986; Office of Science and Technology Policy, 1986; OECD, 1986; EC-Commission, 1986; Committee for proprietary medicinal products, 1987; 1990; FDA, 1987; Ministry of Health and Welfare, 1988; Commission of the European Communities, 1989; FDA, 1990; Health and Welfare Canada, 1990; World Health Organization, 1991; FDA, 1991; Kozak, 1992). Significant differences in the biosynthesis profile for prokaryotic (bacteria) and eukaryotic (fungi, mammalian cells) producer cells are obvious, namely with regard to molecular weight and post-translational modifications.

Typically, genetically engineered bacteria produce their protein product as inclusion bodies, which can easily be separated in a pure form after cell homogenization. The cleavage of the inclusion bodies and the subsequent refolding of the monomer protein in order to obtain the desired product require specific techniques whereas the following downstream processing is comparable to the isolation of secreted proteins in a native form. Complex and – very often – highly glycosylated proteins are typically produced by mammalian cells, and a state of the art downstream process must orientate itself according to the specific features of mammalian recombinant cells, or – in the case of monoclonal antibodies – hybridoma cells. Typical features specific for those mammalian producer cells are concentrations of the native protein in the cell culture fluid in the range of about $0.1-2\,g\,l^{-1}$ and potential hazardous impurities of biological origin, e.g. nucleic acid with an oncogenic potential and viruses (Schiff et al., 1992). Advantages versus disadvantages for the manufacturing of pharmaceutical proteins using mammalian cell cultures are listed in Table 4.2.2.

It is evident that the protein-chemical properties of the product determine the selection of purification procedures. The strategy for the sequence and confirmation of individual unit operations must take into consideration the specific attributes of the producer organism and the required cultivation media.

Table 4.2.2. Features of Pharmaceutical Proteins derived from Mammalian Cell Culture.

Advantages
* native * singular * folded * secreted * high molecular weight * no disulfide bridge limitation * gylcosylated
Disadvantages/Risks
* potential retroviral contamination * potential product-homolog host protein contamination * potential oncogene contamination

4.2.4.2.2 Media Derived Contaminants

Impurities of a proteinaceous nature

The cultivation of mammalian cells requires a complex mixture of nutrients either for the propagation of the cells during growth phase or for the production phase. Typically, the media used for cell mass production and media used for the production

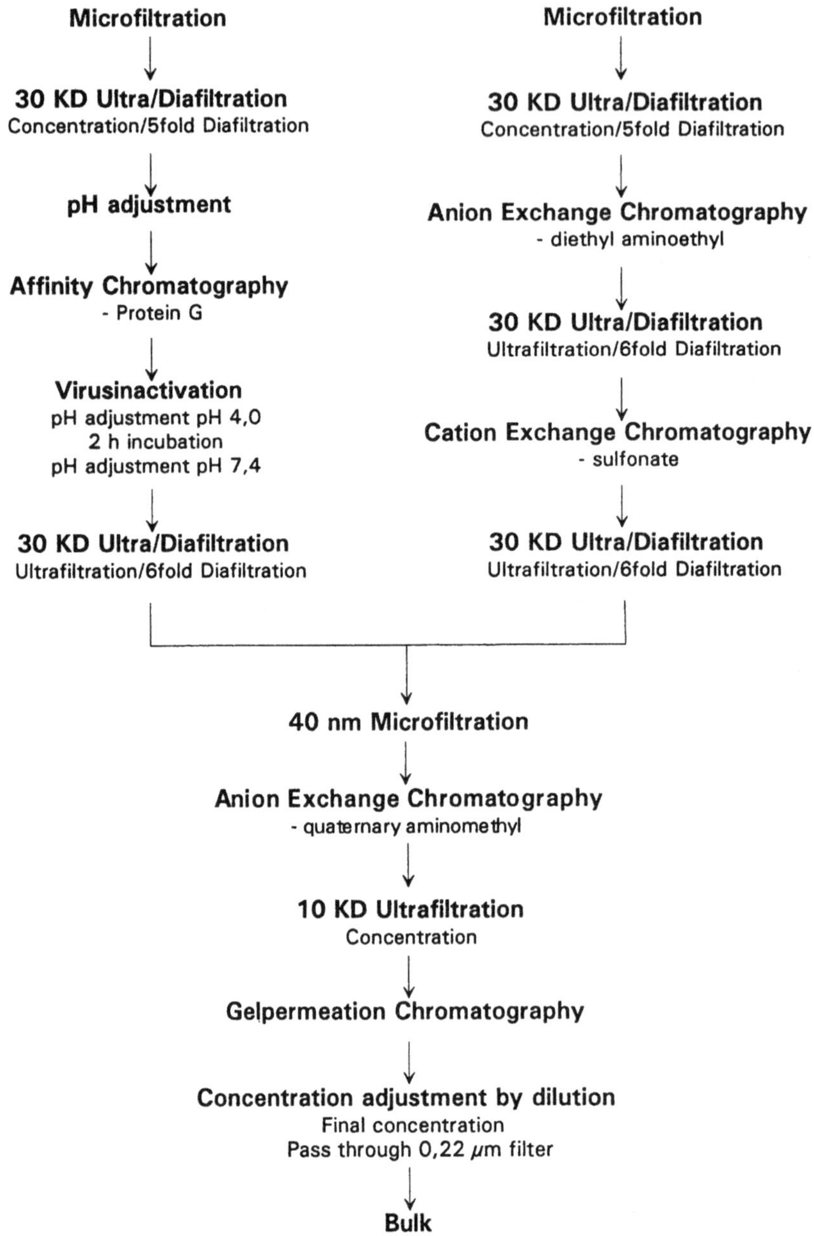

Fig. 4.2.2: Downstream Processing of a Monoclonal Antibody

and secernation of the protein product differ significantly with respect to the supply with physiologically active protein supplements (World Health Organization, 1987). The use of protein-free media still is an exemption for cell culture operations of registered products.

For a number of mammalian producer cell lines the supply with fetal or bovine serum – sometimes up to 10% v/v – is still necessary. Depending on the serum quality and composition with respect to a lot-to-lot consistency, the overall protein pattern may vary and the relation of the serum proteins to each other may differ. Thus, in the case of a contamination of the product with serum, the downstream process has to be designed carefully with respect to a potential variation of the composition of the culture fluid. Well-defined media, i.e. quantitative supplementation with purified proteins such as bovine serum albumin (BSA), transferrin or insulin, allow for easy assessment of their removal during protein purification; sensitive assays like ELISA are available. Figure 4.2.2 shows two strategies for the purification of a monoclonal antibody; the data presented in Figures 4.2.3–4.2.5 are based upon this scheme.

Figure 4.2.3 and Figure 4.2.4 give examples for the monitoring of specified protein contaminants. For the manufacturing of monoclonal antibodies it can be advan-

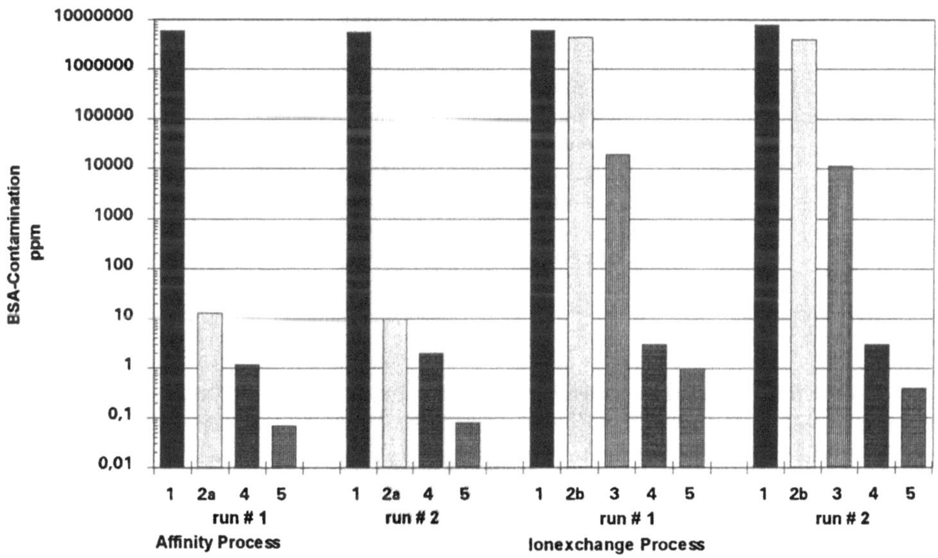

Fig. 4.2.3: Removal of BSA during Downstream Processing of a Monoclonal Antibody in a Generic Process (Protein G Affinity Chromatography) and in an Individual Process (Ionexchange Chromatography). Compare Fig. 4.2.2

BSA concentration given in ppm-values

Process steps:
1) Cellfree culture fluid
2a) Pool after Protein G Affinity Chromatography
2b) Pool after DEAE-Anion exchange Chromatography
3) Pool after S-Cation exchange Chromatography
4) Pool after Q-Anion exchange Chromatography
5) Pool after Gel permeation Chromatography

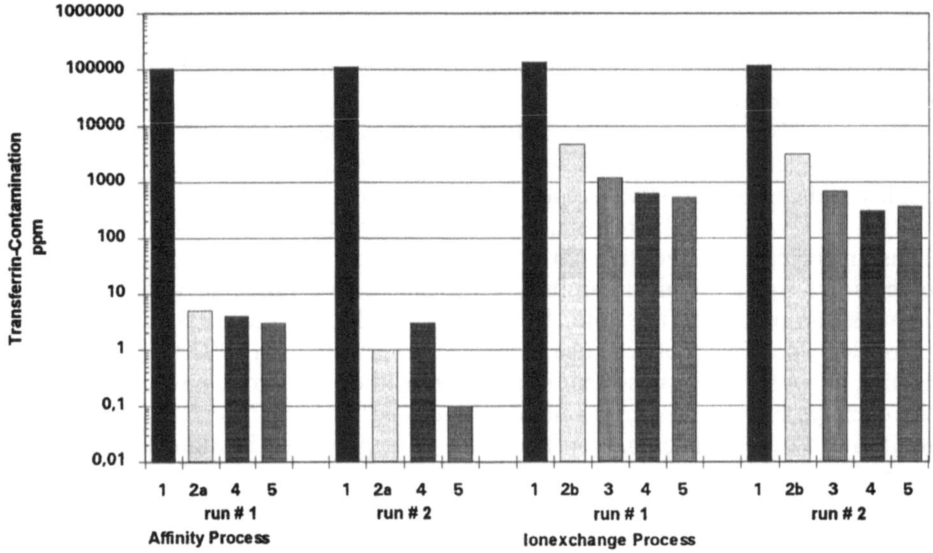

Fig. 4.2.4: Removal of Transferrin during Downstream Processing of a Monoclonal Antibody in a Generic Process (Protein G Affinity Chromatography) and in an Individual Process (Ionexchange Chromatography). Compare Fig. 4.2.2

Transferrin concentration given in ppm-values
Process steps: 1) Cellfree culture fluid
2a) Pool after Protein G Affinity Chromatography
2b) Pool after DEAE-Anion exchange Chromatography
3) Pool after S-Cation exchange Chromatography
4) Pool after Q-Anion exchange Chromatography
5) Pool after Gel permeation Chromatography

tageous to remove residual bovine immunoglobulins (IgG's) from supplementary bovine proteins; Figure 4.2.5 demonstrates the tracing of contaminating bovine IgG's by ELISA for the purification of a murine IgG2a monoclonal antibody (MAb) using a generic process in comparison to a process which is dedicated and optimized for the murine MAb.

Impurities of a non-proteinaceous nature

The complex mixture of ingredients that are required to cultivate mammalian cells contains, in addition to anorganic salts and proteinaceous supplements, additional organic compounds.

Antibiotics might be applied to maintain continuous fermentation over a period of several weeks or even months, and cytotoxic agents like methotrexate are frequently used to keep the cell culture under selective pressure, in order to avoid cell segregation during the time consuming phase of cell propagation. It is evident that these types of additives have to be removed and that appropriate detection assays have to be in place.

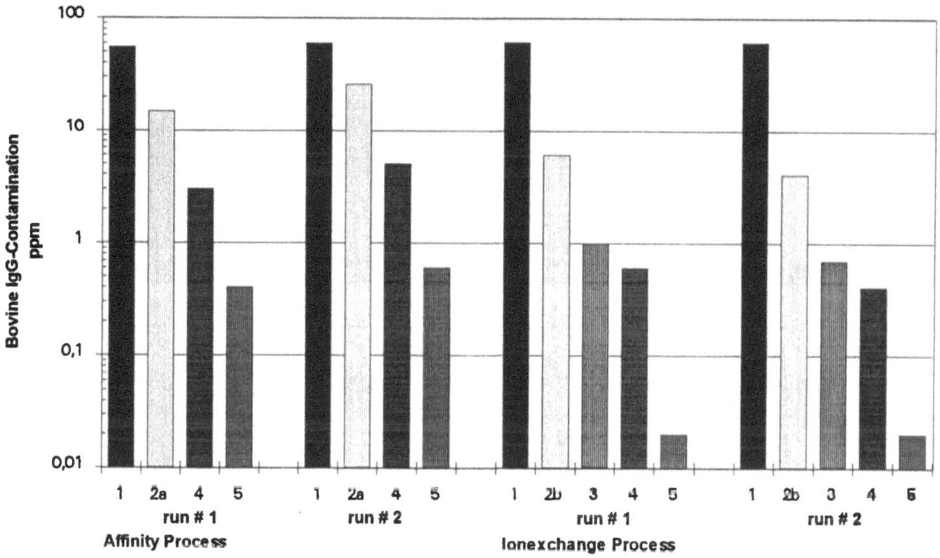

Fig. 4.2.5: Removal of Bovine IgG during Downstream Processing of a Monoclonal Antibody in a Generic Process (Protein G Affinity Chromatography) and in an Individual Process (Ionexchange Chromatography). Compare Fig. 4.2.2

Bovine IgG concentration given in ppm-values

Process steps:
1) Cellfree culture fluid
2a) Pool after Protein G Affinity Chromatography
2b) Pool after DEAE-Anion exchange Chromatography
3) Pool after S-Cation exchange Chromatography
4) Pool after Q-Anion exchange Chromatography
5) Pool after Gel permeation Chromatography

Lipids, fatty acids and antifoaming agents used in fermentation media are likely to cause negative effects on filter membranes and chromatographic matrices due to nonspecific hydrophobic adsorption to the surface; special attention and considerable validation has to be performed for the regeneration procedures.

Amino acids – produced by bacterial fermentation or acidic hydrolysis – and buffer salts are used for the fermentation media as well as for process buffers downstream. These compounds might bear pyrogens derived from gram negative bacteria to which pyrogenic lipopolysaccharides (LPS) are essential integral compounds in their outer cell wall. Ultrafiltration of solutions through membranes (cutoff < 10.000 d) has been proven to remove pyrogens effectively (American Society for Testing and Materials, 1988).

Table 4.2.3. Depyrogenization of Solutions by 5 KD Ultrafiltration.

Solution	Pyrogen Contamination before UF EU ml^{-1}	Pyrogen Contamination after UF EU ml^{-1}
aqua dest.	7.68	0.12
7.5 M urea	3.84	< 0.06
phosphate buffered saline pH 7.2	0.96	0.06

Pyrogens: LPS isolated from E. coli
Membrane: polyethersulfone, cut-off 5.000 dalton, 10.000 cm^2
Holder: stainless steel
Type: tangential flow filtration
TMP: 2 bar (14 psi)
Volume: 50 l

4.2.4.2.3 Host Cell Derived Contaminants

Host cell proteins

During the production phase for the biosynthesis of the desired protein product, the composition of the culture medium is selected such that the product in question is the main protein synthesized by the cell. By supplementation with inducing agents or selective substances like methotrexate, the biosynthesis of the product can be enhanced. Nevertheless, the desired product is by far not the only protein produced during this phase: these proteins, derived from the host cell, in addition contaminate the product solution and have to be separated (American Society for Testing and Materials, 1988).

In a well described reproducible and fully controlled fermentation process, the pattern of the host cell proteins is typical and reproducible for the producer cell. This allows for the development of quantitative assays for contaminating host cell proteins on the basis of immunological methods such as ELISA and immunoblotting. In order to get antibodies for the evaluation of immunological assays, a cell line nearly identical to the producer cell line has to be constructed: the genome of that "blank" cell harbors the genetic information of the producer cell except for the DNA sequence determining the protein product. Cells of this cell line are used to perform a complete fermentation and downstream process ("blank run") under process conditions which are identical to the conditions of the manufacturing process. Samples of intermediate process stages are taken and used as antigen for the immunization of rabbits or goats. The polyclonal antibodies which are produced against the applied antigen are recovered from the animal sera, and are the base molecules for a sensitive bioassay.

The detection limit of a well-developed sensitive ELISA allows for quantitive determinations in a nanogram range of the antigen. The effectiveness of individual purification steps of the recovery process can be determined and proven; for process

Table 4.2.4: Separation of Host Cell Specific Proteins.

Type of Chromatography	Interferon µg ml^{-1}	CHO-Protein µg ml^{-1}	CHO-Protein ppm
Cation Exchange	124	12.50	100725
Anion Exchange	169	0.75	4438
Hydrophobic Interaction	1038	0.59	217

Interferon is synthesized by chinese hamster ovary (CHO) cells. The removal of CHO-Proteins in three consecutive chromatographies is demonstrated by ELISA, using CHO-Protein specific antibodies.

validation it is a "must" to have such sensitive assays available. Furthermore, the release of final product requires such a quantitative and sensitive assay for host cell proteins (Garnick et al., 1988).

Contaminating nucleic acids

Recombinant microbial and mammalian cells are constructed using DNA sequences which might harbor potential harmful genetic information when incorporated into healthy human cells. The DNA sequences are integrated into the genome of the intended producer cell in order to achieve an immortilized cell line by use of oncogenic sequences and with respect to a high producer cell line by use of viral sequences e.g. promoters. For the evaluation of the tumorigenicity of such DNA sequences, investigations with infectious viral nucleic acid bearing oncogenic potential have been performed (Bouchard et al., 1984). As a result of the obtained data the WHO defined a maximal amount of DNA to be as low as 100 pg DNA per single dose of a biotechnologically derived protein drug intended for parenteral application (WHO Study Group,1987). This concentration is even a very conservative approach, since the dose of DNA that is required for tumor induction exceeds the value given by 10 log units, i.e. the concentration of 100 pg DNA per dose remains under the tumor inducing dose by a factor of 10^{10}. Recommendations released by some authorities even suggest DNA concentrations of less than 10 pg DNA per single dose (Löwer, 1990). With respect to these requirements for reasonable drug safety, the downstream operations have to be validated accordingly.

The amount of DNA that contaminates the product fluid after cell separation is highly influenced by the organisms used for production and the type of cell harvest. Microorganisms and mammalian cells that secret their desired product can be separated keeping the cells intact: contamination with DNA is reduced when using cell harvest procedures that avoid cell lysis. Furthermore, the success of a careful cell harvest is influenced by the viability of the cells prior to the process, i.e. fermentation conditions and the timepoint for an adequate harvest have to be considered. In order to prove the required DNA removal, the amount of DNA in the cell-free product fluid is measured and the removal of DNA along the downstream process is determined. As mentioned above, the harvested fermentation fluid, especially for product secreting organisms and mammalian cells already contains only a low amount of DNA. In order to achieve relevant data that demonstrate the required DNA removal, spiking experiments for selected purification procedures have to be performed in addition to the measurements of the native samples derived from the whole downstream processing. Spiking experiments are carried out using appropriate DNA, e.g. DNA which has been isolated from the producer cell. Validation work is done on the designated process step, using product solutions that are supplemented with the prepared DNA sample in sufficient amounts. DNA is effectively removed using chromatographical methods, either binding the product while releasing the DNA or vice versa. Predestinated procedures for efficient DNA removal include affinity chromatography on ligands which selectively bind the desired protein product while leaving the DNA in the effluent, or anion exchange chromatography to which under appropriate physical conditions nucleic acids bind extremely strongly due to the potential high negative charge of their phosphate groups (Bergmann et al., 1993). The accumulation of clearance factors for DNA achieved in different purification steps thus easily yields reduction factors up to 10^{14} and even more.

Two principal detection methods are widely used for the detection and quantification of DNA: a "DNA hybridization assay" and a "total DNA assay". Both assays are appropriate for the measurements in a validation study.

The specific interaction based on biorecognition between complementary DNA sequences is the background for the DNA hybridization assay: a DNA probe which is typically radioactively labeled in order to achieve a reasonable sensitivity binds selectively to a complementary DNA sequence of the sample to be determined. Detection limits that can be achieved using an hybridization assay are in the range of about 5–10 pg DNA (Rigby et al., 1977; Per et al., 1989; Ostrove et al., 1992). The other assay to be mentioned is able to determine DNA with no need for complementary DNA probes. The so-called "total DNA assay" uses two high affinity DNA-binding proteins and a semiconductor sensor for the detection and quantitation of the bound DNA (King et al., 1990). The assay has no need for radioactively labeled probes and thus can be performed in any laboratory. The detection limit of the total DNA assay is as low as about 2 pg DNA (Molecular Devices Corporation, 1988; Smith et al., 1991).

Virus and Virus-Like Particles (VLP's)

A serious concern for rDNA-derived proteins from recombinant mammalian cell cultures and monoclonal antibodies (MAb's) derived from hybridoma cultures is the potential risk of contaminating retroviral particles or adventitious virus. Typically and with rare exceptions retroviral-like particles can be demonstrated for hybridoma cells, but although electron microscopic visible particles have been tested for infectivity, in most cases an infective nature has not been substantiated. Nevertheless, the presence of a virus of unknown origin can not be excluded. Such an unidentified virus is expected to have unknown but potentially harmful physiological effects, but due to the unknown nature of the contaminant a specific assay can not be advised. Therefore it is impossible to monitor the presence and the behavior, the removal and the inactivation of the contaminating virus during the manufacturing process of the pharmaceutical protein. Any drug designated for parenteral

Table 4.2.5. Validation of DNA Removal.

Q-Sepharose FF		
trial No.	Elution Factor	Clearance Factor
# 1	35.700	2.360.000
# 2–5	---	---
# 6	14.500	1.500.000
# 7–10	---	---
# 11	18.500	1.320.000
# 12–15	---	---
# 16	24.500	1.500.000
# 17–20	---	---
# 21	28.600	1.050.000
average	24.360	1.546.000
DEAE-Sepharose FF		
trial No.	Elution Factor	Clearance Factor
# 1	31.700	712.000
# 2–4	---	---
# 5	20.400	750.000
# 6–8	---	---
# 9	23.200	2.600.000
# 10–18	---	---
# 19	18.100	1.300.000
average	23.350	1.340.000
DE-52 Cellulose		
trial No.	Elution Factor	Clearance Factor
average of 7 runs	15.000	700.000

A set of up to 21 consecutive chromatographies, each including a full regeneration process, was performed for 3 different anion exchanger matrices, elution factors and clearance factors were determined in intervals using radiolabeled DNA.
The elution factor determines the total radioactivity which coelutes with the protein and does not discriminate between DNA-molecules of different polymer length.
The clearance factor determines the radioactivity with respect to precipitable DNA: at a 15% trichloroacetic acid (TCA) concentration only DNA-molecules larger than about 50 base pairs precipitate. DNA-molecules containing less than 50 base pairs pose a significantly reduced potential risk: typically a minimum of several hundred base pairs codes for an oncogene and about one hundred base pairs code for a gene promotor.

Column diameter:	1 cm	Protein load:	10 mg ml^{-1} matrix
Column height:	20 cm	DNA load:	7 g ml^{-1} matrix
Column volume:	15.7 cm	linear Flowrate:	150 cm h^{-1}

application must be sterile, i.e. free of any infectious agent, including virus. Taking into account that an absolute sterility can hardly be proven, regulatory authorities have defined a limit of less than one infective particle in $1x10^6$ sterilized units of the final product, independent of its volume. Preventive measures have to be undertaken in order to reduce the probability of the presence of potential viral contamination in the protein drug (Lees and Onions, 1990; Marcus-Sekura, 1991; Walter et al., 1991; 1992; Liptrot and Gull, 1991; Löwer, 1991; Ad Hoc Working Party, 1991; Berthold et al., 1992) (see also Chapter 1.7).

These measures include extensive testing of the master cell bank (MCB) and manufacturer's working cell bank (MWCB) for a great number of specific viruses and assays for the detection of adventitious virus at a number of stages during cell cultivation and fermentation and for biological raw materials.

A preventive measure to avoid virus contamination is to be taken at the stage of cell banking: cell lines engineered and selected to produce a pharmaceutical protein have to be proven to be free of contaminating virus, and the detection of infective virus in the Master Cell Bank (MCB) in general disqualifies such bank for manufacturing. For the production of monoclonal antibodies a selection of 17 viruses which have to be screened has been issued by the Committee for Proprietary Medicinal Products (CPMP) of the European Community (1987). As mentioned earlier, hybridoma cell lines and recombinant cell lines might harbor endogenous retrovirus, at least retroviral particles or even reverse transcriptase activity can be demonstrated. Despite the occurrence of a potential retroviral infection, those cell lines can also be accepted under well-defined circumstances: The removal and the inactivation of retrovirus in the course of downstream processing must be demonstrated by validation.

In the overall consideration of product safety, virus-like particles are considered as if infectious. With respect to an electron microscopic evaluation of the number of particles, a maximal virus contamination can be calculated. Guidelines released by several authorities, e.g. FDA and EC guidelines, recommend the quantitation of particles in harvest samples, more precisely known as "prepurified bulk" samples. Such samples are defined by the CFR as cell culture supernatant after low speed sedimentation of cells. This cell free supernatant is then centrifuged at 100.000 g in order to sediment all particles of virus size. The pellet is screened in an electron microscope and virus-like particles are counted per grid. In this way a number for the average (or maximal) content of retrovirus-like particles is calculated for the virus burden of the starting solution of the downstream processing. The validation of the unit operations of the downstream process has to demonstrate at least their potency to remove this amount of virus (Walker et al., 1992). For safety aspects, the virus accumulated clearance should exceed this level with a margin of 3–5 log units.

According to the expression of the virus titer on a logarithmic scale it has become customary to describe the degree of the removal and inactivation of a virus by the logarithm of the clearance factor as "log reduction", e.g. the reduction in virus titer by 10^6 means a "6 log reduction".

In addition, intensive investigation on the producer cells for retroviral infectivity and reverse transcriptase activity is performed. The downstream process with its purification and recovery operations has to be designed and validated for its ability

Table 4.2.6. Potential Virus Contamination and Preventive Measures for Biotechnologically Manufactured Protein Drugs.

Potential Virus Contamination
* adventitious virus * endogenous host specific virus * endogenous retrovirus * virus like particles
Preventive Measures – Cell biology
* selection of cell line * selection of host/vector construction * MCB testing * choice of biological media compounds * serum/protein free media for > cell expansion > production
Preventive Measures – Protein chemistry
* design of downstream processing with respect to virus removal/inactivation * validation of virus removal/inactivation of individual unit operations in downstream processing > virus removal – chromatography – filtration > virus inactivating – virus inactivating chemicals – chaotropic salts – organic solvents – detergents – extreme pH-values – extreme temperatures – irradiation

to remove and to inactivate viruses. Different process steps contribute to a virus removal/inactivation to a different degree, depending of the physical and chemical conditions. Validaton studies have to be performed for any individual step which claims for a contribution to remove or inactivate virus.

Typically, any process validation has to be performed using the original equipment embedded in the prospective environment of the production facility, at least the equipment has to be of identical type and of the same size of the manufacturing scale and must be operated identically as intended for use in the manufacturing process. An exception to this stringent rule is the validation of the downstream processing with respect to its capability for virus removal and virus inactivation: The deliberate contamination of the designated manufacturing equipment and facility with different infectious viruses – some of them might even be human pathogens – would discriminate for any production due to an irresponsible risk factor for the final product. An alternative for small scale production could be the use of a replica of the manufacturing equipment. For batch sizes up to several thousand liters it is from a practical point of view impossible or at least extremely difficult to provide the considerable amount of virus that would be required for the scale corresponding validation. An appropriate solution to this difficulty is a truly linear down-scaled model of the unit operation in question on conditions that the process parameters can be kept representative for the manufacturing scale. Both strategies have found acceptance with respect to assuring product safety.

The efficacy of removal/inactivation with regard to potential infection with other types of virus of unknown nature must be proven using model viruses. The selection of such model viruses for validation purposes is critical. The nature and origin of the producer cells in question is a benchmark for the selection of appropriate model viruses: a model virus should be close to identical to the virus suspected to be an infectious agent of the cell line, or – more practical – closely related to the potential infective virus, e. g. ecotropic or xenotropic murine leukaemia virus for murine hybridoma cells. With respect to other potential but unknown viruses, the validation for viral clearance has to cover a broad range of virus species exhibiting different physical, biological and physico-chemical properties (Marcus-Sekura, 1991; Ad Hoc Working Party, 1991).

Table 4.2.7. Requirements for Model Viruses used in Validation Studies.

Virus Features
* size of virus * enveloped/non-enveloped * genome structure DNA/RNA * strandedness of genome * resistence to inactivation
Feasibility
* relevance for producer cell * achievable high titer * high sensitivity of detection * ease of detection

In order to fit the requirements for an economic downstream processing, i. e. a design of purification procedures reduced in number but enhanced in selectivity and effectiveness, the model virus should be accessible in high titers and detectable in a reproducible and sensitive assay, otherwise high reduction factors of virus for the individual unit operations can not be demonstrated, and the accumulated virus reduction factor for the complete downstream processing might fail to assure the required safety assessments.

Downstream processing for the purification and recovery of a pharmaceutical protein at industrial scale is based upon two major separation techniques:

Table 4.2.8. Removal of Viruses by Ultrafiltration Membranes.

trial		REO 3			
cutoff	sample	titer (log ml^{-1})	volume (ml)	total virus (log)	clearance factor (log)
200 KD	1. load	5.93	100	7.93	
	1. filtrate	n.d.	163	---	> 5.93
	1. retentate	5.05	50	6.75	
	2. load	6.18	100	8.18	
	2. filtrate	n.d.	158	---	> 6.18
	2. retentate	5.50	50	6.75	
300 KD	1. load	5.68	100	7.68	
	1. filtrate	2.68	165	4.89	2.79
	1. retentate	4.43	50	6.13	
	2. load	5.30	100	7.30	
	2. filtrate	2.30	163	4.51	2.79
	2. retentate	5.18	50	6.88	

trial		PI-3			
cut-off	sample	titer (log ml^{-1})	volume (ml)	total virus (log)	clearance factor (log)
200 KD	1. load	6.80	100	8.80	
	1. filtrate	n.d.	149	---	> 6.80
	1. retentate	6.93	50	8.63	
	2. load	6.93	100	8.93	
	2. filtrate	n.d.	159	---	> 6.93
	2. retentate	6.68	50	8.38	
300 KD	1. load	6.43	100	8.43	
	1. filtrate	< 1.80	176	4.05	4.38
	1. retentate	6.05	50	7.75	
	2. load	6.68	100	8.68	
	2. filtrate	2.05	165	4.27	4.41
	2. retentate	6.05	50	7.75	

	REO 3	PI-3
Titer working bank:	$10^{7.8}$ ml^{-1}	$10^{7.18}$ ml^{-1}
Spike dilution:	1 : 10	1 : 10
Theor. spike virus:	$10^{6.8}$ ml^{-1}	$10^{6.8}$ ml^{-1}
Envelope:	no	yes
Size:	ca. 60 nm	ca. 150 nm

Membrane: polysulfone, Filtron OMEGA™, 70 cm^2 monosheet-cassette
Holder: Filtron VIRUSETTE™, 5 ml dead volume
TMP: 1 bar (7 psi)
Protein: IgG_{2a} monoclonal antibody (MAb), 5 mg ml^{-1}
Buffer: phosphate buffered saline, pH 7.2
Volume: a total volume of 100 ml was processed in each trial
Product Recovery: recovery of the MAb was better than 98 % for both cutoffs

Membrane filtration featuring selective pore sizes is used with regard to the size of the product molecule: an appropriate cutoff of a selected membrane permits the separation of the product from particles, e.g. tangential flow filtration through ultrafiltration membranes or dead-end filtration through filter cartridges, both exhibiting the appropriate cutoff. Validation criteria for filtration steps are the membrane chemistry and pore size, the geometry of the holder and the membranes, geometry, velocity and pressure of the solution streams and transmembrane pressure, as well as product load per area with respect to volume and product concentration.

Chromatographic methods exploiting different physico-chemical features of the protein product are designed and optimized for the separation of contaminating proteins as well as for the removal of contaminants such as nucleic acid, pyrogens and viruses. The nature of viruses as nucleocapsids or enveloped viruses with a number of proteins as an integral part of the envelope makes them accessible to a respective separation. Column chromatography using downscaled equipment requires a careful calculation for the process of parameters. Chromatographic parameters such as matrix chemistry, geometry of the column, bed height, linear flow rate and product load in correlation to matrix volume have to be identical. It is evident that

Table 4.2.9. Removal of Viruses by Chromatographic Procedures.

Protein MAb	Clearance Factor (log 10 units)						
	Enveloped Virus				Non-Enveloped Virus		
Type of Chromatography	X-MuLV	E-MuLV	PI-3	VSV	REO-3	SV-40	Polio
Affinity							
– Protein G	> 5.7	6.5	7.0	3.0	2.0	3.3	3.0
Cation exchange							
– Sulfonate Type 1	n.d.	> 6.3	> 6.3	n.d.	2.8	1.7	n.d.
– Sulfonate Type 2	< 1.0	n.d.	< 1.0	n.d.	< 1.0	1.7	n.d.
Anion exchange							
– Diethyl aminoethyl	4.3	n.d.	6.0	n.d.	6.0	5.9	n.d.
– Trimethyl aminoethyl	n.d.	> 6.3	3.1	n.d.	> 6.8	3.6	n.d.
– Quaternary aminomethyl	4.1	> 5.7	> 6.0	9.1	4.1	> 7.0	< 1.0
Gel permeation							
– fraction MW 3.000–70.000 d	3.2	3.3	3.0	n.d.	2.6	2.6	n.d.

n.d. = not determined

The values of virus log reduction given for the different types of chromatography cover data which have been achieved in different processes demonstrating the significant influence of process buffers e.g. in the case of the example for cation exchange chromatography. These data, in addition, indicate the requirement for virus removal studies for individual process unit operations in chromatography.

in any case the process buffers, which are used for equilibration, the separation process itself, and regeneration processes are identical to the manufacturing process (see also Chapter 4.1).

Virus inactivation can be achieved using harsh environmental conditions such as extreme pH-values, e. g. acid treatment at pH 2–4 for 1–2 hours (Brady et al., 1990), the addition of organic solvents, detergents (Horowitz, 1989) or chaotropic salts in order to dissolve the virus envelope and/or disintegrate the nucleocapsid, or irradiation by UV-light (Prince et al., 1993) or γ-radiation. The stability of the protein product determines the inactivation methods that can be applied.

The virus removal or inactivation efficacy for the unit operations in protein purification is determined by spiking experiments, i. e. the contamination of the product solution with a defined amount of virus. The reduction of virus titer is determined by measuring the residual virus in the product-containing fraction after separation. An obligatory assumption for those spiking experiments is a careful investigation of buffer toxicity for the virus indicator cells and viral interference experiments in order to avoid false experimental approaches and interpretation. A critical issue is the statistical mathematical methods for the calculation of the virus reduction factors (Löwer, 1991; Walter et al., 1992; Berthold et al., 1992). The major limitation factor for spiking experiments is the titer of the virus used and the volume of the test samples. The virus titer is limited by biological criteria, the volume of the test sample is limited by the cell culture test system and is typically in the range of 0.1 ml to 1.0 ml.

In order to maintain the physico-chemical conditions of the process step to be evaluated, a maximum virus spiking of 1 : 10 is recommended, i.e. the achievable titer of the load is at least a 1 : 10 dilution of the virus stock solution. Typically, the volume of a product containing effluent of a chromatography column, or the permeate after filtration is increased in comparison to the load volume due to an additional washing step, which is required in order to save and recover highly valuable product that is in the residual dead volume of the process equipment, thus causing a further dilution. Depending on the type of virus action, the virus titer is expressed either as focus forming units (ffu) e. g. for retrovirus or as plaque forming units (pfu) or tissue culture infectious doses (TCID50) for other viruses.

$$\mathbf{ffu/ml} = \sum \mathbf{n} \times \mathbf{d} \qquad \text{[equation 1]}$$

where: $\sum n$ = number of foci
d = dilution factor

$$\mathbf{pfu/ml} = \sum \mathbf{n} \times \mathbf{d} \qquad \text{[equation 2]}$$

where: $\sum n$ = number of plaques
d = dilution factor

$$\mathbf{TCID_{50}} = \mathbf{x_0} - \frac{\mathbf{d}}{\mathbf{2}} + \frac{\mathbf{d}}{\mathbf{n}} \times \sum \mathbf{xi} \qquad \text{[equation 3]}$$

where: x_0 = positive exponent of the highest dilution with a cytopathic effect (CPE)
d = logarithm of dilution step
n = number of determinations
$\sum xi$ = number of all positive samples up to and including xo

Statistical calculation becomes necessary if the virus titer is in a low range of one infectious particle per ml or even less. The probability of detecting virus at this concentration level in sample volumes of 0.1 ml is fairly low.

The probability **p** for a false negative result is:

$$\mathbf{p} = (\mathbf{V} - \mathbf{v}/\mathbf{V})^{\mathbf{n}} \qquad \text{[equation 4]}$$

where: V = overall volume of the starting test material
v = volume of the test sample
n = number of infectious particles statistically distributed in v

For **V** >> **v** the equation 4 follows the Poisson distribution

$$\mathbf{p} = \mathbf{e}^{-\mathbf{cv}} \qquad \text{[equation 5]}$$

where: c = concentration of infectious particles
v = volume of the test sample

The individual virus reduction factor **Ri** of a purification step is defined as the difference between the virus titer in the load sample and in the product-containing fraction of the spiking experiment, and is expressed as log 10 units.

$$\mathbf{10^{Ri}} = \mathbf{V^1 \times 10^a / V^2 \times 10^b} \qquad \text{[equation 6]}$$

where: Ri = individual reduction factor
V^1 = volume of the starting material
10^a = virus titer of the starting material
V^2 = volume of the processed material
10^b = virus titer of the processed material

The overall reduction factor is the sum of the individual reduction factors for all purification steps (Ad Hoc Working Party, 1991; Walter et al., 1992). With this clearance factor the safety margin of a purification process can be determined. The maximal potential virus contamination in a fermenter batch should be related to the final dose.

The safety margin for virus clearance should be in the range of 10^3 to 10^5 compared to the maximal potential virus contamination.

Pyrogens

The manufacturing of parenteral applied pharmaceutical proteins has to take into consideration the harmful effect of a residual pyrogen contamination of the final product. In addition to the preventive measures that have been outlined in 4.2.4.2.2 for the removal of media derived pyrogens, the purification and recovery process steps have to be validated for their ability to remove the potential content of adventitious pyrogens (Pearson III., 1985; 1987; Parenteral Drug Association, 1985). Pyrogens, i.e. lipopolysaccharides of gram negative bacteria (Hamma et al., 1984), can typically be removed by chromatographical methods similar to those used for the removal of DNA: affinity and anion exchange chromatography (Nolan et al., 1975). Affinity chromatography selective for enterobacterial lipopolysaccharides is described for ligands such as Polymyxin B (Novitsky and Gould, 1985) and nitrogen-containing heterocyclic molecules (Minobe et al., 1983). The use of those types of ligands cannot be recommended in general due to possible negative physiological effects in human. Whenever such ligands are used for depyrogenization, the product fluid has to be monitored accordingly using sensitive detection assays. The LAL-assay is so sensitive that spiking experiments for the validation can be avoided and pyrogenicity is easily determined during the manufacturing process (FDA, 1987).

A typical source and risk for pyrogen contamination with respect to process equipment are those parts that can not be sterilized and – or at least – sanitized, e.g. using caustic solutions. Most prominent examples are chromatographic matrices and ultrafiltration membranes manufactured of materials or polymers sensitive to high pH-values, e.g. silica and cellulose.

Table 4.2.10. Pyrogen Contamination related to the Polymer of Ultrafiltration Membranes.

Polymer	Trial	# 1 EU ml^{-1}	# 2 EU ml^{-1}	# 3 EU ml^{-1}	# 4 EU ml^{-1}	# 5 EUml^{-1}
Cellulose	10 KD UF/DF					
	before	< 0.96	< 0.96	< 0.96	< 0.96	< 0.96
	after	> 7.68	0.48	0.96	1.92	1.92
	10 KD UF*					
	before	1.92	0.48	0.06	3.84	3.84
	after	> 7.68	3.84	30.72	3.84	3.84
Polysulfon	10 KD UF/DF					
	before	0.12	< 0.24	0.12	0.06	
	after	0.24	0.24	0.12	0.12	
	10 KD UF*					
	before	0.12	0.06	0.12	0.06	
	after	0.48	0.24	0.48	0.48	

Sample: protein in phosphate buffered saline, 10 mg/ml
cut off: 10.000 dalton (10 KD)
*: Concentration factor 5–10
EU: Endotoxin Unit
UF/DF: Ultrafiltration/Diafiltration

Filtration membranes manufactured of cellulose derivatives are appreciated because of their significantly low nonspecific binding behaviour to most proteins; table 10 demonstrates the requirement that special attention must be devoted to the membrane maintenance in order to avoid imponderable bacterial/pyrogen contamination.

4.2.4.3 Regeneration and Storage of Chromatographic Matrices and Membranes

Raw materials expenses considerably contribute to the manufacturing costs of a biopharmaceutical. Therefore, validated regeneration methods for resins and membranes with subsequent reuse procedures are necessary for an economic downstream process.

Regeneration procedures must take the potential impurity spectrum of host cell proteins, DNA, viruses, media derived contaminants, pyrogens and microbial contaminations into account. In addition, the stability of resins and membranes to solvents and acid and caustic solutions defined by the resin manufacturer show the frame for development of useful and dedicated regeneration procedures. The European "Notes to applicants" require that the conditions of use and reuse of columns including loading conditions, yields, regeneration and storage between runs and the lifetime of each column be described (Ad Hoc Working Party, 1992). Therefore, small scale validation studies must be carried out with test systems reflecting the large scale conditions in detail.

Before starting the development of regeneration procedures in detail one should bear in mind that regeneration solutions must not only be compatible with the resins but also with all materials used in engineering, that some regeneration solutions may cause environmental problems and that sophisticated procedures with a whole series of regeneration cycles are not easily transferable to large scale manufacturing.

A typical protocol requires from 50 up to 100 equilibration-load-wash-regenerate and storage cycles. Important parameters for the assessment of resin performance in such validation studies are the determination of theoretical plates, the consistency of elution profiles, increasing backpressure of the system indicating biofouling of the matrix, consistency of product yield, product purity and detection of any carry over of impurities from one cycle to another by a number of specific and sensitive assays. With respect to the latter, blank elutions, e.g. after every tenth cycle, can give additional hints for the effectiveness of regeneration procedures. It is self evident that during scale-up and large scale validation of th process these parameters have to be carefully rechecked.

Storage conditions of resins and membranes should be developed mainly with respect to the prevention of microbial growth. The assessment of storage conditions can be facilitated by using the antimicrobial preservatives effectiveness test described in the USP (1990). Microorganisms are kept very well under control by sodium hydroxide solutions of various concentrations with the exception of spore-bearing organisms. Resins instable in alkaline storage conditions like affinity gels with monoclonal antibody ligands can be stored under more physiological solutions as effectively as alkaline conditions. Validation protocols for storage procedures of resins and membranes require after a defined storage period under controlled conditions the application of the same set of assays and methods described above for regeneration procedures.

4.2.4.4 Validation of Product Changeover Procedures

4.2.4.4.1 Multi-Use Manufacturing Facilities for Biologics

The growing success in biotechnology raises the question to companies involved in the development and/or manufacturing of biopharmaceuticals under which conditions and precautions more than one biological product can be handled in the same facility. The concerns regulatory agencies could have with such a situation are founded in potential cross-contamination of a product by another product, cross-contamination by adventitious agents from additional host-vector systems and cross-contamination by additional media- and buffer ingredients. For these reasons, 21 CFR 600.11 e. g. states for example that "all work with spore-bearing microorganisms shall be done in an entirely separate building" (U.S. Government Printing Office, 1991). It is the responsibility of the manufacturer that additional products in a given facility do not affect the purity, safety and efficacy of all products intended for clinical and market supply. The industry perspective of lay-out and design of multi-use facilities for biologics is addressed in a "White Paper" from the Pharmaceutical Manufacturing Association with the title "Multi-use manufacturing facilities for biologicals" with the intention of defining the "rules of the game" for multi-use facilities (Bader et al., 1992).

4.2.4.4.2 Cleaning Validation

Cleaning qualification, validation and changeover procedures have to consider several features of the equipment and the process used to purify active proteins from crude upstream fluids. Equipment could be dedicated, non-dedicated, disposable, product-contacted or non-product-contacted (Allgaier, 1996).

In downstream processing, chromatographic matrices, micro- and ultrafiltration membranes should be product-dedicated. The cleaning and reuse of these raw materials for a certain product is described in chapter 4.2.4.3. Empty chromatographic columns, membrane holders, centrifuges and vessels used to store eluates from downstream steps could be used non-dedicated. The cleaning validation and changeover program has to consider that this equipment comes into direct contact with the manufactured product. Vessels used for preparation and storage of buffer solutions could also be used for more than one product, but unlike vessels used for eluates they have no direct contact with the product. For smaller disposable items like pipettes cleaning validation obviously does not apply.

As important as the definition of dedicated, non-dedicated and disposable equipment is the selection of specific and sensitive assays to detect any kind of impurities that would cause cross-contamination. In general, final rinse samples and surfaces should be assayed by chemical, biochemical and microbiological methods. Water for Injection used for final rinsing in Cleaning-In-Place procedures could be investigated according to the monograph "Purified Water" in the USP XXIII. The conductivity of final rinse samples gives important information concerning the concentrations of dissociated inorganic and organic substances. Non-dissociated organic molecules are very sensitively assayed by Total-Organic-Carbon (TOC) measurements, where Carbon is oxidized to Carbon dioxide and detected by non-dispersive infrared gas analyses (NDIR) (Baffi et al., 1991). The optimized procedure could be as sensitive as down to 200 ppb Carbon. Enzyme-linked Immunosorbent Assays (ELISA's) directed to the active protein are also very sensitive, but one has to consider that hot alkaline cleaning solutions often destroy the epitopes and subsequently the sensitivity of the assay can decrease tremendously. SDS-gel electrophoresis with detection by silver staining in addition is a very powerful method for the detection of proteinaceous impurities.

Microbiological investigations with respect to pyrogens by LAL testing and bioburden by the sterile filtration or direct inoculation method rounds off the assessment of final rinse samples.

Table 4.2.11. Cleaning Validation and Product Changeover with Respect to the Prevention of Cross-contamination.

	Cleaning Validation and Product Changeover							
	Cleaning Validation				Product Changeover			
	Final Rinse			Surface	Final Rinse			Surface
	TOC	SDS/ silver stain	Conduc-tivity	Swab/ Ninhydrin	TOC	SDS/ silver stain	Conduc-tivity	Swab/ Ninhydrin
Vessels Media- and Buffer Preparation	+	–	+	+	–	–	+	–
Fermenters and CCF Hold Vessels	+	–	+	+	+	–	+	–
Vessels Protein Purification	+	–	+	+	+	–	+	–
Vessels Formulation and Sterile Filtration	+	+	+	+	+	–	+	–
Dish Washers	+	–	+	+	–	–	+	–

Surfaces of product equipment can be rubbed off by swabs followed by either extraction and Total-Organic-Carbon measurements or direct detection of amino acids, peptides and proteins with ninhydrine.

Taking all features of the processes and the equipment used, the different assays and the choice of different specifications into account, every manufacturer will end up with his own matrix for cleaning validation and requirements for product change-over. One proposal for such a specific matrix is shown in Table 4.2.1.1.

4.2.4.4.3 Product Changeover Procedures

Independent from the manufacturing strategy in a multi-use-facility, namely campaign manufacturing or concurrent manufacturing, product changeover procedures are of utmost importance to prevent cross-contamination and mix-ups.

Non-dedicated equipment used for the production of more than one product is set into quarantine until final rinse sampling of CIP operations is carried out and the results comply with the in-house specifications. Dedicated equipment from the formerly manufactured protein must be removed and new equipment brought into the manufacturing area. Log books required by 21 CFR 211 give information about the use of every single piece of equipment on the spot. In addition, the documentation of all used equipment in the manufacture of a distinct product in a major equipment list further facilitates the general survey during and after product changeover. Special requalification and revalidation programs may be necessary if major equipment changes have occured.

Standard Operating Procedures (SOP's) and Master Manufacturing Records (MMR's) have to undergo in general the same procedure as described above for manufacturing equipment. Product specific documentation must be removed and new SOP's and MMR's brought into the manufacturing area. Finally, all employees get product specific information and instruction and after a final inspection of the downstream area the next product can be purified according to Good Manufacturing Practice (Allgaier, 1994).

Acknowledgements

All data on the validation of the removal of DNA and the removal/inactivation of virus were elaborated in close collaboration with our colleagues Dr. Klaus Bergemann (Nucleic acid chemistry) and Dr. William Werz (Cell biology), Process Development, Biotechnology Department, Thomae. Analytical assays were performed in the protein chemical laboratories of Dr. Stefan Bassarab and Dr. Helmut Hoffmann, Biotechnology Department, Thomae.

Abbreviations

BSA	Bovine Serum Albumin
CFR	Code of Federal Regulations (USA)
CFU	Colony Forming Units
cGMP	current Good Manufacturing Practice
CIP	Cleaning-in-Place
CPMP	Committee for Proprietary Medicinal Products
d	dalton
DNA	Desoxyribonucleic Acid
EC	European Committee
ELISA	Enzyme-linked Immunosorbent Assay
FDA	Food and Drug Administration
ffu	focus forming unit
IgG	Immunoglobulin G
IQ	Installation Qualification
LAL	Limulus-Amoebocytes-Lysate
LPS	Lipopolysaccharide
MAb	Monoclonal Antibody
MCB	Master Cell Bank
MMR	Master Manufacturing Record
MWCB	Manufacturer's Working Cell Bank
NDIR	Non-Dispersive Infrared gas analysis
OQ	Operational Qualification
P&ID	Piping and Instrumentation Diagram
pfu	plaque forming unit
PQ	Performance Qualification
PW	Purified Water
SDS	Sodium Dodecyl Sulfonate
SIP	Steaming-in-Place
SOP	Standard Operation Procedures
TCID50	Tissue Culture Infectious Dose, at a 50% rate
TOC	Total Organic Carbon
USP	United States Pharmacopeia
VLP	Virus-like-Particle
WFI	Water For Injection
WHO	World Health Organization

References

Ad Hoc Working Party on Biotechnology/Pharmacy, Commission of the European Communities, Biotech headings for Notice to Applicants, Part II, 1992.

Ad Hoc Working Party on Biotechnology/Pharmacy (1991) Note for Guidance: Validation of Virus Removal and Inactivation procedures, EC DG III/8115/89-EN 1991.

Allgaier, H. (1994) Requirements and Validation of a Biotech Multipurpose Plant. Chimia 48, 464–466.

Allgaier, H. (1996) Conversion of a Biotech Facility to "Multi-Use" Operation. In: Pharmaceutical Biotechnology Internationl (Barnacal, P.A., ed.), Sterling Publications Limited, London W2 1XR, pp 47–48.

American Society for Testing and Materials (ASTU) (1988) Draft Standard Guide for Determination of Purity, Impurities, and Contaminants in Biological Drug Products, Philadelphia.

Bader, F.G., Blum, A., Garfinkle, B.D., MacFarlane, D., Massa, T. and Copmann, T.L. (1992) Multiuse Manufacturing Facilities for Biologicals, Biopharm Sept. 1992, 32–40.

Baffi, R., Dolch, G., Garnick, R., Huang, Y.F., Mar, B., Matsuhiro, D., Niepelt, B., Parra, C. and Stephan, M. (1991) A Total Organic Carbon Analysis for Validation Cleaning Between Products in Biopharmaceutical Manufacturing, Journal of Parenteral Science & Technology, *45*, 13–19.

Bergemann, K., Walter, J. and Berthold, W. (1993) Removal of Host Cell DNA from Protein Solutions, presented at: Validation of Pharmaceuticals and Biopharmaceuticals. Parenteral Drug Associates, Inc. (PDA), 2nd International Congress Basel 22–24 February, 1993 and the PDA Workshops, 25–27 February, 1993.

Berthold, W., Walter, J. and Werz, W. (1992) Experimental approaches to guarantee minimal risk of potential virus in purified monoclonal antibodies, Cytotechnology 9, 189–201.

Bill, E., Lutz, U., Sparrmann, M., and Allgaier, H. (1995) Optimization of Protein G Chromatography for Biopharmaceutical Monoclonal Antibodies. Journal of Molecular Recognition 8, 90–94.

Bouchard, L., Gelinas, C., Asselin, C. and Bastin, M. (1984) Tumorigenic activity of Polyoma virus and SV40 DNAs in newborn rodents. Virology *135*, 53–64.

Brady, D., Bonnerjea, J. and Hill, C.R. (1990) Purification of Monoclonal Antibodies for Human Clinical Use: Validation of DNA and Retroviral Clearance. In: Separations For Biotechnology 2, SCI, Elsevier Applied Science, London–New York.

Commission of the European Communities (1989) Guidelines on the Quality, Safety and Efficacy of Medicinal Products for Human Use, Vol. 2 of The Rules Governing Medicinal Products in the European Community.

Commission of the European Communities (1989) Guidelines to Good Manufacturing Practice for Medicinal Products of the Rules Governing Medicinal Products in the European Community, Volume 4.

Committee for proprietary medicinal products (July 1987) Notes to applicants for marketing authorization on requirements for production and quality control of medicinal products derived by recombinant DNA technology.

Committee for Proprietary Medicinal Products Ad Hoc Working Party on Biotechnology/Pharmacy (June 1987) Guidelines on the Production and quality control of monoclonal antibodies of murine origin intended for use in man.

Committee for Proprietary Medicinal Products Ad Hoc Working Party on Biotechnology/Pharmacy (July 1990) Guidelines on the production and quality control of human monoclonal antibodies.

Council of Europe (1990) European Pharmacopeia.

Division of Manufacturing and Product Quality, FDA (1987) Guidelines on General Principles of Process Validation.
Division of Manufacturing and Product Quality, FDA (1987) Guidelines on Sterile Drug Product Produced by Aseptic Processing
EC-Commission, Notes to applicants for marketing authorization, DG III/860/86-EN (1986) Guidelines on the production and quality control of medicinal products derived by recombinant DNA-technology.
FDA, Bethesda, MD, USA (1987) Guidelines on Validation of the Limulus Amoebocyte Lysate Test as an End-Product Endotoxin Test for Human and Animal Parenteral Drugs, Biological Products and Medical Devices.
FDA, Center for Biologics Evaluation and Research, Bethesda, MD, USA (1987) Points to Consider in the Characterization of Cell Lines to Produce Biological Products.
FDA, Center for Biologics Evaluation and Research, Bethesda, MD, USA (1987) Points to Consider in the Manufacture of Monoclonal Antibody Products for Human Use.
FDA, Center for Biologics Evaluation and Research, Bethesda, MD, USA (1990) Cytokine and Growth Factor Pre-Pivotal Trial Information Package with Special Emphasis on Products Identified for Consideration under 21 CFR 312 Subpart E. Drugs Intended to Treat Life-Threatening and Severely Debilitating Illnesses.
FDA, Center for Biologics Evaluation and Research, Bethesda, MD, USA (1991) Points to Consider in Human Somatic Cell Therapy and Gene Therapy.
FDA, Center for Drugs and Biologics, Bethesda, MD, USA (1985) Antiviral Drugs for Non-Life-Threatening Disease: Draft Points to Consider for Safety Evaluation Prior to Phase 1 Studies.
FDA, Office of Biologics Research and Review, 8800 Rockville Rike, Bethesda MD, 20892 USA (1987) Points to consider in the characterization of cell lines used to produce Biologicals.
FDA, Office of Biologics Research and Review, Center for Drugs and Biologics, Bethesda, MD, USA (April 10, 1985) Points to Consider in the Production and Testing of New Drugs and Biologicals Produced by Recombinant DNA Technology.
FDA, Office of Biologics, Bethesda, MD, USA (1983) Interferon Test Procedures: Points to Be Considered in the Production and Testing of Interferon Intended for Investigational Use in Humans.
Garnick, R.L., Solli, N.J. and Papa, P.A. (1988) The Role of Quality Control in Biotechnology: An Analytical Perspective. Anal. Chem. *60*, 2546–2557.
Health and Welfare Canada (1990) Drugs Directorate Guidelines, Manufacture and Testing of Biologics Produced by Recombinant DNA Technology.
Health and Welfare Canada (1990) Drugs Directorate Guidelines, Utilization of Continuous Cell Lines in the Manufacture of Biologics.
Homma, J.Y., Kanegasaki, S., Lüderitz, O., Shiba, T. and Wertphal, O. (eds.) (1984) Bacterial Endotoxin. Verlag Chemie, Weinheim, New York.
Horowitz, B. (1989) Investigations into the application of tri(n-butyl) phosphate/detergent mixtures to blood derivatives. Virus inactivation in plasma products. Curr. Stud. Hematol. Blood Transfus., No. 56 (Mortenthaler, J.-J., ed.), pp 83–96, Karger, Basel.
Kozak, R. (1992) Current FDA Requirements for Phase 1 Studies, Workshop on Preclinical Safety Testing of Monoclonal, Antibodies, Bethesda, MD, USA.
Kung, V.T., Panfili, P.R., Sheldon, E.L., King, R.S., Nagainis, P.A., Gomez, B., Ross, D. A., Briggs, J. and Zuk, R.F. (1990) Picogram Quantitation of Total DNA Using DNA-Binding Proteins in a Silicon Sensor Based System. Anal. Biochem. *187*, 220–227.
Lees, G. and Onions, D. (1990) Validation of Downstream Processes. Technical Bulletin 4, Quality Biotech, Glasgow.

Liptrot, C. and Gull, K. (1991) Detection of Viruses in Recombinant Cells by Electron Microscopy. In: Animal Cell Technology: Developments, Processes and Products (Spier, R. E., Griffiths, J. B. and MacDonald, C., eds.), Butterworth-Heinemann Ltd. Linacre House, Jordan Hill, Oxford OX2 8DP, pp 653–656.

Löwer, J. (1990) Risk of Tumor Induction In Vivo by Residual Cellular DNA: Quantitative Considerations. J. Med. Virol. *31*, 50–53.

Löwer, J. (1991) Virological Aspects of the Quality Control of Biologicals: Quantitative Considerations. In: Symposium on Virological Aspects of the Safety of Biological Products London, England, 1990. Develop. biol. Standard. *75*, pp 221–226, S. Karger, Basel.

Marcus-Sekura, C. J. (1991) Validation of Removal of Human Retroviruses. Center for Biologics Evaluation and Research, FDA, Bethesda, MD, USA.

Ministry of Health and Welfare, Japan (1988) Documents necessary for application for approval of drugs manufactured using cell culture technology.

Minobe, S., Tadashi, S., Tetsuya, T. and Chibata, J. (1983) Characteristics of Immobilized Histamin for Pyrogen Adsorption. J. Chromatogr. *262*, 193–198.

Molecular Devices Corporation, publication 0120–0501A (1988) "Threshold Total DNA Assay".

National Institutes of Health (May 7, 1986) Guidelines for Research Involving Recombinant DNA Molecules. Department of Helth and Human Services.

Nolan, J. G., McDevitt, J. J. and Goldmann, G. S. (1975) Endotoxin binding by charged and uncahrged resin. Proc. Soc. Exp. Biol. Med. *149*, 766–770.

Novitsky, T. J. and Gould, M. J. (1985) Inactivation of Endotoxin by Polymyxin B. in: Technical Report No. 7, Parenteral Drug Association, Inc., Philadelphia.

Office of Science and Technology Policy, Federal Register, Vol. 51, No. 123 (June 26, 1986) 23301–23750, Coordinated Framework for Regulation of Biotechnology.

Organization for Economic Cooperation and Development (OECD) (July 16, 1986) Recombinant DNA Safety Considerations.

Ostrove, J. M., Walsh, W., Vacante, D. and Patel, N. (1992) Molecular Hybridization Techniques and Polymerase Chain Reaction (PCR) as Methods for Safety Assessment of Animal Cells used in Biopharmaceutical Production. In: Animal Cell Technology: Developments, Processes and Products (Spier, R. E., Griffiths, J. B. and MacDonald, C., eds.), Butterworth-Heinemann Ltd. Linacre House, Jordan Hill, Oxford OX2 8DP, pp 689–695.

Parenteral Drug Association, Inc., Philadelphia (1985) Depyrogenation, Technical Report No. 7.

Pearson III, F. C. (1985) Pyrogens. Advances in Parenteral Sciences/2, Marcel Dekker, Inc., New York, Basel.

Pearson III, F. C. (1987) Pyrogens and Depyrogenation, Theory and Practice. In: Aseptic Pharmaceutical Manufacturing (Olson, W. P. and Groves, M. J., eds.), Interpharm Press Inc., Prairie View.

Per, S. R., Aversa, C. R. and Sito, A. F. (1989) Quantification of Residual DNA in Biological Products. Clinical Chemistry *35*, 1859–1860.

Prince, A. M., Stephan, W. and Brotmann, B. (1993) b-Propiolactone/Ultraviolett Irradiation: A Review of Its Effectiveness for Inactivation of Viruses in Blood Derivatives. Rev. Infect. Dis. *5*, 92–107.

Rigby, P. W. J., Dieckmann, M., Rhodes, C. and Berg, P. (1977) Labeling deroxyribonucleic acid to high specific activity in vitro by nick translation with DNA-Polymerase I. J. Mol. Biol. *113*, 237–251.

Schiff, L. J., Moore, W. A., Brown, J. and Wisher, M. H. (1992) Lot Release – Final Product Safety Testing for Biologics. Pharmaceutical Technology International, September 1992.

Schindler, R. and Dinarello, C. A. (1990) Ultrafiltration to Remove Endotoxins and other Cytokine-Inducing Materials from Tissue Culture Media and Parenteral Fluids. Bio Techniques 8, 408–413.

Smith, K.T., Doherty, I., Thomas, J.A., Per, S.R. and Sito, A.F. (1991) Quantitation of Residual DNA in Biological Products: New Regulatory Concerns and New Methodologies. In: Animal Cell Technology: Developments, Processes and Products (Spier, R.E., Griffiths, J.B., MacDonald, C., eds.), Butterworth-Heinemann Ltd. Linacre House, Jordan Hill, Oxford OX2 8DP, pp 696–698.

U.S. Government Printing Office, Washington, D.C. (April 1, 1991) Code of Federal Regulations 21 CFR 600–680.

U.S. Government Printing Office, Washington, D.C. (April 1991) Code of Federal Regulations 21 CFR 210–211.

United States Pharmacopeia Convention, Inc. (1990) U.S. Pharmacopeia XXII.

Walter, J. and Werner, R. G. (1993) Regulatory Requirements and Economic Aspects in Downstream Processing of Biotechnically Engineered Proteins for Parenteral Application as Pharmaceuticals. In: Downstream Processing, Recovery and Purification of Proteins. A Handbook of Principles and Practice (Kroner, K.H., Papamichael, N. and Schütte, H., eds.), Carl Hanser Verlag, München.

Walter, J., Werz, W. and Berthold, W. (1992) Virus Removal and Inactivation – Concept and Data for Process Validation of Downstream Processing, Biotech Forum Europe, *9*, 560–564.

Walter, J., Werz, W., McGoff, P., Werner, R.G., Berthold, W. (1991) Virus Removal/Inactivation in Downstream Processing. In: Animal Cell Technology: Developments, Processes and Products (Spier, R.E., Griffiths, J.B. and MacDonald, C., eds.), Butterworth-Heinemann Ltd. Linacre House, Jordan Hill, Oxford OX2 8DP, pp 624–634.

WHO Study Group (1987) Acceptability of cell substrates for production of Biologicals. Geneva: World Health Organization Technical Report Series 747, pp 1–29.

World Health Organization (1991) Proposed Requirements for Monoclonal Antibodies for Clinical Use in Humans (Requirements for Biological Substances No. 48), Technical Report BS/91.1657.

World Health Organization, Geneva (1987) Acceptability of all substrates for production of biologicals. Report of a WHO Study Group. Technical Report, Series 747.

Index

Bei Fragen zur Produktsicherheit wenden Sie sich bitte an:
If you have any questions regarding product safety,
please contact:

Walter de Gruyter GmbH
Genthiner Straße 13
10785 Berlin
productsafety@degruyterbrill.com